MATHEMATICAL LOGIC

MATHEMATICAL LOGIC

Edited by

Petio Petrov Petkov
Sofia University
Sofia, Bulgaria

PLENUM PRESS • NEW YORK AND LONDON

Library of Congress Cataloging-in-Publication Data

Summer School and Conference on Mathematical Logic (1988 : Chaĭka, Varnenski okrŭg, Bulgaria)
Mathematical logic / edited by Petio Petrov Petkov.
p. cm.
"Proceedings of the Summer School and Conference on Mathematical Logic, honourably dedicated to the ninetieth anniversay of Arend Heyting (1898-1980) held September 13-23, 1988, in Chaika (near Varna), Bulgaria"--T.p. verso.
Includes bibliographical references.
ISBN-13:978-1-4612-7890-0 e-ISBN-13:978-1-4613-0609-2
DOI:10.1007/978-1-4613-0609-2

1. Logic, Symbolic and mathematical--Congresses. I. Petkov, Petio Petrov. II. Heyting, A. (Arend), 1898- . III. Title.
QA9.A1S86 1988
511.3--dc20 90-30549
CIP

Proceedings of the Summer School and Conference on Mathematical Logic, honourably dedicated to the Ninetieth Anniversary of Arend Heyting (1898-1980), held September 13-23, 1988, in Chaika (near Varna), Bulgaria

Softcover reprint of the hardcover 1st edition 1990

A Division of Plenum Publishing Corporation
233 Spring Street, New York, N.Y. 10013

ORGANIZING COMMITTEE OF THE MEETING:

V. Goranko, L. Ivanov, S. Passy, P. Petkov/Chairman, T. Tinchev

PROGRAMME COMMITTEE

D. Bridges (Buckingham)
M. Cresswell (Wellington)
O. Demuth (Prague)
J. de Iongh (Nijmegen)
D. de Jongh (Amsterdam)
A. Ditchev* (Sofia)
A. Dragalin (Debrecen)
K. Fine (Los Angeles)
R. Gandy (Oxford)
N. Georgieva* (Sofia)
J. D. Halpern (Mountain View)
D. Harel (Rehovot)
S. Hayashi (Edinburgh)
M. Kanovich (Kalinin)
A. Kechris (Pasadena)
Ph. Kolaitis (Stanford)
B. Kushner (Moscow)
A. Levy (Jerusalem)
V. Lifschitz (Stanford)
L. Maksimova (Novosibirsk)
G. Mints (Tallinn)
A. Muchnik (Moscow)
D. Normann (Oslo)
H. Ono (Hiroshima)
P. Petkov/Chairman* (Sofia)
G. Sacks (Cambridge, Mass.)
J. Shepherdson (Bristol)
D. Skordev* (Sofia)
C. Smorynski (Utrecht)
R. Soare (Chicago)
I. Soskov (Sofia)
V. Stoltenberg-Hansen (Uppsala)
G. Takeuti (Urbana)
J. Tiuryn (Warsaw)
B. Trakhtenbrot (Tel Aviv)
A. Troelstra (Amsterdam)
D. Vakarelov (Sofia)
D. van Dalen (Utrecht)
W. Veldman (Nijmegen)
A. Visser (Utrecht)
S. Wainer (Leeds)

* Member of the Selection Committee

OTHER REFEREES

G. Delavalette, F.-J. de Vries, G. Gargov, R. Goldblatt, V. Goranko, J. Håstad, C. Haught, J. M. E. Hyland, I. L. Humberstone, D. Israel, L. Ivanov, W. Kowalczyk, S. Kurtz, T. Landholm, O. Lichtenstein, W. Maass, S. Mardayev, E. Munthe-Kaas, D. Niwinski, S. Passy, S. Radev, P. Rodenburg, G. Rosolini, J. Royer, A. Skowron, A. Soskova, V. Sotirov, W. Tait, T. Tinchev, P. Urzyczyn, F. Voorbraak

PREFACE

Heyting'88 Summer School and Conference on Mathematical Logic, held September 13 - 23, 1988 in Chaika, Bulgaria, was honourably dedicated to Arend Heyting's 90th anniversary. It was organized by Sofia University "Kliment Ohridski" on the occasion of its centenary and by the Bulgarian Academy of Sciences, with sponsorship of the Association for Symbolic Logic. The Meeting gathered some 115 participants from 19 countries.

The present volume consists of invited and selected papers. Included are all the invited lectures submitted for publication and the 14 selected contributions, chosen out of 56 submissions by the Selection Committee. The selection was made on the basis of reports of PC members, an average of 4 per submission. All the papers are concentrated on the topics of the Meeting: Recursion Theory, Modal and Non-classical Logics, Intuitionism and Constructivism, Related Applications to Computer and Other Sciences, Life and Work of Arend Heyting.

I am pleased to thank all persons and institutions that contributed to the success of the Meeting: sponsors, Programme Committee members and additional referees, the members of the Organizing Committee, our secretaries K. Lozanova and L. Nikolova, as well as K. Angelov, V. Bozhichkova, A. Ditchev, D. Dobrev, N. Dimitrov, R. Draganova, G. Gargov, N. Georgieva, M. Janchev, P. Marinov, S. Nikolova, S. Radev, I. Soskov, A. Soskova and V. Sotirov, who helped in the organization, Plenum Press and at last but not least all participants in the Meeting and contributors to this volume.

January 1989

Petio Petkov

CONTENTS

HEYTING SESSION (INVITED PAPERS)

* indicates the lecturer invited

CONFERENCE (CONTRIBUTED PAPERS)

H E Y T I N G   S E S S I O N

(invited papers)

ON THE EARLY HISTORY OF INTUITIONISTIC LOGIC

A.S.Troelstra

Faculteit Wiskunde en Informatica
Universiteit van Amsterdam
Roetersstraat 15, 1018 WB Amsterdam, NL

Abstract. We describe the early history of intuitonistic logic, its formalization and the genesis of the so-called Brouwer–Heyting–Kolmogorov interpretation. In particular we discuss at some length whether Heyting's papers contain an anticipation of logic with existence predicate. Finally we publish some source material, in particular letters of Bernays, Glivenko and Kolmogorov to Heyting.

On this topic I have written before (1978,1981,1983), and there is not much news to tell. However, (1978) and (1983) are of limited distribution, and several sources of interest for this topic are unpublished, so I thought it appropriate to collect here my earlier remarks on the topic, together with some additions and a publication of some further sources: some fragments of the correspondence between O. Becker and A. Heyting (3.1), a fragment of a letter from L.E.J. Brouwer to Heyting (3.2), a letter of Brouwer to Th. de Donder (3.3), a letter of P. Bernays (3.4), and the letters of V. Glivenko and A. Kolmogorov to A. Heyting (3.5-6). The correspondence between A. Heyting and H. Freudenthal on the interpretation of intuitionistic logic has already been published in Troelstra (1983).

1. The formalization of intuitionistic logic

1.1. Kolmogorov's first paper on intuitionistic logic. Kolmogorov (1925), in Russian, is the earliest published formalization of a fragment of intuitionistic logic. The paper represented a remarkable achievement, but remained virtually without effect because of the language in which it was written. Gödel was clearly unaware of the paper when he wrote his (1933). The paper is mentioned by Glivenko in his letter of 13-X-1928, and also by Kolmogorov in his third letter to Heyting, but it seems that Heyting still had no idea of its contents when he wrote his survey (1934), since it is missing in the bibliography.

An english translation, with an introduction by H. Wang, may be found in van Heijenoort (1967). Kolmogorov restricts attention to the logical operators $\rightarrow$, $\neg$ $\forall$, $\exists$ (axioms for $\vee$ appear in a footnote); the quantifier rules are not complete, and Kolmogorov rejects the rule ("ex falso sequitur quodlibet") $p \rightarrow (\neg p \rightarrow q)$. The axiom schema $\forall x\, A(x) \rightarrow A(t)$ is not included in Kolmogorov's system (although Wang regards it as part of the system since it is clearly intuitively accepted as true by Kolmogorov). Kolmogorov in his paper also describes an embedding of classical propositional logic into (his fragment of) intuitionistic propositional logic, and shows that this embedding is capable of generalization to stronger systems, thereby anticipating Gödel (1933) and Gentzen (1974, dating from 1933).

1.2. The work of Glivenko and Heyting. In chronological order the next paper is Glivenko (1928), where a philosophical debate is settled by the use of formalization. Glivenko lists ten propositional axioms clearly valid on Brouwer's interpretation of the logical signs, and derives formally from his axioms

Mathematical Logic
Edited by P. P. Petkov
Plenum Press, New York, 1990

$$\neg\neg(\neg p \vee p), \neg\neg\neg p \to \neg p, (\neg p \vee p \to \neg q) \to \neg q,$$

and then uses these theorems to show that "a proposition is tierce" (in the sense of Barzin and Errera (1927)) is false, thus revealing the fallacies in the Barzin-Errera argument which purported to show the contradictoriness of Brouwer's logic (it must be said, however, that the misunderstanding of Barzin and Errera was so tenacious that not even Glivenko's paper could convince them of their mistake). Glivenko does not refer to Kolmogorov's (1925), and the axioms he lists do not seem to be directly inspired by this paper. In a non-formal context, $\neg\neg(\neg p \vee p)$ already appears in Brouwer (1908). A formal approximation of Brouwer's informal reasoning is perhaps most easily given in natural deduction. The principle $\neg\neg\neg p \to \neg p$ is informally derived in Brouwer (1923).

In 1927, the Dutch Mathematical Association ("Het Wiskundig Genootschap") published the following prize question proposed by G. Mannoury (in a free translation):

"By its very nature, Brouwer's set theory cannot be identified with he conclusions formally derivable in a certain pasigraphic system. Nevertheless certain regularities may be observed in the language which Brouwer uses to give expression to his mathematical intuition; these regularities may be codified in a formal mathematical system. It is asked to

1) construct such a system and to indicate its deviations from Brouwer's theories;

2) to investigate whether from the system to be constructed a dual system may be obtained by (formally) interchanging the principium tertii exclusi and the principium contradictionis."

Heyting wrote a prize essay on the topic proposed, and was awarded the prize by the "Wiskundig Genootschap" early in 1928. Though originally publication in the Mathematische Annalen was intended, the essay finally appeared (Heyting 1930, 1930A, 1930B) in revised form,through the intermediacy of L. Bieberbach, in the Sitzungsberichte der Preussisschen Akademie von Wissenschaften (not a good choice from the point of view of the accessibility of the papers). There is no manuscript version of Heytings essay in its original form nor of the revised version among his papers (except for a single page in handwriting preserved by chance, which differs from the final version but cannot otherwise be chronologically located), so we do not know which changes he made for the revision. Clearly he has taken note of Glivenko's work, not known to him while writing the prize essay. Presumably, the references to Hilbert and Ackermann(1928) were inserted by Heyting while preparing his revision.

In his informal argument(1923) for $\neg\neg\neg p \to \neg p$ Brouwer assumes $p \to \neg\neg p$ as obviously valid. Originally Heyting adopted $p \to \neg\neg p$ as an axiom, but Glivenko's proof of $\neg\neg\neg p \to \neg p$ showed him how to derive it from the other principles.

On the other hand, Glivenko (1930), containing the well known Glivenko theorem, has been influenced by Heyting's work; from the correspondence reproduced below, it looks as if Heyting's derivation of $\neg\neg(\neg\neg p \to p)$ suggested to Glivenko his theorem.

The first paper (Heyting 1930) is the most polished one, and had the most immediate effect. Heyting arrived at his axiom system for propositional logic, according to his own account, in a very straightforward and simple-minded way: he went through the axioms and theorems in Principia Mathematica (Whitehead and Russell 1910) and retained those which were intuitionistically valid (cf. the fragments of the correspondence with O. Becker). This is a more thorough way of proceeding (in view of the rather exhaustive treatment of propositional logic in the Principia) than either Kolmogorov (1925) or Bernays (cf. his letter below) appear to have used. It does not strike one as a particularly imaginative method, but one should not forget that the project could be carried out successfully only on the basis of a clear insight in the intuitionistic interpretation of the logical operations (which had not yet been explicity formulated when Heyting wrote the formalization papers). The contents of the later two papers are difficult to compare with the formalism of either the Principia or Hilbert and Ackermann (1925); here more pioneering work was needed, and the second and third paper contain a number of technical flaws and weaknesses. The style of formalization is closer to Peano's "Formulaire".

Soon after the publication of his formalization papers, Heyting must have realized that the formal treatment was unsatisfactory at certain points, at least in (1934) he says, speaking of his own work, "regrettably his presentation is unduly complicated, due to a not completely successful attempt to formalize also the substitution process". In later years he expressed himself even more strongly (Heyting 1978):

"I regret that my name is known today mainly in connection with these papers, which were very imperfect and contained many mistakes. They were of little help in the struggle to which I devoted my life, namely a better understanding and appreciation of Brouwer's ideas. They diverted attention from the underlying ideas to the formal system itself." (We cannot agree with the second sentence in this quotation; here Heyting underestimates the contribution of his own formalization to a better understanding of the basic notions of intuitionism.)

The next two subsections contain some comments on the technical details of the second and third paper. We use modern notation in our discussion.

1.3. Strict identity. The second and third paper deal with predicate logic and mathematical theories based on predicate logic. The individual variables are taken to range over all mathematical objects, and special classes are singled out by introducing axiomatically or defining suitable predicates. The rules are formulated semantically, and it is not always obvious how to interpret them.

A feature of interest is the introduction of a primitive notion of strict identity denoted by ≣. ≡ is reserved for the possibly defined notion of mathematical equality of objects appropriate to the domain under consideration. "p ≣ q" is read as "p is the same object as q". We quote the relevant passages:

"The fundamental relation between objects [Individuen] is p ≣ q, 'p is the same object as q'. We therefore need a special sign for mathematical identity, which has to be defined for each kind of object, in such a way that identity implies that the objects agree w.r.t. all mathematically relevant properties For mathematical identity we choose the sign =. [...]

The formula p ≣ p does not hold for all signs p; rather, we use the formula to express that p denotes an object. Therefore we can explain 6.1 ⊢ ⊢ 1 ≣ 1 as '1 is an object'. Thus we obtain the possibility to introduce objects successively, just as in mathematics."

Of the axioms for ≣ we note in particular

6.11 $\quad p ≣ p \wedge q ≣ q \supset p ≣ q \vee \neg(p ≣ q)$

(identity between objects is decidable).

1.4. Anticipation of logic with existence predicate. Elsewhere I have stated that in Heyting's formalization some of the ideas of Scott's intuitionistic logic with existence predicate E are present. On closer scrutiny I am less certain; due to certain defects of Heytings formalization, it is hard to say to what extent E-logic has been anticipated. The obvious candidate for the E-predicate in Heyting's system is Et := t ≣ t, and below we shall take this as definition.

5.3 stipulates that a true formula must always be well-formed. 5.31-33 state that well-formed formulas are closed under renaming variables, replacing denoting terms by other denoting terms, and variables by denoting terms. There is an unbalance in the stipulations here; we must assume that variables are not assumed to denote (otherwise some of the later axioms would obviously be redundant), but if a variable is replaced by a term, the term is required to denote.

Also substitution cannot be handled as a variable-binding operator in the way proposed by Heyting, and henceforth we shall tacitly correct this and assume substitution to be a syntactical operation in the way which is customary in modern treatments of predicate logic.

In order to interpret the rules 5.4-6 as deduction rules, one should replace the predicate variables a and b by metavariables A and B for formulas, and interpret

$$\binom{p}{x}$$

as syntactically defined substitution. Then 5.4 amounts to (in modern notation)

$$\vdash A \Rightarrow \vdash A[b/B] \text{ (b a propositional variable),}$$

$$\vdash A \wedge Et \Rightarrow \vdash A[x/t] \text{ (x an individual variable),}$$

and 5.5 and 5.6 correspond to

$$\vdash A \to B \Rightarrow \vdash A \to \forall x B \ (x \notin FV(A)),$$

$$\vdash A \to B \Rightarrow \vdash \exists x A \to B \ (x \notin FV(B)).$$

Rule 5.7 becomes redundant if ( ) is interpreted syntactically. The axiom

6.32 $\vdash \forall y(Ey \to \forall y A) \to \forall y A$

may be read as saying that universal quantification runs over objects only (not over "virtual objects"); a corresponding axiom for existential quantification is missing. The axiom

6.3 $\vdash \forall xA \to A[x/t]$

is now problematic; here we would have expected

$$\vdash \forall xA \land Et \to A[x/t].$$

It cannot be that t is tacitly assumed to satisfy E, (as might be suggested by 5.32), because in 6.402 a variable is taken for t. Note that 6.3 with

6.45 $\vdash \forall x\, Ex$

implies $\vdash Ex$ (this would lead to E^+-logic, where variables are assumed to exist, cf.Troelstra and van Dalen 1988, 1.2.3), and this also does not agree very well with certain axioms which appear later (see e.g. Heyting's axiom 7.11), where "existence" for variables is apparently not assumed.

1.5. Arithmetic and analysis. The treatment of arithmetic does not exactly cover what is nowadays called Heyting's arithmetic, since multiplication is not introduced. Also, the definition by recursion of + cannot be treated as an explicit definition but should be regarded as a pair of axioms, which has to be supplemented e.g. by taking 10.4 ($\vdash p \in N \land q \in N \supset p+q \in N$) as an axiom, since the proof of 10.4 is wrong (cf. Henkin 1960).

The third paper deals with spreads in Brouwer's sense, and species; there are few axioms and many definitions, and the main purpose seems to have been to show that with relatively few primitives most of the intuitionistic notions appearing in Brouwer's papers could be defined. It is to be noted that the axioms are extremely weak: thus there is no comprehension principle, and there are no axioms to warrant the existence of a single choice sequence.

There is one particular point of interest in the third paper: it contains the first known statement of the weak continuity axiom (12.22).

1.6. Later developments. Heyting and Glivenko's work on intuitionistic propositional logic **IPC** was soon taken up in other publications; as already said, Kolmogorov (1925) remained without effect until much later. Gentzen (1935) is perhaps the most important early paper on the syntax and proof theory of **IPC**. The semantical characterization of intuitionistic logic started with Jaskowski (1936) and Tarski (1938); see also the references in Troelstra and van Dalen (1988, 13.9.5). Intuitionistic predicate logic **IQC** was also studied by Gentzen (1935), but a (topological) semantics for **IQC** had to wait till Mostowski (1948).

Intuitionistic arithmetic was taken up in Gödel (1933) and Gentzen (1974), but further investigations had to wait till Kleene (1945), where the realizability interpretation was introduced. It took still longer till the formalization of intuitionistic analysis was taken up again by Kleene (1952, 1957), and the formalisms studied by Kleene were not inspired by Heyting (1930B).

2. The Brouwer-Heyting-Kolmogorov interpretation

2.1. By the Brouwer-Heyting-Kolmogorov interpretation of intuitionistic logic (BHK-interpretation for short) we understand the well known explanation of the meaning of "proof of a compound statement A" in terms of what it means to prove the constituents of A.

Characteristic is the explanation of implication:

"A proof of A → B is a construction method c transforming any proof p of A into a proof c(p) of B".

(For more details see Troelstra and van Dalen 1988.) Implicitly such an interpretation is already found in Brouwer's writings, in particular (1908) and (1923).

2.2. Heyting 1930C, 1931. In these papers the clauses of the BHK-interpretation for ∨ and ¬ are explicitly formulated. In (1930C) Heyting associates with a proposition a problem or an expectation ("attente", §2), in (1931) an "Erwartung" or "Intention" (p.113), and distinguishes between a proposition p and the proposition +p ("p is provable"). As a result of correspondence with Freudenthal (published in Troelstra 1983), shortly after the meeting in Königsbergen (Kaliningrad), Heyting observed that his formalization of intuitionistic logic was applicable only to propositions of the form +p. In the correspondence with Freudenthal, Heyting also formulates for the first time explicitly the implicational clause of the BHK-interpretation mentioned above. Kolmogorov, in his first letter to Heyting (3.6), points out that Heyting is mistaken in presenting the Goldbach conjecture as an example of a proposition for which it might be the case that both ⊢ ¬+p and ⊢ ¬¬p.

The possibility of a proposition with a parameter such that one can find a proof for each particular value of the parameter, but not a uniform proof applicable to all values of the parameter (a method), as considered in Heyting (1930C), does not fit into an intuitionistic point of view, as pointed out by Kolmogorov in his second letter; cf. also Heyting's letter to Becker.

2.3. Kolmogorov 1932 and Heyting 1934. Kolmogorov's second letter to Heyting suggests that he sent Heyting a copy of his (1932) before publication. Both Heyting (1934) and Kolmogorov apparently regarded the interpretation of propositions as expectations, and the interpretation as problems, as distinct. Later Heyting came to regard them as essentially the same, cf. Heyting (1958). In his monograph (1934) Heyting not only described his own version of the BHK-interpretation, but also Kolmogorov's calculus of problems, extended with clauses for the quantifiers. Thus we feel justified in saying that Heyting and Kolmogorov have an equal share in the explicit formulation of the proof-interpretation of intuitionistic logic.

2.4. Hypothetical reasoning. This has been discussed at some length in Troelstra (1983), and I do not want to repeat the whole discussion. However, I wish to add a comment. In (1983) it was noted that Brouwer's thesis might leave the reader with the impression that Brouwer did not want to allow hypothetical reasoning. Dirk van Dalen pointed out that the Brouwer-Korteweg correspondence, published in the 1981 reprint of Brouwer (1907), modifies this impression somewhat (though Brouwer is not very clear on this point). In any case, I think that there is little doubt that Brouwer was prepared to allow hypothetical reasoning as soon as he turned his attention to intuitionistic logic in earnest, i.e. from 1908 onwards.

2.5. There exist a number of "implementations" of the BHK-interpretation, of which Kleene's realizability (1945) is perhaps the simplest and most striking example. According to Kleene (1973), part of the idea for this interpretation went back, not to Heyting or Kolmogorov, but rather to remarks in Hilbert and Bernays (1934, page 32) describing the finitist meaning of an existential statement as an incomplete statement which may be completed by presenting the value of the variable verifying the existential statement, and similarly for disjunctions. It is not unlikely that these remarks in Hilbert-Bernays are the result of Brouwer's influence on the formulation of the finitary point of view. As described in his (1973), Kleene tried to use Heyting's clauses for implication for his definition of realizability, but that "did not work"; he then proposed the now familiar clause for "x realizes A → B", apparently without recognizing or viewing this as a "recursive implementation" of Heyting's clause, a view which, in retrospect, seems so natural.

A good deal of discussion has been generated by Kreisel's (1960) proposal for a sharpening of Heyting's clauses; as to this discussion see the references in Troelstra and van Dalen (1988, 1.5.3).

3. Sources

In reproducing the letters we have retained spelling and grammar (errors included), and the layout as far as possible.

3.1. Fragments from letters between O. Becker and A. Heyting. Here Heyting explains how he arrived at his axiomatization of intuitionistic propositional logic, and emphasizes that his logic is applicable only to propositions of the form "+p"; cf. the end of Kolmogorov's second letter below, and the Heyting-Freudenthal correspondence in Troelstra (1983).

Becker to Heyting, 19-III-33:

"... Man kann nämlich fragen: ist Ihr Logikkalkül der einzig mögliche Kalkül, um die Brouwersche intuitionistische Mathematik zu begründen? Dass er dazu hinreicht, glaube ich; aber ist er auch die notwendige logische 'Bedingung' der B'schen Mathematik? Wie sind sie überhaupt auf Ihre Axiome gekommen? Gefühlsmässig oder nach einem eindeutigen logischer Prinzip? Denn aus dem ... Hilbert'schen Kalkül einfach den Satz vom ausgeschlossenen Dritten ($p \vee \neg p$) weglassen, ist doch kein Verfahren, das eindeutig auf ein neues Axiomensystem führt. ... "

Heyting to Becker, 23-IX-33:

"...Sie fragen, wie ich zu meinen Axiomen der Logik gekommen bin. Ich habe die Axiome und Sätze der Principia Mathematica gesichtet und aus den zulässig befundenen ein System von unabhängigen Axiomen gemacht. Bei der relative Vollständigkeit der Principia ist die Vollständigkeit meines Systems m.E. in der best möglichen Weise gesichert. Es ist ja prinzipiell unmöglich mit Gewissheit alle zulässigen Schlußweisen in ein Formales System zu erfassen. eine andere Sache ist es, dass die Anwendung meiner Logik auf konstruktive Fragen beschränkt ist. Was ich damit meine, möge das folgende Beispiel erhellen. Es seien zwei Folgen von reellen Zahlen $\{a_i\}$ und $\{b_i\}$ vorgelegt. Die Aussage 'Für jedes i ist $a_i = b_i$' lässt zweierlei Auffassung zu. a). Sie kann die Aufgabe bedeuten, einen allgemeinen Beweis zu suchen, der sich bei der Wahl eines bestimmtes Index i zu einen Beweis für $a_i = b_i$ spezialisiert; b) man kann darunter die Erwartung verstehen, dass es, wenn man immer wieder einen Index i beliebig wählt, jedesmal gelingen wird, $a_i = b_i$ zu beweisen. Der Unterschied ist klar, wenn man auf a) und b) die Negation anwendet. Es wäre denkbar, dass die Annahme eines Beweises wie unter a) gefördert, als wiederspruchsvoll erwiesen wäre, ohne dass dieser Wiederspruch auch die Annahme unter b) treffen würde. Meine Logik gilt dann, wenn jede Aussage in der Art wie unter a) verstanden wird; die Logik der nicht-konstruktiven Erwartungen b) würde sich viel verwickelter gestalten; ich halte ihre Aufstellung nicht für sehr fruchtbar. ..."

3.2. Letter from P. Bernays to A. Heyting. This letter concerns technical points.

Sehr geehrter Herr Dr. Heyting!

Seit langer Zeit schon habe ich vor, Ihnen für die Übersendung Ihrer zwei Abhandlungen zu danken und Ihnen zu derjenigen , welche den Aussagenkalkül betrifft, einige kleine Bemerkungen mitzuteilen, die ich mir bald nach dem Empfang Ihrer Separata überlegt habe.

Ihre Untersuchung steht mir nicht nur durch die Beziehung zu der axiomatischen Betrachtung der Russellschen Aussagenlogik nahe, – welche Sie in so freundlicher Weise zitiert haben. Diesen Russellschen Aufbau der Aussagenlogik habe ich bald als axiomatisch sehr unbefriedigend empfunden. Und bei der Entstehung der Hilbertschen Beweistheorie ergab es sich sozusagen von selbst, dass die positive Aussagenlogik von der Rolle der Negation abgesondert wurde.

Die Vorträge, die Prof. Brouwer seinerzeit (erstmalig) in Göttingen hielt, regten mich zu der Frage an, wie sich am einfachsten eine Brouwerschen Aussagenlogik aussondern lasse, und ich kam zu dem Ergebnis, dass sich dieses durch weglassen der einen Formel

$\neg\neg a \supset a$ (in Ihrer Symbolik)

bewirken lasse. Ich schrieb auch damals an Prof. Brouwer, als Bemerkung zu seinen Vortrag, eine briefliche Notiz, worin ich – gegenüber der von ihm damals geäusserten Vermutung, dass der Satz von der doppelten Verneinung schwächer sei als der Satz vom ausgeschlossenen Dritten – auf den deduktiven Zusammenhang hinwies, der jetzt als "Satz von der Absurdität der Absurdität des Satzes vom ausgeschlossenen Dritten" benannt zu worden pflegt.)

– Sie finden in Hilberts Abhandlung "Die Grundlagen der Mathematik" (Abh. aus d. Math. Seminar d. Hamburg. Univ., 1928) ein System der Aussagen-Logik, welches bei Weglassung der letzten Formel (12.) mit Ihrem System, bei Weglassung von 4.1., gleichwertig ist. (Insbes. sind die Formeln 1.-7. mit Ihren Formeln 2.1., 2.11., 2.12., 2.13., 2.14., 2.15., gleichwertig.) Für dieses System der Formeln 1.-12. habe ich mir - im Anschluss an Unterhaltungen mit den Herren Prof. Lukasiewicz und Dr. Tarski, (die mir auf dem Kongress zu Bologna ihre Ergebnisse im Gebiete der Aussagenlogik mitteilten), vollständig die Unabhängigkeitsbeweise überlegt. – (N.B. die Formel 2. ist nicht von allen übrigen, wohl aber von den Formeln 1. und 3.-11. unabhängig.)

Diese Beweise sowie einige daran knüpfende Ergebnisse, insbes. über die positive Logik, habe ich nicht publiziert, vielmehr nur mündlich einiges davon in einer Vorlesung vorgetragen. –

Worin sich die Anlage Ihres Systems von den System der Formeln 1.-12. hauptsächlich unterscheidet, das ist die Zusammenfassung der Formeln für die Implikation mit denen für die Konjunktion zu einer Formelgruppe.

Diese Zusammensetzung ist ja von inhaltlichen Standpunkt durchaus zu motivieren; und wenn auch die Absonderung der Formeln der Implikation für sich ein Interesse hat (und es überdies möglich macht, dass in den Formeln, welche die Konjunktion, Disjunktion und die Negation charakterisieren, ausser der betreffende Operation jeweils nur die Implikation auftritt) –, so gibt andrerseits Ihr Verfahren den Axiom-Formeln eine andere Art von grösserer Übersichtlichkeit, – die wie ich annehme von Ihnen beabsichtigt ist, ich meine die Tatsache, dass bei Ihren Axiomen nirgends im Vorderglied oder im Hinterglied einer Implikation wiederum eine Überlagerung von Implikationen vorkommt. Das ist jedenfalls ein Gewinn an formaler Eleganz und erleichtert die Auffassung der Axiome. –

Was ich nun vor allem bemerken wollte, ist dass Ihre Operationsregel "1.2" ("Sind a und b richtige Formeln, so ist a ∧ b eine richtige Formel") entbehrlich ist.

Nämlich es seien a, b richtige Formeln. Aus 2.14. erhält man durch Einsetzung die Formel

$$a \supset .b \supset a.$$

Diese zusammen mit a gibt nach der Regel 1.3. die Formel

$$b \supset a,$$

diese zusammen mit der Formel

$$b \supset a. \supset .b \wedge b \supset a \wedge b,$$

welche durch Einsetzung aus 2.12. hervorgeht, ergibt nach der Regel 1.3. die Formel

(1) $$b \wedge b \supset a \wedge b.$$

Andrerseits entsteht aus 2.1. durch Einsetzung die Formel

$$b \supset b \wedge b;$$

diese zusammen mit b gibt nach der Regel 1.3. die Formel

$$b \wedge b$$

und diese wieder ergibt zusammen mit der abgeleiteten Formel (1) nach Regel 1.3. die Formel

$a \wedge b.$ –

Eine andere Vereinfachung ist, dass die Formel 4.11. ersetzt werden kann durch

$$a \supset \neg a. \supset \neg a.$$

Nämlich aus 4.1. ergibt sich durch Einsetzung

$$\neg b \supset .b \supset \neg a,$$

und aus dieser Formel leitet man mit Hilfe Ihrer sechs ersten Formeln ab:

$$a \supset b. \wedge .a \supset \neg b. \supset .a \supset \neg a.$$

Nimmt man daher

$$a \supset \neg a. \supset \neg a$$

als Axiom, so gelangt man zu 4.11. –
Bei den Unabhängigkeitsbeweisen wäre es vielleicht lohnend, noch gewisse Verschärfungen der Ergebnisse zu erhalten, so z.B. nachzuweisen, dass die Formel 2.14. nicht durch die Formel

$$a \wedge b \supset a$$

vertreten werden kann. – Ich vermute übrigens, dass man in allen diesen Fällen mit endlichen Element-Systemen auskommt.
– Mit besten Grüssen empfiehlt
sich Ihnen Ihr
P. Bernays
5.XI.30.

3.3. Translation of a fragment of a letter of L.E.J. Brouwer to Heyting This letter and the next one are of some interest as revealing Brouwer's reaction to Heyting's work on intuitionistic logic.

17-VII-1928

Dear mr. Heyting,
I found your manuscript extraordinarily interesting, and I regret that you have to urge me to return it. I would appreciate it if, in the future, you would make copies of your manuscripts beforehand, at least if you desire me to read them more than just superficially. Even so, I have come to appreciate your work so much, that I want to ask you to prpare a version in German for the Mathematische Annalen (preferably slightly expanded, instead of shortened). Perhaps you can then distinguish more sharply between primitive symbols and symbols introduced by definition (as abbreviations for other symbols). And perhaps it is also possible (in view of §13) to formalize the notion of a "law". But these remarks are of secondary importance. ...

3.4. Letter of L.E.J. Brouwer to Th. de Donder

9 octobre 1930.

Mon cher collègue

En préparant une note sur l'intuitionisme pour le Bulletin de l'Académie Royale de Belgique, je fus agréablement surpris d'en voir paraître une de mon élève M. Heyting élucidant d'une manière magistrale les points que j'avais voulu mettre en lumière moi-même. Je crois qu'après la note de M. Heyting il ne reste plus grand'chose à dire sur les questions en litige, et que dès à présent le lecteur des éditions de votre Académie saura suffisamment à quoi s'en tenir concernant les idées de M.M. Barzin et Errera, qui, à part du grand intérêt qu'elles présentent, sont néanmoins intenables dans leur tendance essentielle. J'examinerai si à la note de M. Heyting il reste à ajouter quelque chose qui puisse approfondir les notions générales sur la logique intuitionniste, et dans le cas affirmatif je ne tarderai pas à en composer une note et je serai heureux de vous l'envoyer.

Agréez, mon cher collègue, l'expression de mes sentiments cordiaux.

3.5. Letters of V. Glivenko to A. Heyting The letters from 1928 document the discovery of the "Glivenko theorem". Glivenko's claim, in his letter of 4-VII-1928, that $\neg\neg(\neg\neg p \rightarrow p)$ is not provable, might have been proved by him e.g. by using an interpretation in a matrix with three truth values, corresponding to a linear lattice with three elements, where the middle element is assigned to falsehood. Glivenko's letter from 24-X-1933 discusses Kolmogorov's "calculus of problems" interpretation and mentions the possibility of taking falsehood instead of negation as a primitive.

4 Juillet 1928

Monsieur,

comme je suis, cet été, à un village près de Volga, loin de Moscou, c'est seulement aujourd'hui que j'ai reçu votre lettre.

Je suis heureux que nos points de vue sur le problème de formalisation de la logique et mathématique intuitionniste coïncident complètement. Dans ma note polémique de Bruxelles, je n'avais pas pu de poser ce probléme assez nettement, parce que là, conformément au but spécial de cette note, j'ai eu besoin d'un langage qui pourrait être comprise sans peine par des savants qui n'ont pas pénétrés, à mon avis, assez profondement aux idées intuitionnistes. Mais j'ai fait, moi-même, quelques considérations plus essentielles sur ce sujet dans un autre travail qui se trouve dejà sous presse et que, je le crois, je pourrai présenter à vous depuis quelques semaines.

Maintenant, permettez moi de me borner à vous communiquer un résultat qui se rattache encore immédiatement à ce que j'avais discuté dans ma note de Bruxelles, et que j'ai trouvé ces jours derniers. Il est à remarquer que le système d'axiomes I-X qui ont été indiqués dans cette note n'est point un système complet d'axiomes de la logique mathématique classique. D'ailleur, par obtenir un système complet, il suffirait de modifier légèrement ce système-ci, à savoir d'y préciser la correspondance formelle entre implication et multiplication, qui s'exprime par les formules:

$$\text{XI*}.\quad p \,.\supset.\, q \supset r \,:\supset:\, pq \supset r,$$

$$\text{XII*}.\quad pq \supset r \,:\supset:\, p \,.\supset.\, q \supset r,$$

et d'y remplacer le principe du tiers exclu X par ce de réciprocité des espèces complémentaires:

$$\text{X*}.\quad \sim(\sim p) \supset p.$$

Comme les axiomes I-IX et XI*-XII* sont, sans doute, admissibles dans la logique intuitionniste, il est naturel de se demander s'il est, ou non, démontrable, à partir de ces axiomes-ci, la non-fausseté du principe X*:

$$\sim(\sim(\sim(\sim p) \supset p)).$$

J'ai démontré que la réponse est négative.

Que doit-on tirer de ce fait pour la construction de la logique intuitionniste? Au cours de mes recherches, j'ai trouvé quelques indications sur ce sujet, mais pas encore assez claires. Ce que me semble d'être certain, c'est seulement que l'élaboration des principes intuitionnistes exige encore beaucoup de travail qui peut devenir très fecond pour la connaissance de faits logiques importants sur lesquels la logique classique ne fait pas même les indications.

Agréez, Monsieur, l'assurance de ma parfaite considération.

V. Glivenko.

Moscou, le 13 octobre 1928.

Monsieur,

avant de tout, je dois vous remeçier pour que vous avez me présenté les extraits de vos travails. Au village où j'avais passé cet été, je n'avais reçu que votre lettre et même je n'avais su rien sur les extraits qui m'attendaient à Moscou.

Ce que je vous avais écrit, dans la lettre précédente, sur la fausseté de la fausseté du principe $\sim(\sim p) \supset p$, n'est, peut-être, que le résultat d'un malentendu. Dans le t.32 du «Recueil Mathematique» de Moscou (1925) M. Kolmogoroff, un des plus ingénieuses mathématiciens russes, a publié un article intitulé «Sur le principe de tertium non datur» (en russe), où il a proposé une axiomatique de la logique intuitionniste de propositions. Là, il rejette non seulement le principe du tiers exclu, mais aussi le principe $\sim p \,.\supset.\, q \supset p$ (quoique il admet le principe $p \,.\supset.\, q \supset p$). C'est précisement depuis cette restriction que ma conclusion est vraie, c'est-à-dire que la fausseté de la fausseté du principe $\sim(\sim p) \supset p$ n'a pas lieu. Mais maintenant je commence de douter que la restriction faite par M. Kolmogoroff est légitime, c'est-à-dire que, pour la mathématique intuitionniste, il est nécessaire de rejetter le principe $\sim p \,.\supset.\, p \supset q$, parce que je ne peux pas démontrer cette nécessité, tandis que la nécessité de rejetter le principe du tiers exclu a été démontré, plusieurs fois et bien clairement, par M. Brouwer.

Je serais heureux si vous me communiqueraient votre opinion sur ce sujet.

Agréez, Monsieur, l'assurance de ma parfaite considération

V. Glivenko.

Moscou, le 18 octobre 1928.

Monsieur,

hier j'ai reçu votre lettre de 7 octobre qui m'a fournit incidemment la réponse à la question posée dans ma lettre de 13 octobre.

A vrai dire, le rôle de l'axiome $\sim p \,.\supset.\, p \supset q$ n'est pas encore assez clair pour moi. Mais, quoiqu'il en soit, je me suis rendu à vos raisons que la mathématique intuitionniste n'a pas besoin de rejetter cet axiome, de sorte que toutes les reflexions contre cet axiome sortiraient des bornes des matières dont nous occupons actuellement.

Quant à votre démonstration de la formule , basée sur l'axiome en question, il me semble qu'on y pourrait procéder d'une manière bien plus générale. Plus précisement, si l'on admet mes axiomes I-IX et XI*-XII* et vos axiomes

(A) $\quad p \,.\supset.\, q \supset p,$

(B) $\quad \sim p \,.\supset.\, p \supset q,$

on démontre la fausseté de la fausseté non seulement de tous ces axiomes mêmes, mais aussi de l'axiome $\sim p \vee p$. Il en suit, a l'aide de la formule $p \supset q \,.\supset.\, \sim\sim p \supset \sim\sim q$ qu'il est facile à démontrer, la fausseté de la fausseté de tous les formules démontrables à partir des axiomes I-IX, XI*-XII*, (A), (B) et $\sim p \vee p$. Or, ces derniers forment certainement un système complet d'axiomes de la logique classique de propositions (même, peut être, quand on omit les axiomes XI*-XII*). Il en résulte la fausseté de la fausseté non seulement de la formule $\sim\sim p \supset p$, mais aussi de toutes les formules que la logique classique reconnaît pour vraies.

Je remarquerai que, pour la logique intuitionniste, ce résultat positif est, à mon avis, bien plus agréable que le résultat négatif que j'avais vous communiqué cet été.

Agréez, Monsieur, l'assurance de ma parfaite considération

V.Glivenko

Moscou 30.10.1928

Monsieur,

je voudrais publier, dans les «Bulletins» de l'Académie Royale de Belgique, le résultat que je vous avez communiqué dans ma dernière lettre, à savoir que, si un certain principe est démontrable à partir des axiomes de la logique classique des propositions, c'est la fausseté de la fausseté de ce principe qui est démontrable à partir des axiomes de la logique intuitionniste. Bien entendu, cette publication n'aurait aucun sens si votre mémoire qui paraîtra dans les «Math. Ann» contient déjà ce résultat. C'est pourquoi je vous prie de me communiquer s'il en est ainsi ou non. Si non, permettez-moi d'indiquer aussi, dans le «Bulletin» de belgique, que vous avez adopté l'axiome

$$\sim p . \supset . p \supset q$$

dans le mémoire en question.

Agréz, Monsieur, etc.

V. Glivenko

Moscou, le 13 novembre 1928.

Monsieur,

je veux publier mon résultat indépendemment de votre mémoire, car, quoique ce résultat n'est qu'une rémarque presque trivial, sa démonstration rigoureuse est un peu longue. A savoir, il y est à vérifier la règle qui s'exprime par le schème

$$\frac{\begin{array}{l}\sim\sim P\\ \sim\sim(P \supset Q)\end{array}}{\sim\sim Q,}$$

ce qui exige l'emploi de la formule $\sim\sim\sim p \supset \sim p$ et d'autres.

Cependant, il serait bon si vous mentionnait mon résultat dans votre mémoire en rapport avec l'idée d'intégrité. Moi-même, je ne puis rien dire sur ce rapport, et je serai le premier qui lira avec interêt vos explications sur ce sujet.

Agréez, Monsieur, etc.

V. Glivenko

12.10.1933

Monsieur!

J'ai reçu vos Notes sur la logique, et cela était bien opportunément, car, depuis presque quatre années, je me suis ramené aux études de ces questions. De plus, l'Institut des Mathématiques de l'Université m'a chargé, cette année, de la direction du séminaire de la logique mathématique. De fait, les travaux du séminaire doivent embrasser un domaine plus large, de de l'axiomatique moderne, comprenant non pas seulement les travaux sur la logique même, mais aussi des branches de la science où les méthodes élaborées dans la logique mathématique trouvent ses applications: Grundlagenfragen de M. Karl Menger, Theorie der abstrakten Verknüpfungen de M. Fritz Klein, etc. J'ai déjà communiqué votre Note de Zürich à cet séminaire.

Aujourd'hui, j'ai à vous une petite prière. Récemment, j'ai écrit le livre sur l'intégrale de Stieltjes qui sera édite en russe. Avant de recevoir les épreuves, je voudrait savoir comment on prononce le nom «Stieltjes», pour en faire la transcription russe tout-a-fait correcte. Comme je ne connais pas hollandais, je vous prie de m'écrire ce nom comme s'il était écrit en allemagne: c'est cette dernière langue, il me semble, où la phonétique d'un mot peut être exprimée le plus facilement.

Agréez, Monsieur, etc.

V. Glivenko

P.S. Que pouvait-on dire sur le livre de M. Dassen? Je le comprend mal.

V.G.

24.10.1933

Monsieur,

je vous remercie de vos indications qui m'ont été tout-à-fait suffisantes, de sorte que je peux maintenant reproduire sans hésiter la transcription Стультьес déjà employée par certains auteurs russes.

Outre cela, je profite l'occasion de vous communiquer quelques réflexions sur le calcul des problèmes de M. Kolmogoroff.

Sa conception est sans doute très intéressante. Mais il me semble que l'opération

$$\neg a,$$

dans le sens qu'il attribue, fait intervenir dans son calcul un certain élément inévitable de la logique, par exemple de la logique classique de propositions. En effet, le problème

$$\neg a$$

a, chez lui, un sens que voisi:

« étant supposé le problème a résolu, en tirer une contradiction ».

Il semble que ce n'est autre chose que

« démontrer la proposition suivante: la prémisse que le problème a est résolu entraîne une contradiction ».

Or, on peut obtenir un calcul des problèmes entièrement indépendant. A cet effet, admettons vos axiomes 2.1, 2.11-15, 3.1 et 3.11-3.12 et, de plus l'axiome suivant:

« Il existe un élément 0 tel que, quel que soit l'élément b, on a $0 \supset b$ ».

Formellement, ce système d'axiomes sera équivalente à votre. Seulement, au lieu de l'opération

$$\neg a$$

on aura ici l'opération

$$a \supset 0.$$

Les conséquences de cette modification sont manifestes. J'ajouterai seulement un example. Prenons le domaine des problèmes arithmétiques, ça veut dire des problèmes de la forme:

« trouver des entiers $x_1, x_2, \dots, x_n$ satisfaisant à l'équation

$$f(x_1, x_2, \dots, x_n) = 0,$$

où f est une fonction bien déterminée ».

Pour le problème 0, on peut y prendre le suivant:

« trouver un entier x satisfaisant à l'équation

$$x + 1 = x \text{ ».}$$

En effet, si nous supposons ce problème-ci résolu, nous avons successivement, quels que soient les $x_1, x_2, \dots, x_n$ sous le signe f,

$$(x+1)f = xf,$$

$$f = 0.$$

On voit que, malgré l'équivalence formelle, le sens et l'emploi du calcul ainsi modifié devient, lui-aussi, modifié et, en premier lieu, affranchi de tous les éléments de la logique.

Il me serait très intéressante de savoir votre opinion sur ce sujet.

Agréez, Monsieur, etc.

V. Glivenko.

3.6. Letters from A. Kolmogorov to A. Heyting The first letter points to an oversight in Heyting (1930C). In the second letter Kolmogorov observes that the distinction between "P(x) can be solved for each x" and "there is a uniform method for solving P(x) for each x" is non-intuitionistic; the point was accepted by Heyting, as the fragment of his letter to Becker, reproduced above, shows. In the third letter Kolmogorov mentions his anticipation of Gödel (1933) in his (1925).

12-X-31

Monsieur et cher Collège

j'ai lu votre article dans "Erkenntniss" avec le plus vif interêt; l'exposition des principes de la logique intuitionniste est dans cet article extrèmement claire et définitive. Je voudrais seulement vous présenter quelques objections se rattachant à votre distinction entre +p et p.

1. Vous considérez (Académie de Belgique) comme exemple la proposition "tout nombre pair est la somme de deux nombres premiers". Il est cependant connue que la formule $\vdash \sim\sim p \supset p$ subsiste dans ce cas au point de vue intuitionniste aussi bien que classique. Si l'on sait $\vdash \sim\sim p$, on a ici automatiquement "une construction, qui nous donne cette décomposition pour tous les nombres pairs à la fois". C'est pourquoi on a aussi $\vdash \sim\sim p \supset +p$ et le cas

(1) $$\vdash \sim\sim p \wedge \sim +p$$

est impossible.

2. Il me semble que cela n'est pas le défaut de cet example particulier. Chaque "proposition" p dans votre conception est à mon avis de l'une des deux sortes:

(α) p exprime l'esperance que dans telles et telles circonstances un expériment donnera toujours un résultat déterminé (par example, que la tentative de décomposer un nombre pair en somme des deux nombres premiers donnera le résultat positif, si seulement tous les couples (p,q) $p < n$, $q < n$, seront utilisés). Chaque "expériment" doit être naturellement réalisable à l'aide d'un nombre fini d'operations déterminées.

(β) p exprime l'intention de trouver un construction.

3. Nous sommes d'accord, que dans le cas (β) la différence entre p et +p n'est pas essentielle, mais la proposition $\sim\sim p \supset p$ ne doit être considéré comme évidente. Dans le premier cas (α) au contraire p et +p ont une signification distincte, mais l'on a $\vdash \sim\sim p \supset p$ et $\vdash \sim\sim p \supset +p$. C'est pourquoi (1) est toujours impossible dans le cas (α) aussi que (β). On a donc, dans votre tableau des positions différentes d'une proposition seulement deux positions définitives $\vdash p$ et $\vdash \sim p$, puisque la position d' "insolubilité" est impossible.

4. Je préfère de conserver le nom d'une proposition (Aussage) seulement pour des propositions du genre (α) et nommer des "propositions" du genre (β) simplement problèmes (Aufgaben). Pour une proposition p on à des problèmes $\sim p$ (conduire p à une contradiction) et +p (démontrer p).

Veuillez agréer, Monsieur et chere collegue mes meilleurs salutations

A. Kolmogoroff

(undated)

Sehr geehrter Herr Kollege,

für die Aufmerksamkeit, mit welcher Sie meine Arbeit gelesen haben, bin ich ihnen sehr dankbar. Ihre Regel I,2 (oder, was formal dasselbe ist, meine Regel I) ist in der Tat überflüssig; ich lasse sie aber doch im Texte bleiben, um jede Komplikation bei dem Vergleich mit Ihrer Abhandlung zu vermeiden.

In Zürich beabsichtige ich nur eine Mitteilung aus dem Gebiete der Wahrscheinlichheitsrechnung zu machen, und wurde mir nur freuen , wenn Sie in Ihren Vortrage auch meine Gedanken über logische Fragen berühren wollten. Ich habe inzwischen über Ihr Beispiel des Satzes

"Für alle i gilt $a_i < b_i$ "

nachgedacht. Es sei im allgemeinen x eine Variable und P(x) eine von dieser Variable abhängige Aufgabe. Die "Hoffnung" für jedes x eine Lösung der Aufgabe P(x) zu finden ist in meiner Terminologie weder eine "Aufgabe" noch eine "Aussage" . Es wäre sehr interessant zu wissen, ob Sie mit dieser Hoffnung eine positive Erwartung verbinden, dass für jedes x die Aufgabe P(x) wirklich gelöst wird (von wem und wann?). Wenn diese Erwartung dabei nicht gemeint ist, so fürchte ich, dass wir uns der náiven nicht-intuitionistischen Auffassung des Satzes "P(x) ist für jedes x lösbar" zu nahe kommen würden.

Mit besten Grüsse, Ihr sehr ergebener

A. Kolmogoroff

(undated)

Sehr geehrter Herr Kollege,

ich danke Sie herzlich für die Sendung ihres Buches über die Grundlagenforschungen. Dieses Buch betrachte ich erstens als die beste Einführung in die modernen Grundlagenuntersuchungen und zweitens als ein wesentlicher Schritt zu einer Synthese verschiedener Richtungen. Für mich persönlich bildet die Erscheinung dieser zusammenfassender Darstellung einen wesentlichen Impuls um meine eigene Untersuchungen weiter zu verfolgen.

Ich möchte noch Ihnen zeigen, dass die Resultate von Gödel über die Möglichkeit die klassische Logik un die klassische Zahlentheorie im Rahmen der intuitionistischen Mathematik zu interpretieren sehr nahe mit den Ausführungen meiner russischen Arbeit "Sur le principe de tertium non datur" (Recueil Mathématique de Moscou 32 (1925), 646-667) stehen. Ich glaube, dass man noch viel weiter in dieser Richtung gehen könnte und also vom intuitionistischen Standpunkte aus die Widerspruchsfreiheit eines grossen Teiles der klassischen Mathematik zu beweisen.

Mit besten Grüssen,

Ihr sehr ergebener

A. Kolmogoroff.

Bibliography

M. Barzin, A. Errera (1927), Sur la logique de M. Brouwer, Acad. Roy. Belg. Bull. Cl. Sci. (5) 13, 56-71.

L.E.J. Brouwer (1907), Over de Grondslagen der Wiskunde (Dutch) (Maas en van Suchtelen, Amsterdam. Republished (1981) with additional material, edited by D. van Dalen (Mathematiscl Centrum, Amsterdam)

– (1908), Over de onbetrouwbaarheid der logische principes (Dutch), Tijdschrift voor Wijsbegeerte 2, 152-158.

– (1923), Intuïtionistische splitsing van mathematische grondbegrippen (Dutch), Nederl. Akad. Wetensch. Verslagen 32, 877-880. German translation Jahresber. Dtsch. Math.-Ver. 33, 251-256.

V.I. Glivenko (1928), Sur la logique de M. Brouwer, Acad. Roy. Belg. Bull. Cl. Sci. (5) 14, 225-228.

– (1929), Sur quelques points de la logique de M. Brouwer, Bull. Soc. Math. Belg. 15, 183-185.

G. Gentzen (1974), Über das Verhältnis zwischen intuitionistischer und klassischer Arithmetik, Arch math. Logik Grundlagenforsch. 16, 119-132 (the paper is based on a galley proof of the Mathematische Annalen which was withdrawn by Gentzen when he learnt of the results of Göde 1933).

– (1935), Untersuchungen über das logische Schließen I,II, Math. Z. 39, 176-210, 405-431.

K. Gödel (1933), Zur intuitionistischen Arithmetik und Zahlentheorie, Ergebnisse eines mathematischen Kolloquiums 4, 34-38.

J. van Heijenoort (1967), ed., From Frege to Gödel. A Source Book in Mathematical Logic 1879-1931 (Harvard University Press, Cambridge Mass.). Reprinted 1970.

L. Henkin (1960), On mathematical induction, Amer. Math. Monthly 67, 323-338.

A. Heyting (1930), Die Formalen Regeln der intuitionistischen Logik, Sitzungsberichte der Preussischen Akademie von Wissenschaften. Physikalisch Mathematische Klasse, 42-56.

– (1930A), Die Formalen Regeln der intuitionistischen Mathematik II, Sitzungsberichte der Preussischen Akademie von Wissenschaften. Physikalisch Mathematische Klasse, 57-71.

– (1930B), Die Formalen Regeln der intuitionistischen Mathematik III, Sitzungsberichte der Preussischen Akademie von Wissenschaften. Physikalisch Mathematische Klasse, 158-169.

– (1930C), Sur la logique intuitionniste, Acad. Roy. Belg. Bull. Cl. Sci. (5) 16, 957-963.

– (1931), Die intuitionistische Grundlegung der Mathematik, Erkenntnis 2, 106-115.

– (1934), Mathematische Grundlagenforschung. Intuitionismus. Beweistheorie (Springer, Berlin) Reprinted 1974.

– (1958), Intuitionism in mathematics, in: R. Klibansky, editor, Philosophy in the mid-century. A Survey (La Nuova Italia editrice, Firenze), 101-115.

– (1978), History of the foundations of mathematics, Nieuw Archief voor Wiskunde (3), 26, 1-21.

D. Hilbert (1928), Die Grundlagen der Mathematik, Abh. Math. Sem. Hamburg 6, 65-85.

D. Hilbert, W. Ackermann (1928), Grundzüge der theoretischen Logik (Springer, Berlin).

D. Hilbert, P. Bernays (1934), Grundlagen der Mathematik I (Springer, Berlin).

S. Jaskowski (1936), Recherches sur le système de la logique intuitionniste, Actes du Congrès International de Philosophie scientifique, septembre 1935 Paris vol. VI (Hermann, Paris), 58-61.

S.C. Kleene (1945), On the interpretation of intuitionistic number theory, J. Symbolic Logic 10, 109-124.

– (1952), Recursive functions and intuitionistic mathematics, in: L.M. Graves, E. Hille, P.A.Smith O. Zariski ,editors, Proceedings of the Int. Congr. Math., August 1950, Cambridge, MA (Amer. Math. Soc., Providence, RI), 679-685.
– (1957), Realizability, in: Summaries of talks presented at the Summer Institute for Symbolic Logic, July 1957, Ithaca, NY (Institute for Defense Analyses, Communications Research Division), 100-104.
– (1973), Realizability: a retrospective survey, in: A.R.D. Mathias, H. Rogers, eds., Cambridge Summer School in Mathematical Logic (Springer, Berlin), 95-112.
A.N. Kolmogorov (1925), On the principle of the excluded middle (Russian), Mat. Sb. 32, 664-667; translated in van Heijenoort (1967), 414-437.
– (1932), Zur Deutung der intuitionistischen Logik, Math. Z. 35, 58-65.
G. Kreisel (1962), Foundations of intuitionistic logic, in: E. Nagel, P. Suppes, A. Tarski, eds., Proceedings of the first International Congress for Logic, Methodology and Philosophy of Science (Stanford University Press, Stanford), 198-210.
A. Mostowski (1948), Proofs of non-deducibility in intuitionistic functional calculus, J. Symb. Logic 13, 204-207.
A. Tarski (1938), Der Aussagenkalkül und die Topologie, Fundam. Math. 31, 103-134.
A.S. Troelstra (1978), A. Heyting on the formalization of intuitionistic mathematics, in : E.M.J. Bertin, H.J.M. Bos and A.W. Grootendorst, eds., Two Decades of Mathematics in the Netherlands 1920-1940. A Retrospection on the Occasion of the Bicentennial of the Wiskundig Genootschap (Mathematisch Centrum, Amsterdam), 153-175.
– (1981), Arend Heyting and his contibution to intuitionism, Nieuw Archief voor Wiskunde (3) 29, 1-23.
– (1983), Logic in the writings of Brouwer and Heyting, in : V.M. Abrusci, E. Casari and M. Mugnai, eds., Atti del Convegno Internazionale di Storia della Logica. San Gimignano, 4-8 dicembre 1982 (Cooperativa Libraria Universitaria Editrice Bologna, Bologna), 193-210.
A.S. Troelstra, D. van Dalen (1988), Constructivism in Mathematics (North-Holland, Amsterdam).
A.N. Whitehead, B. Russell (1910), Principia Mathematica I (Cambridge University Press, Cambridge).

HEYTING AND INTUITIONISTIC GEOMETRY

D. van Dalen

State University of Utrecht
Mathematical Institute
Budapestlaan 6, P.O. BO.010
350B TA Utrecht, The Netherlands

It may seem strange that the second fully committed intuitionist in mathematics entered his career with a treatise on axiomatic geometry, for axiomatics did have a formalist flavour and one cannot suspect Brouwer, Heyting's teacher, of leanings in that specific direction. There are a number of possible explanations for the choice of this particular topic - which, by the way, had been suggested by Brouwer. One of them is Brouwer's own interest in the foundations of geometry in the Pasch-Hilbert-style; his Ph.D.Thesis contained a good deal of geometry and he regularly lectured on the foundations of geometry. His inaugural address as a "privaat docent" bore the title "The nature of geometry". Hence it is not all that surprising that Heyting choose the intuitionistic foundations as a topic for his Ph.D.thesis.

There is another possible reason for the choice of topic: ever since Hilbert's *Grundlagen der Geometrie* there had been a great deal of activity in this particular area of geometry and the subject had become something of a touchstone for foundational research; the reader must bear in mind that in 1925 geometry was the only reasonably formalized subject in mathematics with a substantial tradition of axiomatic research. The notion of "model" more or less emerged in the wake of the discovery of non-euclidean geometry, and, generally speaking, the foundations of geometry were a focussing point for new trends - algebra (in connection with the coordinatization), and in particular group theory (study of projective transformations). The research activity in other axiom systems (such as Peano's or Whitehead and Russell's) had in no way reached the level and sophistication of that of the foundations of geometry.

Heyting's dissertation "Intuitionistic axiomatics of projective geometry" was the first instance of axiomatization in intuitionistic (or constructive, for that matter) mathematics. The style of the dissertation is informal, that is to say there is no attempt to formalize the matter in any particular system. As a matter of historical fact, the manner of presentation in the foundations of geometry has changed little from Hilbert's Foundations till the present day.

Mathematical Logic
Edited by P. P. Petkov
Plenum Press, New York, 1990

The doctorate was awarded on May 27, 1925 at the University of Amsterdam; it was followed by two papers in the Mathematische Annnalen: *Zur Theorie der linearen Gleichungen in einer Zahlenspezies mit nicht kommentativer Multiplikation* , Heyting (1928) [1],465 - 490, and *Zur intuitionistischen Axiomatik der projektiven Geometrie*, Heyting (1928A).

At the time of writing Heyting was a high-school teacher at Enschede, a small town in the eastern part of the Netherlands where he was rather isolated from the Dutch mathematical community.

2. Heyting on the axiomatic method

The dissertation contains a brief methodological introduction in which Heyting made some remarks on foundational matters and on the geometrical contents. Right in his first major publication he takes the liberal position in the philosophy of mathematics that is so characteristic for all of his later work: "One can view the new conceptions of mathematics from various philosophical positions. It may be immaterial whether one assumes that the notions that are presented by mathematical intuition as immediately clear, correspond to objects in a reality outside of us, or that one considers them exclusively as products of our mind. It only matters that one can reason about a mathematical system only after one has thought it, i.e. constructed it, in the mind."

Heyting rejected the axiomatic method as a foundation of mathematics, but he did not choose to follow Weyl (1921) in discarding it altogether. His position, at the time of the dissertation, on the intuitionistic value of the axiomatic method is somewhat cryptic. Referring to Brouwer's introduction of a hierarchy of species, he stated that "Axiomatics is an application of that principle (comprehension in a typed universe, cf Brouwer (1918)): the defining property of the new species A is that between the elements of its elements the relations specified by the axioms exist". Apparently one has to interpret this statement in the older tradition that an axiom system implicitly defines its basic notions, and hence it would do so in a previously constructed domain.

At a later occasion Heyting returned to this matter in his paper "*L'axiomatique intuitioniste*" Heyting (1951). In this paper the intuitionistic views on the axiomatic method are clearly and forcefully stated: axiomatic reasoning corresponds to an incomplete construction. Whereas the direct activity of a constructive mathematician consists of constructing mathematical systems and proving facts about them (i.e. carrying out proof-constructions), the practice of axiomatics consists of providing hypothetical constructions that can be turned into real constructions by the construction of an actual mathematical domain (the model or structure fitting the axiomatic system) which serves as a point of departure for the hypothetical

[1] I have adopted the dating of the literature from the Ω-bibliography of mathematical logic, as a consequence there are some minor changes compared to earlier accounts of intuitionistic mathematics.

constructions. One might paraphrase this view point as "axiomatic mathematics is instant mathematics". Further comments be found in Heyting (1962).

3. The dissertation

This first investigation into intuitionistic axiomatics had to deal with considerable difficulties. The beginnings of projective geometry are so delicately balanced and the axioms are so minimal, that the first steps towards a theory depend heavily on proof by contradiction. Since this particular tool is denied to the intuitionist, a good deal of ingenuity was required to get the project started at all.

Heyting introduced at this point a novelty that served the purpose and that he exploited with great success in many researches. In order to be able to apply a substitute of the proof by contradiction in the case of equality and incidence he added a positive analogue of 'unequal' and the negation of incidence, this after the example of the apartness relation on the continuum as introduced by Brouwer in Brouwer (1919) under the name "*örtlich verschieden*" (locally distinct) [2]. We will use somewhat anachronistically the terminology "apart from" and symbolize it by $\#$ [3].

The well-known axioms for apartness occur for the first time in the dissertation (p 27)

$$A \# B \rightarrow B \# A$$
$$\neg A \# B \leftrightarrow A = B$$
$$A \# B \rightarrow A \# C \vee B \# C$$

The axiom system for projective geometry used by Heyting is not one designed to bring out the dual character of projective geometry, it is based on treatises by Pieri (1898) and Whitehead (1913). Points are the basic entities, and lines and planes are introduced as species (sets). The first sections of the geometric part show a true mastership of the possibilities of the formal manipulations of intuitionistic axiomatics, once the author got to the introduction of coordinates treatment became more routine.

The geometric part of the dissertation served as the basis of the paper "*Intuitionistische Axiomatik der projektiven Geometrie*" Heyting (1928 A). This paper is an expanded version of the geometric sections of the thesis without the theory of ordering. The coordinatization in the paper is completely general and yields a skew field, whereas the underlying field in the dissertation is that of the reals.

Although there is only a lapse of two years between the dissertation and the Mathematische Annalen paper, there is a remarkable difference in style, one would almost say an increase of self-confidence. From our point of view the choice of the ordering of the material is somewhat unusual, e.g. a theorem like "if P is not apart from l, then P is on l", is shown after a section on the properties of planes and points in space, whereas we would be inclined to prove it right away as a kind of sanctioning of the apartness relation. However, Heyting had a point in that he needed the auxiliary facts to prove the corresponding theorems about points and planes.

There are a number of refinements in the treatment of intuitionistic geometry vis a vis classical geometry. As a rule difficulties arise in those cases where the relative locations of points and lines is unknown. E.g. one has to show Desargues' theorem not only for the cases of triangles in the same plane and triangles in planes apart, but also for the general case in which the relative position of the planes is not known. Heyting has developed a convenient routine by which the general case is reduced to one or more special cases. This routine is by no means a foolproof automatism, it was applied with great elegance by Heyting in widely varying contents.

4. <u>The excursion to algebra</u>

In the dissertation to goal was the characterization of a geometry over Á, so the main non-geometric task was to set up a convenient construction of the continuum that could be mirrored in geometry. The particular approach Heyting chose was that of Brouwer (1921), a general method to construct a continuum out of a countable, densely ordered set without endpoints. In spite of the praise Heyting bestowed on this method, it is a somewhat cumbersome process, that eventually was simplified by Brouwer to a method closer to Dedekind's cuts, Brouwer (1926). The latter, in its turn, is not a miracle of elegance either, the more recent versions of intuitionistic Dedekind reals seem far more natural, cf Troelstra, van Dalen (1988).

In his paper *"Lineare Gleichungen bei nicht kommutativer Multiplikation"* for the *Mathematische Annalen* Heyting developed a linear algebra over an intuitionistic skew field with apartness. In the paper Heyting dealt with the linear algebra required for the treatment of the analytic geometry of projective space (in principle for n dimensions). A novelty that has little to do with intuitionism is his introduction of designants, a version of determinants for skew fields [4]. In a sense the notion is wholly pragmatic, one should note, however, that the role of determinants is rather more essential in constructive than in classical mathematics; they allow a reduction in complexity (they eliminate a quantifier).

Before we return to geometry we will discuss briefly Heyting's paper on intuitionistic algebra, "*Untersuchtungen über intuitionistischen Algebra*", Heyting(1941). The treatment of algebra in this paper follows the by then classical pattern of Van der Waerden's *Moderne Algebra*. Groups, rings, integral domains and fields are introduced with apartness relations and their main properties are derived. The feature that distinguishes Heyting's algebra from the Kronecker-style algebra is the complete generality with respect to equality. No decidability of equality is required (the theory goes beyond discrete structures), in keeping with the idea that algebra should be general; in particular the theory of the reals and their derived structures should be covered.

---

[4] In the literature little attention is paid to "non-commutative determinants"; one version was introduced by Dieudonné [D 1948], cf [A 1957], which, however, does not coincide with Heyting's notion.

The heart of the paper is the theory of polynomial rings, here Heyting demonstrates a complete control over all those intuitionistic tricks that are indispensable to get a viable theory at all. The results are by no means straightforward copies of classical algebra, here is an example: K[x] modulo a prime polynomial f is only an integral domain (K is a field), it is a field if f is regular, i.e. when its leading coefficient is apart from zero. In this paper the "two-step technique" is turned into a powerful tool, by this we mean the proof a general theorem by first dealing with a special case, in which certain elements (coefficients, etc.) are taken to satisfy some convenient apartness conditions, followed by a treatment of the general case by a clever reduction to the special case. Invariably this method uses the axiom

$$a \# b \rightarrow c \# a \vee c \# b.$$

In some cases a chain of steps is required. Heyting's technique of handling factor rings contains all the essential ingredients used in the anti-ideals introduced by Dana Scott (1979); for a recasting of Heyting's arguments cf. Ruitenburg (1982), Troelstra, van Dalen (1988).

5. Later work in geometry.

Heyting was appointed to a chair of geometry at the University of Amsterdam and as such he taught innumerable courses on all kinds of geometry, most of them in the traditional (i.e. classical) presentation. The book Heyting (1963) is the outcome of such activities. His courses were beautifully presented and their contents was systematic to the highest degree. His courses on foundations of geometry bore the traces of similar courses presented by his teacher Brouwer. Among the regular courses on special topics in intuitionistic mathematics, geometry also figured from time to time. In the Heyting archive there are lecture notes from courses in 1948/49 [5] and 1962/63 [6]. The course of 1948/49 is more or less a copy of the material on plane projective geometry from the dissertation, some details have been changed. In particular the presentation of the continuum is simplified following Brouwer (1926).

In 1956/57 another course on intuitionistic projective geometry was offered by Heyting, and this course contained a number of innovations, the most important one being the coordinatization of the plane via Marshal Hall's ternary field. Heyting had already used this approach in his standard course on the foundations of geometry. The course of 1962/63 was more or less a duplicate of the 1956/57 course. At the time of the 1956/57 course Heyting returned to his research in geometry, from which the paper "Axioms for intuitionistic plane affine geometry", Heyting(1959) resulted. The problem considered in this paper is one which is totally routine in classical geometry, but extremely delicate in the intuitionistic case: given an affine plane, find a projective plane into which it can be embedded. The problem turned out to be surprisingly difficult, Heyting had to introduce extra axioms for the affine plane in order to obtain a canonical embedding of the affine plane into a projective plane. The difficulty in the

---

[5] A. Heyting Archief, nr C48 (notes taken by G.W. Decnop).

[6] A. Heyting Archief, nr C62.

absence of non-trivial incidence axioms (such as Desargues) is that one has no control over the points and lines "at infinity". To elaborate this point somewhat, let us consider some of the finer points. In the first place Heyting had to introduce a new class of points, but unlike the the traditional procedure, he could not just introduce "points at infinity" as pencils of parallel lines. He also had to add "points for which it is unknown whether they are at infinity or not" , that is to say , he had to consider pencils in general (called *projective points*), independent of their having a point in common or being parallel. The next step, which clasically is trivial, was to add not just one line at infinity, but a whole host of possibly new lines, namely all the lines through pairs of projective points which are apart from each other. These new lines (*projective lines*) are defined as pairs of mutually apart projective points. For projective points and lines the required notions of "apart", "incidence" and "outside" were introduced without difficulty. The final task was to verify the axioms of the projective plane for the thus constructed system of projective points and lines. It turned out that the axioms concerning the intersection of lines and the properties of the relation "P is outside l" were unmanageable. The behaviour of the points and lines the position of which with respect to "infinity" is unknown proved to be so elusive , that Heyting had to add special cases of the above mentioned axioms as extra affine axioms.

In the dissertation of D. van Dalen (1963) two of the extra axioms were shown to derivable from the axioms of affine geometry .

The 1959 paper was Heyting's last contribution to intuitionistic axiomatic geometry - the subject that he founded as a Ph.D. student.

6. Other axiom systems

The high water mark of Heyting's axiomatic studies was undoubtedly his formalization of intuitionistic logic and mathematics, a topic that is being reported by A.S.Troelstra. In fact most of his original research was axiomatic in nature; the papers Heyting (1966) and Heyting(1962) dealt with the formalization of algebraic structures. The papers on functional analysis, Heyting (1951 A), Heyting (1953), Heyting(1953 B) are the ones among Heyting's oevre that deal with straightforward analysis over the continuum, although even here there is a slight flavour of the axiomatic.

The monograph "Intuitionism" - the first comprehensive treatise on intuitionistic mathematics - contains a number of topics which are treated in an axiomatic framework. Most of this material had already been treated in earlier papers.

Under Heyting's direction two dissertations of an axiomatic nature were written one on geometry van Dalen (1963) and one on topology Troelstra (1966).

Literature

E. Artin, *Geometric Algebra.* Interscience Publ. Inc. New York. 1957.

L.E.J. Brouwer, Begründung der Mengenlehre unabhängig vom logischen Satz vom ausgeschlossenen Dritten. I: Allgemeine Mengenlehre.*Nederlandse Akademie van Wetenschappen, Verhandelingen, Tweede Reeks. Afdeling Natuurkunde.* **12/5**, 1918. 43pp.

___________, Begründung der Mengenlehre unabhängig vom logischen Satz vom ausgeschlossenen Dritten. II: Theorie der Punktmengen. *Nederlandse Akademie van Wetenschappen, Verhandelingen, Tweede Reeks. Afdeling Natuurkunde.* **12/7**, 1919. 33pp.

___________, Besitzt jede reelle Zahl eine Dezimalbruch-Entwicklung? *Mathematische Annalen.* **83**, 1921. 201-210.

___________, Intuitionistische splitsing van mathematische grondbegrippen. *Verslag van de gewone vergadering der wis- ennatuurkundige afdeling.* XXXII, 1923. 877-880.

___________, Intuitionistische Zerlegung mathematischer Grundbegriffe. *Jahresberichte d. Deutschen Mathematiker- Vereinigung.* **33**, 1925. 251-256.

___________, Points and Spaces. *Canadian Journal of Mathematics.* **6**, 1954. 1-17.

___________, Zur Begründung der intuitionistischen Mathematik. II. *Mathematische Annalen.* **95**, 1926. 453-472.

D. van Dalen. Extension Problems in intuitionistic plane projective geometry I,II. *Indag. Math.* **25**, 349 - 383. Also diss., Amsterdam. 1963.

J. Dieudonné, *La géometrie des groupes classiques.* Actualités Sci. Ind. No 1040, Hermann, Paris. 1948.

A. Heyting, Intuitionistische axiomatiek der projectieve meetkunde. (diss.). Amsterdam 1925.

___________, Die Theorie der linearen Gleichungen in einer Zahlenspezies mit nicht kommutativer Multiplikation. *Mathematische Annalen.* **98**, 1927. 491-538.

___________, Intuïtionistische Wiskunde. *Mathematica B.* **5**, 1936. 105-112.

___________, Zur intuitionistischen Axiomatik der projektiven Geometrie. *Mathematische Annalen.* **98** , 1927. 491- 538.

___________, Untersuchungen über intuitionidtische Algebra. *Nederlandse Akademie van Wetenschappen, Afdeling natuurkunde. Eerste sectie,* **18**, no. 2. 1941. 36pp.

___________, L'axiomatique intuitioniste. In R. Bayer (ed.), *Philosophie: XVI. Congres Intern. Phil. Sc. Paris, 1949, I. Logique*. Hermann et Cie. Paris 1951. 81-86.

___________, Espace de Hilbert et intuitionisme. In *Les Méthodes formelles een axiomatique*. Colloques internationaux du C.N.R.S. Paris 1950.59-63.

___________, Note on the Riesz-Fischer Theorem. *Indag. Math.* **13**, 1951. 35-44.

___________, Sur la théorie intuitioniste de la mesure. *Bull. de la Soc. Math. de Belgique*. 6, 1952. 70-78.

___________, *Intuitionism. An Introduction*. North-Holland Publ. Co. Amsterdam 1956.

___________, Axiomatic Method and Intuitionism. In Y. Bar-Hillel, J. Poznanski, M.O. Rabin, A. Robinson (eds.). *Essays on the foundations of Mathematics*. Magnus Press, Hebrew University, Jerusalem/North-Holland Publ. Co. Amsterdam. 1962. 237-247.

___________, Axioms for intuitionistic plane affine geometry. In L. Henkin, P. Suppes, A. Tarski (eds.), *The Axiomatic Method, with special reference to Geometry and physics*. North-Holland Publ. Co. Amsterdam. 1959. 160-173.

___________, *Axiomatic Projective Geometry*. Noordhoff, Groningen/ North-Holland Publ. Co.Amsterdam. 1963.

___________, Remarques sur la théorie intuitioniste des espaces lineaires.*Synthese*, **16**, 1966. 47-52.

G. Pieri, Memorie di Torino.1898.

W. Ruitenburg, Primality and invertibility of polynomials. In A.S. Troelstra, D. van Dalen (eds.) *The L.E.J. Brouwer Centenary Symposium*. North-Holland Publ. Co. Amsterdam.1982. 413-434.

D.S. Scott, Identity and existence in intuitionistic logic. In M.P. Fourman, C.J. Mulvey, D.S. Scott (eds.), *Applications of sheaves. Symposium on Applications of sheaf theory to logic, algebra and analysis*. Springer, Berlin. 1979. 154-193.

A.S. Troelstra, *Intuitionistic Topology*. Diss. Amsterdam. 1966.

A.S. Troelstra, D. van Dalen, *Constructivity in Mathematics*. Vol. I, II. North-Holland Publ. Co. Amsterdam. 1988.

A.N. Whitehead, *The axioms of projective geometry*. Cambridge University Press, Cambridge. 1913.

SUMMER SCHOOL

(invited papers)

PROVABILITY LOGICS FOR RELATIVE INTERPRETABILITY

Dick de Jongh
Department of Mathematics and Computer Science
University of Amsterdam

Frank Veltman
Department of Philosophy
University of Amsterdam

0. INTRODUCTION

In this paper the system **IL** for relative interpretability described in Visser (1988) is studied.[1] In **IL** formulae $A \rhd B$ (read: A *interprets* B) are added to the provability logic **L**. The intended interpretation of a formula $A \rhd B$ in an (arithmetical) theory T is: T + B is relatively interpretable in T + A. The system has been shown to be sound with respect to such arithmetical interpretations (Švejdar 1983, Montagna 1984, Visser 1986, 1988P).

As axioms for **IL** we take, besides the usual axioms $\Box A \rightarrow \Box\Box A$ and $\Box(\Box A \rightarrow A) \rightarrow \Box A$ (*Löb's Axiom*) for the provability logic **L** and its rules, modus ponens and necessitation, the axioms:

(1) $\Box(A \rightarrow B) \rightarrow (A \rhd B)$
(2) $(A \rhd B) \wedge (B \rhd C) \rightarrow (A \rhd C)$
(3) $(A \rhd C) \wedge (B \rhd C) \rightarrow (A \vee B \rhd C)$
(4) $(A \rhd B) \rightarrow (\Diamond A \rightarrow \Diamond B)$
(5) $\Diamond A \rhd A$

With respect to priority of parentheses $\rhd$ is treated as $\rightarrow$.

Furthermore, we will consider the following extensions of **IL**:

ILM = **IL** + **M**, where **M** is the axiom $(A \rhd B) \rightarrow (A \wedge \Box C \rhd B \wedge \Box C)$
ILP = **IL** + **P**, where **P** is the axiom $(A \rhd B) \rightarrow \Box(A \rhd B)$[2]

We will write $\vdash_{IL}$ for derivability in **IL**, similarly for the other systems, but sometimes we may leave the subscript off.

The object of the whole study, undertaken together with Smoryński and Visser is to obtain for the standard formal systems an analogon of Solovay's theorem: which are the interpretability logics corresponding to **PA**, **GB** etc? Solovay's Theorem shows that the provability logics of all these systems are the same. However, their interpretability logics are not. Smoryński and Visser have shown that the interpretability logic of **GB** and other finitely axiomatizable

[1] We want to thank Albert Visser who inspired these investigations by asking us to try and find a useful semantics for the system **ILM**. We also thank Rineke Verbrugge for a number of corrections.

[2] The scheme **M** is named after Franco Montagna who showed its soundness with respect to **PA**, even in the more general case when $\Box C$ is replaced by a Σ-formula. The background of the names of the schemes **P** and **W** is semantic and will be explained in the next section.

Mathematical Logic
Edited by P. P. Petkov
Plenum Press, New York, 1990

systems is **ILP**. It is conjectured that **ILM** is the logic of **PA** and other essentially reflexive systems. A third system

$$\mathbf{ILW} = \mathbf{IL} + \mathbf{W}, \text{ where } \mathbf{W} \text{ is the axiom } (A \rhd B) \rightarrow (A \rhd B \wedge \Box \neg A)$$

is weaker than both other logics, and is conjectured to embody the principles common to all "reasonable" arithmetics. For more details one should consult Visser's paper in this volume.

In this paper we restrict ourselves to purely modal properties of the systems in question. In section 1 the semantics for the different logics is described. In section 2 the fixed point theorem of **L** is extended to **IL**. In the remaining sections modal completeness theorems are proved for the systems **IL**, **ILP** and **ILM**. The logics also turn out to have the finite model property, so decidability is a consequence. We are still working on a completeness proof for **ILW**.

1. SEMANTICS

It is a well-known fact that the modal logic **L** is complete with respect to the **L**-*frames* $\langle W, R\rangle$, which consist of a set of worlds W together with a transitive conversely well-founded relation R.

1.1 Definition. If $\langle W, R\rangle$ is a partially ordered set and $w \in W$, then $W[w]=\{w' \in W \mid w R w'\}$.

1.2 Definition. An *IL-frame* is a **L**-frame $\langle W, R\rangle$ with an additional relation S_w, for each $w \in W$, which has the following properties:

(i) S_w is a relation on $W[w]$,

(ii) S_w is reflexive and transitive,

(iii) if $w', w'' \in W[w]$ and $w' R w''$, then $w' S_w w''$,

We will often write S for $\{S_w \mid w \in W\}$.

1.3 Definition. An *IL-model* is given by an **IL**-frame $\langle W, R, S\rangle$ combined with a forcing relation with the clauses:

$$u \Vdash \Box A \iff \forall v (u R v \Rightarrow v \Vdash A)$$

$$u \Vdash A \rhd B \iff \forall v (u R v \text{ and } v \Vdash A \Rightarrow \exists w (v S_u w \text{ and } w \Vdash B)).$$

1.4 Definition.

(a) We write $\mathbf{F} \models A$ iff $\mathbf{F} = \langle W, R, S\rangle$, and $w \Vdash A$ for every $\Vdash$ on **F** and $w \in W$.

(b) If K is a class of frames, we write $K \models A$ iff $\mathbf{F} \models A$ for each $\mathbf{F} \in K$.

(c) K_W is the class of frames satisfying

(iv) for any w, the converse of $R \circ S_w$ is wellfounded

(d) K_M is the class of frames satisfying

(iv') if $u S_w v R z$, then $u R z$

(e) K_P is the class of frames satisfying

(iv'') if $u S_w v$, then $u S_{w'} v$ for any w' such that $w R w'$, $w' R u$.

The next lemma states that the schemes **W** and **P** characterize the classes of frames K_W and K_P respectively. Their names refer to the character of these classes: in K_P the relation S_w is persistent over R.

1.5 Lemma (*Soundness*).

(a) For each A, if $\vdash_{IL} A$, then $F \models A$.

(b) For $\mathbf{S} = \mathbf{W}, \mathbf{M}, \mathbf{P}$, respectively,

$\mathbf{F} \models \mathbf{ILS} \iff \mathbf{F} \in K_S$ (**ILS** *characterizes* K_S).

(c) For $\mathbf{S} = \mathbf{W}, \mathbf{M}, \mathbf{P}$, respectively, if $\vdash_{ILS} A$, then $K_S \models A$.

Proof. Straightforward. ⊠

In Sections 3 and following completeness will be proved for the three systems **IL, ILP** and **ILM**. Actually, **ILP** will be proved complete with respect to the more restricted class of frames in which S_w and $S_{w'}$ are identical on the intersection of their domains. We will keep writing **ILS** if we want leave open which system we are aiming at.

1.6 Example. For each of the systems above,
$\nvdash \neg((p \rhd \neg p) \wedge (\neg p \rhd p))$.
Proof. The following is a countermodel:

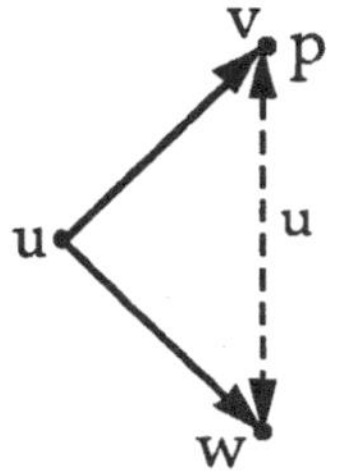

In the above picture only the "extra" arrows for S_u are indicated. Note that in an arithmetical interpretation such a formula would be what is called an *Orey-sentence* (see e.g. Visser 1986). Note also that one could make this model into one in which S_u is antisymmetric; however, the procedure would make the model infinite. ⊠

In the case of provability logic validity on trees is equivalent to validity on **L**-frames. In the case of interpretability logic this is not generally the case.

1.7 Proposition. The formula $\Box(p \to \neg q \wedge \Box\neg q) \wedge (p \rhd q) \to (p \rhd q \wedge \Box\bot)$ is valid on all **ILM**-models on trees, but $\boldsymbol{K}_{\mathbf{ILM}} \nvDash \Box(p \to \neg q \wedge \Box\neg q) \wedge (p \rhd q) \to (p \rhd q \wedge \Box\bot)$ and hence $\nvdash_{\mathbf{ILM}} \Box(p \to \neg q \wedge \Box\neg q) \wedge (p \rhd q) \to (p \rhd q \wedge \Box\neg\bot)$.
Proof. Left to the reader. ⊠

Of course the usual procedure for "stretching out" a partially ordered model into a tree works in this case. The point is that property (iv') will get lost: it will no longer generally hold that, if $w' S_w w'' R u$, then $w' R u$; the only thing one can say of u then is that it will have a forcing relation identical to that of some successor of w', and hence the resulting model will no longer be an **ILM**-model in our sense. For **IL, ILW** and **ILP**, on the other hand, one can restrict oneself to tree models.

2. FIXED POINTS

From the fact that **IL** is an extension of **L** it is obvious that to prove the existence of explicit fixed points in **IL** it is actually sufficient to find a fixed point for $A(p) \rhd B(p)$, i.e. to find a formula C such that $\vdash_{\mathbf{IL}} C \leftrightarrow (A(C) \rhd B(C))$. For, after that we can proceed as in the standard proof for **L** (see Smoryński 1985). One might conjecture that $C = A(T) \rhd B(T)$ would do the trick, and in fact that formula does work for **ILM** (as the reader may check). However, for **IL** a more complicated formula is necessary: $C = A(T) \rhd B(\Box\neg A(T))$. (The even more complicated, but more symmetric formula $A(\Box\neg A(T)) \rhd B(\Box\neg A(T))$ is equivalent to C and therefore works too.) We will give a semantic proof. (The results of this section were reached in cooperation with Visser; see Visser 1988P for a syntactic proof.) Of course, the present proof does need the completeness of **IL** proved in section 3.

2.1 Lemma.
$w \Vdash A(T) \rhd B(\Box\neg A(T)) \iff w \Vdash A(A(T) \rhd B(\Box\neg A(T))) \rhd B(A(T) \rhd B(\Box\neg A(T)))$.
Proof. We first establish some simple general facts, for arbitrary w. If we give them without comment their proof is trivial. We write $u \Vdash_{max} A$ iff $u \Vdash A$ and $\forall v\,(u\,R\,v \Rightarrow v \nVdash A)$, and we write $w\,\underline{R}\,u$ for $w\,R\,u$ or $w = u$.

(1) $w \Vdash D \rhd E \iff \forall u\,(w\,R\,u \wedge u \Vdash_{max} D \Rightarrow \exists v\,(u\,S_w\,v \wedge v \Vdash E))$;
(2) if $w \Vdash \Box D$ and $w\,R\,u$, then $u \Vdash \Box D$;
(3) if $w \Vdash_{max} D$, then $w \Vdash \Box\neg D$;
(4) if $w \Vdash \Box\neg D$, then, if $w\,\underline{R}\,u$, then $u \Vdash D \rhd E$;
(5) if $w \Vdash_{max} D$, then, if $w\,\underline{R}\,u$, then $u \Vdash D \rhd E$;
(6) if $w \Vdash_{max} D$, then $w \Vdash_{max} A(T) \iff w \Vdash_{max} A(D \rhd E)$;
by (5), as $w \Vdash$ can only be depend on $u \Vdash$ for u for which $w\,\underline{R}\,u$, since e.g., $w\,R\,v\,R\,v'\,S_v\,u$ implies $w\,R\,u$ by Def.1.2.(i);
(7) if $w \Vdash_{max} A(T)$, then $w \Vdash_{max} A(A(T) \rhd E)$, by (6);
(8) if $w \Vdash_{max} A(A(T) \rhd E)$, then $w \Vdash A(T)$;
for assume $w \Vdash_{max} A(A(T) \rhd E)$, then, by (7), $w\,R\,u \Rightarrow u \Vdash \neg A(T)$, hence, for all u with $w\,\underline{R}\,u$, $u \Vdash A(T) \rhd E$. As in (6), $w \Vdash A(T)$ follows from $w \Vdash A(A(T) \rhd E)$;
(9) if $w \Vdash \Box\neg E$, then, for all u with $w\,\underline{R}\,u$, $u \Vdash \Box\neg D \iff u \Vdash D \rhd E$;
(10) if $w \Vdash_{max} E$, then, for all u with $w\,\underline{R}\,u$, $u \Vdash B(\Box\neg D) \iff u \Vdash B(D \rhd E)$;
(11) if $w \Vdash_{max} E$, then, for all u with $w\,\underline{R}\,u$,
$u \Vdash_{max} B(\Box\neg D) \iff u \Vdash_{max} B(D \rhd E)$;
(12) if $w \Vdash_{max} B(\Box\neg D)$, then $w \Vdash_{max} B(D \rhd B(\Box\neg D))$;
(13) if $w \Vdash_{max} B(D \rhd B(\Box\neg D))$, then $w \Vdash_{max} B(\Box\neg D)$;
for assume $w \Vdash_{max} B(D \rhd B(\Box\neg D))$, then, by (12), $w \Vdash \Box\neg B(\Box\neg D)$.
So, by (9), for all u with $w\,\underline{R}\,u$, $u \Vdash D \rhd B(\Box\neg D) \iff u \Vdash \Box\neg D$;
so, $w \Vdash_{max} B(\Box\neg D)$.

Now we establish the main claim:

$\Rightarrow$: Let $w \Vdash A(T) \rhd B(\Box\neg A(T))$. Assume $w\,R\,u$ and $u \Vdash_{max} A(A(T) \rhd B(\Box\neg A(T)))$. By (8), $u \Vdash A(T)$. So, for some v with $u\,S_w\,v$, $v \Vdash B(\Box\neg A(T))$. We may just as well assume $v \Vdash_{max} B(\Box\neg A(T))$, as $u\,S_w\,v\,R\,v'$ implies $u\,S_w\,v'$ by def. 1.2 (iii). By (12) this implies $v \Vdash B(A(T) \rhd B(\Box\neg A(T)))$.

$\Leftarrow$: Let $w \Vdash A(A(T) \rhd B(\Box\neg A(T))) \rhd B(A(T) \rhd B(\Box\neg A(T)))$. Assume $w\,R\,u$, $u \Vdash_{max} A(T)$. By (7), $u \Vdash A(A(T) \rhd B(\Box\neg A(T)))$. So, for some v with $u\,S_w\,v$, $v \Vdash B(A(T) \rhd B(\Box\neg A(T)))$. Again we may assume that $v \Vdash_{max} B(A(T) \rhd B(\Box\neg A(T)))$, and (13) gives us $v \Vdash B(\Box\neg A(T))$. ⊠

For completeness' sake we formulate the explicit fixed point theorem which follows from lemma 2.1 by the remarks above.

2.2 Theorem. For each **IL**-formula $A(p, q_1, \ldots, q_n)$ in which p occurs only modalized (i.e. all occurrences of p are under some $\Box$ or $\rhd$) there is a provably unique **IL**-formula $B(q_1, \ldots, q_n)$ such that $\vdash_{IL} A(B(q_1, \ldots, q_n), q_1, \ldots, q_n) \leftrightarrow B(q_1, \ldots, q_n)$.

3. MODAL COMPLETENESS: PRELIMINARIES

The usual method in modal logic for obtaining completeness proofs is to construct the necessary countermodels by taking maximal consistent sets of the logic under consideration as the worlds of the model (without necessarily one consistent set standing for only one world) and providing this set of worlds with an appropriate relation R. This method cannot be applied here, since the logic is not compact: some infinite syntactically consistent sets of formulae are semantically incoherent. The solution is to restrict the maximal consistent sets to subsets of some finite set of formulae. Such a so-called adequate set has to be rich enough to handle the truth definition, and hence has to be closed under the forming of subformulae and single negations. Furthermore, for each particular logic, additional requirements on the adequate set will be needed to be able to apply the axioms.

3.1 Definition. An *adequate* set of formulae is a set Φ which fulfills the following conditions:

(i) Φ is closed under the taking of subformulae
(ii) if $B \in \Phi$, and B is no negation, then $\neg B \in \Phi$
(iii) $\bot \rhd \bot \in \Phi$
(iv) if $B \rhd C \in \Phi$, then also $\Diamond B, \Diamond C \in \Phi$
(v) if B as well as C is an antecedent or a consequent of some $\rhd$-formula in Φ, then $B \rhd C \in \Phi$.

Obviously, each finite set Γ of formulae is contained in a finite adequate set Φ.

3.2 Definition. Let Γ and Δ be two maximal **ILS**-consistent subsets of some finite adequate Φ. Then

$$\Gamma \prec \Delta \iff \text{for each } \Box A \in \Gamma, \Box A, A \in \Delta, \text{ and for some } \Box A \notin \Gamma, \ \Box A \in \Delta.$$

Whenever $\Gamma \prec \Delta$, we say that Δ is a *successor* of Γ.

3.3 Lemma. Let Γ_0 be a maximal **ILS**-consistent subset of some finite adequate Φ, and let W_{Γ_0} be the smallest set such that

(i) $\Gamma_0 \in W$
(ii) if $\Delta \in W$ and Δ' is a maximal **ILS**-consistent subset of Φ such that $\Delta \prec \Delta'$, then $\Delta' \in W$.

Then

(i) $\prec$ is transitive and irreflexive on W_{Γ_0}
(ii) For each $\Gamma \in W_{\Gamma_0}$, $\Box A \in \Gamma \iff A \in \Delta$ for every Δ such that $\Gamma \prec \Delta$.

Proof. As in the case of **L** (i) is trivial, and so is $\Rightarrow$ of (ii). For $\Leftarrow$ of (ii) one needs Löb's axiom. ⊠

One might think that this means that, in essence, the completeness problem for **ILS** reduces to defining relations $\preccurlyeq_\Delta$ on W_{Γ_0} such that

(i) $\preccurlyeq$ has all the properties of the relation S in K_S
(ii) For each Γ in W_{Γ_0}, $B \rhd C \in \Gamma$ iff, for every Δ such that $\Gamma \prec \Delta$ and $B \in \Delta$, there is some Δ' with $\Delta \preccurlyeq_\Gamma \Delta'$ and $C \in \Delta'$.

The situation is not as simple as that. Before we continue with the completeness proofs, we will give an example to make this clear.

3.4 Example. It will be obvious that $\nvdash_{ILS} (p \rhd q \vee r) \to (p \rhd q) \vee (p \rhd r)$. Now, take Γ_0 to be a maximal **ILS**-consistent set in Φ that contains $p \rhd q \vee r$, $\neg(p \rhd q)$, and $\neg(p \rhd r)$, as well as the formulae $\Box\Box\bot$, $\Box(p \vee q \vee r)$, $\Box\neg(p \wedge q)$, $\Box\neg(q \wedge r)$, and $\Box\neg(p \wedge r)$. It is then clear that the resulting W_{Γ_0} will look as follows:

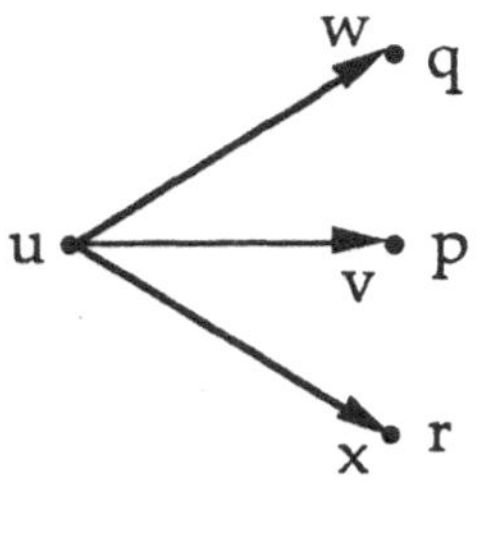

It will also be clear that no S_u can be defined on this model in such a way that $u \Vdash p \rhd q \vee r, \neg(p \rhd q), \neg(p \rhd r)$. By doubling W[u] however an appropriate model can be obtained (the arrows give the additional S_u-relations not given by R):

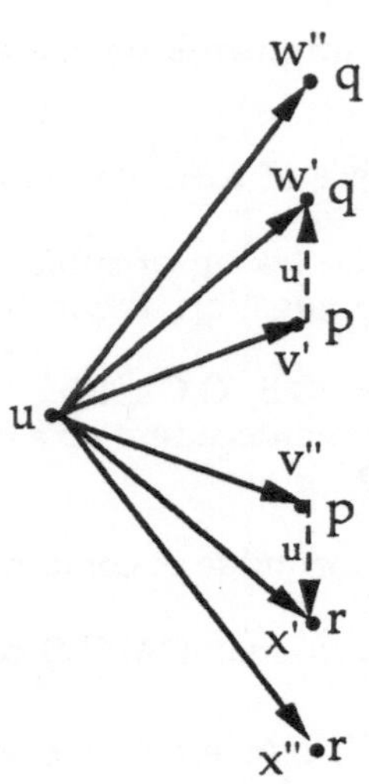

Our strategy in the next section is a generalization of this idea: we will multiply the maximal **ILS**-consistent sets by indexing them with finite sequences of formulae. We write $\tau \subseteq \tau'$ iff the finite sequence τ is a (not necessarily proper) initial segment of the finite sequence τ'; we write $*$ for concatenation, and, if $w = \langle \Gamma, \tau \rangle$, we write $(w)_0$ for Γ and $(w)_1$ for τ.

Using these pairs we set aside, for each world w and each appropriate formula C, a specific set of the successors of w indexed by C (the so-called critical C-successors of w) to provide the counterexamples to the formulae $B \rhd C$ that must be falsified in w. We will restrict the relation S_W so that it does not "leave" this set of C-critical successors. Speaking intuitively, the C-critical successors of w will be the ones that contain no formula A that "asks for" C (where A is an antecedent and C the consequent of a $\rhd$-formula in w). The next two lemmas show that this whole idea is feasible. The first one says that indeed a counterexample can be found, when needed: for each $\neg(B \rhd C)$ in w a C-critical successor with B in it can be found. The second one says that we will need to have S_W lead from C-critical successors of w to C-critical successors of w: if $A \rhd D$ is a member of w, and A is a member of a C-critical successor of w, then yet another C-critical successor of w with D in it can be found.

3.5 Definition. Let Γ and Δ be maximal **ILS**-consistent subsets of some given adequate Φ. Then Δ is a C-*critical* successor of Γ iff

(i) $\Gamma \prec \Delta$

(ii) $\neg A, \Box\neg A \in \Delta$ for each A such that $A \rhd C \in \Gamma$.

Note that successors of C-critical successors of Γ are C-critical successors of Γ.

3.6 Lemma. Suppose Γ is maximal **ILS**-consistent in Φ and $\neg(B \rhd C) \in \Gamma$; then there exists a C-critical successor Δ of Γ, maximal **ILS**-consistent in Φ, such that $B \in \Delta$.

Proof. Take Δ to be a maximal **ILS**-consistent extension of

$$\{D, \Box D \mid \Box D \in \Gamma\} \cup \{\neg A, \Box\neg A \mid A \rhd C \in \Gamma\} \cup \{B, \Box\neg B\}$$

Note first that the adequacy of Φ insures that all the formulae of Δ are indeed available. Secondly, note that if such a Δ exists, it is indeed a C-critical successor of Γ: the fact that

$$\{D, \Box D \mid \Box D \in \Gamma\} \cup \{\Box\neg B\} \subseteq \Delta$$

makes it a successor of Γ, and the fact that

$$\{\neg A, \Box\neg A \mid A \rhd C \in \Gamma\} \subseteq \Delta$$

makes it C-critical.

Now, if no such Δ exists, then there are $A_1, \ldots, A_m$ and $D_1, \ldots, D_k$ with

$$D_1, \ldots, D_k, \Box D_1, \ldots, \Box D_k, \neg A_1, \ldots, \neg A_m, \Box\neg A_1, \ldots, \Box\neg A_m, B, \Box\neg B \vdash \bot.$$

Or, equivalently:

$$D_1, \ldots, D_k, \Box D_1, \ldots, \Box D_k, \neg(A_1 \vee \ldots \vee A_m), \Box\neg(A_1 \vee \ldots \vee A_m), B, \Box\neg B \vdash \bot$$

This would mean that:

$$D_1, \ldots, D_k, \Box D_1, \ldots, \Box D_k, B, \Box\neg B \vdash A_1 \vee \ldots \vee A_m \vee \Diamond(A_1 \vee \ldots \vee A_m).$$

In other words:

$$D_1, \ldots, D_k, \Box D_1, \ldots, \Box D_k \vdash B \wedge \Box\neg B \rightarrow A_1 \vee \ldots \vee A_m \vee \Diamond(A_1 \vee \ldots \vee A_m).$$

Since **IL** contains **L**:

$$\Box D_1, \ldots, \Box D_k \vdash \Box(B \wedge \Box\neg B \rightarrow A_1 \vee \ldots \vee A_m \vee \Diamond(A_1 \vee \ldots \vee A_m))$$

By axiom (1):

$$\Box D_1, \ldots, \Box D_k \vdash B \wedge \Box\neg B \rhd A_1 \vee \ldots \vee A_m \vee \Diamond(A_1 \vee \ldots \vee A_m)$$

In view of particularly the axioms (5) and (3) we have that

$$\vdash A_1 \vee \ldots \vee A_m \vee \Diamond(A_1 \vee \ldots \vee A_m) \rhd A_1 \vee \ldots \vee A_m.$$

So, by axiom (2):

$$\Box D_1, \ldots, \Box D_k \vdash B \wedge \Box\neg B \rhd A_1 \vee \ldots \vee A_m$$

Given that $A_1 \rhd C, \ldots, A_m \rhd C \in \Gamma$, we also have $\Gamma \vdash A_1 \vee \ldots \vee A_m \rhd C$ (apply axiom (3)), and so by axiom (2):

$$\Gamma \vdash B \wedge \Box\neg B \rhd C$$

Now, it is not difficult to see that

$$\vdash B \rhd B \wedge \Box\neg B$$

(To that purpose, note first that $\vdash (B \wedge \Box\neg B) \vee \Diamond(B \wedge \Box\neg B) \rhd B \wedge \Box\neg B$. Secondly, since **ILS** contains **L**, $\vdash \Box(B \rightarrow (B \wedge \Box\neg B) \vee \Diamond(B \wedge \Box\neg B))$. So by axiom (1), it is provable in **ILS** that $B \rhd (B \wedge \Box\neg B) \vee \Diamond(B \wedge \Box\neg B)$. Combining these two facts we find $\vdash B \rhd B \wedge \Box\neg B$.)

Finally, by applying axiom (2) once more, it follows from $\Gamma \vdash B \wedge \Box\neg B \rhd C$ and $\vdash B \rhd B \wedge \Box\neg B$ that

$$\Gamma \vdash B \rhd C$$

This contradicts the consistency of Γ. ⊠

3.7 Lemma. Suppose $B \rhd C \in \Gamma$ and let Δ be an E-critical successor of Γ with $B \in \Delta$. Then there is an E-critical successor Δ' of Γ with $C \in \Delta'$.

Proof. Suppose there is not such a Δ'. Then there would be $\Box D_1, \ldots, \Box D_n \in \Gamma$, and $F_1 \rhd E, \ldots, F_k \rhd E \in \Gamma$ such that

$$D_1, \ldots, D_n, \Box D_1, \ldots, \Box D_n, \neg F_1, \ldots, \neg F_k, \Box\neg F_1, \ldots, \Box\neg F_k, C \vdash \bot$$

and, therefore,

$$D_1, \ldots, D_n, \Box D_1, \ldots, \Box D_n \vdash C \rightarrow F_1 \vee \ldots \vee F_k \vee \Diamond(F_1 \vee \ldots \vee F_k)$$

which as before implies:

$$\Box D_1, \ldots, \Box D_n \vdash C \rhd F_1 \vee \ldots \vee F_k.$$

By axiom (2), $B \rhd C \in \Gamma$ implies that $\Gamma \vdash B \rhd F_1 \vee \ldots \vee F_k$ and, by axiom (3), $\Gamma \vdash B \rhd E$. Given the adequacy conditions, this can be strengthened to $B \rhd E \in \Gamma$. Since Δ is an E-critical successor of Γ, this implies $\neg B \in \Delta$, and we have arrived at a contradiction, since it is assumed that $B \in \Delta$. ⊠

4. THE MODAL COMPLETENESS OF IL

In this section we just have to carefully adjoin sequences to the maximal **IL**-consistent sets and see that the intuitive ideas of the previous section can be set to work properly.

4.1 Theorem (*Completeness and decidability of* **IL**). If $\nvdash_{IL} A$, then there is a finite **IL**-model **K** such that $\mathbf{K} \nvDash A$.

Proof. Take some finite adequate set Φ containing $\neg A$. Let Γ be a maximally consistent subset of Φ containing $\neg A$.

Now, set W_Γ to be the smallest set of pairs $<\Delta,\tau>$, where τ is a finite sequence of formulae from Φ, that fulfills the following requirements:

(i) $<\Gamma,<>> \in W_\Gamma$

(ii) If $<\Delta,\tau> \in W_\Gamma$, then $<\Delta',\tau> \in W_\Gamma$ for every successor Δ' of Δ

(iii) If $<\Delta,\tau> \in W_\Gamma$, then $<\Delta',\tau * <C>> \in W_\Gamma$ for every C-critical successor Δ' of Δ.

W_Γ is finite. (For every Δ, the number of successors of Δ is finite. Moreover, if $\Delta \prec \Delta'$, the number of successors of Δ' is smaller than the number of successors of Δ.)

Observation: If $<\Delta,\tau> \in W_\Gamma$ and E occurs in τ, then $\neg E \in \Delta$.

Proof: Show with induction on the construction of W_Γ that if $<\Delta,\tau> \in W_\Gamma$ and E occurs in τ then $\neg E, \Box\neg E \in \Delta$.

Define R on W_Γ as follows:

$$w R w' \text{ iff } (w)_0 \prec (w')_0 \text{ and } (w)_1 \subseteq (w')_1 .$$

It is easy to check that R has all the properties required.

Finally, let $u S_w v$ apply if (I) and (II) hold:

(I) $u, v \in W_\Gamma[w]$

(II) $(w)_1 = (u)_1 \subseteq (v)_1$, or $(u)_1 = (w)_1 * <C> * \tau$ and $(v)_1 = (w)_1 * <C> * \sigma$ for some C, σ and τ.

We leave it to the reader to check that under this definition S_w will have the required properties:

We are now ready to define

$$w \Vdash p \text{ iff } p \in (w)_0$$

and prove that

$$\text{for each } A \in \Phi,\ w \Vdash A \text{ iff } A \in (w)_0 .$$

Given (ii) it is clear from lemma 3.3 that the model treats $\Box$-formulae properly.[3] So, the only interesting case to look at in the inductive proof is the one that A is $B \rhd C$, i.e. we have to show that

$$B \rhd C \in (w)_0 \Leftrightarrow \forall u\,(w R u \wedge B \in (u)_0 \Rightarrow \exists v\,(u S_w v \wedge C \in (v)_0)):$$

$\Leftarrow$: Suppose $B \rhd C \notin (w)_0$. Then $\neg(B \rhd C) \in (w)_0$. We must show that $\exists u\,(w R u \wedge B \in (u)_0 \wedge \forall v\,(u S_w v \rightarrow \neg C \in (v)_0))$. Let Δ be as in lemma 3.6 with $(w)_0$ as Γ, and take u to be $<\Delta, (w)_1 * <C>>$. Consider any v such that $u S_w v$. Then C occurs in $(v)_1$. By the observation above, $\neg C \in (v)_0$.

$\Rightarrow$: Suppose $B \rhd C \in (w)_0$. Consider any u such that wRu and $B \in (u)_0$. Let us first assume that $(u)_1 = (w)_1 * <E> * \tau$. In that case we can apply lemma 3.7 for $\Gamma = (w)_0$ and $\Delta = (u)_0$ to obtain an E-critical successor Δ' of Γ with $C \in \Delta'$. It suffices now to take $v = <\Delta', (w)_1 * <E>>$. It is clear that v fulfills all requirements to make $u S_w v$.

---

[3] An alternative, perhaps more elegant, set up would be to do without (ii). Then one has to use the equivalence of $\Box A$ with $\neg A \rhd \bot$ and to adapt the definition of adequate set. In the present proof this equivalence is used in the treatment of $\rhd$. (See the end of the proof which can be deleted in the alternative set up.)

If $(u)_1 = (w)_1$, then all we know is that $(w)_0 \prec (u)_0$. Note, however, that every successor of Γ is a $\bot$-critical successor of Γ. (By axiom (4), $\vdash F \rhd \bot \rightarrow \Box\neg F$; hence if $F \rhd \bot \in \Gamma$, then $\Box\neg F \in \Gamma$, and therefore $\neg F, \Box\neg F \in \Delta$ for every Δ such that $\Gamma \prec \Delta$.) So we can apply lemma 3.7 for $\Gamma = (w)_0$, $\Delta = (u)_0$, and $E = \bot$, in order to obtain a ($\bot$-critical) successor Δ' of Γ with $C \in \Delta'$. Take $v = \langle \Delta', (w)_1 \rangle$. ⊠

5. THE MODAL COMPLETENESS OF ILP

5.1 Definition. A set Φ of formulae is **ILP**-*adequate* iff

(i) Φ is adequate in the sense of definition 3.1

(ii) if $B \rhd C \in \Phi$, then also $\Box(B \rhd C) \in \Phi$.

Clearly, each finite set Γ of formulae is contained in a finite **ILP**-adequate set Φ.

5.2 Theorem (*Completeness and decidability of* **ILP**). If $\nvdash_{\mathbf{ILP}} A$, then there is a finite **ILP**-model **K** such that $\mathbf{K} \nVdash A$.

Proof. Take some finite adequate set Φ containing $\neg A$. Let Γ be a maximally consistent subset of Φ containing $\neg A$.

In constructing the model, we multiply the maximal **ILP**-consistent sets similarly as with **IL** while at the same time transforming the model into a tree in the standard manner. The purpose of making the model into a tree is insuring that a unique immediate predecessor exists for each world. A world in the model will be a sequence of pairs $\langle\langle \Gamma_0, \tau_0 \rangle, \ldots, \langle \Gamma_{n-1}, \tau_{n-1} \rangle, \langle \Gamma_n, \tau_n \rangle\rangle$.

More precisely, W_Γ is built up according to the following clauses:

(i) $\langle\langle \Gamma, \langle\rangle \rangle\rangle \in W_\Gamma$

(ii) If $\langle\langle \Gamma_0, \tau_0 \rangle, \ldots, \langle \Gamma_n, \tau_n \rangle\rangle \in W_\Gamma$, and Δ is a successor of Γ then also $\langle\langle \Gamma_0, \tau_0 \rangle, \ldots, \langle \Gamma_n, \tau_n \rangle, \langle \Delta, \tau_n \rangle\rangle \in W_\Gamma$;

(iii) If $\langle\langle \Gamma_0, \tau_0 \rangle, \ldots, \langle \Gamma_n, \tau_n \rangle\rangle \in W_\Gamma$ and Δ is a C-critical successor of Γ, then also $\langle\langle \Gamma_0, \tau_0 \rangle, \ldots, \langle \Gamma_n, \tau_n \rangle, \langle \Delta, \tau_n * \langle C \rangle \rangle\rangle \in W_\Gamma$

If $w = \langle\langle \Gamma_0, \tau_0 \rangle, \ldots, \langle \Gamma_n, \tau_n \rangle\rangle \in W_\Gamma$, we write $\Delta_w = \Gamma_n$ and $\tau_w = \tau_n$.

We next define R on W_Γ as follows: $w R w'$ iff w is a proper initial segment of w'. Thus, R is transitive and irreflexive. More importantly, every world different from $\langle\langle \Gamma, \langle\rangle \rangle\rangle$ has precisely one immediate R-predecessor.

Note that that the model will treat $\Box$ properly.

We are now ready to define $u S_w v$ as applying if (I) and (II) hold:

(I) $w R u$, and for every w', if $w' R u$ then $w' R v$

(II) $\tau_u \subseteq \tau_v$

It is easy to check that under this definition S_w will have the required properties.

Next we define

$$w \Vdash p \text{ iff } p \in \Delta_w,$$

and prove that

$$\text{for each } A \in \Phi,\ w \Vdash A \text{ iff } A \in \Delta_w.$$

Again, the only interesting case to look at in the inductive proof is the one that A is $B \rhd C$, i.e. we have to show that

$$B \rhd C \in \Delta_w \Leftrightarrow \forall u (w R u \wedge B \in \Delta_u \Rightarrow \exists v (u S_w v \wedge C \in \Delta_v)).$$

$\Leftarrow$: Suppose $B \rhd C \notin \Delta_w$. Then $\neg(B \rhd C) \in \Delta_w$. We must show that $\exists u (w R u \wedge B \in \Delta_u \wedge \forall v (u S_w v \rightarrow \neg C \in \Delta_v))$.

Assume $w = \langle\langle \Gamma_0, \tau_0 \rangle, \ldots, \langle \Gamma_n, \tau_n \rangle\rangle$. Let Δ be as in lemma 3.6 with Γ_n as Γ. Take u to be $\langle\langle \Gamma_0, \tau_0 \rangle, \ldots, \langle \Gamma_n, \tau_n \rangle, \langle \Delta', \tau_n * \langle C \rangle \rangle\rangle$ with the Δ' given by that lemma. Consider any v such that $u S_w v$. Then C occurs in τ_v. As in the previous case, it is easy to see that this means that $\neg C \in \Delta_v$.

$\Rightarrow$: Suppose $B \rhd C \in \Delta_w$ and wRu with $B \in \Delta_u$. Let w' the(!) immediate predecessor of u. Note that axiom **P** and the **ILP**-adequacy of Φ insure that $B \rhd C \in \Delta_{w'}$.

Let us first assume that $\tau_u = \tau_{w'} * <E>$. In that case we can apply lemma 3.7 with $\Gamma = \Delta_{w'}$ and $\Delta = \Delta_u$ to obtain an E- critical successor Δ' of Γ with $C \in \Delta'$. It suffices now to take $v = w' * < \Delta', \tau_u>$. It is clear that v fulfills all requirements to make $u S_w v$.

If, on the other hand, $\tau_u = \tau_{w'}$, then all we know is that $\Delta_{w'} \prec \Delta_u$. Recall however that every successor is a $\bot$-critical successor. So, here too, we apply lemma 3.7 for $\Gamma = \Delta_{w'}$, $\Delta = \Delta_u$, and $E = \bot$, in order to obtain a ($\bot$-critical) successor Δ' of Γ with $C \in \Delta''$. Take $v = w' * <\Delta'', \tau_u>$. ⊠

5.3 Corollary (to the proof of theorem 5.2). **ILP** is complete with respect to the frames in which, if wRw', then $S_{w'} = S_w \upharpoonright W[w']$.

Proof. It is clear from the proof that, in the model constructed $u S_w v$ iff $u S_{w'} v$ for the immediate predecessor w' of w. ⊠

The corollary means that we can take the S-relation in **ILP** to be a rigid relation, essentially independent of w.

6. THE MODAL COMPLETENESS OF ILM

The completeness proof for **ILM** is rather more complicated than the ones for the completeness of **IL** and **ILP**. The first problem arises from the fact that to be able to apply the characteristic axiom $(A \rhd B) \rightarrow (A \wedge \Box C \rhd B \wedge \Box C)$ we are forced to add the consequent of this formula to the adequate set, whenever we have the antecedent.

6.1 Definition. An **ILM**-*adequate* set of formulae is a set Φ which fulfills the conditions:

(i) Φ is closed under the taking of subformulae

(ii) if B and $C \in \Phi$, then for each Boolean combination D of B and C there is a formula **ILM**-equivalent to D in Φ

(iii) $\bot \rhd \bot \in \Phi$

(iv) if $B \rhd C \in \Phi$, then also formulae **ILM**-equivalent to $\Diamond B$, $\Diamond C$ in Φ

(v) if both B and C are antecedent or consequent of some $\rhd$-formula in Φ, then $B \rhd C \in \Phi$

(vi) if $B \rhd C, \Box D \in \Phi$, then there is in Φ a formula **ILM**-equivalent to $B \wedge \Box D \rhd C \wedge \Box D$.

With this definition it is, of course, not at all obvious that each finite set is contained in a finite adequate one. The problem in keeping things finite is that with $B \wedge \Box D \rhd C \wedge \Box D$ also $\Diamond(B \wedge \Box D)$ and $\Diamond(C \wedge \Box D)$ will have to be an element of Φ and these will via clause (vi) generate new formulae in the adequate set, e.g. $B \wedge \Box D \wedge \Box \neg(B \wedge \Box D) \rhd C \wedge \Box D \wedge \Box \neg(B \wedge \Box D)$. What we have to show is that this does not lead to an infinite regress: after a while the process starts delivering formulae equivalent to ones which have occurred previously. A little thought will convince the reader that the next lemma shows just that.

6.2 Lemma. Starting with a finite set of formulae $\Diamond B_1, \ldots, \Diamond B_n$, and and closing off under the operation of taking $\Diamond(B_i \wedge \Box \neg B_j)$ (adding each new $\Diamond$-formula to the stock) leads to a finite set of **L**-equivalence classes of formulae.

Proof. By induction on n. In the case that there is only one formula $\Diamond B$ the process stops immediately, because $\Diamond(B \wedge \Box \neg B)$ is **L**-equivalent to $\Diamond B$.

Assume the validity of the lemma for n starting formulae and apply the closing off procedure to $\Diamond B_1, \ldots, \Diamond B_{n+1}$. The formulae obtained will be of the forms $\Diamond(B_i \wedge \Box \neg D_1 \wedge \ldots \wedge \Box \neg D_k)$ $(1 \leq i \leq n+1)$. For each of these classes we have to show that they contain only a finite number of equivalence classes. Without loss of generality we restrict ourselves to the case that i=1.

By the induction hypothesis there can be only finitely many formulae $\Diamond(B_1 \wedge \Box\neg D_1 \wedge \ldots \wedge \Box\neg D_k)$ in which the formula B_1 has not been used in the construction of $D_1, \ldots, D_k$. Now consider a formula $\Diamond(B_1 \wedge \Box\neg D_1 \wedge \ldots \wedge \Box\neg D_k)$ in which B_1 has been used. This formula is **L**-equivalent to $\Diamond(B_1 \wedge \Box\neg B_1 \wedge \Box\neg D_1 \wedge \ldots \wedge \Box\neg D_k)$. We now use the fact that

$$\vdash_L \Box\neg B_1 \rightarrow \Box(B_1 \leftrightarrow \bot) \text{ and } \vdash_L \Box\neg B_1 \rightarrow \Box \ldots \Box(B_1 \leftrightarrow \bot).$$

From this it easily follows that B_1 can, in each of the $D_1, \ldots, D_k$ occurring in $\Diamond(B_1 \wedge \Box\neg B_1 \wedge \Box\neg D_1 \wedge \ldots \wedge \Box\neg D_k)$, be **L**-equivalently replaced by $\bot$, since $\Box\neg B_1$ occurs in that formula. Now, each of the D_i is built up in such a manner that B_1 occurs only in the context $\Box\neg(B_1 \wedge \ldots)$. This means that after replacing B_1 by $\bot$ we get a tautology, which can be left out altogether. We end up with a formula $\Diamond(B_1 \wedge \Box\neg E_1 \wedge \ldots \wedge \Box\neg E_m)$ in which each of the E_i has been constructed according to procedure from $B_2, \ldots, B_{n+1}$. We already concluded that there can be only finitely many such formulae. ⊠

6.3 Theorem (*Completeness and decidability of* **ILM**) If $\nvdash_{ILM} A$, then there is a finite **ILM**-model **K** such that $K \nVdash A$.

Proof. Take some finite **ILM**-adequate set Φ containing $\neg A$. Let Γ be a maximal **ILM**-consistent subset of Φ containing $\neg A$. Unfortunately, we need more worlds than present in the W_Γ used in the proofs for **IL** and **ILP**.

This time we set W_Γ to be the collection of all pairs $<\Delta, \tau>$, with

- (i) $\Gamma \prec \Delta$ or $\Gamma = \Delta$
- (ii) τ is a finite sequence of formulae from Φ, the length of which does not exceed the the depth[4] of Γ minus the depth of Δ. (So, Γ is only paired off with the empty sequence.)

Clearly, W_Γ is finite. Note that the sequence τ in a pair $<\Delta, \tau>$ provides no longer sufficient information on the "C-critical" status of Δ.

Define R on W_Γ as follows:

$$w R w' \text{ iff } (w)_0 \prec (w')_0 \text{ and } (w)_1 \subseteq (w')_1 .$$

It is easy to check that R has all the properties required.

We say that u is a C-critical R-successor of w if $(u)_0$ is a C-critical successor of $(w)_0$ and $(u)_1 = (w)_1 * <C> * \tau$.

Let $u S_w v$ apply if (I)–(IV) hold:

- (I) $u, v \in W_\Gamma[w]$
- (II) $(u)_1 \subseteq (v)_1$
- (III) for each A such that $\Box A \in (u)_0$ also $\Box A \in (v)_0$
- (IV) if u is a C-critical R-successor of w, then v is a C-critical R-successor of w.

Let us check right away that under this definition S_w will have the required properties:

- (i) that $u, v \in W[w]$ if $u S_w v$, is instantaneous.
- (ii) reflexivity and transitivity of S_w are easy to check.
- (iii) if $u, v \in W[w]$ and $u R v$, then (I), (II) and (III) are immediate. As for (IV) it suffices to recall that successors of C-critical successors are C-critical.
- (iv) Suppose $w' S_w w'' R u$. We must show that $w' R u$. That $(w')_1 \subseteq (u)_1$ is immediate. That $(w')_0 \prec (u)_0$ follows from $(w'')_0 \prec (u)_0$ combined with (III) for w', w''.

We are now ready to define $w \Vdash p$ iff $p \in (w)_0$ and prove that in that case $w \Vdash A$ iff $A \in (w)_0$, holds for each $A \in \Phi$. Again, we restrict ourselves to the case that A is $B \rhd C$, i.e. we have to show that

---

[4] Γ has *depth* n if the maximal length of a complete chain $\Gamma = \Gamma_0 \prec \ldots \prec \Gamma_m$ is n+1.

$$B \rhd C \in (w)_0 \Leftrightarrow \forall u(wRu \wedge B \in (u)_0 \Rightarrow \exists v(uS_w v \wedge C \in (v)_0)):$$

$\Leftarrow$: Suppose $B \rhd C \notin (w)_0$. Then $\neg(B \rhd C) \in (w)_0$. We must show that

$$\exists u(wRu \wedge B \in (u)_0 \wedge \forall v(uS_w v \to \neg C \in (v)_0)).$$

Let Δ be as in lemma 3.6 with $(w)_0$ as Γ, and take u to be $<\Delta, (w)_1 * <C>>$. Consider any v such that $uS_w v$. Since u is a C-critical R-successor of w, v will be one too. Therefore, $\neg C \in (v)_0$.

$\Rightarrow$: Suppose $B \rhd C \in (w)_0$ and let u be such that wRu and $B \in (u)_0$. Let $\{\Box D_1, \ldots, \Box D_n\} = \{\Box D \mid \Box D \in (u)_0\}$. Note that axiom **M** and the adequacy of Φ insure that $(w)_0$ contains a formula equivalent to

$$B \wedge \Box D_1 \wedge \ldots \wedge \Box D_n \rhd C \wedge \Box D_1 \wedge \ldots \wedge \Box D_n.$$

Let us first assume that u is an E-critical R-successor of w. Then, for some τ, $(u)_1 = (w)_1 * <E> * \tau$. In that case we can apply lemma 3.7 with $\Gamma = (w)_0$, $\Delta = (u)_0$ and a formula equivalent to $B \wedge \Box D_1 \wedge \ldots \wedge \Box D_n \rhd C \wedge \Box D_1 \wedge \ldots \wedge \Box D_n$, rather than $B \rhd C$ itself, as input. In so doing, we obtain an E-critical successor Δ' of Γ with (i) $C \in \Delta'$ and (ii) $\Box D \in \Delta'$ for each D such that $\Box D \in \Delta$. It suffices now to take $v = <\Delta', (u)_1>$. Given that each $\Box$-formula in Δ is also an element of Δ', the depth of Δ' cannot be larger than the depth of Δ. Therefore $v \in W_\Gamma$. It is clear that v fulfills all requirements to make $uS_w v$.

If, on the other hand u is not an E-critical R-successor of w, then all we know is that $(w)_0 \prec (u)_0$. Recall once more that every successor of Γ is a $\bot$-critical successor of Δ. So, an application of lemma 3.7 with $\Gamma = (w)_0$, $\Delta = (u)_0$, $E = \bot$, and $B \wedge \Box D_1 \wedge \ldots \wedge \Box D_n \rhd C \wedge \Box D_1 \wedge \ldots \wedge \Box D_n$ as input, yields a ($\bot$-critical) successor Δ' of Γ with $C \in \Delta'$ and $\Box D \in \Delta'$ for each D such that $\Box D \in \Delta$. Take $v = <\Delta', (u)_1>$. ⊠

BIBLIOGRAPHY

F. Montagna, 1984, Provability in finite subtheories of PA and Relative Interpretability: a Modal Investigation, *Rapporto Matematico 118*, Dipartimento di Matematica, Universita di Siena.

V. Švejdar, 1983, Modal Analysis of generalized Rosser sentences, *Journal of Symbolic Logic 48*, p. 986-999.

C. Smoryński, 1985, *Modal Logic and Self-reference*, Springer-Verlag, New York

A. Visser, 1986, Peano's Smart Children, a provability-logical study of systems with built-in consistency, *Logic Group Preprint Series* No. 14, Department of Philosophy, University of Utrecht, to be published in *The Notre Dame Journal of Formal Logic*.

A. Visser, 1988P, Preliminary Notes on Interpretability Logic, *Logic Group Preprint Series No. 29*, Department of Philosophy, University of Utrecht.

A. Visser, 1988, Interpretability Logic, This Volume.

CONSTRUCTIVE MATHEMATICS AND COMPUTER-ASSISTED REASONING SYSTEMS

Susumu Hayashi [1]

Research Institute for Mathematical Sciences
Kyoto University
Sakyo, Kyoto, Japan

1. Introduction

The relationship between constructive mathematics and computer science has been noticed for a long while. But it is rather recent that live interactions between constructive mathematics and computer science began. In theoretical aspects, metamathematics of constructive mathematics are utilized to give semantics of higher order functional programming languages. Researches of constructive mathematics in computer science are not just theoretical. Quite a few systems based on constructive mathematics have been designed and implemented. They are program verification systems, program extraction systems, and high-level programming languages. These developments of systems need theoretical investigations as well and so give new insights into constructive mathematics. Applications of constructive mathematics to semantics of programming languages have begun to produce significant feedbacks to constructive mathematics and related areas, see Hyland (1987), Longo and Moggi ($1988+\alpha$). Although researches on constructive systems have not yet been used to unvail new aspects of constructive mathematics, there are clues of possible applications of computer systems to theoretical researches in constructive mathematics Coquand (1986), Howe (1987). It is reasonable to imagine that constructive mathematicians a decade later will be helped by computer systems as mathematicians are now helped by computer algebra systems like REDUCE. In this paper, we will give a survey of existing systems based on constructive mathematics to show the readers how constructive theories are used in actual systems.

Unfortunately my knowledge about works in Eastern bloc countries is quite limited, I regret that works in Eastern bloc countries are rather neglected in this survey.

We would like to thank Mariko Yasugi for comments on the paper.

2. Curry-Howard isomorphism

The basic principle of constructive mathematics in computer science is Curry-Howard isomorphism. We will explain it briefly. Heyting emphasized that activities in constructive mathematics are activities to construct mathematical entities. We are not God. We are finite human beings. Don't care about the paradise in the air which belongs not to us but to God. This philosophy underlies in every constructive philosophy in mathematics. A mathematician who is a finite human being must be able to construct entities when we say they exist. In Brouwer's philosophy, such a construction *must* take place in a mathematician's mind. This gives a mysterious flavor to intuitionism. Science existing only in one's mind? How subjective it is! Hetying partly blew off this mysterious mist of Brouwer's philosophy by the concept "construction." But, in his age, constructions in mathematics, which are highly abstract and intricated, could take place only in human's minds. Modern digital computers have changed the situation. Actual constructions may take place on physical machines.

---

[1] This paper was written while the author was visiting LFCS, Department of Computer Science, University of Edinburgh. This work is supported by Japanese Organization of Promotion of Science.

Heyting (1971) explains that the implication $A \supset B$ is asserted, if and only if we possess a construction r, which, joined to any construction proving A, would automatically effect a construction proving B. In other words, a construction of $A \supset B$ is a function mapping a construction of A to a construction of B. If P^* is the set of constructions of proposition P, then we can identify $(A \supset B)^*$ with function space $A^* \rightarrow B^*$. Similarly $(A \wedge B)^*$ is identified with cartesian product $A^* \times B^*$. Furthermore, the constructions of quantified formulas $(\forall x : A.B(x))^*$ and $(\exists x : A.B(x))^*$ are identified with *dependent types* $\Pi x \in A^*.B(x)^*$ and $\Sigma x \in A^*.B(x)^*$ defined as

$$\Pi x \in T_1.T_2(x) = \{f | f(x) \in T_2(x) \text{ holds for all } x \in T_1\},$$
$$\Sigma x \in T_1.T_2(x) = \{< x, y > | x \in T_1,\ y \in T_2(x)\}.$$

These function spaces, cartesian products, and dependent types are used in programming languages and module systems as data types. This correspondence between types and formulas (or proposition) are called Curry-Howard isomorphism, or "propositions as types" or "formulas as types" principle. Since a formula corresponds to a type, a program which computes a data of type A^* corresponds to a construction of formula A. For example, a construction of an implication may be regarded as a program which computes a data of B from a data of A. Since a construction of a formula is identified with a proof of the formula, we have the equation

$$\text{proofs} = \text{constructions} = \text{programs}.$$

Hence Curry-Howard isomorphism is also referred as "proofs as programs" principles. All in all, programming corresponds to constructive mathematics, hence the principle is also called as "constructive mathematics as programming language." Detailed explanations of Curry-Howard isomorphisms can be found in Howard (1980), Martin-Löf (1982).

3. Early developments

We will mainly concern the recent studies, but we will mention some early developments both in computer science and logic. The historical account below will give the reader a general idea how the present systems evolved, although it is far from being complete.

3.1. Early developments in computer science

The researches related to constructive mathematics in computer science began, when de Bruijn started AUTOMATH project. In the project, many versions of typed lambda calculus are designed and implemented to check mathematics on machines (see a survey de Bruijn (1980) on the project). A version of AUTOMATH has checked the entire book of Landau's "Grundlagen der Analysis". Although, it is used as a proof checker for classical mathematics, it is a system of Curry-Howard isomorphism and so it is considered as a forerunner of recent constructive type theories, e.g., Calculus of Construction.

The deductive approach to program synthesis of Manna and Waldinger (1971) is another forerunner. They proposed to synthesize a program from a specification by proving the specification mechanically. Although they used classical logic, this idea was taken by Constable (1971) as a basis of "constructive mathematics as a programming language." It is apparent that some constructivists were aware of the possibility of using constructive mathematics as a programming language, but Constable seems the first person who took this idea seriously. Later, he and his team implemented Nuprl (1986) based on Martin-Löf's type theory, which is one of the most sophisticated proof checkers ever built.

Influenced by Constable (1971), Goto began to search for procedures which can realize program extraction. (Constable (1971) did not include any description of the working algorithm.) Constable (1971) criticized classical approach in Manna and Waldinger (1971). Goto (1975) tried to apply realizability interpretations to solve some anomaly of the classical approach. Later he shifted to Gödel interpretation and implemented a system based on HA (Goto, 1977, 1979). His system extracted some small LISP programs from proofs of HA. Although the extracted programs were tiny, they were rather natural programs. Sato (1979) also investigated program extraction by Gödel interpretation.

In Soviet, Nepeĭvoda (1978) gave an algorithm of program extraction for realistic programming languages, e.g., LISP, Algol, for a class of proofs.

C. Goad's thesis (Goad, 1980) is one of the most significant early works. Goad implemented an interpreter which execute first order proofs. His interpreter avoided to reduce proofs of lemmata which are Harrop formulas. His interpreter could also avoid unnecessary case distinctions looking up conditions. The second optimization increased efficiency dramatically for some cases. He reported that an application of his method to the bin packing problem increased efficiency dramatically. This could be done, because the interpreter could use logical informations in a proof In execution (reduction) of the proof. This showed "proofs as programs" approach may be used for not only verification but also optimization. Goad also used "recursive proofs" or "circular proofs" A recursive proof is a proof whose lemma may be the same as its conclusion. Such a lemma is specially labeled to indicate that it is a recursive call of the conclusion. Since recursive proofs represent recursive definitions of functions, it may fail to terminate in a normal form. So total correctness was not guaranteed unlike the other systems of "proofs as programs." But, even if a circular argument is used in a proof, a proof may terminate in a normal form which has no recursive calls of conclusion. In a sense, his system guranteed partial correctness of recursive proofs as recursive programs. But he left the problem of verification of termination unsettled. (Cf. 4.3. below.)

Aside from these studies which are explicitly and intentionaly related to constructive mathematics, some very important studies which are implicitly related to constructive mathematics were done by Reynolds and Milner. Reynolds (1974) coined a system of second order lambda calculus for study of types in module programming languages, and Milner introduced a metalanguage ML for a proof checker Edinburgh LCF (Gordon, et al., 1979). These languages had "unusual" function spaces known as "polymorphic function spaces" which are not interpretable by classical function spaces. Later these works turned out to be closely related to Girard's functional system F^{ω}described below.

3.2. Early developments in logic

In the last two decades, studies of constructive mathematics made a great advance. Bishop (1967) showed that a fairly big portion of analysis can be carried out in a constructive way without exotic principles like Church's axiom and the continuity principle. After his work, logicians tried to find out correct formal theories in which Bishop's mathematics can be formalized. Theories used to formalize Bishop's mathematics can be classified in three groups, Martin-Löf's type theory (Marin-Löf, 1973), Myhill-Friedman's constructive set theories (Myhill, 1975; Friedman, 1977) and Feferman's theories of operations and classes (Feferman, 1975, 1979). (Martin-Löf's theory was not designed to formalize Bishop's constructive mathematics, but, at least, it was one of the main applications of his theory. Furthermore, it is Bishop's work that makes possible to claim that Martin-Löf's theory is enough to formalize a great portion of mathematics.) These theories were predicative so that their proof theoretic strength were very weak. They demonstrated that Bishop's constructive analysis is formalizable in those predicative theories. Even theories which are conservative extensions of constructive first order arithmetic HA were enough to formalize it (cf. Beeson, 1985). (Warning: This result depends on particular ways of formalization of mathematical concepts. Constructive logic and predicative theories cannot prove some equivalent definitions of basic concepts like compactness of topology.)

On the other hand, metamathematical study of impredicative theories also made a great advance. Girard (1972) introduced a higher order functional system F^{ω} and extended Gödel interpretation to intuitionistic higher order logic (IHOL). He proved strong normalization theorem of F^{ω}. Girard and Troelstra gave a model of F^{ω}, which was later used by Moggi to give a semantics of polymorphic lambda calculus (Longo and Moggi, 1988+α). Friedman and Myhill, e.g., Friedman (1973), investigated metamathematics of intuitionistic Zermelo-Frankel set theory (IZF).

Aside from these, even earlier, some attempts formulating the concept of construction were made by Scott (1970) and some others. These are regarded as forerunners of Martin-Löf's type theory.

All of these works have computational significance. In the rest of the paper, we will explain how they are related to computer science.

4. The new generation systems

4.1. Martin-Löf's type theory as a programming language

Martin-Löf (1982) introduced a new version of his type theory and related it to programming languages. Martin-Löf's type theory is a logic free theory of typing expressions. An expression denotes a value of a data type. When an expression e has a value of data type A, it is denoted as $e \in A$. This form is called a judgement. (Strictly speaking, A itself is an expression which denotes a type.) Data types of Martin-Löf's type theory is so rich that constructive logic can be coded by means of Curry-Howard isomorphism. As Bishop identified a set as a base set with an equality on it, every type is equipped with an extensional equality on it. When two expressions e_1 and e_2 of type A are extensionally equal, Martin-Löf writes as $e_1 = e_2 \in A$. This is called an equality judgement. Besides this, Martin-Löf's type theory has a type called propositional equality or equality type. If two expressions e_1 and e_2 are of a type A, then $Eq(A, e_1, e_2)$ is a type which has a value $\mathbf{r}$ if and only if $e_1 = e_2 \in A$ holds. Namely the rules

$$\frac{\mathbf{r} \in Eq(A,a,b)}{a = b \in A}, \qquad \frac{a = b \in A}{\mathbf{r} \in Eq(A,a,b)}$$

hold for the equality type. By this type, equality of each type can be interpreted as a type. So constructive logic can be interpreted by Curry-Howard isomorphism. Martin-Löf pointed out that his theory can be considered as a type inference system for a functional programming language. Actually, his typing rules are closely related to typing rules of some typed functional languages, e.g., ML, Pebble (Burstall and Lampson, 1984), and second order lambda calculus. Furthermore, Martin-Löf's equality type enabled to represent a specification of a program as a type. For example, a specification of a program which finds a prime larger than a natural number m is given by

$$\Pi m \in N.\Sigma p \in N.Eq(N, test(p, m), 1), \tag{1}$$

where $test(p, m)$ is 1 if and only if p is prime greater than m. If f has this type, $f(m)$ is defined and its value is $\langle p, \mathbf{r}\rangle$ where p is a prime greater than m. Namely verifying that f satisfies the specification can be done by type checking of f. Furthermore, a program development of f can be done by a proof development of the formula

$$\forall m \in N.\exists p \in N.test(p, m) = 1.$$

This synthesis of type checking, verification, and rigorous program development received attentions from computer scientists. We will look at two efforts of development of Martin-Löf's idea in the next two sections.

4.2. Göteborg type theory

A group at Göteborg, Sweden, is developing a type theory and program methodology based on Martin-Löf's type theory as well as its computer implementation. (We will call their type theory as Göteborg type theory.) They follow Martin-Löf rather faithfully. One important departure from the original Martin-Löf type theory is the introduction of a subset type. In the example of "there is a greater prime" above, a function f of the type (1) returns a value $\langle p, \mathbf{r}\rangle$, whose second component $\mathbf{r}$ is redundant. So the type (1) is not quite the specification we have in our mind. Furthermore, if $Eq(N, test(p, m), 1)$ were a very complicated type, it would be costly to compute the unnecessary second component. To solve this problem, Nordström and Petersson (1983) introduced a subset type If A is a type and $B(a)$ is a type depending on a variable a of the type A, then the subset of A of the elements a for which $B(a)$ is inhibited is denoted by

$$\{a \in A | B(a)\}.$$

By this concept of subset type, we modify the specification of (1) as

$$\Pi m \in N.\{p \in N | Eq(N, test(p, m), 1)\}. \tag{2}$$

A function of this type computes a prime which is greater than m.

The rules for the subtype are

$$\frac{a \in A \qquad b \in B(a)}{a \in \{x \in A | B(x)\}}, \qquad \frac{a \in \{x \in Z | B(x)\} \qquad c(x) \in C(x)[x \in Z, y \in B(x)]}{c(a) \in C(a)},$$

where y must not occur free in $c(x)$ or in $C(x)$. The first one is an introduction rule and the second one is an elimination rule. (Besides these a formation rule is necessary. We omitted it, since it is obvious.)

Note that an element of a subset type $\{x \in A | B(x)\}$ does not contain an element of $B(x)$. So $\forall x \in \{x \in A | B(x)\}.B(x)$ is not always provable. A sufficient condition of provability is that $B(x)$ is stable, i.e., $\forall x \in A.((\neg\neg B(x)) \supset B(x))$. (See Salvesen and Smith, 1988.)

Göteborg type theory has been used for some program developments. A detailed introduction to their type theory and program development methodology will appear in Nordström, Petersson, and Smith (1988+α).

4.3. Nuprl

The most sophisticated computer system related to constructive mathematics ever built is Nuprl by Constable's PRL group at Cornell University. After his pioneering paper, Constable seemed to shift to program verification rather than "proofs as programs." Constable and his students implemented a programming logic PL/CV2 which is a verification system for a dialect PL/I programming language. It was similar to dynamic logic so that programs explicitly appear in proofs. So it was a system of "programs as proofs". Namely, the main objects which users concern were programs and proofs were their annotations. PL/CV2 was based on first order logic and did not have enough types like function spaces, so it suffered from weak expressive power (p.271, Constable, Johnson, and Eichenlaub, 1982). Constable and his coworkers tried to cure this and began to design and implement a proof checker using Martin-Löf's type theory. The result of the effort was Nuprl.

Nuprl is renowned for its sophisticated environment. Actually its window system oriented refinement logic (backward reasoning logic) is one of the most novel features of Nuprl. But we will not explain it here, since it is irrelevant to constructive logic. Basically, Nuprl's type theory is an extension of Martin-Löf's type theory. But there are some significant departures from Martin-Löf's type theory. Nuprl used quotient types as well as subset type of Göteborg type theroy. Bishop identified a set with a base set A with an equivalence relation E on it. Theoretically, it is sufficient to assume that a set is always equipped with an equivalence relation. But, in practice, it will be more convenient to make it to a quotient type. Nuprl allows to make it as a quotient type $(x, y){:}A//E$ with its own equality rather than an equivalence relation $E(x, y)$. So the equality rules for equality judgement of the quotient type can be used. The quotient type is useful only when $E(x, y)$ is a "degenerated" or "squashed" type as was pointed out in Constable et al. (1986). Namely it is useful only when $E(x, y)$ has at most one element. This is because of that the equality judgement and the equality type forget about the proof of equality. Let a and b be elements of A. Let's denote their cosets with respect to E by $[a]$ and $[b]$. When $[a] = [b] \in (x, y){:}A//E$ is derivable, then we hope retrieve an element of $E(a, b)$. Since $[a] = [b] \in (x, y){:}A//E$ is equivalent to $\mathbf{r} \in Eq((x, y){:}A//E, [a], [b])$ and $\mathbf{r}$ is the only one possible element of $Eq((x, y){:}A//E, [a], [b])$, the equality judgement does not carry any information with respect to E except that it is inhabited. So we cannot retrieve an element of E unless E is degenerated. This restriction on quotient type causes no problem in almost all cases. For example, if a pre-real number is defined as a pair of Cauchy sequence a and its modulus of convergence r, then the type representing its equivalence relation has at most one element and so its quotient type is useful.

Another departure of Nuprl type theory from Martin-Löf's type theory is inductive types and partial function spaces. Martin-Löf's type theory has some basic built-in recursive types like natural numbers and well-founded trees. Theoretically, these are enough to implement other data types. But this is not enough in practise. So Nuprl has inductive types. Furthermore, it has partial function types as well. Nuprl's semantics is based on a computation system which consists a *type free* term model and reduction rules on the terms . But not all terms in the term model is normalizable as they are untyped. Nuprl has a rule of computation by which user can evaluate his term. The evaluation is done by an evaluator using call-by-name strategy. So it is natural to extend Martin-Löf's type theory to include partial functions.

Considering partial functions, we can define functions using arbitrary recursions in the conventional way. The lack of arbitrary recursion obstacles to develop conventional and efficient programs via constructive mathematics. Although program development via constructive mathematics is neat, developed algorithms are not always natural from algorithmic point of view. For some very algorithmic problems, neatness of mathematical proofs and naturality of the extracted terms are often opposite to each other. (See 4.3.2.5 in Hayashi and Nakano, 1988.) Since program constructors and their verification rules of Martin-Löf's type theory are quite natural from mathematical point of view, this means Martin-Löf's type theory may fail to develop natural programs in some cases. So it is necessary to tune up Martin-Löf's type theory so that the conventional programming can be done by sacrificing mathematical neatness when necessary. Inductive types and partial function types of Nuprl can serve this purpose.

For example, searching a root of $f(n) = 0$ for $n = 0, 1, \ldots$ is defined by

$$\begin{aligned} search() &= search1(0), \\ search1(x) &= if\ f(x) = 0\ then\ x\ else\ search1(x+1). \end{aligned}$$

The recursion of $search1$ would not be able to be obtained by ordinary induction principles of Martin-Löf's type systems. But Nuprl can do this. Since propositions are identified with types, one can define the domain of the termination of $search1(x)$ as $\{x \in N | dom(x)\}$ by defining dom as

$$dom(x) = \begin{cases} true & \text{if } f(x) = 0, \\ dom(x+1) & \text{if } f(x) \neq 0. \end{cases}$$

Assuming $dom(0)$ is inhabited, $\exists x \in N. f(x) = 0$ is provable and the extracted term uses the recursion above. As subset type, recursive type does not have constructor for its elements. So the elements of $dom(0)$ is not used to compute the extracted term of the proof of $\exists x \in N. f(x) = 0$. As we will explain later in the section of PX, this gives a powerful methodology to extract efficient programs and do termination proof separately. But it seems that this technique is not so used for program development in Nuprl.

The recursion associated with inductive types is *not* an arbitrary recursion. Nuprl has partial function type $A \rightharpoonup B$ so that arbitrary recursions are allowed.[2] Functions defined by arbitrary recursions

[2] PRL group uses another notation for partial function space.

may fail to terminate and so they do not belong to total function types of Martin-Löf's type theory. In Constable et al. (1986), a partial function f from A to B was considered as a total function on its domain of convergence. Inductive types are useful to characterize termination domains and graphs of a kind of recursions.

Constable and Smith (1987) have presented another approach to partial functions in Nuprl. This approach uses a type of computation $\overline{A}$. $\overline{A}$ is a type of computation whose possible value belongs to the type A. So the typing rule for the fixpoint operator fix is

$$\frac{f \in A \rightarrow A}{fix(f) \in \overline{A}.}$$

Partial function type $A \rightharpoonup B$ is defined by $A \rightarrow \overline{B}$. Since a formula $\exists y \in B.P(x,y)$ is understood as a type, $\overline{\exists y \in B.P(x,y)}$ is at the same time a type and a formula. So $\forall x \in A.\overline{\exists y \in B.P(x,y)}$ is a formula and this is a specification of partially correct programs of the problem $\forall x \in A.\exists y \in B.P(x,y)$, i.e., if a function of the type terminates for a value of A, then the value satisfies the specification. Namely, if A is a formula, $\overline{A}$ is the type of "partial proofs" of A. The semantics of bar type $\overline{A}$ is a collection of terms whose values belong to type A if they exist. Equality on bar types are Kleene's equalities. So all divergent terms belong to any bar type and are equal. Note that this approach resembles Goad's "recursive proofs". Nuprl's approach has a significant superiority. Nuprl can reason about the properties of "proofs" which are Nuprl's terms. So Nuprl can reason about terminations of "partial proofs". Goad had to argue about termination metamathematically, since proofs are not terms in his system. Basin (1988) has shown how to develop partial programs using bar types.

4.4. PX

PX is a system developed by Hayashi and Nakano at Kyoto University (Hayashi, 1987: Hayashi and Nakano, 1988). The underlying theory of PX is a constructive formal theory based on Feferman's T_0. In type theories, logic is defined by means of Curry-Howard isomorphism. In PX, logic is fixed in advance and programs are extracted from proofs by Curry-Howard isomorphism, and data types called classes are defined by means of logic.

PX is a first oreder two sorted theory based on logic of partial terms (LPT). LPT is a logic whose terms may fail to denote values (cf. Beeson, 1985; Hayashi and Nakano, 1988; Moggi, 1988). In PX, arbitrary recursive definition is allowed. This does not cause a contradiction, for terms may fail to denote values. By a definedness predicate $E(t)$, we can express "term t has a value." The equality of PX is Kleene's equality. LPT of PX axiomatizes LISP. It is an axiomatization of call-by-value intensional partial recursive functions over S-expressions. The axiomatization resembles that of axiomatic recursion theory.

The sorts of PX are the sorts of individuals and classes. Individuals are S-expressions in the sense of LISP. Since the sort of classes is a subsort of individuals, every value in PX is an individual. PX is thus untyped. Data types are values in PX, since data types are defined as classes.

Furthermore, every function has a canonical representation as an individual.

Data types of Martin-Löf's type theory except universes can be interpreted as classes. Since classes are objects, it is easy to define a class of all classes generated from given finitely many classes by finite many class forming operators. For example, a cartesian closed category generated from finite many classes is again a class. Total and partial intensional function spaces can be defined as classes. The partial function space $A \rightharpoonup B$ is defined as

$$\{f | \forall a : A.(E(app*(f,a)) \supset app*(f,a) : B)\},$$

where $app*(f,a)$ is the result of the application of f as a function to a. E is the definedness predicate mentioned above. This reads that "if the application of f to a from A has value, then it belongs to B." Bar type $\overline{A}$ is defined as $1 \rightharpoonup A$ or

$$\{p | E(app*(p)) \supset app*(p) : A\}.$$

A least fixpoint operator with the same typing rule as that of Nuprl is easily definable.[3]

Since the equality of PX is the intensional equality on S-expressions, these function spaces are intensional. Extensional types must be defined as partial equivalence relations (PER). A PER is a binary class C such that (i) $[x,y] : C \supset [y,x] : C$, (ii) $[x,y] : C \supset [y,z] : C \supset [x,z] : C$. (See Mitchell,

[3] MGA in Hayashi and Nakano (1988) is necessary to show the typing rule.

1986.) PER C is interpreted as equality on the class $\{x|[x,x]:C\}$ and extensional spaces can be defined by using PER, e.g., an extensional total function space is defined as

$$\{f|\forall[a,a']:A.[app(f,a),app(f,a')]:B)\}.$$

Curry-Howard isomorphism in PX is given by a formalized realizability. If A is a theorem of PX, then there is a term e for which the fact "e realizes A" is provable in PX again. It is different from ordinary realizability so that if A is realizable, then A holds. Namely, if A is realizable, then it holds in the standard model of PX. Since the standard model is given *classically*, this makes the following difference from Marin-Löf's and Nuprl's approach. Nuprl contradicts classical logic (see Theorem 3, Constable and Smith, 1987). Contrary to this, PX is consistent with classical logic. Hence, in Martin-Löf's type theory and Nuprl, if $\forall x:A.\exists y:B.F(x,y)$ is proved and a term f is extracted from it, then f meets the specification $\forall x:A.F(x,f(x))$ in the sense of constructive logic. In PX, f meets the specification in the sense of classical logic. (Note that it meets the specification in the sense of constructive logic of PX as well.) Namely, PX allows to interpret the input-output predicate $F(x,y)$ classically. Furthermore, this compatibility allows to prove many lemmata classically. A family of formulas, called rank 0 formulas, obeys classical logic in PX. The properties expressible by rank 0 formulas are properties about data types and termination conditions of extracted programs. Furthermore, PX has a modal operator $\Diamond$ which resembles the squash operator of Nuprl. If A is a formula $\Diamond A$ holds iff A holds classically. Namely, putting $\Diamond$, every formula can be classicalized and loose all informations except A holds *classically*. In the case of squash operator, a squashed proposition holds, if and only if the original proposition holds constructively.

The realizability and the extractor of PX are optimized so that efficient and natural programs are extracted. For example, if A is a rank 0 formula, a realizer of $A \wedge B$ is just a realizer of B. Note that the interpretation of logical operators of the usual Curry-Howard isomorphism is defined uniformly from the types of realizers of arguments of logical operators. In PX, the interpretation of a formula depends not only on its outermost logical operator but also on the *ranks* of immediate subformulas. (The rank of a formula is the length of realizers of the formula.)

We have done some program developments with PX, e.g., a solution of the Chinese remainder theorem, a deterministic theorem prover for classical propositional sequent calculus (see Hayashi and Nakano, 1988). A technique which resembles the one used for the example of *search* in Nuprl described above were extensively used in these program developments. We define termination domains as classes and so termination problems are stated as rank 0 formulas. We can then extract programs before we settle termination problems. Those termination problems may be proved by classical logic or even metamathematical considerations on the standard semantics. PX can prove these by a classical proof checker based on higher order logic in which the semantics of PX is embedded. This resembles Cook's idea of the relative completeness of Hoare logic. We have given a translation of Hoare logic to PX using this idea.

4.5. Calculus of Constructions

Calculus of Construcion (CC) is a system developed by Coquand and Huet at INRIA, Roquencourt. It is a decent of AUTOMATH, Martin-Löf's type theory, and Girard's F^ω. It uses de Bruijn notation to avoid renaming of variables. It is a impredicative system as F^ω. Its types are defined simultaneoulsy with terms as in Martin-Löf's type theory. In these sense, CC is an impredicative extension of Martin-Löf's type theory. Basic differences besides impredicativity from Martin-Lóf's type theory are

1. CC does not have equality judgement in the sense of Martin-Löf's type theory,
2. CC does not have basic types as equality types, natural number type, and finite set types, etc,
3. CC does not have either of quotient types or subset types,
4. CC does not have strong dependent sums.

The only type constructor of CC is the dependent product and the only basic type is *Prop.*[5] Impredicativity allows to define many types by these. Equality type is defined as Libnitz's equality. A type of natural number can be defined, although ordinary properties cannot proved in the basic CC. Empty type can be defined as $\forall\alpha \in Prop.\alpha$. The lack of basic types is not quite important, for CC is an open system so that such types are added when they are necessary.

Quotient types would be treated as PER's as we did in PX. Subset types would be treated as propositional functions. These types are meaningful when equivalence relations and subset conditions are "degenerated." Namely, a concept as squashed types or rank 0 formula is necessary. So these would need distinction of *Prop* and *Spec* by Pauling-Mohring described below.

[5] CC has many versions. In some versions, CC has more basic types or kinds.

The lack of strong dependent sum is essential. A dependent sum $\Sigma x : A.B(x)$ is called strong, if the first projection of elements of the type is available. Martin-Löf's dependent sum is strong. It is known that this contradicts to impredicativity (see Coquand, 1986).

A good point of CC is a beautiful combination of minimality and strong expressive power. Since CC has a small numbers of basic concepts, it looks as if it is an abstract machine. It can express many complicated types from small numbers of basic concepts. The higher order impredicative types seems enough to formalize almost all concepts of abstract mathematics like abstract algebra and topology. Since it extends F^ω, all of provably computable functions of higher order arithmetic are representable as terms of CC without recursions.

An early implementation and experiments with it was described in Coquand and Huet (1985). Now there are some different implementations of CC at INRIA, Cornell University, and Edinburgh University. Griffin (1987) has impelmented CC in his EFS system. Randy Pollack at Edinburgh University has implemented a proof checker of CC called LEGO and has done many interesting examples on it. The implementation of LEGO is based on higher order unification and so resembles Isablle described below. J. Despeyroux (1988) has implemented a window oriented proof checker for CC on CENTAUR. Coquand is implementing a new version of his proof checker using LCF-like tactics.

So far CC has been mainly used as a framework to implement mathematics. It is interesting how CC serves as a program extraction system. Pauling-Mohring (1986, 1987) is developing program extraction methodologies with CC and a new version of CC for program extraction. She extended CC so that there are two kinds of types. One is *Prop* and another is *Spec*. *Spec* corresponds to *Prop* of the original CC. Types of *Prop* resemble Harrop formulas. So types of *Prop* are used as squashed types of Nuprl and rank 0 formulas of PX to avoid unnecessary informations.[6] Typing rules for *Spec* and *Prop* are isomorphic to those of original CC, so the extension still retains CC's minimality. This was possible since CC has only dependent product as type constructors.

There are some attempts to extend CC to stronger type theories. Luo (1988, 1988a) extended CC using Martin-Löf like predicative universes. His type system has strong dependent sum at higher predicative types. He proved that his system is strongly normalizable and gave a mathematical model. Although the predicative hierarchy resembles hierarchies of small models of ZF set theory, its proof theoretic strength is much weaker than ZF. It seems that Luo's system is a right framework to formalize category theory. His type theory has been implemented in LEGO.

4.6. ICOT-QJ

QJ is a system introduced by Sato (1985). It is a type theory based on a concurrent functional language Quty based on unification. QJ has a minimal fixpoint operator by which even inductive types are definable. Implementation of QJ is being carried at ICOT by Takayama (1987). Takayama (1988) has presented a new idea to avoid unnecessary informations automatically by "proof analysis" which resembles program analysis.

4.6. ZFR

As we saw above, Martin-Löf's type theory and Feferman's untyped theory have been implemented and used from computational point of view in Göteborg type theory, Nuprl, and PX. But there has not been such an attempt for Myhill and Friedman's theories mentioned in 3.2. ZFR by Beeson (1987) aims to fill this gap. ZFR is a mixture of lambda calculus and ZF set theory. Basic idea is to use sets as data types as infinite S-expressions and give an interpreter of lambda calculus on it. Since set recursion can be defined by lambda calculus, some recursive structures like transitive closure of a set are easily defined. ZFR has not been implemented or used for program development.

4.7. LF, Isabelle, λProlog

Curry-Howard isomorphism shows that type theories can define constructive logic. LF (logical framework) group at Edinburgh university has extended the "formulas as types" notion to the "judgements as types" notion, so that many logics can be defined by means of a Martin-Löf-like type theory called Edinburgh LF (Harper, R., Honsell, F., and Plotkin, G., 1987). LF enables us to "code up" logics easily without changing the basic proof checker. All systems of Despeyroux, Griffin, and Pollack can be used as systems for LF.

It is not necessary to keep constructions (proofs) to code up logics, if the provability alone is concerned. Paulson's Isabelle system (Paulson, 1987) uses intuitionistic higher order logic to code up logics. A similar approach is used by Felty and Miller (1988) to code up logics in λProlog. λProlog is a higher order Prolog based on hereditary Harrop formulas, which form a fragment of Harrop formulas of

[6] Salvesen and Smith (1987) presented a similar idea.

intuitionistic higher order logic. In λProlog, it is possible to code up logics so that proofs are preserved. An interesting point is that an extension of the existence property under Harrop formulas is used to make search space tractable in λProlog.

References

Basin, D.A., 1988, An Environment for Automated Reasoning About Partial Functions, in: Springer Lecture Notes in Computer Science 310.

Beeson, M.J., 1985, "The Foundations of Constructive Mathematics", Springer, Berlin.

Beeson, M.J., 1987, Towards a Computation System Based on Set Theory, preprint.

Bishop, E., 1967, "Foundations of Constructive Analysis," McGraw-Hill.

de Bruijn, N.G., 1980, A survey of the project AUTOMATH, in: "Essays in Combinatory Logic, Lambda Calculus, and Formalism," Seldin, J.P. and Hindley, J.R., eds., Academic Press, New York.

Burstall, R., and Lampson, B., 1984, A kernel language for abstract data types and modules, in: Springer Lecture Note in Computer Science 173.

Constable, R.L., 1971, Constructive mathematics and automatics programs writers, in: "Proceedings of IFIP Congress," Ljubljana.

Constable, R.L., et al., 1986, "Implementing Mathematics with the Nuprl Proof Development System," Prentice-Hall.

Constable, R.L., Johnson, S.D., and Eichenlaub, C.D., 1982, *Introduction to the PL/CV2 Programming Logic*, Springer Lecture Notes in Computer Science 135.

Constable, R.L., and Smith, S.F., 1987, Partial Objects in Constructive Type Theory, in: "Second Annual Symposium on Logic in Computer Science," IEEE.

Coquand, T., 1986, An analysis of Girard's paradox, in: "First Annual Symposium on Logic in Computer Science," IEEE.

Coquand, T., and Huet, G., 1985, Constructions: A higher order proof system of mechanizing mathematics, in: Springer Lecture Notes in Computer Science 203.

Coquand, T., and Huet, G., 1988, The Calculus of Constructions, *Information and Computation 76*.

Despeyroux, J., 1988, First experiments with theorem proving in Centaur: the Calculus of Construction, the Edinburgh Logical Framework, and Theo, in: "GIPE, ESPRIT project 348, Third annual review report."

Feferman, S., 1975, A language and axioms for explicit mathematics, in: Springer Lecture Notes in Computer Science 450.

Feferman, S., 1979, Constructive theories of functions and classes, in: "Logic Colloquium '78," North-Holland, Amsterdam.

Felty, A. and Miller, D., Specifying Theorem Provers in a Higher-order Logic Programming Language, in: Springer Lecture Notes in Computer Science 310.

Friedman, H., 1973, Some applications of Kleene's methods for intuitionistic systems, in: Springer Lecture Note in Mathematics 337, Springer, Berlin.

Friedman, H., 1977, Set theoretic foundations for constructive analysis, *Annals of Mathematics 105*.

Girard, J.Y., 1972, "Interprétation fonctionnelle et élimination des coupures de l'arithmétique d'ordre supérieur", Ph.D. thesis, University of Paris, VII.

Goad, C., 1980, "Computational uses of the manipulation of formal proofs", Ph.D. thesis, Stanford University.

Gordon, M.J., Milner, R., and Wadsworth, C.P., 1979, "Edinburgh LCF", Springer Lecture Notes in Computer Science, Springer-Verlag 78.

Goto, S., 1975, Recursive Realizability and Automatic Program Synthesis (in Japanese), IECE (The Institute of Electrical and Communication Engineers, Japan), Shingaku-Giho AL (Automata and Language) AL-75-54.

Goto, S., 1977, Foundations of Automatic Program Synthesis: an Application of Godel's Interpretation (in Japanese), Shingaku-Giho AL (Automata and Language) AL-77-16, English version appeared in: Springer Lecture Note in Computer Science 75 (1979).

Goto, S., 1979, Program synthesis from natural deduction proofs, in: "Proceedings of the Sixth International Joint Conference on Artificial Intelligence, vol. 1."

Griffin, T.G., 1987, An environment for formal systems, Technical Report 87-846, Department of Computer Science, Cornell University.

Harper, R., Honsell, F., and Plotkin, G., 1987, A Framework for Defining Logics, in: "Second Annual Symposium on Logic in Computer Science," IEEE.

Hayashi, S., 1987, **PX**: a system extracting programs from proofs, in "Formal Description of Programming Concepts-III," Wirsing, M., ed., North-Holland, Amsterdam.

Hayashi, S., and Nakano, H., 1988, **PX**: A Computational Logic, The MIT Press, Cambridge, Massachusetts.

Heyting, A., 1971, *Intuitionism, An Introduction, third revised edition*, North-Holland, Amsterdam.

Howard, W., 1980, The formulas-as-types notion of construction, in: "Essays on Combinatory Logic, Lambda Calculus and Formalism," Seldin, J.P. and Hindley, J.R., eds., Academic Press, New York.

Howe, D.J.., 1987, The Computational Behaviour of Girard's Paradox, in: "Second Annual Symposium on Logic in Computer Science," IEEE.

Hyland, J.M.E., 1987, A small complete category, preprint.

Longo, J., and Moggi, E., 1988+α, Constructive Natural Deduction and its "modest" Interpretation, in: "Semantics of Natural and Computer Languages," Meseguer et al. eds., The MIT Press, Cambridge, Massachusetts.

Luo, Z., 1988, A Higher-order Calculus and Theory Abstraction, LFCS Report ECS-LFCS-88-57, LFCS, Edinburgh University.

Luo, Z., 1988a, CC^{ω} and its Meta Theory, LFCS Report ECS-LFCS-88-58, LFCS, Edinburgh University.

Manna, Z., and Waldinger, R., 1971, Towards automatic program synthesis, *Communications of ACM 14*.

Martin-Löf, P., 1974, An intuitionistic theory of types: Predicative Part, in: "Logic Colloquium 73," Rose, H., and Shepherdson, J., eds., North-Holland, Amsterdam.

Martin-Löf, P., 1982, Constructive mathematics and computer programming, in: "Logic, Methodology, and Philosophy of Science VI," Cohen, L.J., et al., eds., North-Holland, Amsterdam.

Mitchell, J.C., 1986, A type-inference approach to reduction properties and semantics of polymorphic expressions, in: "1986 ACM Symposium on Lisp and Functional Programming."

Moggi, E., 1988, "The Partial Lambda-Calculus," Ph.D. thesis, University of Edinburgh.

Myhill, J., 1975, Constructive set theory, J. Symbolic Logic 40.

Nepeĭvoda, N.N., 1978, A relation between the natural deduction rules and operators of higher level algorithmic languages, *Soviet Mathematics Doklady 19*.

Nordström, B., and Petersson, K., 1983, Types and specifications, in: *Proceedings IFIP '83*, Elsevier, Amsterdam.

Nordström, Petersson, and Smith, 1988+α, *Programming in Martin-Löf's type theory. An introduction*, outcoming from Oxford University Press.

Paulson, L.C., 1987, The representation of logics in higher-order logic, Cambridge Technical Report 113.

Paulin-Mohring, C., 1986, Algorithm development in the calculus of constructions, in: "First Annual Symposium on Logic in Computer Science," IEEE.

Paulin-Mohring, C., 1987, Extraction de programmes dans le Calcul des Constructions, preprint.

Reynolds, J.C., 1974, Towards a theory of type structure, in: Springer Lecture Note in Computer Science 19.

Salvesen, A. and Smith, J.M., 1988, The strength of the subset type in Martin-Löf's type theroy, in: "Third Annual Symposium on Logic in Computer Science," IEEE.

Sato, M., 1979, Towards a mathematical theory of program synthesis, in: "Proceedings of the Sixth International Joint Conference on Artificial Intelligence, vol. 2."

Sato, M., 1985, Typed Logical Calculus, Department of Information Science, Faculty of Science, University of Tokyo, Technical Report 85-13.

Scott, D., 1970, Constructive validity, in: Springer Lecture Notes in Computer Science 125, Springer, Berlin.

Takayama, Y., 1987, Writing Programs as QJ-Proofs and Compiling into PROLOG Programs, in: "Proceedings of 4th Symposium on Logic Programming."

Takayama, Y., 1988, Proof Theoretic Approach to the Extraction of Redundancy-free Realizer Codes, preprint.

MARKOV'S CONSTRUCTIVE MATHEMATICAL ANALYSIS: THE EXPECTATIONS AND THE RESULTS

Boris Kushner

Computing Center, Akademia Nauk
Vavilov str.40
Moscow, 117333, USSR

1. The main aim of this work is a concise review of constructive mathematics in Markov's sense. Soviet constructivism originated in the late 40-s or early 50-s of this century. This trend of mathematical thought is similar in many of its features to its immediate precursor, intuitionism, which has been in the same radical opposition to the rest of the mathematical world. History shows that revolutionary ideas are always connected with brilliant and original personalities, from whose temperament and spiritual power these ideas take their energy and convincing might. For intuitionism, the crucial person was L.Brouwer (1881-1966); for constructivism, it was A.A.Markov (Jr., 1903-1979).

2. Below the word "constructive" will be used in the sense of Markov's school only. As is known, constructive mathematics can be characterized by the following main features (cf. [1-2]). a. The objects of study are constructive processes and constructive objects arising as results of these processes. One deals practically always with a special case of constructive objects - words in one or another alphabet. b. One uses a special constructive logic which takes into account the specific nature of constructive processes and objects. In particular, the law of the excluded middle and the law of double negation cannot be accepted in any real sense. Assertions about the existence of constructive objects are understood as assertions about the possibility of constructing the required objects. c. One admits the abstraction of potential realizability but the abstraction of actual infinity is completely excluded. d. The intuitive concept of "effectivity", "computability", etc. are connected with a precise concept of algorithm.

The question arises what kind of mathematics one can develop on such restricted basis. Is it realizable, the initial revolutionary intention of rebuilding mathematics in the constructive way? We shall attemt to examine this question through constructive mathematical analysis. We believe that it is mathematical analysis which is the most important and the most characteristic part of the above mentioned programme. Let us note, as well, that a distinguishing feature of con-

Mathematical Logic
Edited by P. P. Petkov
Plenum Press, New York, 1990

structive mathematics as opposed to classical and intuitionistic mathematics is the purely syntactic nature of the constructive mathematical univers. Such an approach seems, perhaps, very restrictive and even strange, but it is quite in agreement with contemporary tendencies of computer science. The constructive orientation on effectiveness, which is in the very nature of constructivism, is also in agreement with those tendencies.

3. All unexplained notions of constructive analysis can be found in [2] . The author believes that constructive mathematics has achieved great success in the development of analysis of the real numbers and functions over real numbers. Let us recall some well-known facts.

4. The theory of the constructive continuum has been developed. The elements of this continuum are so-called constructive real numbers (c.r.n.), i.e. pairs of algorithms (coded in the proper way), of which the first algorithm determines a sequence of rational numbers and the second effectively estimates its rate of convergence in itself. Below we shall denote the constructive continuum by D. By D_1 we denote the set of pseudonumbers. As is known, D is strictly included in D_1 and D_1 can be regarded as an analog of D when we are passing to ϕ-computability.

Because of its atomic nature, the constructive continuum has some resemblence to the classical one. Yet it has, in contrast, purely syntactic nature since c.r.n. are words in some fixed alphabet. It should be noted that the constructive continuum has turned out to be quite adequate for the introduction of basic functions and operations of elementary analysis. In particular, the completeness theorem gives the opportunity to develop the convergrnce theory in practically its usual strength. Yet the theorem about the convergence of a bounded monotone sequence and compactness theorems are not valid for D.

5. Starting from Markov's definition of a constructive function (c.f.) the constructive theory of functions of a real variable has been developed. In its applied aspects this theory is quite adequate to meet the claims which one usually makes of mathematical analysis. The very notion of a constructive function is a natural algorithmic variant of the classical notion of function in Dirichlet's sense. In terms of constructive functions one can without any difficulties introduce all the usual elementary and special functions. Differential and integral calculus for constructive functions can also be developed to the full. On the other hand, some property o. constructive functions have no classical analogues. The most famous result of this kind is Tseitin's theorem about the continuity of constructive functions: any c.f. is constructively continuous at any point where it is defined. One should also mention the well-known examples of constructive functions with unusual properties. For example, there exists a c.f., which is defined everywhere on the constructive unit segment and which is effectively nonuniformly continuous on that segment (I.D.Zaslavskiĭ). Some progress has been achieved lately in the study of such peculiarities

of constructive functions connected with the non-compactness of the constructive continuum. With this purpose one consides the behaviour of constructive functions on various extensions of the constructive continuum, on D_1 in particular (V.Lifshits, O.Demuth, B.Kushner et al.). Let us present some relevant results of the author [3]: 1) if a partial-recursive operator computing a c.f. f on D can be extended onto D_1, then f is constructively uniform continuous on the unit segment; 2) there is a c.f. which has a continuous extension on D_1 and which is effectively nonuniformly continuous on the unit segment; 3) if a c.f. can be continuously extended onto D_2 than it is a pseudo-uniformly continuous function on the unit segment (by D_n we denote the system of $\emptyset^{(n)}$-computable real numbers).

6. In fact, in the last section we have already mentioned the method of relativisation. The method enables us to generalize results onto various computable structures, constructing in terms of relative computability. But consistent constructive interpretation of that method is not an easy thing at all. The corresponding theory is developed in the splendid work of O.Demuth, R.Kryl and A.Kucera [4] which contains a number of original results.

7. One should point out the rather developed topology of the constructive plane. Not to mention V.P.Orevkov's famous achievements concerning the Brouwer's fix-point theorem, there is a cycle of Manukyan's works about plane constructive curves. Let us note the very difficult proof of a constructive version of Jordan's theorem for the constructive plane (in co-operation with I.D.Zaslavskiĭ) and the following surprising counterexample [6]: there exists a pair of plane curves f, g such that: 1) f and g lie inside the unit square; 2) f connects points (0,0) and (1,1), g connects points (0,1) and (1,0); 3) f, g do not intersect.

8. The possibility of development of basic parts of the theory of differential equations and the theory of functions of complex variable is quite evident. But, regretfully, works on these topics are not numerous. Let us note here Manukyan's work [7] containing a constructive version of Cauchy's integral theorem.

9. The theory of complexity of algorithms and calculations has also had, regretfully, only a few applications in constructive analysis (V.A.Shurygin, V.P.Orevkov, M.I.Kanovich, B.A.Kushner et al.). As an example, we present the following result of the author [8]. Let h be a general recursive function. An algorithm q is said to be a calculation of a c.r.n.x if for every i q(i) is a rational number and condition $|q(i) - x| \leqslant 1/i+1$ holds. A calculation q is said to be h-simple if for any i there exists some k such that ($k \geqslant i$ & $[q(k)] \leqslant h(k)$) where $[q(k)]$ denotes the number of steps in which q finishes its operations on k. Let us call a c.r.n. x h-complex if it does not admit h-simple calculation.

A c.r.n. x is said to be h-simple if x is not h-complex. The following assertion holds: for any general recursive function h the set of h-simple c.r.n. has constructive measure 0 (and, so, the set of h-complex c.r.n. has full constructive measure).

10. The development of constructive functional analysis was started by N.A.Shanin in his famous work [9]. In particular, the concept of constructive metric space introduced in [9] turned out to be very usefull. Many results concerning the real line acquired their natural generality in terms of constructive metric spaces. Let us note, first of all, Tseitin's famous theorem on the continuity of algorithmic operators over constructive metric spaces. The notions of normed and Hilbert's spaces can be successfully treated in the constructive way as well. It is more difficult to develop theories which premise an essential freedom in dealing with sets. In this connection we could mention abstract topology and the theory of Lebesgue measure and integral. We would like to note the investigations of Phan Dinh Dieu on constructive generalised functions and linear topological spaces which has been summed up in his monograph [10] and a large cycle of works of O.Demuth and his students on various problems of measure theory, function theory, Lebesgue integral theory, general integration theory etc (see, f.e., [11-15]). The Prague school of O.Demuth is now the most fruitful working group in constructive analysis. As an illustration of constructive distinction we mention the following result proved independently by O.Demuth [13] and M.Khachatryan [16] which is of interest in connection with the famous Lebesgue theorem: there is an increasing constructive function which is not differentiable at any point. At the same time many classical constructions and results received their constructive analogues in works of Demuth's school. In particular, O.Demuth has constructed a developed theory of the Lebesgue integral including theory of L_r and S spaces. It is worth noting that these spaces turned out to be complete and separable ones and that a function f is an indefinite integral af an element of L_1 if and only if it is absolutely continuous. There are also constructive analogues of Fubini's theorem and Lusin's theorem about measurable functions. The author believes that it is now time for a monographical exposition of the afore results. The complicated language and notations of journal publications could thus, perhaps, be simplified.

11. Considerable progress has been made by constructive mathematics in the study of algorithmic problems of analysis. In particular, in many cases we can now clarify the question as to which initially given data suffice to find various objects of analysis. A number of results of this kind can be found, f.e., in [2].

12. Summing up this short and incomplete review one can say that although the most revolutionary claims of the constructive programme are now, perhaps, behind us, constructive mathematics has its worthy place among other mathematical trends. Many of the results obtained here are of undoubded all-mathematical worth. On the other hand, the very develop-

ment of mathematics on a constructive basis can be regarded as a bold and valuable experiment whether or not we agree to consider constructive mathematics as the only one. In particular, the purely syntactic nature of the constructive mathematical univers is, in the opinion of the author, of exceptional interest from this point of view.

Biblioghraphy

1. А.А. Марков, О конструктивной математике, Труды Матем. ин-та АН СССР им. В.А.Стеклова, 67, 8-14, (1962). English transl. in Amer.Math.Soc. Transl. (2) 98, (1971).
2. Б.А. Кушнер, "Лекции по конструктивному математическому анализу", Наука, Москва (1973). English transl. B.A.Kushner "Lectures on Constructive Mathematical Analysis",Amer.Math.Soc., vol.80, (1984).
3. B.A. Kushner, Some extensions of Markov's constructive continuum and their applications to the theory of constructive functions, The L.E.J.Brouwer Cent. Symp., Amsterdam, 261-273, (1982).
4. O.Demuth,R.Kryl,A.Kucera, Об использовании теории частично-рекурсивных относительно числовых множеств в конструктивной математике,Acta universitatis Carolinae, Math.et Phys.,19 N1, 15-60, (1978).
5. И.Д. Заславский, С.Н. Манукян, О разбиении плоскости конструктивными кривыми, Математические вопросы кибернетики и вычислительной техники, Ереван, 26-138, (1968).
6. С.Н. Манукян, О некоторых топологических особенностях конструктивных простых дуг, Исследования по теории алгорифмов и математической логике, Изд. ВЦ АН СССР, Москва, 122-129, (1976).
7. С.Н. Манукян, О конструктивных кривых и криволинейных интегралах от функции комплексной переменной, Известия АН Армянской ССР, Математика,№4, №2, 137-143, (1969).
8. Б.А. Кушнер, Сложно-вычислимые действительные числа, Zeitschr. f. Math.Logic und Grundlagen d. Math.,19, 447-452, (1973).
9. Н.А. Шанин, Конструктивные вещественные числа и конструктивные функциональные пространства, Труды Матем. ин-та АН СССР им. В.А.Стеклова, 67, 15-294, (1962). English transl. Transl. of math. monograph, Amer. Math. Soc.,vol. 21, (1968).
10. Фан Динь Зиеу, Некоторые вопросы конструктивного функционального анлиза, Труды Матем. ин-та им. В.А.Стеклова АН СССР, 113, (1970).English transl., Proc. Steklov Inst., Math., 114, (1970).
11. О. Демут, Интеграл Лебега в конструктивном анализе, Зап.научн. семинаров Ленингр.отд. Матем. ин-та АН СССР им.В.А.Стеклова,№4, 30-43, (1967).
12. О. Демут, Пространства L_r и S в конструктивной математике, Commentationes Math. Universitatis Carolinae 10, 2, 261-284, (1969).
13. О.Демут, О дифференцируемости конструктивных функций,Ibid. 10,2, 167-175, (1968).
14. О. Демут, Некоторые вопросы теории конструктивных

функций действительной переменной, Acta Universitatis Carolinae, Math. et Phys., 19, N1, 61-96, (1978).
15. O. Demuth, О конструктивном интеграле Перрона, Ibid., 21, N1, 3-57, (1980).
16. М.А. Хачатрян, Пример конструктивной недифференцируемой монотонной функции, Известия АН Армянской ССР, Математика, 4, №4, 269-299, (1969).

NORMALIZATION THEOREMS FOR THE INTUITIONISTIC SYSTEMS WITH CHOICE PRINCIPLES

Grigorii Mints

Institute of Cybernetics
Estonian Academy of Sciences
200108 Tallinn, USSR

INTRODUCTION

We review here some intuitionistic systems with choice principles

$$\forall x(Ey)A(x,y) \to (Ef)\forall xA(x,f(x)) \qquad (1)$$

for which normalization theorems have been established. These are mainly first order systems or systems close to the first order ones in their deductive power. This is not accidental, since in higher order intuitionistic logic with extensionality choice seems to imply excluded third [1].

We remind the treatment of the intuitionistic predicate calculus in [2], and refer the reader there for review of other work ([3]-[7]) comment on the treatment of equality in general ([8],[9],[10]) and of decidable equality [11].

Then we consider first-order arithmetic with choice and, at last, the system with quantifiers for functionals of all finite types [12]. The latter system derives not only the axiom of choice AC, but also relativised dependent choice RDC, and we propose more elegant natural deduction formulation which is presumably equivalent and allows easy treatment of extensionality. Even more simple system of the intuitionistic arithmetic with choice was proposed by H.Schwichtenberg [13], but due to the lack of some restrictions it is not conservative over first order intuitionistic arithmetic: it derives the so-called Uniformity Principle.

HEYTING PREDICATE CALCULUS WITH EPSILON-SYMBOL

A close connection exists between the choice principles and the use of Hilbert's epsilon-symbol εxA with the epsilon-axiom $A_x[t]\to A_x[\varepsilon xA]$, where $A_x[t]$ denotes the result of substituting t for free occurrences of x (with renaming of bound variables to avoid collisions). Sometimes we will not show subscript x explicitly and, for example, write the epsilon-axiom as

$$A[t] \to A[\varepsilon xA] , \qquad (2)$$

so we begin with several systems containing the epsilon-symbol.

Mathematical Logic
Edited by P. P. Petkov
Plenum Press, New York, 1990

A reformulation of the choice principle (1) in the first order case, where quantifiers over functions f are not available, takes the form of the choice rule

$$\frac{X \to \forall x(Ey)A; \quad \forall x A_y[f(x)] \to Y}{X \to Y} \tag{3}$$

where f is a new function variable, i.e., it does not occur in X, A, Y. It is easy to derive choice rule (3) by substituting $\varepsilon y A_x[t]$ for f(t) throughout the whole derivation of the right premise in (3), and using epsilon-axiom in the form $(Ey)A \to A[\varepsilon yA]$.

It is known that the unrestricted use of the epsilon-axiom is not conservative for the intuitionistic predicate logic: such formulas as $(C \to (Ex)A) \to (Ex)(C \to A)$ become derivable. We followed in [2] the suggestion of A.Dragalin [3] and treated the epsilon-symbol as partially defined: the use of εxA is allowed only if (Ex)A is established.

Formulas of the Heyting predicate calculus HPCe with the epsilon-symbol are constructed as usual: to the standard definitions of terms and formulas one extra clause is added: if A is a formula and x is (individual) variable, then εxA is a (epsilon-)term. Recall that quantifiers are not definable here in terms of the epsilon-symbol. The derivable objects of HPCe are sequents $X \to Y$ with the usual intuitionistic restriction: antecedent X is an arbitrary list of formulas and succedent Y is a formula or an empty symbol (understood as falsity). The rules of HPCe are the same as for the usual Gentzen-type formulation of the HPC, with the exception of (E→), i.e., (E) on the left, which becomes

$$\frac{A[\varepsilon xA], Y \mid\!- Z}{(Ex)A, Y \mid\!- Z} \ (E\mid\!-) \tag{4}$$

An essential difference from the classical logic with epsilon-symbol is in the notion of derivation. A tree constructed according to the rules is a derivation only if for every occurrence E of an epsilon-term εxA in any sequent S the antecedent of S (or some sequent below S) contains a member $\forall x_1 \ldots \forall x_n (Ex)A$, where $x_1 \ldots x_n$ is the complete list of variables free in εxA but bound in the occurrence E. So, for example, instead of $A_x[\varepsilon yB] \to (Ex)A$ one can prove only $(Ey)B \;\&\; A_x[\varepsilon yB] \to (Ex)A$.

Theorem 1. Cutelimination theorem holds for HPCe. If a formula containing no ε is derivable in HPCe, it is derivable already in HPC (conservativeness of HPCe over HPC).

The cutelimination in [2] is done by the familiar Gentzen-type induction on degree and rank, and some ideas from [14] are used. The conservativeness is proved by an analysis of cutfree derivations.

Another proof-theoretic demonstration of conservativeness is implicitly contained in the remark made at the end of [11]: the justification of the rule (3) presented there is also valid for the intuitionistic rules. A model-theoretic proof of conservativeness was presented in [3].

HEYTING PREDICATE CALCULUS WITH EQUALITY AND EPSILON-SYMBOL

It is known that in the presence of equality adding of the epsilon-symbol (even with restriction on the derivation described above) is not conservative. Denote by HPCe= the result of adding the epsilon-symbol and the corresponding rules to HPCe (with the restriction on derivation stated above for HPCe).

The first example of non-conservativeness has been given by the author in [8], but a more perspicuous one is by Osswald [15]:
$(\forall x)(Ey)((x=0 \& y=1) \vee (x=1 \& y=1) \vee (x=c \& y=c)) \& 0 \neq 1 \vdash c=0 \vee c \neq 0$
is not derivable in the Heyting predicate calculus with equality HPC=, but is derivable in HPCe= . Indeed, denote the scope quantifier (Ey) by $P(x,y)$, instantiate variable x above to 0 and c, and consider the cases according to the disjunctions present. In all cases except $\varepsilon y P(0,y)=1 \& \varepsilon y P(c,y)=c$ one has $c=0$ or contradiction with $0 \neq 1$. In the remaining case the assumption $c=0$ together with the equality axiom for epsilon-symbol produces contradiction with $0 \neq 1$, so we have $c \neq 0$ in the latter case.

The reason for non-conservativeness lies in the equality axioms for the epsilon-symbol. As noted by D.Miller [10] one possible solution (allowing to preserve conservativeness and to have Skolemisation even for higher types) is to drop these axioms.

The amount of non-conservativeness is estimated in [9] by Smorynski who realised a suggestion made by the present author.

Theorem 2. Let theory T based on the intuitionistic predicate calculus with equality contain the axiom $\forall x(Ey)R(x,y)$. Let T' be the result of Skolemisation of that axiom, i.e., replacement of $\forall x(Ey)R(x,y)$ by $\forall x R(x,f(x))$ for a new function symbol f. Then the f-free fragment of T' is axiomatised by adding to T all the formulas of the form

$$\forall x_1(Ey_1)\ldots\forall x_n(Ey_n)[\&_i R(x_i y_i) \; \& \; (\&_{i<j}(x_i=x_j \to y_i=y_j)] \; . \tag{5}$$

The proof in [9] is model-theoretic, but the author, while making his suggestion, had in mind the following syntactic proof based on lemmas 2,4 in [11].

Let a proof in HPC= of a f-free formula F from the axiom $\forall x R(x,f(x))$ be given. Turn it into a proof in HPC by adding as premises the equality axioms EQ for all predicate and functional symbols of the theory T occurring in F, as well as equality axioms EQ(f) for f. We have HPC proof of the sequent $EQ, EQ(f), \forall x R(x,f(x)) \vdash F$.

Now replace $\forall x R(x,f(x))$ by $\forall x(Ey)R(x,y)$, delete all ancestors of EQ(f) containing quantifiers and replace all the terms beginning with f by new distinct individual variables. In order to transform the resulting figure into a proof in HPC one has to deal with the ancestors of $\forall x R(x,f(x))$ which have the form $R(t,a)$ and with the remaining ancestors of EQ(f) which have the form $t=t' \to a=a'$. Here a,a' are the variables replacing $f(t), f(t')$, respectively. Now add as the premise formula (5) with n equal to the number of replaced terms $f(t)$, and have the proof in HPC of EQ, (5), $\forall x(Ey)R(x,y) \vdash F$. Passing to HPC= allows to drop EQ and to get desired result.

Another situation occurs when we have a decidable equality. Then HPCe= is conservative over HPC.

Theorem 3. If a sequent of the form

$$\forall xy(x=y \vee x \neq y), X \vdash Y \tag{6}$$

with ε-free X Y is derivable in HPCe=, then it is derivable in HPC=.

The easiest proof is via Theorem 2. Simply use the fact that (5) is derivable in HPC= from $\forall xy(x=y \vee x \neq y)$ and $\forall x(Ey)R(x,y)$, using case distinction $(x_i=x_j) \vee (x_i \neq x_j)$ before applying the rule (E|-) to analyse the formula $(Ey)R(x_j,y)$ for $i<j$. In the case $x_i=x_j$ simply do not apply that (E|-) and take y_j to be the same as y_i. In the case $x_i \neq x_j$ do apply the rule, but have in mind that the corresponding implication in (5) is valid since its premise is false.

A more informative proof uses cutelimination.

Theorem 4. Cutelimination can be done in HPCe= for the sequents of the form (6).

The cutelimination proof given in [2] still goes through. The only addition is treatment of the case when the cut-formula is of the form $(Ey)R(t,y)$ and $\varepsilon yR(t,y)$, $\varepsilon yR(t',y)$ for some t' simpler than t occur essentially in the derivation. The most difficult case arises when the equality axiom is applied to $t=t'$ and the two epsilon-terms. We essentially work in the same way as in the previous proof. Apply the case distinction $(t=t') \vee (t \neq t')$ below the cut in question. In the first branch replace the cut formula $(Ey)R(t,y)$ by $(Ey)R(t',y)$, so it will be simplified. In the other branch the equality axiom is not needed at all. This concludes the proof of cutelimination.

FIRST ORDER HEYTING ARITHMETIC WITH EPSILON-SYMBOL

This system (denoted here by HAe) treated in section 1 of part II of [12] is obtained by adding the epsilon-symbol to the language of the first order Heyting arithmetic HA and changing the (E)-rule and the notion of derivation exactly as for the predicate calculus.

Theorem 5. HAe is a conservative extension of HA.

This theorem is not an immediate consequence of Theorem 4, since the induction axioms in HAe can contain the epsilon-symbol. It is derived from cutelimination.

An infinitary system HAe_{inf} is obtained from HAe by replacing the induction axiom by the omega-rule

$$\frac{X|-A[0];\ X|-A[1];\ldots\ X|-A[N];\ldots}{X|-\forall xA}\ (\forall inf)$$

adding similar infinitary computation rule

$$\frac{\ldots t=N, X|-Y\ldots}{X|-Y}\ (Comp)$$

and changing the equality axioms in a suitable way (cf. [12]). The (E)-rule has the same form as for HPCe. Derivations in HAe_{inf} are defined as for HPCe with an additional requirement: increasing ordinals less than a fixed ordinal less than ε_0 should be assigned to the nodes.

Theorem 6. There is a primitive recursive cutelimination operator for HAe. This can be proved in the Primitive Recursive Arithmetic PRA.

The proof of this theorem in [12] uses some ideas from the preceding proofs but it is slightly simpler combinatorially as it is usual for the infinitary formulations.

Theorem 5 is derived from theorem 6 in [12] almost in the same way as similar result is established for HPCe. The main difference is that the result of elimination of the epsilon-symbol from a cut-free HAe_{inf}-derivation is a cut-free derivation d_{inf} in the infinitary version of HA. The derivability in HA itself is proved by Kreisel's trick: transfinite induction on (the ordinal of) d_{inf} is applied to prove that all the sequents in d_{inf} are true. Truth definition exists since all the formulas in a cut-free derivation are subformulas of the final formula, and transfinite induction is derivable in HA since the ordinal is less than ε_0.

HEYTING ARITHMETIC WITH THE AXIOM OF CHOICE IN ALL FINITE TYPES

This system $HA+AC^{\omega}$ is treated in the section 10 of [12]. It is obtained from the Heyting Arithmetic HA^{ω} with the (free and bound) variables for the functionals of all finite types by adding the axiom of choice

$$\forall x^t(Ey^s)A \rightarrow (EY^{(t\rightarrow s)})\forall x^t A_y[Y(x)] \qquad (AC^{(t\rightarrow s)})$$

The conservativeness of $HA+AC^{\omega}$ over HA was first proved by N.Goodman [16]. It also follows from the cutelimination for infinitary version $HA+AC^{\omega}_{inf}$ of $HA+AC^{\omega}$ established in [12]. That infinitary system uses the formulas-as-types functionals [17] to make explicit choice functionals Y in the axiom AC. The Howard functionals contain free variables a^A for any formula A, to be interpreted as constructions realising A. Main new means of definition are primitive recursion functional (denoted here by R) for all types, pairing <u,v> and inverse functionals $(p)_0, (p)_1$. In particular the type of ∀xA is treated as (type(x) → type(A)), and terms of type (Ex)A are pairs <u,x'> such that u is of the type $A_x[x']$.

This reminds of various functional interpretations, in particular, of the Kleene recursive realisability and Kreisel's modified realisability (cf. [18]). The cutelimination proof from [12] was obtained in the following way. Take the derivation in $HA+AC^{\omega}$, apply HRO interpretation (replace the higher-type functionals by indices of recursive operations) and the Kleene-realisability. We obtain a derivation in Heyting arithmetic, from which the cut can be eliminated (after transition to infinitary formulation). Now observe that a sufficient amount of structure of original derivation can be preserved during cutelimination, so that everything could be done without destroying the formulas by HRO and the Kleene-realisability. The presentation in [12] does not mention these heuristical considerations.

The Gentzen-type formulation of $HA+AC^{\omega}_{inf}$ in [12] is rather involved and it is not clear how to extend it to a system with extensionality so that cutelimination be preserved. We outline here a simpler natural deduction formulation. It probably allows normalisation (which is slightly simpler than that for the formulation in [12]) as well as the proof of conservativeness over HA along the same lines as in [12] and the treatment of extensionality.

The new system (denoted by $NHA+AC^{\omega}_{inf}$) has familiar natural deduction rules for introduction and elimination of &,→,V, and the introduction of (E) (with some distinctions for the quantifier rules to be stated below). Each sequent in a derivation is assigned some term according to the rules familiar for the formulas-as-types. (E)-elimination has the form

$$\frac{Z|-(Ey)A \qquad q}{Z|-A_y[(q)_1] \qquad (q)_0} \tag{7}$$

Here q is a term assigned to the premise (upper sequent). Then $(q)_0$ is assigned to the conclusion of the rule (lower sequent) and $(q)_1$ is substituted for the existential variable y. The derivability of a sequent $Z|-\forall x^s(Ey^t)A$ with the assigned term q justifies in a similar way the use of the term $\lambda x(q(x))_1$ of the type $(s\to t)$. The condition imposed on the derivations is similar to one for HPCe: each term used in the derivation should have a justification. The details are as in [12]. In particular, quantifiers of a type different from 0 are instantiated only by variables, and quantifiers of the type 0 also by numerals. So the computation rule similar to (Comp) above is very important.

We expect that equivalence of the infinitary system in [12] to $NHA+AC^{\omega}_{inf}$ can be proved by the familiar Prawitz translation between natural deduction and Gentzen's L-type systems. Normalisation proof for $NHA+AC^{\omega}_{inf}$ (including construction of the computed derivation, cf. [12]) is probably even easier than one in [12], as is the proof of conservativeness over HA. Adding of the extensionality rule

$$\frac{Z|-u(a)=v(a)}{Z|-w(u)=w(v)}$$

(for the type of u,v different from 0, and the corresponding infinitary rule for arguments of the type 0) probably has little influence on the normalisation proof.

Full use of the formulas-as-types in our systems allows to derive not only the axiom of choice AC and the axiom of dependent choice DC (which can be obtained from AC by using induction), but also the schema RDC of relativised dependent choice

$$\forall x(A[x] \to (Ey)(A[y]\&B[x,y])),A[z] \;|-(Ef)(f(0)=z\&\forall nB[f(n),f(n+1)]) .$$

In order to prove it we denote $\forall x(A[x] \to (Ey)(A[y]\&B[x,y]))$ by I and the (formulas-as-types) variables which will correspond to the assumptions I and A[z] in RDC, by a^I and a^z , respectively. We will derive RDC from the sequent

$$I,A[z] \;|-(Ef)(f(0)=z \;\&\; \forall n(A[f(n)]\& \;B[f(n),f(n+1)]) \tag{8}$$

which will be obtained from the sequent

$$I,A[z] \;|- q(0)=z \;\&\; \forall n(A[q(n)] \;\&\; B[q(n),q(n+1)]) \tag{9}$$

for some term q containing a^I and a^z. Term q will be defined by primitive recursion with the base q(0)=z, and (9) will be derived by &-introduction, so we have to determine only the derivation of the second conjunct in (9) and the values of q(n+1). This will be done by means of derivation of the sequent

$$I,A[z] \;|- \forall n(Ey')(Ey)(A[y] \;\&\; B[y,y']) \quad ; r \tag{10}$$

with the assigned term r: we put $q(n+1)=(r(n))_1$. Let us define r again by primitive recursion. Since r is intended to be the realisation of the conclusion in (10), we expect it to have the form

$$r = \lambda n <<<u_n , \quad v_n>, \quad f_n>, \quad f_{n+1}> , \tag{11}$$

where f_n , f_{n+1} are the values of the function f at n, n+1; u_n realises $A[f_n]$, v_n realises $B[f_n,f_{n+1}]$. So it remains only to put

$$u_{n+1} = ((a^I(f_n)) (u_n))_{00}$$

$$v_{n+1} = ((a^I(f_n)) (u_n))_{01}$$

$$f_{n+1} = ((a^I(f_n)) (u_n))_1$$

$$u_0 = a^z , v_0 = ((a^I(z))(a^z))_{01} , f_0 = z .$$

OPEN QUESTIONS

1. Is our system with the formulas-as-types equivalent to HA+RDC for all finite types?

2. (G. Kreisel). Can one restrict instances of AC needed for derivations of formulas containing quantifiers only for a given set of types?

3. The same as question 2, for RDC.

REFERENCES

1. N. Goodman and J. Myhill, Choice implies excluded middle, Z.math.Log. und Grundl.Math. 24:461 (1978).
2. G. E. Mints, Heyting predicate calculus with epsilon-symbol, Zap. Nauch.Sem. Leningrad.Otd.Mat.Inst. Akad. Nauk SSSR 40:101 (1974). English translation: J.SovMath. 8:317 (1977).
3. A. Dragalin, Intuitionistic logic and Hilbert's epsilon-symbol, in: "History and Methodology of Natural Sciences," Moscow Univ. Press, Moscow 16:78 (1974) (in Russian).
4. D. Leivant, "Existential Instantiation in a System of Natural Deduction for Intuitionistic Arithmetic," Stichting Math. Centrum, Amsterdam (1973).
5. S. Maehara, A general theory of completeness proofs, Ann.Jap. Ass. Philos.Sci. 3:54 (1970).
6. K. Shirai, Intuitionistic predicate calculus with epsilon-symbols, Ann.Jap.Ass.Philos.Sci. 4:49 (1971).
7. V. Smirnov, Elimination des termes epsylon dans le logique intuitioniste, Revue Intern. de Philosophie 98:512 (1971).
8. G. Mints, Skolem's method of elimination of positive quantifiers in sequential calculi, Soviet Math. Dokl. 7:861 (1966).
9. C. Smorynski, On axiomatizing fragments, J.Symbol.Log. 42:530 (1977).
10. D. Miller and G. Nadatur, Higher order logic programming, in: "Proc. Third Int. Logic Progr. Conf.," London, June 1986, 448.
11. G. Mints, The Herbrand theorem, in: "Mathematical Theory of Logical Deduction," Nauka, Moscow (1967), 311 (in Russian).
12. G. Mints, Finite investigation of infinite derivations, Zap.Nauch.Sem. Leningrad.Otd.Mat.Inst. Akad. Nauk SSSR 49:57 (1975). English translation: J.Sov.Math. 10:548 (1978).
13. H. Schwichtenberg, A normal form for natural deduction in a type

theory with realizing terms, in: Atti del Congresso, "Logica e Filosofia della Scienza," Logica. CLUEB, Bologna (1986), I:95.
14. A. Leisenring, "Mathematical Logic and Hilbert's Epsilon-symbol," MacDonald, London (1969).
15. H. Osswald, Über Skolemerweiterungen in der intuitionistischen Logik mit Gleichheit, Lect. Notes Math. 500:264 (1975).
16. N. Goodman, The faithfulness of the interpretation of arithmetic in the theory of constructions, J. Symbolic Logic 38:453 (1973).
17. P. Howard, The formulae-as-types notation of construction, in: "To H. B. Curry. Essays on Logic, Lambda Calculus and Formalism," Academic Press, London (1980), 479-490.
18. A. Troelstra, ed., Mathematical investigation of intuitionistic arithmetic and analysis, Lect. Notes Math. N344 (1973).

P. S. After presentation of the paper two more references have been pointed out, viz. papers by Renardel de Lavalette and by L. Gordeev in Annals of Pure and Aplied Logic.

FORMALIZING THE NOTION OF TOTAL INFORMATION

Dag Normann

Institute of mathematics , University of Oslo

P.O. Box 1053, 0316 Blindern, Oslo 3 , Norway

1. *Introduction*

Suppose that Λ is a model for untyped λ-calculus and that T is some typed λ-calculus (transfinite, second order or whatever). One standard procedure for obtaining a model for T is to regard all terms in T as untyped terms, interpret each type as a subset of Λ and then prove that each typed term is interpreted into the interpretation of the type.
When the types are inductively defined it is possible to isolate the total objects of the type, while if T is second order type theory this is not obvious. If T permits certain recursive definitions of types, the isolation of the total or well-founded objects becomes more vital. As mentioned, it is possible by straightforward induction on the type-formation to isolate the total (or well-founded) objects of the type, and the proofs that λ-terms define total objects are often simple though the result is essential for the applicability of the theory in question. It does not, however, involve a systematic analysis of the concept of totality itself.

Girard-85 has given a semantics for second order λ-calculus based on the concept of a qualitative domain. He introduces a totality-domain as a qualitative domain with certain objects as the total ones. He uses this to interpret each closed type as one totality-domain and proves that each closed λ-term is total. His results give nice information about second order λ-calculus but we do not learn much about the structure of the total objects from them.
In this paper we will focus on the concept of totality in defining the structures we will investigate. We will see a totality-domain as a set of *total processes* giving definite answers to a fixed class of questions. We will isolate certain properties the class of total objects might have, and we will discuss which properties of the "processes" these reflect.
The inspiration for this work came from two sources. In an unpublished paper, Normann 87, I have worked on a transfinite version of Kleene's countable functionals, the so called *Kleene-spaces*. The primitives in a Kleene-space are the *extension-maps* , demonstrating that each neighbour-hood in question is non-empty, and the *trace-functions* , a technical gadget used (originally by Kleene) to define extension maps of higher types.Based on the considerations of this paper, a revised version of the Kleene spaces are found in Normann 88.
The other source is the progressing work by Lill Kristiansen where she uses Girard-type totality domains to define hierarchies of domains suitable as interpretations of type-terms. She discovered that in order to preserve some reasonable properties of totality, other more

technical and seemingly more ad hoc properties had to be assumed as well. Our analysis showed that these properties also are founded on a reasonable conception of totality.
This paper will consist of two parts, a conceptual part and a technical part. In the conceptual part we discuss the notion of totality, choose a class of structures called *totality-domains* and define a notion of *embedding* between such structures. We briefly discuss possible additional properties, but do not touch upon the applications for Kleene-spaces or qualitative domains that originally inspired the definition. In the technical part we will show how we may consider inductive closure of certain operators as a direct ω-limit in the chosen category. We will isolate certain functors called *strictly positive* and use them in the analysis of inductively defined domains.
In preparing the paper I had several discussions with Lill Kristiansen, and her remarks have helped me in forming the final concepts.

2. *Totality domains*

In this section we will see how we can build mathematical models for the following situation:
We have a set S of questions (S for "spørsmål") , a set A of possible answers and a set $\mathcal{P}$ of *processes* that give us answers to the questions.
The first we will discuss is:

What is a possible answer?

We will accept answers like 'yes', 'no', '17', 'true', 'false' etc., e.g. what we call atomic answers. We will not, however, accept answers that are reals, functions on **N** (**N** is the set of natural numbers) or functions on **R** (the reals), since in order to get accurate information we must ask about decimal expansions or values for certain inputs.
In our models the answers are *atomic* i.e. finite entities that cannot or needn't be resolved through further questioning.
We will not, however, insist that the questions are finite entities. We will accept questions like

Determine g(a) with an accuracy of 10^{-n}

where $g:\mathbf{R} \to \mathbf{R}$.
In order to answer such a question we must have access to some information about g (e.g. local modulus of continuity) and the real a , and we cannot utilize more than a finite amount of this information in answering the question. Thus we will assume that the questions can be "approximated" by finite entities and that each process will answer the question on the basis of one of these approximations. We need not go into detail here.
Our main task is to build mathematical models for the processes. Here we will be liberal on what we consider to be a process. We will accept processes in the range from

"Left for the reader"

to

"Deterministic, algorithmic process".

In the first case the "process" is just sufficient amount of information about the question to determine the answer, with no reference to how the answer is obtained.
Examples are:

1. A process is a finite set of pairs of rational intervals

$$(\langle q_1, q_2 \rangle, \langle r_1, r_2 \rangle).$$

If the question is

determine f(a) up to n decimals

the process answers this question if we from the assumption that

$$<q_1,q_2> \subseteq f^{-1}(<r_1,r_2>)$$

for each pair in the process can find the answer.

2. A process is a finite valuation of a set of atomic propositions. If B is a formula of propositional calculus, the question might be:

What is the truth value of B?

The process answers this question if the valuation leaves us with only one possible truth value.

The deterministic, algorithmic processes will tell the full story about how information about a question is collected and used in order to find the answer.

In these cases and in all cases we have in mind the *real processes* are finite entities and they code in some way information about one or several questions, how this information is obtained (e.g. given or sought for) and how this information is used to obtain the answers. The *ideal processes* are collections of compatible real processes . Again we choose to be vague, compatible means that they can bee seen as parts of one common process.

We will use partial orderings to model our concept.

2.1 *Definition*

a) A *processing domain* P is a partial ordering $P, <$ satisfying

i) For each $p \in P$ we have that

$$\{ q \in P \mid q \leq p \}$$

is finite.

ii) If p_1 and p_2 are in P and there is a $q \in P$ such that $p_1 \leq q$, $p_2 \leq q$, then there is a least $q = p_1 \vee p_2$ with that property.

b) If P is a processing domain and $\alpha \subseteq P$, then α is *consistent* if each finite subset of α is bounded in P.

c) Let $\alpha \subseteq P$. $\alpha \in \mathbb{P}$ if α is consistent.

2.2 *Definition*

a) A *pre-totality domain* is a quadruple

$$(P, S, A, \{A_p\}_{p \in P})$$

where P is a processing domain, $A_p \subseteq S \times A$ for each $p \in P$, such that

i) $p_1 \leq p_2 \Rightarrow A_{p_1} \subseteq A_{p_2}$.

ii) If $(s,a_1) \in A_{p_1}$ and $(s,a_2) \in A_{p_2}$ and p_1 and p_2 are consistent, then $a_1 = a_2$.

b) If $(P, S, A, \{A_p\}_{p \in P})$ is as above, then $\alpha \in \mathbb{P}$ is *total* if

$$\forall s \in S\ \exists a \in A\ \exists p \in \alpha\ ((s,a) \in A_p)$$

2.3 *Remark*

We think of A_p as the set of pairs (s,a) such that p answers 'a' to the question s. We have not yet justified the finitary aspect of P. Our further intuition is tied up with the *minimal question - answerers*, described in the next trivial lemma left without a proof:

2.4 *Lemma*

Let $(s,a) \in A_p$. Then there is a (possibly several) $p' \leq p$ *being minimal for* s, i.e.

$(s,a) \in A_{p'}$

$(s,a) \in A_{p''} \wedge p'' \leq p' \Rightarrow p'' = p'$.

2.5 *Remark*

Given s and p there is at most one $a \in A$ such that $(s,a) \in A_p$. Thus a can be suppressed from the definition of 'minimal for'.
The p's that are minimal for some s are the building-blocks of our processes. We do not, however, want to use them as atomic entities, since we will accept the situation that $p \vee q$ may answer more questions than p or q individually.

We will now discuss a few properties pre-totality domains may have and how they model certain classes of processes.

Property I

If $p_1 \not\leq p_2$, there is a question s and a $p \leq p_1$ that is minimal for s such that $p \not\leq p_2$.

This property reflects that when $p_1 \not\leq p_2$ then at least some angle to a question is taken by p_1 but not by p_2. One consequence is that we will not extend a p without adding some real information.

Property II

If p_1 and p_2 are inconsistent then there is an s, a $q_1 \leq p_1$ and a $q_2 \leq p_2$ such that q_1 and q_2 are inconsistent and minimal for s.

This property reflects that two processes are inconsistent only if some question is treated unacceptably different by the two processes.

Property III

If $p \in P$ and $s \in S$ there is a $q \in P$ and $a \in A$ such that $p \leq q$ and $(s,a) \in A_q$.

This property reflects that we do not accept approximations to processes that for trivial reasons cannot be extended to some total object. In a way, if p blocks out all answers to s, then p is implicitly inconsistent and should be ruled out. Indeed, if we outrule such p we will be left with exactly the same total objects.

We were tempted to include I, II and III in our definition but decided not to do so in order to keep the basic concepts as simple as possible. These and the three following properties are seen as suggestions for further axioms that would be satisfied for large classes of actual domains.
The next property is a natural strengthening of III and need not be motivated further:

Property IV

All $p \in P$ are elements of some total $\alpha \in \mathbb{P}$ (depending on p).

The final two properties are of a less obvious character and are not shared by all the examples we have in mind:

Property V
If p and q are minimal for s then p and q are either equal or inconsistent.

We call this the *stability property*. It reflects the view that p and q are consistent only if they can be seen as parts of the same *deterministic* procedure; then the same procedure cannot answer s in two different ways.

Property VI
Given s, consistency is an equivalence relation on the set of p that are minimal for s.

This property may seem a bit more far-fetched than the others. The property will however model the situation that when p is minimal for s then any other question t answered by p is so similar to s that the answering process is the same. If in addition consistency just means that the same questions will have the same answers, then the equivalence-classes will just be those p that are minimal for s for some fixed answer a.
In the more general situation p might contain some minimal information about s and also code the way this information is transformed into an answer, where the various "ways" may be clearly distinguished or identified.

It is clear from the above discussion that the p's that are minimal for s play an important role in our description. We will in fact use this set as the "interpretation" of the question.

2.6 *Definition*
Let P be a processing domain.
A *chain* on P is a set $C \subseteq P$ such that if $p \neq q$ are in C then
$p \not\preceq q$ and $q \not\preceq p$.

2.7 *Definition*
Let $(P, S, A, \{A_p\}_{p \in P})$ be a pre-totality domain.
For each $s \in S$, let C_s be the set of p's that are minimal for s.

2.8 *Observation*
Each C_s is a chain and $\alpha \in \mathbb{P}$ is total if and only $c(\alpha) \cap C_s \neq \emptyset$ for each $s \in S$, where

$$c(\alpha) = \{ q \in P \mid \exists p \in \alpha (q \leq p) \}$$

We will use this observation while formulating our concept of a totality domain:

2.9 *Definition*
a) A *totality domain* is a triple

$$\mathcal{P} = (P, S, \{C_s\}_{s \in S})$$

where P is a processing domain, S is a (possibly empty) set and for each $s \in S$, C_s is a chain in P.
b) If $\mathcal{P}$ is a totality-domain as above we define the total objects

$$\mathcal{P}_{tot} = \{ \alpha \in \mathbb{P} \mid c(\alpha) \cap C_s \neq \emptyset \text{ for all } s \in S \}.$$

c) We call a totality-domain a I, II, III, IV, V or VI - domain if the corresponding properties are satisfied:

I For $p, q \in P$ we have

$$p \leq q \Leftrightarrow \forall s \in S \, \forall p' \in C_S \, (p' \leq p \Rightarrow p' \leq q)$$

II If p and q are inconsistent there is an $s \in S$ such that C_S separates p and q, i.e. such that there are inconsistent $p' \leq p$ and $q' \leq q$ in C_S.

III For each $s \in S$,

$$\{ p \mid \exists q \leq p \, (q \in C_S) \}$$

is dense in P.

IV Each $p \in P$ is the element of some $\alpha \in \mathcal{P}_{tot}$.

V The elements of each C_S are pairwise inconsistent.

VI Consistency is an equivalence relation on each C_S.

2.10 *Remarks*

As mentioned before we will not be working with the additional properties here. We included the discussion of them because this discussion illuminates the original intuition behind the concepts.

It is our aim to throw some light also on less reflected concepts of totality. To this end we introduce the class of weak totality domains. There the total objects are just given, not indirectly defined via a set of chains.

2.11 *Definition*

A *weak totality-domain* is a pair

$$\mathcal{P} = (P, X)$$

where P is a processing domain, $X \subseteq \mathbb{P}$ and

$$\alpha \in X, \beta \in \mathbb{P} \text{ and } \alpha \subseteq \beta \Rightarrow \beta \in X.$$

The elements of X are called *the total objects.*

3. *Embeddings*

In this section we will define what will be the natural notion of an embedding from a structure

$$\mathcal{P} = (P, S, \{ C_S \}_{s \in S})$$

to a structure

$$Q = (Q, T, \{ D_t \}_{t \in T})$$

First we will require that each process in P can be identified with a process in Q ,i.e that there is an embedding from P to Q.

3.1 *Definition*

Let $\phi: P \to Q$ be 1 - 1.

We call ϕ a *PRO - embedding* if ϕ is an order isomorphism from

$$\{ p' \mid p' \leq p \}$$

to

$$\{ q' \mid q' \leq \phi(p) \}$$

for each $p \in P$.

3.2 *Remark*

This definition reflects that our structure is not just some partial ordering, but that the order-type of the predecessors is seen as a basic property of an element.

Secondly we must decide what the natural connections between the question-sets might be. Our definition is based on the following considerations:

1. We may extend a structure by demanding answers to fewer questions.

2. We may extend a structure by accepting new procedures giving a new kind of answers to the old questions. These new answers may not be atomic, so in order to make the answers atomic the old question may be split up into a family of questions, thought of as supplementing the original question with alternative additional questions.

We will illustrate 2. with the following example:

$$N \rightarrow (N \oplus \varnothing)$$

is a substructure of

$$N \rightarrow (N \oplus (N \rightarrow N))$$

In the first case the questions are

s_n: What is the value on n?

In the second case the questions are

t_{nm}: What is the value on n, and if the answer is of the form right(f), what is f(m)?

Thus we will require that there is a map $\pi: T \rightarrow S$, bearing the meaning that t corresponds to $\pi(t)$ in the sense that it is a refinement of $\pi(t)$.

The relation between D_t and $S_{\pi(t)}$ is then obvious. The processes in P can answer the refined question in the same way as they answer the original question since the refinement only is needed when we don't get an atomic answer. These considerations lead us to

3.3 *Definition*

Let

$$\mathcal{P} = (P, S, \{C_s\}_{s \in S})$$

and

$$\mathcal{Q} = (Q, T, \{D_t\}_{t \in T})$$

be two totality-domains.

A *TOT - embedding* from P to Q is a pair (ϕ, π) such that

i) $\phi: P \rightarrow Q$ is a PRO - embedding

ii) $\pi: T \rightarrow S$

iii) For each $t \in T$ we have

$$\phi^{-1}(D_t) = C_{\pi(t)}.$$

3.4 *Definition*

a) We let TOT be the category with totality-domains as objects and the TOT - embeddings as morphisms

b) We let PRO be the category with processing-domains as the objects and with PRO - embeddings as morphisms.

When the category under consideration is clear from the context, we will just use the terms *embedding* and *morphism* .

We will now show that the category TOT is closed with respect to the formation of direct limits.

3.5 *Theorem*
Let $I, \angle$ be a directed set, $\{ \mathcal{P}_i \}_{i \in I}$, $\{ (\phi_{ij}, \pi_{ij}) \}_{i \angle j}$ be a directed system from TOT. Then the direct limit , $\lim(i \in I)\, \mathcal{P}_i$, exists in TOT.

Proof
Let $\mathcal{P}_i = (P_i , S_i , \{ C^i{}_s \}_{s \in S_i})$
Let $P = \lim(i \in I) P_i$ with limit embeddings ϕ_i.
Let s be a *tower* if

$$s = \{ s_i \}_{i \in I}$$

such that $s_i \in S_i$ for $i \in I$ and $\pi_{ij}(s_j) = s_i$ when $i \angle j$.
Let S be the set of towers,

$$\pi_i(\{ s_j \}_{j \in I}) = s_i.$$

If

$$s = \{ s_i \}_{i \in I}$$

is a tower, then

$$\{ C_{s_i} \}_{i \in I}$$

is a directed set of chains. Let C_s be the limit of those chains.

The unique factorisation property for this suggested limit-structure is easy to establish, and is left for the reader.

The category will also contain pullbacks We will, however, not prove this here since we will only be interested in pullbacks in the categories PRO and WTOT , the category of *weak totality - domains* to be defined below. In these categories the characterization of the pullbacks are easy to find.

We see the chains as a means for analyzing the total objects. Now we will see what the embeddings will do for the total objects of a domain:

3.6 *Lemma*
Let $(\phi , \pi): \mathcal{P} \rightarrow \mathcal{Q}$ be an embedding and let $\alpha \in \mathcal{P}_{TOT}$.
Then $\phi(\alpha) \in \mathcal{Q}_{tot}$. (Where $\phi(\alpha) = \{ \phi(p) \mid \pi \in \alpha \}$)

Proof
Let S , T , C_s , D_t be as they use to.
Let $t \in T$.
Then $\quad \exists p \in \alpha\ \exists p' \leq p\ (p \in C_{\pi(t)})$.
Then $\quad \phi(p) \in \phi(\alpha)$, $\phi(p') \leq \phi(p)$, $\phi(p') \in D_t$.
Thus $\quad \phi(\alpha) \in \mathcal{Q}_{tot}$.

This corresponds to the natural notion of embedding between weak totality-domains:

3.7 *Definition*

a) Let (P, X) and (Q, Y) be two weak totality domains. A WTOT - *embedding* from (P, X) to (Q, Y) is an embedding $\phi: P \to Q$ mapping X into Y.

b) We let WTOT be the category with:
   -- objects: weak totality - domains
   -- morphisms: WTOT - embeddings.

4. *Operators and Functors*

In this section we will investigate a class of operators on totality-domains. It will be an advantage to formulate these operators as functors for two reasons:

1. The functorial representation can be used to describe the uniformity and complexity of the operator.

2. We want to consider operators that are essentially but not set-theoretically monotone, and to take iterative limits of such. We then need to bring along the embeddings.

We will isolate a class of operators called *strictly positive*. They may not seem as a natural class to study as given, but they have the following nice technical properties:

-- Operators given by strictly positive definitions will be strictly positive in our sense.
-- They commute with direct limits.
-- They have canonical extensions to functors on WTOT commuting with arbitrary pullbacks (but not with direct limits)
.-- The ω - limit in TOT corresponds to the least fixed-point in WTOT.
-- The class of operators is closed under composition and under the formation of the least fixed-point.

In this section we will work with functors of one variable, while we will extend this to finitely many variables in section 6.

4.1 *Definition*

Let Γ:TOT $\to$ TOT be a functor.

We say that Γ is *separable* with PRO - part Γ_0 if Γ_0 is a functor on PRO such that

$$\Gamma(P, S, \{C_s\}_{s\in S}) = (\Gamma_0(P), T, \{D_t\}_{t\in T})$$

$$\Gamma(\phi, \pi) = (\Gamma_0(\phi), \nu) \quad \text{when } (\phi, \pi) \text{ is an embedding.}$$

4.2 *Discussion*

Let Γ:PRO $\to$ PRO be a functor commuting with pullbacks and direct limits. We will show how we can define a *denotation system* for Γ.

Let $Q = \Gamma(P)$. We write P as the direct limit of its finite substructures. Then every element q in Q will have a finite substructure of P as a *basis*, the unique minimal finite substructure P' such that q is in the image of $\Gamma(\mathrm{id}: P' \to P)$. In all our examples we will have that Γ will preserve the property of being an inclusion map. Then the basis for q is the minimal substructure P' of P with $q \in \Gamma(P')$.

Let $\{H_i\}_{i\in N}$ be a sequence of finite processing-domains containing exactly one element of each isomorphism-class of such.

A Γ- *denotation - index* will be a pair (H, q) where H is one of the H_i given above, $q \in \Gamma(H)$ and H is minimal with this property. Let $G_{H,q}$ be the group of automorphisms τ on H such that $\Gamma(\tau)(q) = q$. We may identify Q with the set of $\approx$-equivalence classes of triples (H, q, ϕ) where $\phi: H \to P$ is an embedding ; where $(H, q_1, \phi_1) \approx (H, q_2, \phi_2)$ if for some automorphism τ on H we have that $\phi_2 = \phi_1 \circ \tau$ and $\Gamma(\tau)(q_1) = q_2$.

The identification will be dependent of a choice of isomorphisms $\psi : H_i \to P'$, one for each finite substructure P' of P We will however observe that each denotation index will define a subfunctor $\Gamma_{H,q}$ of Γ. such that $\Gamma_{H,q}(P)$ will be the elements of Q with denotations based on the index (H, q).

The equivalence relation $\approx$ is generated by the corresponding equivalence - relation on the set of denotation indices, where (H, q) is equivalent to (H, p) when there is an automorphism τ on H such that $p = \Gamma(\tau)(q)$.

We will from now on assume that we have chosen one denotation-index from each equivalence class. Then the various $\Gamma_{H,q}$ will be disjoint, and Γ can be seen as the disjoint union of the $\Gamma_{H,q}$. Following Girard we will call a set of denotation-indices like this a *trace* for Γ. Any subset of a trace will define a subfunctor of Γ. We will call two subfunctors like this *disjoint* if the corresponding traces are disjoint.

Since a processing domain is supposed to be closed under taking suprema of finite consistent sets, each finite H will contain a set of pairwise disjoint maximal elements. The number of these elements is called the *arity* of H.

Likewise a denotation-index (H,q) will have an arity, the arity of H. We will in particular be interested in indices with arity 0 or 1, as seen in the next definition.

4.3 *Definition*

Let Γ: TOT $\to$ TOT be a separable functor with PRO-part Γ_0. We call Γ *strictly positive* if

i) Γ_0 commutes with direct limits and pullbacks.

ii) There is a set Δ and families

$$\{I_\delta\}_{\delta \in \Delta}, \{p_{i,\delta}\}_{i \in I_\delta{}^*, \delta \in \Delta},$$

where $I_\delta{}^* = I_\delta \cup \{*\}$ for some additional point $*$, satisfying iii) - v) below.

iii) Each I_δ is a set such that

-- If $\Gamma(P, S, \{C_s\}_{s \in S}) = (Q, T, \{D_t\}_{t \in T})$ then τ is in a 1-1 correspondence with the set

$$\{(\delta,\sigma) \mid \delta \in \Delta \text{ and } \sigma: I_\delta \to S\}$$

via a "denotation-system"

$$t = (t_{(\delta,\sigma)})_S$$

(we will drop the S when it is clear from the context).

-- If $\pi: S' \to S$, $\phi: P \to P'$ and $\Gamma(\phi, \pi) = (\psi, \nu)$ then ν can be defined by

$$\nu((t_{(\delta,\sigma)})_{S'}) = (t_{(\delta,\pi\circ\sigma)})_S$$

iv) If $Q = \Gamma_0(P)$ then for each $\delta \in \Delta$ and $i \in I_\delta{}^*$

-- If $i = *$ then $\rho_{i,\delta}$ is a subfunctor of Γ_0 such that all indices have arity 0.

-- If $i \in I_\delta$ then $\rho_{i,\delta}$ is a subfunctor of Γ_0 such that all indices have arity 1.

-- If

$$\alpha \subseteq \bigcup\{\rho_{i,\delta}(P) \mid i \in I_\delta{}^*\}$$

is a consistent set in Q , then there is one and only one $i \in I_\delta{}^*$ such that we have

$$\alpha \subseteq \rho_{i,\delta}(P)$$

and then we have that

$$\alpha \subseteq \rho_{i,\delta}(\beta)$$

for some consistent subset β of P.

If $i \in I_\delta$, if q has a $\rho_{i,\delta}$ -basis p and if $p' \leq p$ then p' is the $\rho_{i,\delta}$ -basis for some $q' \leq q$.

If α is just a consistent subset of Q we will let the δ - predecessor of α be the set of $\rho_{i,\delta}$ -basises for elements in $c(\alpha)$. We will be more specific later.

v) The chains $D_{t(\delta,\sigma)} = \bigcup\{\rho_{i,\delta}(C_{\sigma(i)}) \mid i \in I_\delta\}$ where $\rho_{i,\delta}(C_{\sigma(i)})$ is the set of elements in $\rho_{i,\delta}(P)$ where the basis, when the arity is 1 , is in $C_{\sigma(i)}$. In particular $\rho_{*,\delta}(P) \subseteq D_{t(\delta,\sigma)}$

This ends the definition.

4.4 *Examples*

a) Let A be a set, $\mathcal{P}$ a totality-domain.

Define $A \times \mathcal{P}$ as follows

$Q = \{(a,p) \mid a \in A \text{ and } p \in P\}$

$(a,p) \leq (b,q)$ when $a = b$ and $p \leq q$.

$T = \{\sigma \mid \sigma : A \to S\}$

$D_\sigma = \{(a,p) \mid p \in C_{\sigma(a)}\}$

Here Δ is just a singleton $\{\delta\}$ with $I_\delta = A$.

We have $\rho_{a,\delta}(P) = \{(a,p) \mid p \in P\}$ for $a \in A$, $\rho_{*,\delta}(P) = \emptyset$.

The verifications of the properties, including the extension to a functor, are left for the reader. Note that the predecessor essentially is the second coordinate.

In the example above we really defined the disjoint union of A copies of $\mathcal{P}$. As we will see later, if A is finite, we may take the disjoint union of various domains. If we are not interested in seeing the product as a disjoint union of independent copies of $\mathcal{P}$, we may use the following construction instead:

b) We let Q with the ordering $\leq$ be as in a).

Let T = S.

Let $D_s = \{(a,p) \mid a \in A \text{ and } p \in C_s\}$

We still let Δ be a singleton set $\{\delta\}$ but now I_δ is also a singleton $\{i\}$ and

$\rho_{i,\delta}(P) = \{(a,p) \mid p \in P \text{ and } a \in A\}$, i.e $\rho_{i,\delta} = \Gamma_0$.

We will now give an example of an operator where $\rho_{*,\delta}$ is nonempty. This occurs when we have an operator in several variables and fix one of the variables. The example will reflect this.

c) Let $\mathcal{N} = (\mathbf{N}, \{ * \}, C_*)$ where $\mathbf{N}$ is the set of natural numbers with the identity-relation as the ordering, and $C_* = \mathbf{N}$.

We define $\mathcal{N} \oplus \mathcal{P}$ as follows:

$Q = \{ (1,n) \mid n \in \mathbf{N} \} \cup \{ (2,p) \mid p \in P \}$

We let $(i,p) \leq (j,q)$ if $i = j$ and $p \leq q$ in the relevant ordering.

We let $T = S$, and we let $D_s = \{ (1,n) \mid n \in \mathbf{N} \} \cup \{ (2,p) \mid p \in C_s \}$

Here $\rho_{*,i}(P) = \mathbf{N} \oplus \varnothing$.

We leave the rest of the construction and the verifications for the reader. If an element comes from $\mathbf{N}$ then it is called *initial* while if it comes from $\mathbb{P}$ then the predecessor is just the original from $\mathbb{P}$.

4.5 *Example*

Let A be a set, $\mathcal{P}$ a totality-domain.

We define $\mathcal{Q} = A \to \mathcal{P}$ as follows:

The elements of Q are nonempty finite sets

$$\{ (a_1,p_1), \ldots, (a_n,p_n) \}$$

with $a_1, \ldots, a_n \in A$ and $p_1, \ldots, p_n \in P$ such that if $a_{i_1} = \ldots = a_{i_k}$ then

$$\{ p_{i_1}, \ldots, p_{i_k} \}$$

is a consistent set in P.

We let

$$\{ (a_1,p_1), \ldots, (a_n,p_n) \} \leq \{ (b_1,q_1), \ldots, (b_m,q_m) \}$$

if for all $i \leq n$ there is a $j \leq m$ such that $a_i = b_j$ and $p_i \leq q_j$.

Let $T = A \times S$.

Let $D_{(a,s)} = \{ \{ (a,p) \} \mid p \in C_s \}$.

Here $\Delta = A$ and each I_a is a singleton, say $\{ a \}$.

$\rho_{a,a}(P) = \{ \{ (a,p) \} \mid p \in P \}$

$\rho_{*,a}(P) = \varnothing$

Essentially $\mathcal{Q}_{tot} = A \to \mathcal{P}_{tot}$. For each $\alpha \in \mathbb{Q}$ the a-predecessor of α will essentially be $\alpha(a)$.

We leave the details for the reader.

Example 4.5 gives us the set-theoretical product over the index-set A. In the next example we will see how we can construct a modified product over a structured set A.

4.6 *Example*

Let $(A,\leq)$ be a partial ordering, Δ a family of directed subsets of A.

We say that some elements in A are *consistent* if they have a mutual extension.

Let $\mathcal{P}$ be a totality-domain.

We will construct $\Delta \to \mathcal{P}_{tot}$ as the set of order-preserving maps from $\mathbb{A}$ to $\mathbb{P}$ sending Δ into $\mathcal{P}_{tot}$.

Let Q be the set of nonempty

$$\{ (a_1, p_1), \ldots, (a_n, p_n) \}$$

such that whenever

$$\{ a_{i_1}, \ldots, a_{i_k} \}$$

is consistent then

$$\{ p_{i_1}, \ldots, p_{i_k} \}$$

is consistent, where each a_i is from A and each p_i is from Q.
We use the same ordering of Q as in 4.5.
For each $B \in \Delta$ and $s \in S$ we define the chain
$D_{(B,s)} = \{ \{ (a,p) \} \mid a \in B \text{ and } p \in C_s \}$.
Here Δ is given. For each $B \in \Delta$ we let I_B be a singleton $\{ B \}$ and
$\rho_{B,B}(p) = \{ \{ (a,p) \} \mid a \in B \text{ and } p \in P \}$.

Now let $f : \mathbb{A} \to \mathbb{P}$ be monotone and commute with directed unions.
We define α_f as the set of non-empty subsets of

$$\{ (a,p) \mid p \in f(\{ b \mid b \leq a \}) \}$$

If $f : \Delta \to \mathcal{P}_{tot}$ then $c(\alpha_f)$ will intersect all the chains.
Conversely, if α is in $\mathbb{Q}$ and $c(\alpha)$ intersect all the chains, then f_α defined by

$$f_\alpha(B) = \{ p \mid \exists \alpha \in B (\{ (a,p) \in c(\alpha) \}$$

will be as required.
Again we leave the details for the reader.

We will now start proving results about these functors.

4.7 *Lemma*
Let Γ^1 and Γ^2 be two strictly positive functors on TOT. Then the composition is strictly positive.

Proof
Let $\Gamma^3 = \Gamma^2 \circ \Gamma^1$. We will use the superscripts 1,2 and 3 throughout this proof, and we will use all letters as they are used in definition 4.2.
Clearly $\Gamma^3{}_0 = \Gamma^2{}_0 \circ \Gamma^1{}_0$ will be the PRO-part of Γ^3 and will commute with pullbacks and direct limits.

$$\text{Let } \Delta^3 = \{ (\delta, \sigma) \mid \delta \in \Delta^2 \text{ and } \sigma : I^2{}_\delta \to \Delta^1 \}$$

$$I^3{}_{(\delta,\sigma)} = \{ (i,j) \mid i \in I^2{}_\delta \text{ and } j \in I^1{}_{\sigma(i)} \}$$

$$\rho^3{}_{(i,j),(\delta,\sigma)} = \rho^2{}_{i,\delta} \circ \rho^1{}_{j,\sigma(i)})$$

$$\rho^3{}_{*,(\delta,\sigma)}(P) = \rho^2{}_{*,\delta}(Q_1) \cup \bigcup \{ \rho^2{}_{(i,\delta)}(\rho^1{}_{*,\sigma(i)}(P)) \mid i \in I^2{}_\delta \}$$

It is essentially trivial to verify that we have defined the composition and that it has the desired properties. We leave some of the details for the reader, but indicate the argument for two of the properties.

Let $\mathcal{P} = (P, S, \{C_s\}_{s \in S})$

$$\mathcal{Q}_1 = \Gamma^1(\mathcal{P}) = (Q_1, T_1, \{D^1{}_t\}_{t\in T_1})$$
$$\mathcal{Q}_2 = \Gamma^2(\mathcal{Q}_1) = (Q_2, T_2, \{D^2{}_t\}_{t\in T_2})$$

Then $T_2 = \{ t_{(\delta,\sigma)} \mid \delta \in \Delta^2 \text{ and } \sigma: I^2{}_\delta \to T_1 \}$

Now each $\sigma(i)$ will be of the form (τ, σ') where $\tau \in \Delta^1$ and $\sigma': I^1{}_\tau \to S$, so

$$T_2 \approx \{ t_{(\delta,\sigma,\{\sigma_i\}_{i\in I^2{}_\delta})} \mid \delta \in \Delta^2 \text{ and } \sigma: I^2{}_\delta \to \Delta^1 \text{ and } \sigma_i:\sigma(i) \to S \}$$
$$\approx \{ t_{(\delta,\sigma,\tau)} \mid \delta \in \Delta^2 \text{ and } \sigma: I^2{}_\delta \to \Delta^1 \text{ and } \tau: I^3{}_{(\delta,\sigma)} \to S \}.$$

This shows that T_2 is of the right form.

Now let

$$\alpha \subseteq \bigcup \{ \rho^3{}_{(i,j),(\delta,\sigma)}(P) \mid i \in (I^2{}_\delta)^* \text{ and } j \in (I^1{}_{\sigma(i)})^* \}$$

be consistent. By the property of ρ^2 we get that there is a consistent subset β of Q_1 and a fixed $i \in (I^2{}_\delta)^*$ such that $\alpha \subseteq \rho^2{}_{i,\delta}(\beta)$.

If $i = *$ then we may choose β to be empty and correspondingly we may use the empty subset of P.

If $i \in I_\delta$ then by the existence of a unique basis in Q_1 and then in P for every element of α, we may assume that

$$\beta \subseteq \bigcup \{ \rho^1{}_{(j,\sigma(i))}(P) \mid j \in I^1{}_{\sigma(i)}{}^* \}$$

But then we may use the corresponding property of ρ^1 to obtain the result.

We leave the proof that we define the chains $D^2{}_t$ as we shall, for the reader.

4.8 *Lemma*

Let Γ be a strictly positive functional on TOT.

Then Γ commutes with direct limits.

Proof

Let $\{ \mathcal{P}_e \}_{e \in E}, \{ (\phi_{e,d}, \pi_{e,d}) \}_{e \leq d}$ be a directed system with limit $\mathcal{P}, \{ \phi_e, \pi_e \}_{e \in E}$

Let $\mathcal{P}_\varepsilon = \Gamma(\mathcal{P}_e), (\psi_{ed}, \nu_{ed}) = \Gamma(\phi_{ed}, \pi_{ed})$

$$\mathcal{Q} = \Gamma(\mathcal{P}), (\psi_e, \nu_e) = \Gamma(\phi_e, \pi_e)$$

We will use the letters $P, S, C_s, S_e, C^i{}_e, Q, T, D_t, \ldots$ etc. as usual. We will also use the letters Δ, I_δ, etc. as in the definition of strictly positive.

Since Γ_0 commutes with direct limits, we only need to concentrate on T and D_t.

Let $t \in T$. Then $t = t_{(\delta,\sigma)}$ for some $\delta \in \Delta$, $\sigma: I_d \to S$.

Then for each $i \in I_\delta$ we have that $\sigma(i)$ is a tower $\{ s_{i,e} \}_{e\in E}$.

Let $\sigma_e(i) = s_{i,e}$. Then

$$t_e = t^e{}_{(\delta,\sigma_e)} \in T_e$$

and $\{ t_e \}_{e \in E}$ is a tower that can be identified with $t_{(\delta,\sigma)}$. Conversely, if

$$\{ t^e(\delta_e,\sigma_e) \}_{e \in E}$$

is a tower in $\lim(e\in E)T_e$, then $v_{ed}(t^d(\delta_d,\sigma_d)) = t^e(\delta_e,\sigma_e)$.

It follows that there is some δ such that $\delta_e = \delta$ for all $e \in E$ and that for all $i \in I_\delta$ we have that

$$\sigma_e(i) = \pi_{ed}(\sigma_d(i)).$$

Thus , if we define

$$\sigma(i) = \{t^e{}_{\sigma_e(i)} \}_{e \in E}$$

we get a tower $\sigma(i)$ for each $i \in I$. Then $t_{(\delta,\sigma)}$ may be identified with the original tower.

Now we show that the definition of D_t is preserved through the limit :

$$\begin{aligned} D_{t(d,s)} &= \bigcup\{ \rho_{i,\delta}(C_{\sigma(i)}) \mid i \in I_\delta{}^* \} \\ &= \bigcup\{\rho_{i,\delta}(\lim(e\in E)C^e{}_{\sigma_e(i)} \mid i \in I_\delta{}^* \} \\ &= \lim(e\in E) \bigcup\{\rho_{i,\delta}(C^e{}_{\sigma_e(i)}) \mid i \in I_\delta{}^* \} \\ &= \lim(e\in E)D^e{}_{t^e(\delta,\sigma_e)}. \end{aligned}$$

Here we have made use of the fact that the ρ - functions themselves commute with direct limits and that they are disjoint for various i. In this argument we have let the formally meaningless expression

$$\rho_{*,\delta}(C^e{}_{\sigma_e(*)})$$

stand for $\rho_{*,\delta}(P_e)$.

We will now give the precise definition of the predecessors:

4.9 *Definition*

Let Γ be a strictly positive operator on TOT with PRO - part Γ_0.

Let P be a processing domain and let $Q = \Gamma_0(P)$. Let $\alpha \in \mathcal{Q}$.

a) For each $\delta \in \Delta$ we define the δ - *predecessor* as follows:

 $\text{pre}(\delta,\alpha) =$
 $\{ p \mid p$ is the $\rho_{i,\delta}$ basis for an element of $c(\alpha)$ for some $i \in I_\delta \}$

b) We call α δ - *initial* if $c(\alpha)$ intersects $\rho_{*,\delta}(P)$.

c) We let the *support of* α be

$$\sup(\alpha) = \{ \text{pre}(\delta,\alpha) \mid \delta \in \Delta \text{ and } \alpha \text{ is not } \delta \text{ - initial} \}$$

4.10 *Lemma*

Let Γ be as above, $\mathcal{P}$ a totality-domain , $\mathcal{Q}= \Gamma(\mathcal{P})$ and $\alpha \in \mathcal{Q}$.
Then the following are equivalent:

i) $\alpha \in \mathcal{Q}_{tot}$

ii) $\sup(\alpha) \subseteq \mathcal{P}_{tot}$

Proof

i) $\Rightarrow$ ii)

Suppose $\alpha \in \mathcal{Q}_{tot}$, and assume that α is not δ - initial.

Let $\beta = \mathrm{pre}(\delta, \alpha)$ and let $s \in S$.
Consider $t_{(\delta,\underline{s})}$ where $\underline{s}$ is the function with constant value s. Then

$$D_{t(\delta,\underline{s})} = \bigcup\{ \rho_{i,\delta}(C_s) \mid i \in I_\delta^* \}$$

Let $q \in c(\alpha)$ such that $q \in D_{t(\delta,\underline{s})}$.

Since α is not δ - initial there will be some $i \in I_\delta$ such that q has $\rho_{i,\delta}$ - basis p for some $p \in C_s$. But then $p \in \beta$ so $\beta \cap C_s \neq \varnothing$.

ii) $\Rightarrow$ i):

Suppose that $\alpha \notin Q_{tot}$. Then there is $\delta \in \Delta$ and a $\sigma: I_\delta \to S$ such that

$$c(\alpha) \cap D_{t(\delta,\sigma)} = \varnothing.$$

We claim that α is not δ - initial and that $\beta = \mathrm{pre}(\delta, \alpha) \notin \mathcal{P}_{tot}$.

If α is δ - initial, then $c(\alpha)$ will intersect $\rho_{*,\delta}(P) \subseteq D_{t(\delta,\sigma)}$ which is contradicting the assumption.

The condition on the $\rho_{i,\delta}$ - functors that they are "closed downwards" will guarantee that the predecessors β will satisfy $\beta = c(\beta)$.

For the rest of the proof there are two cases:

Case 1

There is no $i \in I_\delta$ such that

$$\rho_{i,\delta}(P) \cap c(\alpha) \neq \varnothing$$

In this case $\mathrm{pre}(\delta,\alpha)$ is empty.

But we exclude the empty set from the total objects by insisting on S to be non-empty.

Case 2

There is an $i \in I_\delta$ and a $p \in P$ such that p is the $\rho_{i,\delta}$ - basis for some

$$q \in c(\alpha).$$

Now assume that there is a $p_0 \in C_{\sigma(i)} \cap \beta$.

Since $p_0 \in \beta$ we have for some $j \in I_\delta$ and q_0 with $\rho_{j,\delta}$ - basis p_0 that $q_0 \in c(\alpha)$.

But then q and q_0 are consistent so $i = j$.

Thus

$$\rho_{j,\delta}(C_{\sigma(i)}) = \rho_{j,\delta}(C_{\sigma(j)}) \subseteq D_{t(\delta,\sigma)}.$$

But then $q_0 \in c(\alpha) \cap D_{t(\delta,\sigma)} = \varnothing$, a contradiction.

Thus $C_{\sigma(i)} \cap \beta = \varnothing$ and $\beta \notin \mathcal{P}_{tot}$.

This ends the proof.

The main consequence of these results is that the strictly positive functors have canonical extensions to functors on weak totality - domains:

4.11 *Definition*

Let Γ with PRO - part Γ_0 be strictly positive.

The *companion* of Γ is the functor Γ_W: WTOT $\to$ WTOT defined by:

$$\Gamma_W(P, X) = (\Gamma_0(P), Y)$$

where

$$Y = \{ \alpha \in \Gamma_0(\mathbb{P}) \mid \sup(\alpha) \subseteq X \}.$$

If $\phi: P_1 \to P_2$ is a morphism of (P_1, X_1) to (P_2, X_2) then

$$\Gamma_W(\phi) = \Gamma_0(\phi).$$

4.12 *Lemma*

$\Gamma_W(\phi)$ as defined in 4.11 is indeed a morphism.

Proof

We must show that

$$\alpha \in Y_1 \Rightarrow \Gamma_0(\phi)(\alpha) \in Y_2$$

where

$(Q_1, Y_1) = \Gamma_W(P_1, X_1)$

$(Q_2, Y_2) = \Gamma_W(P_2, X_2)$

But since the following diagram commutes for every $\delta \in \Delta$ and $i \in I_\delta^*$

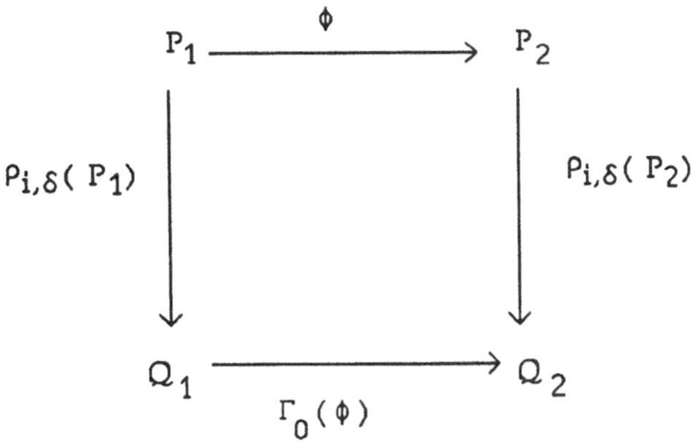

we have

$$\phi(\text{pre}(\delta, \alpha)) = \text{pre}(\delta, \Gamma_0(\phi)(\alpha))$$

so

$$\sup(\Gamma_0(\phi)(\alpha)) = \{ \phi(\beta) \mid \beta \neq \sup(\alpha) \}$$

Then

$$\sup(\Gamma_0(\phi)(\alpha)) \subseteq X_2 \text{ and } \phi(\alpha) \in Y_2.$$

As a direct consequence of 4.10 we get

4.13 *Corollary*

If Γ is strictly positive, then Γ is *chain independent*, i.e. $(\Gamma(\mathcal{P}))_{tot}$ only depends on P and $\mathcal{P}_{tot}$, not on the chains.

4.14 *Lemma*

Let Γ^1 and Γ^2 be two strictly positive functors, $\Gamma^3 = \Gamma^2 \circ \Gamma^1$.

Let $\sup_1$, $\sup_2$ and $\sup_3$ be the three support-set-operators.

For given P, $Q_1 = \Gamma^1{}_0(P)$, $Q_2 = \Gamma^2{}_0(Q_1)$ and $\alpha \in Q_2$ we have
$\beta \in \sup_3(\alpha) \Leftrightarrow \exists \beta_1 \in \sup_2(\alpha)(\beta \in \sup_1(\beta_1))$.

Proof

$\Rightarrow$: Let $\beta = \mathrm{pre}^3((\delta,\sigma), \alpha)$ for some $\delta \in \Delta^2$, $\sigma: I^2{}_\delta \to \Delta^1$.
Then by the argument of lemma 4.7 there is an $i \in I^2{}_\delta$ and $j \in I^1{}_{\sigma(i)}$ such that each element of β is a $\rho^3{}_{(i,j),(\delta,\sigma)}$ - basis of an element in α.
Let $\beta_1 = \mathrm{pre}^2(\delta,\alpha)$. Then $\beta = \mathrm{pre}^1(\sigma(i),\beta_1)$ and the proof is complete.

$\Leftarrow$: Let $\delta_1 \in \Delta^1$ and $\delta_2 \in \Delta^2$ such that a is not δ_2 - initial over Q_1 and $\mathrm{pre}(\delta_2, \alpha)$ is not δ_1 - initial. Let σ be the function on Δ^2 with constant value δ_1. Then

$$\mathrm{pre}^1(\delta_1, \mathrm{pre}^2(\delta_2, \alpha)) = \mathrm{pre}^3((\delta_2, \sigma), \alpha).$$

4.15 *Corollary*

Let Γ_1 and Γ_2 be two strictly positive functors on TOT and let $\Gamma_3 = \Gamma_2 \circ \Gamma_1$.
Then $(\Gamma_W)_3 = (\Gamma_W)_2 \circ (\Gamma_W)_1$.

Proof

Immediate from lemma 4.14.

4.16 *Lemma*

Let Γ be strictly positive on TOT.
Then Γ_W commutes with arbitrary pullbacks on WTOT.

Proof

Let $\phi_i:(P_i, X_i) \to (P, X)$ for all $i \in I$.
The pullback is given by

$P' = \bigcap\{\phi_i(P_i) \mid i \in I\}$

$\alpha \in X' \Leftrightarrow \forall i \in I\, \exists \beta \in X_i\, (\alpha = \phi_i(\beta))$

We leave the proof of this fact for the reader.
Now let

$$(Q_i, Y_i) = \Gamma_W(P_i, X_i)$$
$$(Q, Y) = \Gamma_W(P, X)$$
$$\psi_i = \Gamma_W(\phi_i) = \Gamma_0(\phi_i).$$

Let $\Gamma(P', X') = (Q', Y')$ where $Q' = \cap\{\psi_i(Q_i) \mid i \in I\}$
By general category theory

$$Y' \subseteq \bigcap\{\psi_i(Y_i) \mid i \in I\}$$

Now, let $\alpha \in \{\psi_i(Y_i) \mid i \in I\}$. Then for each $i \in I$
$\sup(\alpha) \subseteq \phi_i(X_i)$.
So $\sup(\alpha) \subseteq X'$. But then $\alpha \in Y'$.

We are now ready to investigate inductive definitions based on strictly positive operators. This will be the theme of the next section.

5. *Inductive definitions*

An inductively defined set is normally obtained by iterating some monotone operator on a bottom element until a fixed-point is reached. We will do so using strictly positive operators as the basis for our inductions.

5.1 *Definition*

Let $\mathcal{B}$ be the totality - domain with empty processing - domain, a one-point set $\{ * \}$ as the question - set and $\varnothing$ as the only chain.

For each $\mathcal{P}$ there is a unique morphism from $\mathcal{B}$ to $\mathcal{P}$.

5.2 *Definition*

Let Γ be a strictly positive functor on TOT.

a) Let $(\mathcal{P}^{\Gamma})_0 = \mathcal{B}$

Let $(\mathcal{P}^{\Gamma})_{n+1} = \Gamma((\mathcal{P}^{\Gamma})_n)$

Let $((\phi^{\Gamma})_0, (\pi^{\Gamma})_0) : \mathcal{B} \to \Gamma(\mathcal{B})$

Let$((\phi^{\Gamma})_{n+1}, (\pi^{\Gamma})_{n+1}) = \Gamma((\phi^{\Gamma})_n, (\pi^{\Gamma})_n)$

Let $(\mathcal{P}^{\Gamma})_{\omega} = \lim(n \to \infty)\, (\mathcal{P}^{\Gamma})_n$.

b) Let $(P^{\Gamma})_0 = \varnothing$, $(P^{\Gamma})_{n+1} = \Gamma_0((P^{\Gamma})_n)$ where Γ_0 is the PRO - part of Γ.

Let $((P^{\Gamma})_{\alpha}, (X^{\Gamma})_{\alpha})$ be defined for any ordinal α by

$$((P^{\Gamma})_0, (X^{\Gamma})_0) = (\varnothing, \varnothing)$$

$$((P^{\Gamma})_{\alpha+1}, (X^{\Gamma})_{\alpha+1}) = \Gamma_W((P^{\Gamma})_{\alpha}, (X^{\Gamma})_{\alpha})$$

$((P^{\Gamma})_{\lambda}, (X^{\Gamma})_{\lambda}) = \lim(\alpha \to \lambda)((P^{\Gamma})_{\alpha}, (X^{\Gamma})_{\alpha})$ when λ is a limit ordinal.

5.3 *Remark*

From now on in this section we will let Γ be fixed. For notational reasons we will drop the superscript Γ (and then one set of parentheses). We will also assume that Γ_0 is set-theoretically monotone.

We can make the following observations:

-- For $\alpha \geq \omega$ we have that $P_{\alpha} = P_{\omega}$.

-- For $n \leq \omega$ we have that $X_n = (\mathcal{P}_n)_{tot}$.

-- There is an α such that $X_{\alpha+1} = X_{\alpha}$., we call this value X_{∞}.

-- $\mathcal{P}_{\omega}$ is a fixed-point for Γ and thus $(P_{\omega}, (\mathcal{P}_{\omega})_{tot})$ is a fixed-point for Γ_W.

-- $X_{\alpha} \subseteq (\mathcal{P}_{\omega})_{tot}$ for all α so $X_{\infty} \subseteq (\mathcal{P}_{\omega})_{tot}$.

5.4 *Theorem*

Let Γ be a strictly positive operator and use the notation and assumptions of 5.2 and 5.3. Then $X_\infty = (\mathcal{P}_\omega)_{tot}$.

Proof

Choose $\alpha \notin X_\infty$. We will show that $\alpha \notin (\mathcal{P}_\omega)_{tot}$.

Since $\alpha \notin X_\infty$, the predecessor - tree of α cannot be well - founded.
Thus there are sequences

$$\{ \alpha_i \}_{i \in N} , \{ \delta_i \}_{i \in N}$$

such that each $\alpha \notin X_\infty$, $\delta_i \in \Delta$, $\alpha_1 = \alpha$ and $\alpha_{i+1} = \mathrm{pre}(\delta_i , \alpha_i)$.
We will construct a tower of chains reflecting the following facts

$\alpha_1 \notin X_1$

$\alpha_1 \notin X_2$ because $\alpha_2 \notin X_1$

$\alpha_1 \notin X_3$ because $\alpha_2 \notin X_2$ because $\alpha_3 \notin X_1$ and so on.

Now let $\mathcal{P}_i = (P_i , S_i , \{ C^i{}_s \}_{s \in S_i})$

Recall that $S_0 = \{ * \}$, $S_{i+1} = \{ (\delta,\sigma) \ \delta \in \Delta$ and $\sigma : I_\delta \to S_i \}$

For any object x , let c_x denote the function with constant value x, and with the domain given by the context

Let $s_{0,i} = *$

Let $s_{n+1,i} = (\delta_i , c_{s_{n,i+1}})$

Claim 1

i) $s_{n,i} \in S_n$

ii) $\pi_{n+1,n}(s_{n+1,i}) = s_{n,i}$
where $\pi_{n+1,n}$ is the projection from S_{n+1} to S_n.

Proof

Both i) and ii) are proved by a trivial induction on $n \geq 0$.

It follows that $\{ s_{n,1} \}_{n \in N}$ is a tower. We will show that the corresponding chain

$$C^\omega{}_{\{s_{n,1}\}_{n \in N}}$$

will avoid $c(\alpha)$.

Claim 2

$C^n{}_{s_{n,i}}$ will not intersect $c(\alpha_i)$

Proof

We use induction on n, and for $n = 0$ this is trivial.
For $n > 0$ we have that

$$C^n{}_{s_{n,i}} \cap c(\alpha_i) = \varnothing$$

because

$$C^{n-1}{}_{s_{n-1,i+1}} \cap c(\alpha_{i+1}) = \emptyset$$

since $s_{n,i} = (\delta_i, c_{s_{n-1,i+1}})$ and $\alpha_{i+1} = \mathrm{pre}(\delta_i, \alpha_i)$.
For further details see the proof of lemma 4.10. This ends the proof of the claim.

In particular $c(\alpha)$ will not intersect $C^n{}_{s_{n,1}}$ for any n, and thus not the limit chain

$C^{\omega}{}_{\{s_{n,1}\}_{n\in N}}$

We conclude that $c(\alpha) \notin (\mathcal{P}_{\omega})_{tot}$, and the theorem is proved.

6. *Multivariable operators*

In this section we will extend the notion of strictly positive functionals to functionals of several (finitely many) variables. The main result is that the fixed point operator described in section 5 in a fair sense will be strictly positive itself.
We will drop , or just give indications of, proofs when the corresponding proofs of sections 4 and 5 essentially will work.
As a convention we will also use the letters S , C in connection with a mentioned P etc. as we have used to do, but without stating this explicitly for each case.

6.1 *Definition*

Let $\Gamma:(\mathrm{TOT})^n \rightarrow \mathrm{TOT}$ be a functor.

a) We call Γ *separable* with PRO - part Γ_0 if $\Gamma_0:(\mathrm{PRO})^n \rightarrow \mathrm{PRO}$ and whenever
$\Gamma((P_1, S_1, \{C^1{}_s\}_{s\in S_1}), \ldots, (P_n, S_n, \{C^n{}_s\}_{s\in S_n}))$
$= (Q, T, \{D_t\}_{t\in T})$ then $\Gamma_0(P_1, \ldots, P_n) = Q$.
In addition we assume the corresponding separation for the morphisms.

b) Γ is called strictly positive if Γ is separable with PRO - part Γ_0,
Γ_0 commutes with pullbacks and direct limits, and we have
-- A set Δ
-- Disjoint set $I_{1,\delta}, \ldots, I_{n,\delta}$ for each $\delta \in \Delta$.
Let $I_\delta = I_{1,\delta} \cup \ldots \cup I_{n,\delta}$
-- Functors $\{\rho_{i,\delta}\}_{i\in I_\delta{}^*}$

such that
-- The elements in T can be identified with
$$\{ t(\delta,\sigma_1,\ldots,\sigma_n) \mid \delta\in\Delta, \sigma_1:I_{1,\delta}\rightarrow S_1, \ldots, \sigma_n:I_{n,\delta}\rightarrow S_n \}$$
For short we will write $t(\delta,\sigma)$ for $t(\delta,\sigma_1,\ldots,\sigma_n)$ assuming that σ can be decomposed to $\sigma_1, \ldots, \sigma_n$ as above.
-- If $i\in I_{k,\delta}$ then $\rho_{i,\delta}$ is a subfunctor of Γ_0 of arity 1 with base in P_i.
-- $\rho_{*,\delta}$ is as before.
-- We assume morphism - conditions on $t(\delta,\sigma)$ as for dimension 1.
-- If
$$\alpha \subseteq \bigcup\{ \rho_{i,\delta}(P_1, \ldots P_n) \mid i\in I^*{}_\delta \}$$
is consistent then either

$$\alpha \subseteq \rho_{*,\delta}(P_1, \ldots, P_n)$$

or there is a unique $k \leq n$ and $i \in I_{k,\delta}$ such that

$$\alpha \subseteq \rho_{i,\delta}(P_1, \ldots P_n)$$

Moreover there is a consistent $\beta \subseteq P_i$ such that $\alpha \subseteq \rho_{i,\delta}(\beta)$ with the obvious meaning of that expression.

-- The "domain" of each $\rho_{i,\delta}$ for $i \in I_\delta$ is closed downwards as in the 1 - dimensional case.

-- $D_{t(\delta,\sigma)} = \bigcup\{\rho_{i,\delta}(C_{\sigma(i)}) \mid i \in I^*_\delta\}$.

6.2 *Lemma*

Let $\mathcal{Q}$ be a fixed totality - domain.

Then $\Gamma_{\mathcal{Q}}(P_1, \ldots, P_n) = \mathcal{Q}$

$\Gamma_{\mathcal{Q}}((\phi_1,\pi_1), \ldots, (\phi_n,\pi_n)) = id_{\mathcal{Q}}$

is strictly positive.

Proof

Let $\Delta = T$, $I_{k,t} = \emptyset$ for $k \leq n$.

Then the empty function is the only function $\sigma: I_{k,t} \to S_i$, so we may identify t with the only possible denotation based on t.

Let $\rho_{*,t}(P_1, \ldots, P_n) = D_t$ (independently of $(P_1, \ldots, P_n)$)

The verification of the correctness of this definition is trivial and is left for the reader.

6.3 *Lemma*

$\Gamma_i(P_1, \ldots, P_n) = \mathcal{P}_i$

$\Gamma_i((\phi_1,\pi_1), \ldots, (\phi_n,\pi_n)) = (\phi_i,\pi_i)$

is strictly positive.

The proof is trivial and is left for the reader.

6.4 *Lemma*

Let $\Lambda: (TOT)^m \to TOT$ and $\Gamma_1, \ldots \Gamma_m: (TOT)^n \to TOT$ be strictly positive.

Then the composition $\Lambda \circ (\Gamma_1, \ldots, \Gamma_m)$ is strictly positive.

Proof

The construction and the proof follow the same line as in the one - dimensional case and is left for the reader.

We can easily extend the examples of section 4 to give us

$$\Gamma_\oplus(\mathcal{P}_1, \ldots, \mathcal{P}_n) = \mathcal{P}_1 \oplus \ldots \oplus \mathcal{P}_n$$
$$\Gamma_\times(\mathcal{P}_1, \ldots, \mathcal{P}_n) = \mathcal{P}_1 \times \ldots \times \mathcal{P}_n$$

This gives the following result

6.5 *Corollary*

Any operator defined by finite iteration of the following basic operators:

- -- Cartesian product
- -- Disjoint union
- -- Modified function - space: $(A,<) \to \mathcal{P}$
- -- Product with an arbitrary set: $A \times \mathcal{P}$,

and using fixed totality - domains as parameters, will generate a strictly positive functor.

Proof

Trivial from the preceding lemmas.

Let $\Gamma_0(P_1, \ldots, P_n) = Q$ and let $\alpha \in \mathbb{Q}$. We will see how to obtain the α - predecessors and the support of α.

Let $\delta \in \Delta$. Assume that $\rho_{i,\delta}(P_1, \ldots, P_n) \neq \emptyset$ for some i.
Then i is unique and $i = *$ or $i \in I_{k,\delta}$ for some $k \leq n$.
If $i = *$ then α is δ - initial. Otherwise, let

$$\text{pre}(\delta,\alpha) = \{ p \in P_k \quad p \text{ is the } \rho_{i,\delta} \text{ - base for some } q \in c(\alpha) \}$$

If for all $i \in I^*_\delta$ we have that $\rho_{i,\delta}(P_1, \ldots, P_n) \cap c(\alpha) = \emptyset$
then $\text{pre}(d,\alpha) = \emptyset$.
We define $\sup(\alpha)$ as before, and as before we have that
$\alpha \in \mathcal{Q}_{tot} \Leftrightarrow \sup(\alpha) \cap \mathbb{P}_k \subseteq (\mathcal{P}_k)_{tot}$ for $k = 1, \ldots, n$.

We are now ready to prove the main theorem of this section.

6.6 *Theorem*

Let $\Gamma(\mathcal{P}, \mathcal{Q}_1, \ldots, \mathcal{Q}_n)$ be a strictly positive functor in n+1 variables.
For each $\mathcal{Q}_1, \ldots, \mathcal{Q}_n$, let

$$\text{Fix}_\Gamma(\mathcal{Q}_1, \ldots, \mathcal{Q}_n)$$

be the least fixed - point in variable $\mathcal{P}$, i.e the least $\mathcal{P}$ such that

$$\mathcal{P} = \Gamma(\mathcal{P}, \mathcal{Q}_1, \ldots, \mathcal{Q}_n)$$

Then Fix_Γ is a strictly positive functor.

Proof

For simplicity we will assume that $n = 1$ in the proof.
The proof will be given in several stages.

1. Fix_Γ is a functor.

Proof

Let $(\psi,\nu) : \mathcal{Q}_1 \to \mathcal{Q}_2$.

Let $\mathcal{P}^1{}_0 = \mathcal{P}^2{}_0 = \mathcal{B}$.

Let $\mathcal{P}^1_{n+1} = \Gamma(\mathcal{P}^1_n, Q_1)$

$\mathcal{P}^2_{n+1} = \Gamma(\mathcal{P}^2_n, Q_2)$

Let (ϕ^i_0, π^i_0) be the unique morphism from $\mathcal{B}$ to $\mathcal{P}^i_1$ $i = 1,2$.

Let $(\phi^i_{n+1}, \pi^i_{n+1}) = \Gamma((\phi^i_n, \pi^i_n), \mathrm{id}_{Q_i})$ $i = 1,2$

Then $(\phi^i_{n+1}, \pi^i_{n+1}) : \mathcal{P}^i_{n+1} \rightarrow \mathcal{P}^i_{n+2}$; $n \in \mathbf{N}$ are the morphisms used in the directed system defining $\mathrm{Fix}_\Gamma(Q_i)$.

Claim: The following diagram commutes for each n:

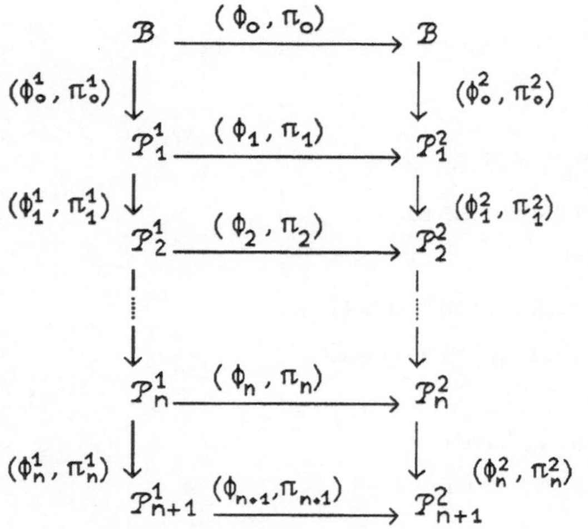

Diagram 1

The proof is by elementary category - theory; clearly the top - diagram commutes, and the commutation of the rest of the diagram follows by induction.

Let the limits be $\mathcal{P}^1_\omega$ and $\mathcal{P}^2_\omega$ and let

$\mathcal{P}^i_j = (P^i_j, S^i_j, \{C^{i,j}_s\}_{s \in S^i_j})$

for $i = 1,2$ and $j \in \mathbf{N} \cup \{\omega\}$.

Then we define

$\mathrm{Fix}_\Gamma(\pi, \nu) = (\phi_\omega, \pi_\omega)$

where ϕ_ω is the obvious limit of $\{\phi_n\}_{n \in \mathbf{N}}$.

If $s \in S^2_\omega$, then $s = \{s_n\}_{n \in \mathbf{N}}$ where $\pi^2_n(s_{n+1}) = s_n$ for all n.

Let $t = \{t_n\}_{n \in \mathbf{N}} = \{\pi_n(s_n)\}_{n \in \mathbf{N}}$.

By the commutation of the diagram we see that $\pi^1_n(t_{n+1}) = t_n$ for all n.

Thus $t = \{t_n\}_{n \in \mathbf{N}}$ is a tower and we let $\pi_\omega(s) = t$.

2. It is clear that in the above case, P^1_ω, P^2_ω and ϕ_ω only depends on Q_1, Q_2, ψ and the original PRO - part Γ_0.

Thus Fix_{Γ} is separable with the fixed - point operator for Γ_0 as PRO - part. We will consequently denote this PRO - part Fix_{Γ_0}.

3 Fix_{Γ_0} will commute with direct limits.

Proof

Let $\{ Q_i \}_{i\in I}$, $\{\psi_{i,j} \}_{i\leq j}$ be a directed system. We construct a diagram, by for each $i \leq j$ constructing the PRO - part of diagram 1. The final diagram will be a directed system and the limit of the full system will be the joint value of

$$\lim(n\rightarrow\infty)\lim(i\in I)P^i{}_n = \lim(n\rightarrow\infty)P^I{}_n$$

and

$$\lim(i\in I)\lim(n\rightarrow\infty)P^i{}_n = \lim(i\in I)P^i{}_\omega.$$

4. Fix_{Γ_0} commutes with pullbacks.

Proof

In the category PRO it is easy to see that the direct limit of a directed system of pullbacks - diagrams is itself a pullback - diagram. The property follows.

We will now define the items used in defining the chain - denotations and the chains, and we will then verify the properties.

Definition

A Δ - *tree* is a tree τ of sequences $(r_1, \ldots, r_k)$ satisfying

i) r_1 is a fixed element of Δ called the *head* of τ.

ii) All $r_{2i+1} \in \Delta$

iii) All $r_{2i+2} \in I_{1,r_{2i+1}}$

iv) If $\gamma\delta \in \tau$ for some $\delta \in \Delta$, then $\gamma\delta i \in \tau$ for all $i \in I_{1,\delta}$.

v) If $\gamma i \in \tau$ then $\gamma i\delta \in \tau$ for exactly one $\delta \in \Delta$.

Here γ denotes an arbitrary sequence $(r_1, \ldots, r_k)$ and $\gamma\delta$ etc. denotes the result of adding δ to the end of γ.

When τ is a Δ - tree, let I_τ be the disjoint union of all $I_{2,\delta}$ for occurrences of δ in τ, where the occurrence in a subsequence is identified with the corresponding occurrence in a sequence.

5. S_ω can be put in a 1 - 1 - correspondence with

$$\{ (\tau,\sigma) \mid \tau \text{ is a } \Delta\text{ - tree and } \sigma:I_t \rightarrow T \}$$

Proof

Let $t \in S_\omega$. Then t is a tower

$$\{ t_n \}_{n\in \mathbf{N}}$$

From $\{ t_n \}_{n\in \mathbf{N}}$ we will determine a head δ, a map $\sigma_\delta:I_{2,\delta} \rightarrow T$ and for each $i \in I_{1,\delta}$, a subtower $\{ t^i{}_n \}_{n\in \mathbf{N}}$ which we will use for the further construction of (τ,σ).

t_0 will be the fixed element of S_0, where $\mathcal{P}_0 = \mathcal{B}$.

For $n > 0$, there is a δ_n, $\sigma_{1,n}: I_{1,\delta_n} \to S_{n-1}$, $\sigma_{2,n}: I_{2,\delta_n} \to T$ such that

$$t_n = (\delta_n, \sigma_{1,n}, \sigma_{2,n}).$$

Since $\pi_{n+1}(t_{n+1}) = t_n$ we must have

$\delta_n = \delta_{n+1} = \delta$ for some fixed δ

$\sigma_{2,n} = \sigma_{2,n+1} = \sigma_\delta$ for some fixed σ_δ.

For each $i \in I_{1,\delta}$ we have that $\pi_n(\sigma_{1,n+1}(i)) = \sigma_{1,n}(i)$.

Thus $\{ t^i_n \}_{n \in \mathbf{N}}$ defined by $t^i_n = \sigma_{1,n+1}(i)$ for $n \geq 0$, $i \in I_{1,\delta}$

is a tower, based on which we will continue our construction of τ and σ.

Conversely let (τ, σ) be given. By induction on n we define t_n:

$n = 0$: Let $t_n = *$; the unique element in S_0.

$n > 0$: Let δ be the head of τ, $\sigma_\delta: I_\delta \to T$ be the part of σ mapping the relevant copy of I_δ into T.

For each $i \in I_\delta$, let τ_i be the subtree of τ defined by

$$\tau_i = \{ \gamma \mid \delta i \gamma \in \tau \}$$

where γ is a sequence as above.

Let σ_i be the restriction of σ to those $I_{\delta'}$ whose occurrences are in elements of τ_i.

By the induction - hypothesis (τ_i, σ_i) defines an element t^i_{n-1} in S_{n-1}.

Let

$$\sigma'(i) = t^i_{n-1}$$

and let $t_n = (\delta, \sigma', \sigma_\delta)$.

By induction on n we prove that

$$\{ t_k \}_{k \leq n}$$

satisfy the tower - property, using the property for each

$$\{ t^i_k \}_{k < n}$$

as an induction hypothesis.

It is clear that the two "conversions" described above are inverses of each other. This ends the proof of the claim.

In order to give a complete proof we must show that this "denotation - system" commutes with the morphisms $\nu: T_2 \to T_1$, i.e. that

$\pi_\infty(t_{(\tau,\sigma)}) = t_{(\tau, \sigma \circ \nu)}$

This is not hard and is left for the reader.

6. Let τ be a Δ - tree. We will define the operators $\rho_{i,\tau}$ for $i \in I_\tau^*$.

Let $i \in I_\tau$

Then there is a unique minimal sequence

$$(\delta_0, i_0, \ldots, \delta_k, i_k, \delta) \in \tau$$

where the occurrence of δ gives the copy of $I_{2,\delta}$ where i is.

Let $i' \in I_{2,\delta}$ be the element corresponding to i.

Let $\rho_{i,t} = \rho_{i_0,\delta_0} \circ \ldots \circ \rho_{i_k,\delta_k} \circ \rho_{i',\delta}$

We define $\rho_{*,\tau}$ by

$$\rho_{*,\tau} = \bigcup\{ \rho_{i_0,\delta_0} \circ \ldots \circ \rho_{i_k,\delta_k} \circ \rho_{*,\delta} \mid (\delta_0, i_0, \ldots, \delta_k, i_k, \delta) \in \tau \}$$

It is tedious but trivial to verify that the ρ - operators thus defined have the required properties.

7. Finally we will show that the chains are correctly defined, i.e. that

$$C_{(\tau,\sigma)} = \bigcup\{\rho_{i,\tau}(D_{\sigma(i)}) \mid i \in I_\tau^* \}$$

Let $C^n_{(\tau,\sigma)}$ be the n'th approximation to $C_{(\tau,\sigma)}$ defined via the tower $\{ t_n \}_{n\in \mathbf{N}}$ obtained from (σ,δ) in the second half of the argument 5. By induction on n we prove

$$C^n_{(\tau,\sigma)} = \bigcup\{\rho_{i,\tau}(D_{\sigma(i)}) \mid i \in I_\tau^* \text{ and the depth of i is} \leq n \}$$

where the depth is given by the number of δ's in the minimal sequence witnessing that $i \in I_\tau^*$. The depth of the i in argument 6 is k+1.
We also assume that $C^n_{(\tau,\sigma)} \subseteq C^{n+1}_{(\tau,\delta)}$ for all n.

For $n = 0$ this is trivial.
For $n > 0$ let δ be the head of τ. Then we have

$$\begin{aligned} C^n_{(\tau,\sigma)} = \; & \rho_{*,\delta}((\emptyset,Q)) \\ & \cup \bigcup\{ \rho_{i,\delta}(D_{\sigma(i)} \mid i \in I_{2,\delta} \} \\ & \cup \bigcup\{ \rho_{i,\delta}(C^{n-1}_{(\tau_i,\sigma(i))}) \mid i \in I_{1,\delta} \} \end{aligned}$$

where τ_i is the tree we obtain from τ by removing δi from the start of every sequence in τ.
By the induction hypothesis we get for each $i \in I_{1,\delta}$ that

$\rho_{i,\delta}(C^{n-i}_{(\tau_i,\sigma(i))}) =$

$\rho_{i,\delta}(\bigcup\{ \rho_{j,\tau_i}(D_{\sigma(j)}) \mid j \in I_{\tau_i}^* \text{ and the depth of j in } \tau_i \text{ is } < n \}) =$

$\bigcup\{ \rho_{j,\tau}(D_{\sigma(j)}) \mid j \in I_{\tau_i}^* \text{ and the depth of j in } \tau \text{ is } \leq n \}$

The full equation then follows.
This completes the proof of Theorem 6.6.

The immediate predecessors are given by the functions pre(δ , α), i.e. Δ can be used as an index set for the predecessors.
In the proof of lemma 4.14 we can use the product $\Delta_2 \times \Delta_1$ as an index - set for the predecessors instead of the rather complex set Δ_3 of the composition.
We have a similar simplification here.

6.7 *Theorem*

Let Γ be a strictly positive functional of two variables with PRO - part Γ_0.

Let Q be a processing - domain , and let P be the least solution to the equation

$$P = \Gamma_0(P , Q).$$

Let $\alpha \in \mathbb{P}$ and let τ be a Δ - tree as defined in the proof of theorem 6.6.

Let $\beta = \text{pre}(\alpha, \tau)$ and assume that β is nonempty.

Then for some sequence $(\delta_1 , \ldots , \delta_{n-1})$ from Δ we have that

$$\beta = \text{pre}(\delta_1 , \text{pre}(\ldots \text{pre}(\delta_{n-1} , \alpha) \ldots))$$

Proof

Assume that α is not τ - initial.

Choose one $q \in Q$ such that q is the $\rho_{i,\tau}$ - basis for some $p \in c(\alpha)$.

Then i is unique and is obtained from a unique minimal sequence in τ.

If we take just the δ's of that sequence we get what we want.

This result might suggest that we may use finite sequences from Δ as an index - set for the predecessors. If we do so, we must, however, be open for the alternative

There is no τ - predecessor for α but α is not τ - initial

Thus we choose Δ^ω as the index - set for the predecessors, and we may give precise definitions for

-- α is $\{\delta_n\}_{n\in \mathbf{N}}$ - initial if when we iterate the α - predecessors along $\delta_1 , \delta_2 , \ldots$ we find a δ_n - initial element in $\mathbb{P}$.

-- α has β as the $\{\delta_n\}_{n\in \mathbf{N}}$ -predecessor if $\beta \in \mathbb{Q}$ and we hit β the first time while iterating the α - predecessors along $\delta_1 , \delta_2 , \ldots$ we find a predecessor in $\mathbb{Q}$.

We omit the details here.

References

Girard - 85: The system F of variable types fifteen years later; Preprint.

Normann 87: Kleene spaces, Informal manuscript

Normann - 88: Kleene spaces, To appear in the proceedings from Logic Colloquium '88, Padova 1988.

STRUCTURAL RULES AND A LOGICAL HIERARCHY

Hiroakira Ono

Faculty of Integrated Arts and Sciences
Hiroshima University
Hiroshima

1. INTRODUCTION

Gentzen-type sequent calculi usually contain three structural rules, i.e., exchange, contraction and weakening rules. In recent years, however, there have been various studies on logics that have not included some or any of these structural rules. The motives or purposes of these studies have been so diverse that sometimes close connections between them have been overlooked. Here we will make a brief survey of recent results on these logics in an attempt to make these interrelationships clearer.

We will start by giving a brief historical overview of the study of logics without structural rules. Lambek(1958) introduced a Gentzen-type system for the implicational fragment of the intuitionistic logic without any structural rule, now called the Lambek calculus, which serves as a formal system of type changing in categorial grammar. This is probably the first major paper on logics without structural rules, though until recently it was not well-known. By proving the cut elimination theorem for that logic, Lambek demonstrated the existence of a decision procedure for testing whether or not a given syntactic type can be deduced from a given set of syntactic types. The linguistic side of this work has been developed recently by van Benthem(1986), Buszkowski(1986), etc..

Lambek(1961) alluded to relationships between Gentzen-type formal systems and categories. This paper became a starting point for his later works (Lambek, 1968, 1969, 1972) on relationships between proof theory and category theory and for works by Minc(1977), Szabo(1978), etc.. It was stated, for example, that cartesian closed categories correspond to the intuitionistic logic, while symmetric monoidal closed categories correspond to the intuitionistic logic with neither contraction nor weakening rules.

Wang(1963) first noticed that the classical predicate logic without contraction rules is decidable (see p.228). Grišin studied the logic extensively in the mid-70s, though his works have been relatively unknown until recently. Computer scientists have also been interested in this logic. Ketonen and Weyrauch(1984) considered it to be a subsystem of the classical predicate logic, which is decidable but still encompasses some mode of reasoning in mathematics. Recently, Girard(1987) developed a study of linear logic, i.e. the classical logic with neither contraction rules nor weakening rules. It contains a lot of interesting and important ideas for both logic and computer science. See also Girard(1987a).

Mathematical Logic
Edited by P. P. Petkov
Plenum Press, New York, 1990

In the author's joint paper with Komori(1985), we developed a semantical study of the intuitionistic propositional logic without contraction rule. Our semantics, using partially ordered monoids, can cover a wide class of logics, including Łukasiewicz's many-valued logics and relevant logics. Closely related results were obtained independently by Došen(1986). On the other hand, some difficulties will occur when we will try to extend our semantics to one for predicate logics. Some attempts in this direction have been made by Komori(1986) and the author(1985). Abrusci(1987) introduced another type of semantics for the intuitionistic linear logic, which is an extension of the phase semantics by Girard.

Finally, we will mention a connection between these logics and lambda calculus. It is well-known that any proof of the natural deduction system of (the implicational fragment of) the intuitionistic logic can be expressed by a lambda-term, and conversely that type-assignment of a given lambda-term will give a proof of the type-scheme assigned to the lambda-term (see e.g. Hindley and Seldin, 1986). Komori(1987) and Hindley(1987) recently studied this relation also for logics without structural rules. By making use of this relation, Hirokawa(1987) proved that every *minimal* formula in BCK logic has a unique normal proof. Here BCK logic is the implicational fragment of the intuitionistic logic without contraction rule.

A major source of inspiration for the present paper was discussions the author had on these topics with J. van Benthem and A. Wroński.

2. FULL LAMBEK CALCULUS AND ITS EXTENSIONS

We will introduce a Gentzen-type system for the intuitionistic logic without any structural rule, which we call *full Lambek calculus* FL, after the syntactic calculus introduced by Lambek(1958). Our language contains two logical constants 0 and 1, four logical connectives $\wedge$, $\vee$, $\supset$ and $*$ (multiplicative, or intensional conjunction), and two quantifiers $\forall$ and $\exists$, as logical symbols. The negation $\neg A$ of a formula A is defined as abbreviation of $A \supset 0$. A sequent of FL has the form $\Gamma \longrightarrow A$, where Γ is a finite (possibly empty) sequence of formulas and A is a formula.

An initial sequent of FL is a sequent in one of the following forms;

1) $A \longrightarrow A$, 2) $\longrightarrow 1$, 3) $\Gamma, 0, \Delta \longrightarrow A$.

Rules of inferences of FL are, roughly speaking, those of the intuitionistic logic LJ minus all structural rules. Attention is drawn to the fact that the order of formulas in a sequent is essential, since FL lacks the exchange rule. Thus, rules of inferences for implication, for example, would be;

$$\frac{\Gamma, A \longrightarrow B}{\Gamma \longrightarrow A \supset B} \ (\rightarrow \supset) \qquad \frac{\Gamma \longrightarrow A \quad \Delta, B, \Sigma \longrightarrow C}{\Delta, A \supset B, \Gamma, \Sigma \longrightarrow C} \ (\supset \rightarrow)$$

Moreover, FL has the following rules for 1 and $*$;

$$\frac{\Gamma, \Delta \longrightarrow C}{\Gamma, 1, \Delta \longrightarrow C} \ (\text{1w})$$

$$\frac{\Gamma \longrightarrow A \quad \Delta \longrightarrow B}{\Gamma, \Delta \longrightarrow A*B} \ (\rightarrow *) \qquad \frac{\Gamma, A, B, \Delta \longrightarrow C}{\Gamma, A*B, \Delta \longrightarrow C} \ (* \rightarrow)$$

Exchange, contraction and weakening rules will be abbreviated to e, c and w, respectively, and each extension of FL with some of these structural rules will be expressed by adding corresponding letters to 'FL' as its subscripts. Thus, FL_{ew} would be the formal system obtained from FL by adding exchange and weakening rules. It is easily observed that FL_{ew} and FL_{ecw} are essentially the same as L_{BCK} and the intuitionistic logic, respectively.

Theorem 2.1 (Wroński, Ono-Komori, Došen, Komori, etc.) The cut elimination theorem and Craig's interpolation theorem hold for FL, FL_e, FL_w FL_{ew} and FL_{ecw}.

Theorem 2.2 (Komori) FL, FL_e, FL_w and FL_{ew} are decidable.

We will next introduce Gentzen-type systems for the classical logics without some or all of the structural rules. In this case, though, there are many ways of defining systems without exchange rules, creating unnecessary complications. As a basic system, therefore, we will take the classical logic with neither contraction nor weakening rules, GL, which is essentially the same logic as linear logic by Girard(1987). We will add also another logical connective, +, (multiplicative disjunction) to our language. A sequent of GL has the form $\Gamma \longrightarrow \Sigma$, where both Γ and Σ are finite (possibly empty) sequences of formulas. An initial sequent of GL is defined similarly to that of FL, but 3) is replaced by the following;

3') $0, \Gamma \longrightarrow \Sigma$.

Rules of inferences of GL are those of the classical logic LK minus contraction and weakening rules. In addition, GL has the following six rules;

$$\frac{\Gamma \longrightarrow \Sigma}{1, \Gamma \longrightarrow \Sigma}\ (1w) \qquad \frac{\Gamma \longrightarrow \Sigma}{\Gamma \longrightarrow \Sigma, 0}\ (0w)$$

$$\frac{\Gamma \longrightarrow \Sigma, A \quad \Delta \longrightarrow \Pi, B}{\Gamma, \Delta \longrightarrow \Sigma, \Pi, A*B}\ (\rightarrow *) \qquad \frac{A, B, \Gamma \longrightarrow \Sigma}{A*B, \Gamma \longrightarrow \Sigma}\ (* \rightarrow)$$

$$\frac{\Gamma \longrightarrow \Sigma, A, B}{\Gamma \longrightarrow \Sigma, A+B}\ (\rightarrow +) \qquad \frac{A, \Gamma \longrightarrow \Sigma \quad B, \Delta \longrightarrow \Pi}{A+B, \Gamma, \Delta \longrightarrow \Sigma, \Pi}\ (+ \rightarrow)$$

We can define extensions GL_c, GL_w and GL_{cw} of GL, similarly as before. Clearly, GL_{cw} is the classical logic and GL_w the classical logic without the contraction rules studied by Grišin(1974,1982).

Theorem 2.3 (Grišin, Wroński-Krzystek) The cut elimination theorem holds for GL, GL_c, GL_w and GL_{cw}.

Theorem 2.4 (Wang, Grišin) Both GL and GL_w are decidable.

The following figure shows the logical hierarchy determined by the logics introduced above.

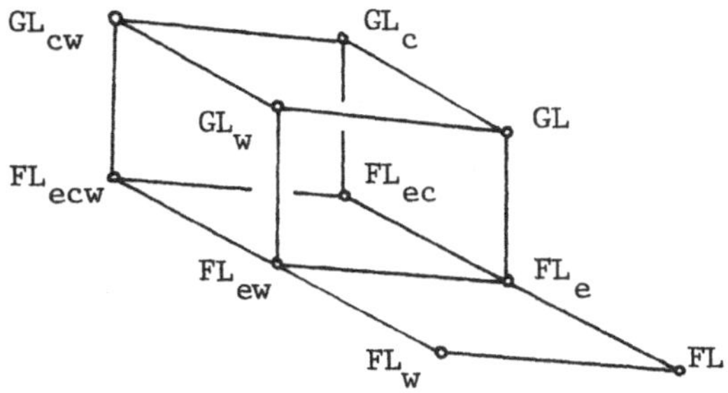

Theorem 2.5 $GL_x = FL_{ex} + \neg\neg A \supset A$, where x is either empty, or any of c, w and cw.

Theorem 2.6 Let x be either empty or c or w. For any set $\Gamma \cup \{A\}$ of formulas containing neither + nor 0, or for any set $\Gamma \cup \{A\}$ of formulas containing neither + nor ,

$$GL_x \vdash \Gamma \longrightarrow A \quad \text{if and only if} \quad FL_{ex} \vdash \Gamma \longrightarrow A.$$

We will introduce extensions of logics in the above, with two additional unary operators, ! and ?, as proposed by Girard(1987). See also Girard and Lafont(1987), Lafont(1987) and Abrusci(1987). For each logic GL_x, let GL_x^+ be the logic obtained from GL_x by adding the following rules of inference;

$$\frac{!\Gamma \longrightarrow A, ?\Delta}{!\Gamma \longrightarrow !A, ?\Delta}\ (\rightarrow !) \qquad \frac{\Gamma, A \longrightarrow \Pi}{\Gamma, !A \longrightarrow \Pi}\ (! \rightarrow)$$

$$\frac{\Gamma \longrightarrow \Pi}{\Gamma, !A \longrightarrow \Pi}\ (!w) \qquad \frac{\Gamma, !A, !A \longrightarrow \Pi}{\Gamma, !A \longrightarrow \Pi}\ (!c)$$

$$\frac{\Gamma \longrightarrow A, \Delta}{\Gamma \longrightarrow ?A, \Delta}\ (\rightarrow ?) \qquad \frac{!\Gamma, A \longrightarrow ?\Pi}{!\Gamma, ?A \longrightarrow ?\Pi}\ (? \rightarrow)$$

$$\frac{\Gamma \longrightarrow \Delta}{\Gamma \longrightarrow ?A, \Delta}\ (?w) \qquad \frac{\Gamma \longrightarrow ?A, ?A, \Delta}{\Gamma \longrightarrow ?A, \Delta}\ (?c)$$

Similarly, for each logic FL_{ex}, let FL_{ex}^+ be the logic obtained from FL_{ex}^+ by adding the above rules except (?w) and (?c), with the requirement that Δ is empty and Π consists of a single formula. Logics with operators ! and ? will be discussed in the next section.

3. TRANSLATIONS BY GRIŠIN AND GIRARD

Grišin(1974) showed that the provability of a formula in LK (the classical logic) can be reduced to the provability of some corresponding formulas in GL_w. To be precise, for each formula A of LK and each natural number n, define formulas $A^{(+n)}$ and $A^{(-n)}$ inductively as follows;

1) $A^{(+n)} = A^{(-n)} = A$, when A is an atomic formula,
2) $(\neg A)^{(+n)} = \neg(A^{(-n)})$, $(\neg A)^{(-n)} = \neg(A^{(+n)})$,
3) $(A \supset B)^{(+n)} = A^{(-n)} \supset B^{(+n)}$,
   $(A \supset B)^{(-n)} = \neg(A^{(+n)}) \vee B^{(-n)}$,
4) $(A \wedge B)^{(+n)} = A^{(+n)} \wedge B^{(+n)}$,
   $(A \wedge B)^{(-n)} = A^{(-n)} * B^{(-n)}$,
5) $(A \vee B)^{(+n)} = A^{(+n)} + B^{(+n)}$,
   $(A \vee B)^{(-n)} = A^{(-n)} \vee B^{(-n)}$,
6) $(\forall xA)^{(+n)} = \forall x(A^{(+n)})$,
   $(\forall xA)^{(-n)} = (\forall x(A^{(-n)}))^n$,
7) $(\exists xA)^{(+n)} = n(\exists x(A^{(+n)}))$,
   $(\exists xA)^{(-n)} = \exists x(A^{(-n)})$,

where for any formula D, D^n and nD are formulas defined by

$$D^1 = 1D = D, \quad D^{n+1} = D^n * D \quad \text{and} \quad (n+1)D = nD+D.$$

Let Γ be a sequence of formulas $A_1, \ldots, A_m$. Then, $\Gamma^{(-n)}$ denotes the sequence of formulas $A_1^{(-n)}, \ldots, A_m^{(-n)}$. $\Gamma^{(+n)}$ is defined similarly.

Theorem 3.1 (Grišin) For any sequent $\Gamma \longrightarrow \Delta$ of LK, $LK \vdash \Gamma \longrightarrow \Delta$ if and only if $GL_w \vdash \Gamma^{(-n)} \longrightarrow \Delta^{(+n)}$ for some n.

When A is a propositional formula, 1 can always be taken for n. It is possible to modify Grišin's translation so that it can be applied to the logic GL. Let us define formulas $A^{[+n]}$ and $A^{[-n]}$ in the same way as $A^{(+n)}$ and $A^{(-n)}$, except

1') $A^{[+n]} = 0 \vee A$ and $A^{[-n]} = 1 \wedge A$, when A is an atomic formula.

The following theorem can be shown in the same way.

Theorem 3.2 For any sequent $\Gamma \longrightarrow \Delta$ of LK, $LK \vdash \Gamma \longrightarrow \Delta$ if and only if $GL \vdash \Gamma^{[-n]} \longrightarrow \Delta^{[+n]}$ for some n.

But what does the number n in Theorems 3.1 and 3.2 represent? To explain this, let us take a Gentzen-type calculus LC of the classical logic by Kanger(1957) which does not contain any structural rule explicitly. Rules ($\forall \rightarrow$) and ($\rightarrow \exists$) of LC have however the following form;

$$\frac{\Gamma'', \Gamma', A(t), \forall x A(x), \Gamma \longrightarrow \Delta}{\Gamma'', \forall x A(x), \Gamma', \Gamma \longrightarrow \Delta} \quad (\forall \rightarrow)$$

$$\frac{\Gamma \longrightarrow \Delta'', \Delta', A(t), \exists x A(x), \Delta}{\Gamma \longrightarrow \Delta'', \exists x A(x), \Delta', \Delta} \quad (\rightarrow \exists)$$

Let Π be a proof of a formula A in LC. For each path in Π, let us count the total number of applications of ($\forall \rightarrow$) and ($\rightarrow \exists$) in that path. Let m be the maximum number among them. We can then show that $GL_w \vdash A^{(+m+1)}$ and $GL \vdash A^{[+m+1]}$. The number n in Theorems 3.1 and 3.2 will therefore denote a measure of the *complexity* of proofs, and thus it has an intrinsic meaning both for proof theory (e.g. Herbrand's theorem) and for mechanical theorem proving.

Several possible translations of LK into linear logic were discussed by Girard(1987). Here we take an example for comparison with Grišin's translation. For each formula A of LK, define formulas $[A]^+$ and $[A]^-$ of GL^+ as follows;

1) $[A]^+ = [A]^- = A$, when A is an atomic formula,
2) $[\neg A]^+ = \neg [A]^-$, $[\neg A]^- = \neg [A]^+$,
3) $[A \supset B]^+ = \neg [A]^- \vee [B]^+$, $[A \supset B]^- = ?[A]^+ \supset ![B]^-$,
4) $[A \wedge B]^+ = ?[A]^+ * ?[B]^+$, $[A \wedge B]^- = [A]^- \wedge [B]^-$,
5) $[A \vee B]^+ = [A]^+ \vee [B]^+$, $[A \vee B]^- = ![A]^- + ![B]^-$,
6) $[\forall x A]^+ = \forall x ?[A]^+$, $[\forall x A]^- = \forall x [A]^-$,
7) $[\exists x A]^+ = \exists x [A]^+$, $[\exists x A]^- = \exists x ![A]^-$.

Theorem 3.3 (Girard) For any sequent $\Gamma \longrightarrow \Delta$ of LK, $LK \vdash \Gamma \longrightarrow \Delta$ if and only if $GL^+ \vdash ![\Gamma]^- \longrightarrow ?[\Delta]^+$, where $![\Gamma]^-$ is a sequence of formulas $![A_1]^-, \ldots, ![A_m]^-$ if Γ is $A_1, \ldots, A_m$, and $?[\Delta]^+$ is defined similarly.

Girard's translation will suggest the following translation, which is naturally induced from Grišin's one. For each formula A of LK, the definition of formulas A^+ and A^- (of GL^+) are obtained from that of $A^{[+n]}$ and $A^{[-n]}$ by replacing $A^{[+n]}$ and $A^{[-n]}$ by A^+ and A^-, respectively, except cases

6) and 7) for which we take the following;

6") $(\forall xA)^+ = \forall x(A^+)$, $(\forall xA)^- = !\forall x(A^-)$,

7") $(\exists xA)^+ = ?\exists x(A^+)$, $(\exists xA)^- = \exists x(A^-)$.

Theorem 3.4 For any sequent $\Gamma \longrightarrow \Delta$ of LK, LK $\vdash \Gamma \longrightarrow \Delta$ if and only if $GL^+ \vdash \Gamma^- \longrightarrow \Delta^+$.

For any formulas A and B, A => B means that $GL^+ \vdash A \longrightarrow B$. The following theorems will clarify relations between results mentiond in above theorems (cf. Theorem 5.2 of Girard, 1987).

Theorem 3.5 (approximation theorem) For each formula A of LK, the following holds;

$$A^- \Rightarrow \dots \Rightarrow A^{[-(n+1)]} \Rightarrow A^{[-n]} \Rightarrow \dots \Rightarrow A^{[-1]}$$
$$\Rightarrow A^{[+1]} \Rightarrow \dots \Rightarrow A^{[+n]} \Rightarrow A^{[+(n+1)]} \Rightarrow \dots \Rightarrow A^+.$$

Recall here Girard's slogan in his paper(1987) which says that *usual logic is obtained from linear logic by a passage to limit.* Recall also our discussion on the meaning of the number n in Theorems 3.1 and 3.2.

Theorem 3.6 For each formula A of LK, the following holds;

Girard(1987) introduced a simple translation between the intuitionistic logic and the logic FL_e^+ (see also Girard and Lafont, 1987). At present, there is no nontrivial Grišin-type translation corresponding to it, but a Grišin-type translation between the classical logic and FL_{ew} was studied by Dardžania(1977).

4. SEMANTICS

There have been various studies of semantics of FL and its extensions, which we can have quick look at now.

Grišin(1982) introduced an algebraic semantics for GL_w and obtained some interesting results on algebraic properties of GL_w-algebras. Among other things, he proved that the predicate logic GL_w is complete with respect to semantics determined by complete GL_w-algebras. Other papers on algebraic semantics are mostly concerned with propositional logics. Došen (1986) and Suzuki(1987) showed that residuated lattice-ordered monoids serve for algebraic structures for FL. Algebraic structures for the implicational fragment of FL_{ew} are traditionally called BCK-algebras, of which many studies have already been made. Some algebraic results on FL_w- and FL_{ew}-algebras were obtained by Ono and Komori(1985) and Ono(1985a). It was pointed out also that in some respects, FL_{ew}-algebras behave like Heyting algebras and GL_w-algebras like Boolean ones. Girard(1987) introduced and studied algebraic structures for GL and GL^+, which are called *phase structures* and *topophase structures*, respectively. This idea was generalized by Abrusci

(1987) who introduced intuitionistic phase and topophase structures for FL_e and FL_e^+, obtained from commutative monoids, and proved completeness theorems with respect to them.

Kripke-type semantics for propositional logics in FL-series is introduced by Ono and Komori(1985), and also independently by Došen(1986), using semilattice-ordered monoids (or partially ordered monoids, if we treat disjunctionless fragments of these logics). Buszkowski(1986) proved that for the { $*, \supset$ }-fragment of FL or FL_e, partially ordered monoids can be also replaced by monoids. There may be some corelation between our semantics and Abrusci's, but at present it is not so clear. Our semantics is an extension of p.o. set semantics for the intuitionistic logic and of Urquhart semantics, which is popular among relevant logicians. Because of this it will be able to be applied to a variety of nonclassical logics. Details of this are given in Ono and Komori(1985). The connection of our semantics with Urquhart semantics can be seen in Bull(1987). However, there are some problems involved in extending our semantics to one for predicate logics, since usual Henkin construction, essential for proof of completeness, does not work in these cases. Some attempts in this direction have been made by Komori(1986) and the author(1985).

5. LAMBDA CALCULI AND IMPLICATIONAL FRAGMENTS OF EXTENSIONS OF FL

In the rest of this paper, we will mainly concern ourselves with implicational fragments of extensions of FL. Hereafter FL, FL_e, etc. will denote their implicational fragments.

By modifying the introduction rule of $\supset$ of the natural deduction system NJ of the intuitionistic logic, we can get natural deduction systems for FL_e and FL_{ew} as follows;

$$\frac{\begin{matrix}[A]\\ \vdots \\ B\end{matrix}}{A \supset B}$$

1) for FL_e, exactly one occurrence of A in assumptions is cancelled,

2) for FL_{ew}, at most one occurrence of A in assumptions is cancelled.

It is well-known that to each proof of NJ we can attach a lambda-term expressing the proof, and conversely that for each lambda-term or combinator X, a proof of type-assignment to X gives a proof of the type-scheme assigned to X in NJ (see Hindley and Seldin, 1986). Here we will identify a type-scheme with the implicational formula corresponding to it. These facts suggest the existence of a strong connection between proofs and lambda-terms (or combinators). We should recall here that (the implicational fragment) of FL_{ew} is usually called BCK-logic, whose name comes from combinators B, C and K. Actually, type-scheme assigned to these combinators constitute three axiom schemes for FL_{ew} in Hilbert-style formulation.

What kind of lambda-terms corresponds to proofs of FL_{ew} or FL_e, then? We will define here ew- and e- (lambda-) terms as follows;

1) each variable is an ew-term,
2) if M and N are ew-terms and the set of free variables in M is disjoint from that in N, then (MN) is an ew-term,
3) if M is an ew-term, then $\lambda x.M$ is also an ew-term.

Similarly, e-terms are defined, but we take the following 3') instead of 3).

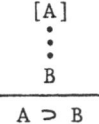

3') if M is an e-term and x is a free variable in M, then $\lambda x.M$ is also an e-term.

Theorem 5.1 For any implicational formula A, A is provable in FL_{ew} (and FL_e) if and only if there exists a closed ew-term (and a closed e-term, respectively) M such that $TA_\lambda \vdash M \in A$.

Theorem 5.2 1) Every ew-term is stratified.
2) Every ew-term has a β-normal form. Moreover, every reduction of it to a β-normal form terminates in linear time.

For details, see Hindley(1987), Komori(1987) and Lafont(1987), and for general information, see Hindley and Seldin(1986). To get back to similarities between proofs and lambda-terms, it can be assured that normal proofs in a natural deduction system correspond exactly to lambda-terms in normal form. By Church-Rosser theorem, two normal forms of a given lambda-terms are equal (modulo congruent). So when does a given provable formula have a unique normal proof? Komori(1987) conjectured that if a formula A is *minimal* in FL_{ew} it has a unique normal proof. Here, we say that a formula A is minimal in FL_{ew} if 1) it is provable in FL_{ew} and 2) if B is any substitution instance of A which is provable in FL_{ew} then conversely A is a substitution instance of B. Hirokawa(1987) solved it affirmatively by showing that if a type-scheme A which is minimal in FL_{ew} is assigned to two closed ew-terms they will be identical. On the other hand, this result can be obtained from a categry-theoretic result by Babaev and Solovjov(1979) (see also Minc, 1984), which says that if a formula is balanced (i.e, every variable occurs at most twice) it has at most one normal proof, since Jaśkowski(1963) proved that every minimal formula in FL_{ew} is balanced. (This proof is essentially due to Wroński.)

6. CATEGORIAL GRAMMAR

We will show some of the relationships between the study of logics without structural rules as described so far and the study of flexible categorial grammar. The following explanation owes much to van Benthem's papers (1986), (1987), (1987a) and (1987b). The study of categorial grammar was started by Ajdukiewicz and Bar-Hillel. The basic idea of the study is to clarify the relation between syntactic categories of expressions such as proper name, transitive verb etc., and semantic types. Here the set of types is defined recursively;

1) e and t are types,
2) if a and b are types then (a,b) is a type,

where e and t denote individual objects and truth values, respectively, and (a,b) denotes functions which associate a b-object to each a-object. Types of proper name and transitive verb are, for instance, e and (e,(e,t)), respectively.

There have been various attempts to incorporate the flexibility of types and to add some *correct* rules of type changing. For example, Geach proposed the following rule of type changing;

(a,b) derives ((c,a),(c,b))

to explain the *polymorphism* of the negation. The negation is usually considered to have the type (t,t), while the negation of predicates must have the type ((e,t),(e,t)). It was recognized then that Lambek(1958) had already introduced a calculus for this flexible categorial grammar which can produce correct rules for type changing. (Note that in the original system of Lambek, two directed slashes a \ b and a / b were used, instead of (a,b).)

To define a modal predicate logic with equality (m.p.l.=) just replace here MF by $MF^=$ and add the words "with equality" to (1).

A superintuitionistic predicate logic (s.p.l.) is a set $L \subseteq IF$ such that:

it contains all intuitionistic predicate tautologies; (6)

$A \in L \Rightarrow xA \in L$; (7)

$A, (A \rightarrow B) \in L \Rightarrow B \in L$; (8)

every IF = substitutional instance of a formula in L is also in L. (9)

Superintuitionistic predicate logics with equality (s.p.l.=) are defined in the obvious way.

The smallest m.p.l. (respectively, m.p.l.=, s.p.l., s.p.l.=) is denoted by QS4 (respectively, by QS4=, QH, QH=). Let $[L + \Gamma]$ denote the smallest m.p.l. containing an m.p.l. L and $\Gamma \subseteq MF$. This notation is obviously extended to other cases (m.p.l.=, s.p.l., s.p.l.=).

2. ALGEBRAIC SEMANTICS

Let $\Omega = \langle \Omega, \cap, \cup, \rightharpoonup, \emptyset, 1 \rangle$ be a complete Heyting algebra, $D \neq \phi$. Recall that an Ω-set with the domain D is a function $E: D \times D \rightarrow \Omega$ such that

$$\forall a,b \in D \quad E(a,b) = E(b,a); \tag{10}$$

$$\forall a,b,c \in D \quad (E(a,b) \cap E(b,c)) \leqslant E(a,c); \tag{11}$$

$$(\bigcup_{a \in D} E(a,a)) = 1 . \tag{12}$$

The triple $F = (\Omega,D,E)$ is then called a Heyting-valued structure (H.v.s.). For any $\Gamma \subseteq (MF^= \cup IF^=)$ consider the derived set Γ_D of D-valued formulas (they are obtained by replacing free variables in formulas from Γ by elements of D). For a D-valued formula, let $E(A(a_1,\dots,a_n) = (\bigcap_{i=1}^{n} E(a_i,a_i))$.

A valuation on $F = (\Omega,D,E)$ is a function $\phi: AF_D \rightarrow \Omega$ such that:

$$E(a_1,b_1) \cap \dots \cap E(a_n,b_n) \cap \phi(P(b_1,\dots,b_n)) \leqslant \phi(P(a_1,\dots,a_n)) \leqslant E(P(a_1,\dots,a_n)) \text{ for any } a_1,b_1,\dots,a_n,b_n \in D,\ P \in PRL_n. \tag{13}$$

It can be uniquely extended to $\phi_F: IF_D^= \rightarrow \Omega$ satisfying

$$\phi_F(a = b) = E(a,b); \tag{14}$$

$$\phi_F(\exists x A(x)) = \bigcup_{a \in D} \phi_F(A(a)); \tag{15}$$

$$\phi_F(\bot) = 0; \tag{16}$$

$$\phi_F(A \vee B) = E(A \vee B) \cap (\phi_F(A) \cup \phi_F(B)); \tag{17}$$

$$\phi_F(A \wedge B) = \phi_F(A) \cap \phi_F(B); \tag{18}$$

$$\phi_F(A \rightarrow B) = E(A \rightarrow B) \cap (\phi_F(A) \rightharpoonup \phi_F(B)); \tag{19}$$

$$\phi_F(\forall x A(x)) = E(\forall x A(x)) \cap \bigcap_{a \in D} (E(a,a) \to \phi_F(A(a))), \tag{20}$$

(cf. Dragalin, 1979; Goldblatt, 1979). A formula $A(x_1,\dots,x_n)$ is called valid in F (notation: $F \mid= A(x_1,\dots,x_n)$) iff $\phi_F(A(a_1,\dots,a_n)) = E(A(a_1,\dots,a_n))$ for any valuation ϕ on F and any $a_1,\dots,a_n \in D$. $IL^=(F) = \{A \in IF^= | F| = A\}$ is called the logic of F (with equality). We set: $IL(F) = IL^=(F) \cap IF$ (the logic without equality).

Now let $\mathcal{O}l = \langle Ol, \cap, \cup, -, I, 0, 1 \rangle$ be a complete topo-Boolean (= interior) algebra; its open elements constitute a complete Heyting algebra Ol_0. The triple $F = (Ol,D,E)$ is called a topo-Boolean valued structure (t.v.s.) iff (Ol_0,D,E) is an H.V.S. A valuation on F is a function ϕ: $AF_D \to Ol$ satisfying (13). It can be uniquely extended to ϕ_F: $MF_D \to Ol$ satisfying (14), (15), (16) and

$$\phi_F(A \supset B) = E(A \supset B) \cap (-\phi_F(A) \cup \phi_F(B)), \tag{21}$$

$$\phi_F(\Box A) = I\phi_F(A). \tag{22}$$

As in the intuitionistic case define: $F| = A(x_1,\dots,x_n)$ iff $\phi_F(A(a_1,\dots,a_n)) = E(A(a_1,\dots,a_n))$ for any valuation ϕ and for any $a_1,\dots,a_n \in D$; $ML^=(F) = \{A \in MF^= | F| = A\}$; $ML(F) = ML^=(F) \cup MF$.

An algebraic semantics (with equality) associated to a class Ω of t.v.s. (respectively, of H.v.s.) is defined as $S^=(\Omega) = \{ML^=(F) | F \in \Omega\}$ (respectively, $\{IL^=(F) | F \in \Omega\}$); $ML^=(\Omega) = \cap S^=(\Omega)$ is called the modal logic of Ω (being unprecise, we identify $S^=(\Omega)$ with Ω). If logics without equality are considered, definitions of $S(\Omega)$, $ML(\Omega)$ are given in the obvious way. By default, we consider only semantics for m.p.l.='s leaving other cases to the reader.

A semantics Ω gives rise to the logical consequence relation:

$$L \models A \text{ iff } \forall F \in \Omega\ (L \subseteq ML^=(F) \Rightarrow A \in ML^=(F)). \tag{23}$$

The Ω-completion of a logic L is $C_\Omega(L) = \{A \in MF^= | L| = A\}$. L is called Ω-complete iff $L = C_\Omega(L)$ or, equivalently, iff $L = ML^=(\Omega')$ for some $\Omega' \subseteq \Omega$. A semantics Ω is reducible to Ω' ($\Omega \leqslant \Omega'$) iff $C_{\Omega'}(L) \subseteq C_\Omega(L)$ for any m.p.l. = L. It can be easily seen that $\Omega \leqslant \Omega'$ iff every Ω-complete logic is Ω'-complete. Semantics Ω and Ω' are called equivalent ($\Omega \approx \Omega'$) iff $\Omega \leqslant \Omega'$ and $\Omega' \leqslant \Omega$. We write: $\Omega < \Omega'$ iff $\Omega \leqslant \Omega'$ but not $\Omega' \leqslant \Omega$.

Some of semantics we consider have the following collection property:

$$\forall \Omega' \subseteq \Omega \quad \exists F \in \Omega \quad ML^=(\Omega') = ML^=(F). \tag{CP}$$

It follows immediately that equivalent semantics satisfying (CP) are equal; although this fact can be disproved in the general case.

3. PRESHEAVES

Let Ol be a complete topo-Boolean algebra; its non-zero open elements constitute a category $Ol_\Box$ in which (u,v) is a unique morphism from u to v iff $u \geqslant v$. A presheaf over Ol is defined as a presheaf over $Ol_\Box$, i.e., as a functor $\mathcal{F}$: $Ol_\Box \to$ SETS. It means that $\mathcal{F}(u)$ is a set for any non-zero open u and $\mathcal{F}(u,v)$ is a function from $\mathcal{F}(u)$ to $\mathcal{F}(v)$ whenever $u \geqslant v$, such that

$$\mathcal{F}(u,u) = 1_{\mathcal{F}(u)}; \tag{24}$$

$$\mathcal{F}(v,w) \times \mathcal{F}(u,v) = \mathcal{F}(u,w) \text{ whenever } u \geqslant v \geqslant w.$$

$\mathcal{F}$ is called <u>inhabited</u> iff $\bigcup \{u \mid \mathcal{F}(u) \neq \phi\} = \mathbb{1}$.

Consider also the following <u>disjointness property</u>:

$$(u \neq v) \Rightarrow (\mathcal{F}(u) \cap \mathcal{F}(v) = \phi). \tag{DP}$$

Every presheaf over 01 is isomorphic to a presheaf satisfying (DP). We set: $\mathcal{F}^* = \bigcup_u \mathcal{F}(u)$. If $\mathcal{F}$ satisfies (DP) and $a \in \mathcal{F}^*$, the single u such that $a \in \mathcal{F}(u)$ is called the life-zone of a and denoted by $|a|$. Define the restriction of a to $u \leqslant |a|$ as $(a|u) = \mathcal{F}(|a|,u)(a)$.

<u>Proposition 1</u>

For an inhabited presheaf $\mathcal{F}$ satisfying (DP) set:

$$E(a,b) = \quad \{u|u \leqslant (|a| \quad |b|) \text{ and } (a|u) = (b|u)\}.$$

Then $F(\mathcal{F}) = (01,\mathcal{F}^*,E)$ is a t.v.s., and $F(\mathcal{F})$, $F(\mathcal{F}')$ are isomorphic whenever $\mathcal{F}$ and $\mathcal{F}'$ are.

Thus, we can define $ML^=(\mathcal{F})$, the <u>modal logic</u> (with equality) <u>of an inhabited presheaf</u> $\mathcal{F}$, as $ML^=(F(\mathcal{F}'))$ for some presheaf $\mathcal{F}'$ satisfying (DP) and isomorphic to $\mathcal{F}$.

<u>Proposition 2</u>

For any t.v.s. over 01 there exists a presheaf over 01 with the same modal logic.

Therefore, every algebraic semantics is generated by a class of presheaves.

4. SOME EXAMPLES OF ALGEBRAIC SEMANTICS

<u>Example 1</u>

The <u>algebraic semantics with constant domains</u> $\mathcal{CA}$ corresponds to the class of all constant presheaves or to the class of t.v.s. (01,D,E) such that for any $a,b \in D$, $E(a,b)$ is either $\mathbb{0}$ or $\mathbb{1}$.

<u>Example 2</u>

The <u>mono-presheaf semantics</u> $\mathcal{MA}$ is generated by the class of all presheaves $\mathcal{F}$ such that $\mathcal{F}(u,v)$ is always injective, or by the class of t.v.s (01,D,E) such that for any $a,b \in D$ $E(a,b)$ is either $\mathbb{0}$ or $(E(a,a) \cap E(b,b))$.

<u>Example 3</u>

A presheaf over a topological space X is a presheaf over 01^X, the topo-Boolean algebra of all subsets of X. Presheaves over topological spaces give rise to the general <u>topological semantics</u> $\mathcal{T}$. Constant presehaves and mono-prehseaves over topological spaces generate semantics $\mathcal{CT}$ and $\mathcal{MT}$.

Recall that a presheaf F over Ol is a sheaf iff:

$$([(u = (\bigcup_{i \in J} u_i)) \text{ and } (a,b \in F(u)) \text{ and } i \in J((a|u_i) = (b|u_i))] \Rightarrow (a = b)); \tag{26}$$

$$([(u = (\bigcup_{i \in J} u_i)) \text{ and } \forall i \in J(a_i \in F(u_i)) \text{ and } i,j \in J((a_i|(u_i \cap u_j)) = (a_j|(u_i \cap u_j)))] \Rightarrow \exists a \in F(u) \forall i \in J((a|u_i) = a_i)). \tag{27}$$

Proposition 3

For any presheaf F over a topological space X there exists s sheaf over X (namely, the canonical sheaf $\tilde{F}$, c.f., Godement, 1958) having the same modal logic. Therefore, every topological semantics is generated by a class of sheaves.

There is quite another way to obtain T. Consider a set $W \neq \phi$ whose elements are called possible worlds. A system of domains over W is a triple $G = (W,D,V)$, with a set of individuals $D \neq \phi$ and a function $V: W \to (P(D)\setminus\{\phi\})$. A valuation on G is $\xi \in \prod_{x \in W} 2^{MF_{V(x)}}$ (i.e., ξ is a family of functions $\xi(x): MF_{V(x)} \to 2$ indexed by $x \in W$) such that for any $x \in W$:

$$\xi(x)(\perp) = 0; \tag{28}$$

$$\xi(x)(A \supset B) = 0 \text{ iff } \xi(x)(A) = 1 \text{ and } \xi(x)(B) = 0; \tag{29}$$

$$\xi(x)(\exists y A(y)) = 1 \text{ iff } \xi(x)(A(a)) = 1 \text{ for some } a \in V(x). \tag{30}$$

In this case the pair $\Phi = (G,\xi)$ is called a model on G. A formula $A(x_1,\ldots,x_n)$ is said to be true in Φ ($\Phi \models A(x_1,\ldots,x_n)$) iff $\xi(x)(A(a_1,\ldots,a_n)) = 1$ for any $x \in W$, $a_1,\ldots,a_n \in V(x)$.

Now, having a presheaf F over a topological space W, consider the system of domains $G(F) = (W,F^+,V)$, in which $F^+ = \bigcup_{x \in W} V(x)$, $V(x) = \{\alpha_x | \alpha \in F^*, x \in |\alpha|\}$, α_x (the germ of α in x) being the class of α modulo the equivalence relation $\tilde{=}_x = \{(\alpha,\beta) \mid x \in E(\alpha,\beta)\}$ ($V(x)$ is also denoted by F_x and called the fiber in x, cf. Godement, 1958). A model $(G(F),\xi)$ is called fiberwise iff for any $x \in W$:

$$\xi(x)(a = b) = 1 \text{ iff } a = b; \tag{31}$$

$$\xi(x)(A(a\ ,\ldots,a\ ) = 1 \text{ iff there exist an open } U,\ldots, \text{ such that } x \in U,\ \forall i \leqslant n(a_i = (\alpha_i)_x) \text{ and } \forall y \in U(\xi(y)(A((\alpha_1)_y,\ldots,(\alpha_n)_y) = 1). \tag{32}$$

Proposition 4

A modal formula is valid in a presheaf F iff it is true in every fiberwise model on F.

Example 4

A pre-ordered set (W,R) gives rise to a topological space $\mathbf{T}(W,R)$ on W whose open sets are just all R-stable subsets ("right topology"). So we obtain Kripke spaces generating Kripke semantics K. Its "frames" (i.e., T.V.S. over topo-Boolean algebras of Kripke spaces) are called

Kripke frames with equality (KFE). A KFE $F = (\mathcal{O}l^{\mathbf{T}(W,R)}, D, E)$ (denoted by (W,R,D,E)) gives rise to the system of domains (W,D,V), in which $V(x) = \{a \in D \mid x \in E(a,a)\}$. A Kripke model on F is a pair (F,η), with a valuation η on (W,D,V) such that

$$\eta(x)(a = b) = 1 \text{ iff } x \in E(a,b); \tag{33}$$

$$\eta(x)(\Box A) = 1 \text{ iff } \forall y \in R(x)\eta(y)(A) = 1; \tag{34}$$

$$\eta(x)(P(a_1,\dots,a_n)) = 1 \text{ and } x \in \bigcap_{i=1}^{n} E(a_i,b_i) \Rightarrow \Rightarrow \eta(x)(P(b_1,\dots,b_n)) = 1. \tag{35}$$

It is well-known that a modal formula is valid in a KFE F iff it is true in all Kripke models on F.

5. KRIPKE-TYPE SEMANTICS

An interior onto-mapping between Kripke spaces $\pi\colon \mathbf{T}(D,\rho) \to \mathbf{T}(W,R)$ is called a Kripke bundle. This is in fact a p-morphism, i.e., a mapping $\pi\colon D \to W$ such that

$$(a\ \rho\ b) \Rightarrow (\pi(a) R \pi(b)); \tag{36}$$

$$(\pi(a) R x) \Rightarrow \exists b(a\ \rho\ b \text{ and } \pi(b) = x). \tag{37}$$

In this context elements of W and D are treated as worlds and individuals respectively, $\pi(a)$ as the "home" for a, R as an accessibility relation between worlds, and $a\ \rho\ b$ as "b interits a". A Kripke bundle satisfying

$$(\pi(a) = \pi(b) \text{ and } a\ \rho\ b) \Rightarrow (a = b) \tag{38}$$

is called a Kripke quasi-sheaf. A Kripke sheaf is a Kripke bundle such that

$$(\pi(a) R x) \Rightarrow \exists! b(a\ \rho\ b \text{ and } \pi(b) = x) \tag{39}$$

(in this case the single b is denoted by $(a|x)$).

A Kripke bundle π gives rise to the system of domains $G = (W,D,V)$, in which $V(x) = \pi^{-1}(x)$. A Kripke model over π is a model (G,ξ) satisfying (28) - (31) and

$$\text{for any different } a_1,\dots,a_n \ D\colon \xi(x)(\Box A(a_1,\dots,a_n)) = 1 \text{ iff } \forall y \in R(x) \forall b_1,\dots,b_n \in V(y)((\mathop{\&}_{i=1}^{n} (a_i\ \rho\ b_1)) \Rightarrow \Rightarrow \xi(y)(A(b_1,\dots,b_n)) = 1). \tag{40}$$

(Note that $\xi(x)(\Box(a = a)) = 1$ because of the definition.)

A modal formula A is called valid in π (notation: $\pi \models A$) iff it is true in all Kripke models on π. A is called strongly valid in π ($\pi \models\!\!\!\equiv A$) iff all its substitutional instances are valid in π. An intuitionistic formula is called (strongly) valid iff its Tarski translation is.

Proposition 5

1) $ML^=(\pi) = \{A \in MF^= | \pi \models A\}$ is a m.p.l. = (called the modal logic of π).

2) $\pi \models A$ iff $\pi \models A$ for any Kripke sheaf π and $A \in MF^=$.

Remark. The condition 2) can fail in the general case; e.g., there exists a Kripke quasi-sheaf π such that $\pi \models (\Diamond\Box p \supset \Box\Diamond p)$, but $\pi \not\models$ $\not\models (\Diamond\Box P(x) \supset \Box\Diamond P(x))$.

So we introduce two new semantics:

$KB = \{ML^=(\pi) | \pi$ is a Kripke bundle$\}$,

$KQ = \{ML^=(\pi) | \pi$ is a Kripke quasi-sheaf$\}$.

Proposition 6

$K = \{ML^=(\pi) | \pi$ is a Kripke sheaf$\}$.

Consider also two properties of a KFE (W,R,D,E):

$$\forall a,b \in D((a = b) \vee (E(a,b) = \phi)); \quad (41)$$

$$\forall a,b \in D((E(a,b) = \phi \vee (E(A,b) = E(a,a) \cap E(b,b))). \quad (42)$$

Frames satisfying (41) (respectively, (42)) are called strictly monic (respectively, monic), and they generate semantics PRF (respectively, MK).

6. BIBLIOGRAPHICAL REMARKS

Algebraic semantics for modal and intuitionistic first-order logics (but only with constant domains) is considered in Rasiowa and Sikorski, 1963. Algebraic semantics for intuitionistic first-order logic with variable domains is introduced in Dragalin, 1979; Fourman and Scott, 1979; Goldblatt, 1979 (there have also been earlier publications on this topic). However, in Fourman and Scott, 1979 and Goldblatt, 1979, the existence predicate E is included in the language.

Equivalence of categories of presheaves over Heyting algebras and of Heyting-valued sets was proved in Higgs, 1973 (c.f., also Goldblatt, 1979). The t.v.s. $F(F)$, defined above, can be obtained from the corresponding functor (denoted by $\mathbf{A}_F$ in Goldblatt, 1979).

The definition (28) - (32) of "forcing-at-a-point" (i.e., of a fiberwise model) corresponding to a presheaf over a topological space seems to be new; the well-known notion of Kripke-Joyal forcing (c.f., Reyes, 1978; Goldblatt, 1979) is quite different since here "worlds" are open sets.

There exists a notion rather close to that of a fiberwise model, namely, a Montague model of a pragmatic language (Montague, 1970). But his language also contains the existence predicate, and quantifiers range over all individuals, so there is no need in germs.

The original Kripke semantics for intuitionistic logic (Kripke, 1965) is equivalent to PKF. PKF for modal logics is introduced in Schütte, 1968. KFE's for intuitionistic logic are described in Thomason, 1968.

Kripke sheaves appeared in Ellerman, 1974, as functors from poset categories to SETS. A connection between KFE's and Kripke sheaves and hints to a proof of our Proposition 6 can be found in Goldblatt, 1979.

Kripke-type semantics in which domains of different worlds are a priori independent are considered, e.g., in Kripke, 1963; Fine, 1978. However, corresponding predicate logics differ from ours; for example, the formula $(\exists x\Diamond P(x)\supset\Diamond\exists xP(x))$ is contained in QS4 but can be non valid in these semantics.

7. DIAGRAM OF SEMANTICS

Here are semantics we study:

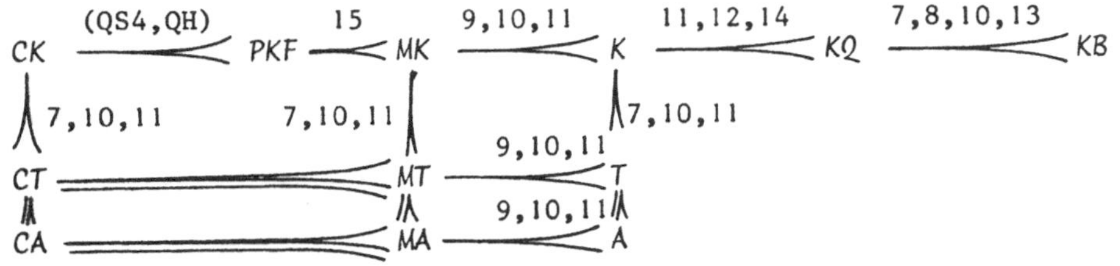

Clue. *A* is generated by all t.v.s. (or H.v.s.), and *CK* is the constant domain Kripke semantics. Numbers refer to propositions proving "<". In fact four diagrams should be drawn here according to four types of logics we consider. We do not know if $\leqslant$ can be replaced by < here, and if all the relations between semantics are displayed (e.g., is it true that $KQ \leqslant A$? or $PKF \leqslant CA$?). Note also that $KB \leqslant CJ$ is not true both in modal and intuitionistic cases, and $CJ < CA$, $MJ < MA$, $T < A$ are true in the modal case; these facts follow from incompleteness results for propositional logics (Gerson, 1975; Shehtman, 1980), but corresponding counter-examples are rather complicated.

8. LIST OF FORMULAS

Modal Formulas

Ba. $\Diamond\exists xP(x)\supset\exists x\Diamond P(x)$ (Barcan formula)

Ba_0. $\Diamond\exists x\Box P(x)\supset\exists x\Diamond P(x)$

Ba_1. $\Diamond\exists xP(x)\supset\exists x\Diamond\exists y(P(y)\wedge\Diamond(x=y))$

Ba*. $\Box\Diamond\exists xP(x)\supset\Diamond\Box\exists xP(x)$

CE. $\Diamond(x=y)\supset(x=y)$

Grz. $\Box(\Box(p\supset\Box p)\supset p)\supset p$ (Grzegorczyk formula)

M. $\Box\Diamond p\supset\Diamond\Box p$

N. $\Diamond(p_1\wedge\Box q_1)\wedge\Diamond(p_2\wedge\Box q_2)\wedge\Box(q_1\wedge q_2\supset\exists xR(x))\supset$
$\supset\exists x(\Diamond(p_1\wedge\Diamond R(x))\wedge\Diamond(p_2\wedge\Diamond R(x))$

TD. $\Box\forall x(\Box P(x)\vee\Box q)\supset\Box\forall x\Box P(x)\vee\Box q$

TJ. $\Diamond\Box p\supset\Box\Diamond p$

Y. $\neg\exists x(P(x)\wedge\Box\forall x(P(x)\supset\Diamond\neg P(x)\wedge\exists xP(x)))$

Intuitionistic formulas

CD. $\forall x(P(x) \vee q) \to \forall xP(x) \vee q$

DE. $\forall x\forall y((x = y) \vee \neg(x = y))$

E. $\neg\neg \exists xP(x) \to \exists x\neg\neg P(x)$

F. $\forall x\neg\neg P(x) \to \neg\neg \forall xP(x)$

G. $\exists x\forall y(P(y) \to P(x))$

IN. $(\neg\forall xR(x)) \wedge \forall x((p_1 \to q_1 \vee R(x)) \vee (p_2 \to q_2 \vee R(x)) \to$
$\to (p_1 \to q_1) \vee (p_2 \to q_2)$

J. $\neg q \vee \neg\neg q$

U. $\forall x\forall y(P(x) \to P(y))$

U_n. $\exists x_1,\dots,\exists x_n \bigwedge_{1 \leqslant i < j \leqslant n} \neg(x_i = x_j)$

V_0. $\forall x(P(x) \vee U_3) \to \forall xP(x) \vee U_2$

V_1. $(p \to U_3) \vee (\neg p \wedge \neg U_3 \to q \vee \neg q)$

V_2. $U_2 \vee \neg U_3 \vee (U_2 \wedge \neg U_3 \to q \vee \neg q)$

Z. $(p \to q) \vee (q \to p)$

9. INCOMPLETENESS

It is well-known that $QS4^=$, and $QH^=$ are *K*-complete but *MA*-incomplete, and QS4, QH are *PKF*- and *CT*-complete but *CK*-incomplete (Rasiowa and Sikorski, 1963; Dragalin, 1979); to prove incompleteness use formulas CE, DE, Ba, CD.

Proposition 7

Set LTD = [QS4 + TD]. Then $Ba \in C_{KQ}(LTD)\setminus(C_{CT}(LTD) \cup C_{KB}(LTD))$. Note that TD is the Tarski translation of CD, and [QH + CD] is known to be Kripke-complete (Görnemann, 1971)!

Proposition 8

1) Let L be a modal propositional logic which contains Grz but does not contain $(p \supset \Box p)$, QL = [QS4 + L]. Then $Y \in C_{KQ}(QL)\setminus C_{KB}(QL)$.

2) Set QGB = [QS4 + Grz + Ba]. Then $Y \in C_{KQ}(QGB)\setminus C_{KB}(QGB)$.

We do not know any completeness results for these logics.

Remark. Cf., Montagna, 1984: QGL = $[QK + (\Box(\Box p \supset p) \supset \Box p)]$ is *K*-incomplete. Cf., also Ono, 1983: [QH + L] is *K*-incomplete for any non-classical propositional intermediate logic L of a finite slice; on the other hand, every logic $(QH + LP_n + CD)$ is *CK*-complete (LP_n is the least logic of n-th slice).

Proposition 9

Set QHE = [QH + E]. Then $(U \vee J) \in C_{MA}(QHE)\setminus C_K(QHE)$.

Remark. In Ono, 1973, it is proved that $CD \in C_{PKF}(QHE)\setminus C_{CT}(QHE)$. In fact $CD \notin C_K(QHE)$ as well.

Proposition 10

Set QHG = [QH + G]. Then $CD \in C_{KQ}(QHG)\backslash(C_{CT}(QHG) \cup C_{KB}(QHG))$, $(U \vee Z) \in C_{MA}(QHG)\backslash C_K(QHG)$, $(CD \vee Z) \in (C_{MA}(QHG) \cap C_{KQ}(QHG))\backslash$ $\backslash(C_T(QHG) \cup C_{KB}(QHG))$.

Proposition 11

Set QB_i = [QS4 + B_i], i = 0,1. Then $Ba \in C_{MA}(QB_1)\backslash C_K(QB_1)$; $Ba_1 \in C_K(QB_0)\backslash(C_{KQ}(QB_0) \cup C_{CT}(QB_0))$.

Proposition 12

Set Q2B = [QS4 + TJ + Ba]. Then $N \in C_K(Q2B)\backslash C_{KQ}(Q2B)$.

Proposition 13

Set Q1.2B = [Q2B + M]. Then $Ba^* \in C_{KQ}(Q1.2B)\backslash C_{KB}(Q1.2B)$.

Proposition 14

Set QJD = [QH + J + CD]. Then $IN \in C_K(QJD)\backslash C_{KQ}(QJD)$.

Remark. S. Ghilardi and G. Corsi proved that QJ = [QH + J] is *PKT*-complete. The proof can be transferred also to Q2 = [QS4 + TJ]. Logics QJDF = [QJD + F] and Q1.2B* = [Q2B + Ba*] are *CK*-complete.

The last proposition also provides a counter-example, although it is not too simple:

Proposition 15

Set QHV = [QH + V_0 + V_1]. Then $V_2 \in C_{PKF}(QHV)\backslash C_{MK}(QHV)$.

REFERENCES

van Benthem, J. F. A. K., 1978, Two simple incomplete logics, Theoria, 44:No. 1, 25-37.
Dragalin, A. G., 1979, "Mathematical Intuitionism. Introduction to Proof Theory", Nauka, Moscow (in Russian).
Ellerman, D., 1974, Sheaves of structures and generalized ultraproducts, Ann. Math. Log., 7:No. 2-3, 163-195.
Fine, K., 1974, An incomplete logic containing S4, Theoria, 40:No. 1, 23-29.
Fine, K., 1978, Model theory for modal logics, I: De re/de dicto distinction, J. Phil. Log., 7:125-156.
Fourman, M.P., and Scott, D.S., 1979, Sheaves and logic, in: "Lect. Notes Math.", 753:302-401.
Gerson, M. S., The inadequacy of neighborhood semantics for modal logics, J. Symb. Logic., 40:No. 2, 141-147.
Godement, R., 1958, "Topologie Algebrique et Theorie de Faisceaux", Hermann, Paris.
Goldblatt, R., 1979, "Topoi", Studies in Logic and Found. Math., v.98.
Görnemann, S., 1971, A logic stronger than intuitionism, J. Symb. Logic, 36:No. 2, 249-261.
Higgs, D., 1973, "A Category Approach to Boolean-Valued Set Theory", Lect. Notes, Univ. of Waterloo.
Kripke, S. A., 1963, Semantical considerations on modal logic, Acta Philos. Fennica, 16:83-94.

Kripke, S. A., 1965, Semantical analysis of intuitionistic logic, in: "Formal Systems and Recursive Functions", J. N. Crossley and M. A. E. Dummett, eds., North-Holland, pp 92-130.

Montagna, F., 1984, The predicate modal logic of provability, Notre Dame J. of Form. Log., 25:No. 2, 179-189.

Montague, R., 1970, Pragmatics and intensional logic, Synthese, 22:No. 112, 68-94.

Ono, H., 1973, Incompleteness of semantics for intermediate predicate logics. I: Kripke's semantics, Proc. Jap. Acad., 49:No. 9, 711-713.

Ono, H., 1983, Model extension theorem and Craig's interpolation theorem for intermediate predicate logics, Reports in Math. Logic, 15:41-58.

Rasiowa, H., and Sikorski, R., 1963, "The Mathematics of Metamathematics", Warsaw.

Reyes, G., 1978, Théorie des modeles et faisceaux, Adv. Math., 30:No. 2, 156-170.

Schütte, K., 1968, Vollständige Systeme modaler und intuitionistischer Logik, Ergebn. Math., v. 42.

Shehtman, V. B., 1977, On incomplete propositional logics, Dokl. AN SSSR., 235:No. 3, 542-545 (in Russian).

Shehtman, V. B., 1980, Topological models of propositional logics, Semiotika i informatika, No. 15:74-98 (in Russian).

Thomason, R. H., 1968, On the strong semantical completeness of the intuitionistic predicate calculus, J. Symb. Logic, 33:1-7.

Thomason, S. K., 1972, Semantic analysis of tense logic, J. Symb. Logic, 37: No. 1, 150-158.

Thomason, S. K., 1974, An incompleteness theorem in modal logic, Theoria, 40:No. 1, 30-34.

ON THE COMPUTATIONAL POWER OF THE LOGIC PROGRAMS*)

Ivan N. Soskov

Laboratory for Applied Logic at Sofia University
blvd. Anton Ivanov 5
Sofia 1126, Bulgaria

0. Introduction

The aim of the present paper is a comparative study of the computational power of the logic programs and search computability [1].

It is a well known fact that in the standard structure of the natural numbers the partial recursive functions are the computable by means of logic programs. More precisely, the sets of natural numbers, definable by means of logic programs coincide with the recursively enumerable ones. Here the notion of definability by means of logic programs or more general definability by means of arbitrary first order formulae is used in the sense of Fraisse [2]. The exact definitions are formulated in section 1.

In the paper definability by means of logic programs in arbitrary first order partial structures is considered. Two results will be shown. First, it will be proved that in every partial structure with a denumerable domain definability by means of logic programs is equivalent to ∀-weak admissibility [3], and hence, that the computational power of the logic programs is greater than this of search computability.

The second result is that definability by means of arbitrary recursively enumerable sets of first order formulae is uniformly equivalent to definability by means of logic programs.

This result shows that the recent attempts to capture more features of the whole predicate calculus in logic programming will hardly increase the expressive power of the language to a considerable degree.

This work is not the first to dwell on this topic. I shall mention two

*) Research partially supported by the Ministry of Culture, Science and Education, Contract # 933, 1988

Mathematical Logic
Edited by P. P. Petkov
Plenum Press, New York, 1990

other works which have influenced the investigations presented here. In his paper [4] Moschovakis compares search computability and Fraisse computability and proves that in partial structures with equality, both concepts are equivalent. In his book [5], Fitting proves the equivalence between search computability and R - definability in the case of structures with equality and effective pairing function. The notion of R-definabilityis very close to the notion of definability by means of logic programs.

I am finishing the introduction with some words concerning the method used in the proofs. This method is based on an external characterization of the Ranges of the search computable functions, based on the notion $\forall$-weak admissibility [3]. Because of the method only structures with denumerable domains will be considered. Recent investigations show that the proposed method can be modified for structures with arbitrary domains but I shall not consider this case here.

Speaking about the method, it is worth mentioning Lacombe's paper [6], where the equivalence between Fraisse computability and Lacombe's notion of $\forall$-recursiveness is proved once again for partial structures with equality.

1. Preliminaries

Let $\mathfrak{A} = (B; \theta_1, \ldots, \theta_n; \Sigma_1, \ldots, \Sigma_k)$ be a partial structure, where B - the domain of $\mathfrak{A}$ - is a denumerable set, $\theta_1, \ldots, \theta_n$ are partial functions of many arguments on B and $\Sigma_1, \ldots, \Sigma_k$ are partial predicates of many arguments on B.

In the first part of the paper - up till section 6 - we shall assume that the structure $\mathfrak{A}$ is fixed.

Moreover, we shall assume that the partial predicates $\Sigma_1, \ldots, \Sigma_k$ obtain only the value "true" whenever are defined.

This assumption is made for the following two reasons. First, it is not restrictive with respect to search computability. On the other hand, the logic programs can not use the negative part of the basic predicates of the structure because of their syntax.

In addition we shall assume that $k \geq 1$ and that Σ_1 is the unary totally defined predicate $\lambda s.$"true". By means of the predicate Σ_1 we can consider the domains of the basic functions $\theta_1, \ldots, \theta_n$ as predefined predicates.

The relational type of $\mathfrak{A}$ is the ordered pair $\langle\langle a_1, a_2, \ldots, a_n\rangle, \langle b_1, b_2, \ldots, b_k\rangle\rangle$ where each θ_i is a_i-ary and each Σ_j is b_j-ary.

We shall use the following notations. The letters s, t, p will denote arbitrary elements of B; x, y, z, u, v -- elements of the set of all natural numbers - N. We shall identify the partial predicates with the partial mappings which obtain values in {0, 1}, writing 0 for true and 1 for false.

Let $\mathcal{L}$ be the first order language corresponding to the structure $\mathfrak{A}$ i.e. $\mathcal{L}$ consists of n functional symbols $f_1, \ldots, f_n$ and k predicate symbols $T_1, T_2, \ldots, T_k$ where each f_i is a_i-ary and each T_j is b_j-ary.

Let $\{\underline{X}_1, \underline{X}_2, \ldots\}$ be a denumerable set of variables. We shall use the capital letters X, Y, Z to denote variables.

If τ is a term in the language $\mathcal{L}$, then we shall write $\tau(X_1, X_2, \ldots, X_a)$ to denote that all of the variables in τ are among $X_1, X_2, \ldots, X_a$. If $\tau(X_1, X_2, \ldots, X_a)$ is a term, $s_1, \ldots, s_a$ are arbitrary elements of B, then with $\tau_{\mathfrak{A}}(X_1/s_1, \ldots, X_a/s_a)$ we shall denote the value, if it exists, of the term τ in the structure $\mathfrak{A}$ over the elements $s_1, \ldots, s_a$.

Analogously, if $\Pi(X_1, \ldots, X_a)$ is an atomic predicate (atom) in $\mathcal{L}$ with variables among $X_1, \ldots, X_a$ then with $\Pi_{\mathfrak{A}}(X_1/s_1, \ldots, X_a/s_a)$ we shall denote the value of Π over the elements $s_1, \ldots, s_a$ in $\mathfrak{A}$.

Suppose that G is a finite set of atoms in $\mathcal{L}$. Let $X_1, \ldots, X_a$ be all variables which occur in the elements of G and $s_1, \ldots, s_a$ are arbitrary elements of B. Then, with $G_{\mathfrak{A}}(X_1/s_1, \ldots, X_a/s_a)$ we shall denote the value of the conjunction of the elements of G in $\mathfrak{A}$. In other words, $G_{\mathfrak{A}}(X_1/s_1, \ldots, X_a/s_a) \cong 0$ iff for each element Π of G, $\Pi_{\mathfrak{A}}(X_1/s_1, \ldots, X_a/s_a) \cong 0$. A formal constant c_s is introduced for every element s of B .

Denote with C the set $\{c_s : s \in B\}$ and with $\mathcal{L}_C$ the language $\mathcal{L} \cup C$. Let $\mathcal{T}$ denote the set of all ground terms, i. e. terms without variables, in $\mathcal{L}_C$ and $\mathfrak{A}^*$ denote the enrichment of $\mathfrak{A}$ to the language $\mathcal{L}_C$, where each c_s is interpreted as the respective s.

The <u>diagram</u> $\partial(\mathfrak{A})$ of the structure $\mathfrak{A}$ is defined by

$$\partial(\mathfrak{A}) = \{T_j(\tau^1, \ldots, \tau^{b_j}) : 1 \le j \le k \ \&\ \tau^1, \ldots, \tau^{b_j} \in \mathcal{T} \ \&$$
$$\Sigma_j(\tau^1_{\mathfrak{A}^*}, \ldots, \tau^{b_j}_{\mathfrak{A}^*}) \cong 0\}.$$

Let F be an arbitrary first order formula and H be an a-ary predicate symbol which does not belong to $\{T_1, \ldots, T_k\}$.

A subset A of B^a is said to be <u>definable</u> in $\mathfrak{A}$ by means of F and H iff the following equivalence is true:

$(s_1,\ldots, s_a) \in A \leftrightarrow \exists\tau^1 \ldots \exists\tau^a((\tau^i \in \mathcal{T} \;\&\; \tau^i \underset{\mathfrak{A}^*}{\cong} s_i),\ i = 1,\ldots, a \;\&$

$\partial(\mathfrak{A}) \cup \{F\} \vdash H(\tau^1,\ldots, \tau^a))$.

Here the sign "$\vdash$" means deducibility in the sense of the first order predicate calculus.

A subset A of B^a is called logically definable in $\mathfrak{A}$ if it is definable bymeans of some first order formula F and some predicate symbol H.

The notion of definability by means of logic programs is a particular case of the notion above, where the formula F is a logic program.

Logic programs are called formulae of the form $F^1\&\ldots\&F^l$, where each F^i is an universal closure of Horn clause, i. e. F^i is formula of the form $\forall X_1\ldots\forall X_r(\Pi \vee \neg\Pi_1 \vee \cdots \vee \neg\Pi_n)$, where $n \geq 0$, and $\Pi, \Pi_1,\ldots, \Pi_n$ are atomic predicates.

We shall use for such sentences the usual notation $\Pi \leftarrow \Pi_1,\ldots, \Pi_n$, omitting the quantifires.

In next section we introduce the notion of $\forall$-weak admissibility. The equivalence between $\forall$-weak admissibility and definability by means of logic programs will be proved in section 4. This equivalence together with the results from [3] allow us to establish in section 5 the relationships between logically definability, definability by means of logic programs and search computability.

From now on we shall consider only subsets of B, i. e. the case a = 1. The proofs in the general case can be made in the same way.

2. Weak Admissibility

Let $\mathfrak{B} = (N; \varphi_1, \varphi_2, \ldots, \varphi_n; \sigma_1, \sigma_2,\ldots,\sigma_k)$ be a partial structure over the set of all natural numbers. A subset W of N is said to be partial recursive (p.r.) in $\mathfrak{B}$ iff $W = \Gamma(\varphi_1, \varphi_2,\ldots, \varphi_n, \sigma_1, \sigma_2,\ldots,\sigma_k)$ for some enumeration operator Γ, see Rogers, [7].

Enumeration of the structure $\mathfrak{A}$ is called every ordered pair $\langle\alpha, \mathfrak{B}\rangle$, where $\mathfrak{B} = (N; \varphi_1, \varphi_2, \ldots, \varphi_n; \sigma_1, \sigma_2, \ldots,\sigma_k)$ is a partial structure of the same relational type as $\mathfrak{A}$ and α is a partial surjective mapping of N onto B such that the following conditions hold:

(i) The domain of α ($Dom(\alpha)$) is partial recursive in $\mathfrak{B}$ and closed with respect to the partial operations $\varphi_1, \varphi_2, \ldots, \varphi_n$.

(ii) $\alpha(\varphi_i(x_1, \ldots, x_{a_i})) \simeq \theta_i(\alpha(x_1), \ldots, \alpha(x_{a_i}))$ for all natural numbers $x_1, \ldots, x_{a_i}$, $1 \leq i \leq n$;

(iii) $\sigma_j(x_1, \ldots, x_{b_j}) \simeq \Sigma_j(\alpha(x_1), \ldots, \alpha(x_{b_j}))$ for all natural numbers $x_1, \ldots, x_{b_j}$, $1 \leq j \leq k$.

Let $A \subseteq B$ and $\langle\alpha, \mathfrak{B}\rangle$ be an enumeration of $\mathfrak{A}$. The set A is said to be weak admissible in $\langle\alpha, \mathfrak{B}\rangle$ iff for some p.r. in $\mathfrak{B}$ subset W of N,
$s \in A \leftrightarrow \exists y(y \in W \ \&\ \alpha(y) \cong s)$.

A set A is said to be $\forall$-weak admissible in $\mathfrak{A}$ iff it is admissible in every enumeration $\langle\alpha, \mathfrak{B}\rangle$ of $\mathfrak{A}$.

Proposition 1. Every logically definable in $\mathfrak{A}$ subset A of B is $\forall$-weak admissible in $\mathfrak{A}$.

Proof. Let $A \subseteq B$ and let A be logically definable in $\mathfrak{A}$ by means of the formula F and the predicate symbol H. In other words,
$s \in A \leftrightarrow \exists\tau(\ \tau \in \mathcal{T} \ \&\ \tau_{\mathfrak{A}^*} \cong s \ \&\ \mathcal{D}(\mathfrak{A}) \cup \{F\} \vdash H(\tau))$.

Suppose that $\langle\alpha, \mathfrak{B} = (N; \varphi_1, \ldots, \varphi_n; \sigma_1, \ldots, \sigma_k)\rangle$ is an enumeration of $\mathfrak{A}$. We shall define a p.r. in $\mathfrak{B}$ subset W of N such that $\alpha(W) = A$.

For every natural x denote with $x^\sim$ a constant symbol - name for x. We shall assume that $\mathcal{L}_C \cap \{x^\sim : x \in N\} = \emptyset$ and that there is an effective way to find for a given x the constant $x^\sim$ and for a given $x^\sim$ the respective natural number x.

Let $K = \{x^\sim : x \in Dom(\alpha)\}$. It is clear that K is p.r. in $\mathfrak{B}$.

Let $\mathcal{L}_K = \mathcal{L} \cup K$ and let $\mathfrak{B}^*$ be the enrichment of $\mathfrak{B}$ to the language $\mathcal{L}_K$, where each $x^\sim$ is interpreted as x.

Denote with $\mathcal{T}_K$ the set of all ground terms of the language $\mathcal{L}_K$.

For every element τ of $\mathcal{T}_K$, define the term (τ) of $\mathcal{T}$ by the inductive clauses:

(i) if $\tau = x^\sim$ for some $x^\sim \in K$ then $(\tau) = c_{\alpha(x)}$, i.e. the name of $\alpha(x)$ in C;

(ii) if $\tau = f_i(\tau^1, \ldots, \tau^{a_i})$ then $(\tau) = f_i((\tau^1), \ldots, (\tau^{a_i}))$.

Lemma. For every element τ of $\mathcal{T}_K$, the following assertions are true:

(i) $\tau_{\mathfrak{B}^*}$ is defined $\leftrightarrow (\tau)_{\mathfrak{A}^*}$ is defined;

(ii) $\alpha(\tau_{\mathfrak{B}^*}) \cong (\tau)_{\mathfrak{A}^*}$.

Let $\partial(\mathfrak{B}) = \{T_j(\tau^1,\ldots,\tau^{b_j}) : 1 \le j \le k \ \&\ \tau^1,\ldots,\tau^{b_j} \in \mathcal{T}_K \ \&$

$T_j((\tau^1),\ldots,(\tau^{b_j})) \in \partial(\mathfrak{A})\}$.

The set $\partial(\mathfrak{B})$ is p.r. in $\mathfrak{B}$. Indeed, let $1 \le j \le k$ and let $\tau^1,\ldots,\tau^{b_j}$ be elements of $\mathcal{T}_K$. Then,

$$T_j(\tau^1,\ldots,\tau^{b_j}) \in \partial(\mathfrak{B}) \leftrightarrow T_j((\tau^1),\ldots,(\tau^{b_j})) \in \partial(\mathfrak{A}) \leftrightarrow$$

$$\Sigma_j((\tau^1)_{\mathfrak{A}^*},\ldots,(\tau^{b_j})_{\mathfrak{A}^*}) \cong 0 \leftrightarrow \Sigma_j(\alpha(\tau^1_{\mathfrak{B}^*}),\ldots,\alpha(\tau^{b_j}_{\mathfrak{B}^*})) \cong 0 \leftrightarrow$$

$$\sigma_j(\tau^1_{\mathfrak{B}^*},\ldots,\tau^{b_j}_{\mathfrak{B}^*}) \cong 0.$$

Let $c_{s_1},\ldots,c_{s_r}$ be the elements of C which occur in F. Let $m_1,\ldots,m_r$ be elements of Dom(α) such that $\alpha(m_i) \cong s_i$, $i = 1,\ldots,r$. Replace every occurrence of c_{s_i} in F by $m_i^\sim$ and denote the obtained formula by $F^\sim$. Define the subset W of N by the equivalence

$x \in W \leftrightarrow \exists\tau(\tau \in \mathcal{T}_K \ \&\ \tau_{\mathfrak{B}^*} \cong x \ \&\ \partial(\mathfrak{B}) \cup \{F^\sim\} \vdash H(\tau))$. It is clear that Wis p.r. in $\mathfrak{B}$.

It remains to prove that $\alpha(W) = A$. This fact follows easy from the definitions of W and A and some simple properties of the first order predicate calculus, namely the reduction theorem and the theorem on constants [8].

3. Normal Enumerations

Our next goal is to prove that every $\forall$-weak admissible in $\mathfrak{A}$ set is definable in $\mathfrak{A}$ by means of logic programs. As in [3], we shall use in the proof a special kind of enumerations - the normal enumerations. In this section the definition of the normal enumerations is given and some properties are proved. The omitted proofs can be found in [3].

For the sake of simplicity from now on we shall assume that all basic functions $\theta_1, \ldots, \theta_n$ and all basic predicates $\Sigma_1, \ldots, \Sigma_k$ of the structure $\mathfrak{A}$ are unary.

Let $\langle,\rangle$ be an effective coding of the ordered pairs of natural numbers. Denote by φ_i^*, $1 \le i \le n$, the unary recursive function $\lambda x.\langle i,x\rangle$. Let N^0 be the set $N \setminus (\mathrm{Range}(\varphi_1^*) \cup \ldots \cup \mathrm{Range}(\varphi_n^*))$ and α^0 be a surjective mapping of N^0 onto B.

The partial mapping α of N onto B is defined by the inductive clauses:

If $x \in N^0$, then $\alpha(x) \cong \alpha^0(x)$;

If $x = \langle i,y\rangle$, $\alpha(y) \cong s$ and $\theta_i(s) \cong t$, then $\alpha(x) \cong t$.

Proposition 2. For every natural number x and every i, $1 \leq i \leq n$, $\alpha(\langle i,x\rangle) \cong \theta_i(\alpha(x))$.

Let $D_1,\ldots, D_n$ be unary partial predicates in N such that

$$D_i(x) \cong \begin{cases} 0, & \text{if } \theta_i(\alpha(x)) \text{ is defined,} \\ \text{undefined,} & \text{otherwise;} \end{cases}$$

and let $\varphi_1,\ldots, \varphi_n$ be unary partial functions such that

$$\varphi_i(x) \cong \begin{cases} \varphi_i^*(x), & \text{if } D_i(x) \cong 0, \\ \text{undefined,} & \text{otherwise.} \end{cases}$$

It is clear that every φ_i is partial recursive in $\{D_i\}$ and every D_i is partial recursive in $\{\varphi_i\}$, $i = 1,\ldots, n$.

Define the partial predicates $\sigma_1,\ldots, \sigma_k$ on N, using the equalities $\sigma_j(x) \cong \Sigma_j(\alpha(x))$, $j = 1,\ldots, k$.

Let $\mathfrak{B}$ denote the partial structure $(N;\ \varphi_1, \varphi_2, \ldots, \varphi_n;\ \sigma_1, \sigma_2,\ldots,\sigma_k)$.

Proposition 3. The pair $\langle\alpha, \mathfrak{B}\rangle$ is an enumeration of $\mathfrak{A}$.

Every enumeration $\langle\alpha, \mathfrak{B}\rangle$ which is obtained by the method,described above, is called normal enumeration. The mapping α^0 is called basis of the enumeration $\langle\alpha, \mathfrak{B}\rangle$. It is clear that α^0 completely determines the enumeration $\langle\alpha, \mathfrak{B}\rangle$.

Let $\langle\alpha, \mathfrak{B}\rangle$ be a normal enumeration. Denote with $R_{\mathfrak{B}}$ the subset of N which consists of codes of ordered pairs and $\langle j, x\rangle \in R_{\mathfrak{B}} \Longleftrightarrow (1 \leq j \leq k\ \&\ \sigma_j(x) \cong 0)$ or $(k+1 \leq j \leq k+n\ \&\ D_{j-k}(x) \cong 0)$. It is clear that for every $W \subseteq N$, W is p.r. in $\mathfrak{B}$ iff W is p.r. in $R_{\mathfrak{B}}$.

Every partial mapping with a finite domain we shall call finite. If α_1 and α_2 are partial mappings of N into B then by $\alpha_1 \subseteq \alpha_2$ we shall denote that $Dom(\alpha_1) \subseteq Dom(\alpha_2)$ and for every x in $Dom(\alpha_1)$, $\alpha_1(x) \cong \alpha_2(x)$.

Let α' be a finite mapping of N^0 into B and let Γ be an enumeration operator. A subset A of B is called sufficient for α' and Γ iff the following two conditions hold:

(i) For every normal enumeration $\langle\alpha, \mathfrak{B}\rangle$ of $\mathfrak{A}$, if $\alpha' \subseteq \alpha$ then $\alpha(\Gamma(R_{\mathfrak{B}})) \subseteq A$;

(ii) For every element s of A, there exists a finite extension α'' of α', such that $Dom(\alpha'') \subseteq N^0$ and for every normal enumeration $\langle\alpha, \mathfrak{B}\rangle$ of $\mathfrak{A}$, if $\alpha'' \subseteq \alpha$ then $s \in \alpha(\Gamma(R_{\mathfrak{B}}))$.

A family $\mathcal{P}$ of subsets of B is said to be <u>sufficient</u> if for every finite mapping α' of N^0 into B and every enumeration operator Γ, there exists a sufficient for α' and Γ set A in $\mathcal{P}$.

<u>Proposition 4.</u> Every sufficient family of subsets of B contains all $\forall$-weak admissible subsets of B.

Proof. Let $\mathcal{P}$ be a sufficient family of subsets of B. Let $D \subseteq B$ and $D \notin \mathcal{P}$. We shall prove that D is not $\forall$-weak admissible in $\mathfrak{A}$. To make the proof, we shall construct a normal enumeration $\langle\alpha, \mathfrak{B}\rangle$ such that D is not weak admissible in it. The basis α^0 of this enumeration will be constructed by steps. In each step k we shall define a finite mapping α_k of N^0 into B so that for every k, $\alpha_k \subseteq \alpha_{k+1}$, and take $\alpha^0 = \bigcup_{k=0}^{\infty} \alpha_k$.

With the even steps we shall ensure that α^0 is totally defined over N^0 and $\mathrm{Range}(\alpha^0) = B$. With the odd steps $k = 2n+1$ we shall ensure that if Γ_n is the n-th enumeration operator and $\langle\alpha, \mathfrak{B}\rangle$ is a normal enumeration such that $\alpha_k \subseteq \alpha$, then $\alpha(\Gamma_n(R_{\mathfrak{B}})) \neq D$.

Let $s_0, s_1, \ldots$ be an arbitrary enumeration of the elements of B and $x_0, x_1, \ldots$ be an enumeration of N^0.

Let $\alpha_0(x_0) \cong s_0$ and let $\alpha_0(x)$ be undefined for $x \neq x_0$.

Let $k > 0$ and let α_q be defined for $q < k$. We have to consider the following two cases.

I. $k = 2n$. Let z be the first element of the sequence $x_0, x_1, \ldots$ which does not belong to $\mathrm{Dom}(\alpha_{k-1})$ and s be the first element of the sequence $s_0, s_1, \ldots$ which does not belong to $\mathrm{Range}(\alpha_{k-1})$. If such s does not exist then let s be an arbitrary element of B. Define $\alpha_k(z) \cong s$ and $\alpha_k(x) \cong \alpha_{k-1}(x)$ for $x \neq z$.

II. $k = 2n+1$. Let Γ_n be the n-th enumeration operator and A be an element of $\mathcal{P}$ which is sufficient for α_{k-1} and Γ. Apparently $A \neq D$. There are two possibilities.

1. For some t in B, $t \in D$ and $t \notin A$. Define $\alpha_k = \alpha_{k-1}$. Let $\langle\alpha, \mathfrak{B}\rangle$ be a normal enumeration and $\alpha_k \subseteq \alpha$. Since A is sufficient for α_k and Γ_n, $\alpha(\Gamma_n(R_{\mathfrak{B}})) \subseteq A$, and hence, $t \notin \alpha(\Gamma_n(R_{\mathfrak{B}}))$.

2. For some t in B, $t \in A$ and $t \notin D$. Let α'' be a finite mapping of N^0 into B, such that $\alpha_{k-1} \subseteq \alpha''$ and for every normal enumeration $\langle\alpha, \mathfrak{B}\rangle$, $\alpha'' \subseteq \alpha$ implies $t \in \alpha(\Gamma_n(R_{\mathfrak{B}}))$. Define $\alpha_k = \alpha''$.

Let $\langle\alpha, \mathfrak{B}\rangle$ be the normal enumeration with basis α^0. We shall prove that D is not weak admissible in $\langle\alpha, \mathfrak{B}\rangle$. Toward a contradiction suppose that $D = \alpha(W)$ for some p.r. in $\mathfrak{B}$ subset W of N. Let Γ_n be an enumeration operator such that $W = \Gamma_n(R_{\mathfrak{B}})$. Since $\alpha_{2n+1} \subseteq \alpha^0$, and hence, $\alpha_{2n+1} \subseteq \alpha$, $\alpha(W) = \alpha(\Gamma_n(R_{\mathfrak{B}})) \neq D$.

In next section we shall prove that the family of all subsets of B which are definable by means of logic programs is sufficient.

4. $\forall$-weak admissibility implies definability by means of logic programs

Let α' be a finite mapping of N^0 into B, Γ be an enumeration operator and H be an unary predicate symbol, $H \notin \{T_1, \ldots, T_k\}$. We shall define a logic program P so that the subset A of B which is definable by means of P and H in $\mathfrak{U}$ is sufficient for α' and Γ.

Let $\mathrm{Dom}(\alpha') = \{w_1, \ldots, w_r\}$ and $\alpha'(w_i) \cong s_i$, $i = 1, \ldots, r$. Let $c_{s_1}, \ldots, c_{s_r}$ be the names in C for $s_1, \ldots, s_r$, respectively.

Let $\underline{0}$ and $\underline{nil}$ be two new constant symbols, let f_0 be a new unary functional symbol and g be a new binary functional symbol. Denote by $\mathbb{F}$ the set $\{c_{s_1}, \ldots, c_{s_r}, \underline{0}, \underline{nil}, f_0, g, f_1, \ldots, f_n\}$ and with $\mathcal{S}$ the set of all terms built by means of the elements of $\mathbb{F}$ and the variables $\underline{X}_1, \underline{X}_2, \ldots$ We shall use the script letters a, b, c, d to denote arbitrary elements of $\mathcal{S}$.

For each element a of $\mathcal{S}$, with $\mathrm{var}(a)$ we shall denote the set of all variables which occur in a.

All logic programs, which will be considered in the sequel, will use functional and constant symbols among these of $\mathbb{F}$.

Let P be a logic program with functional and constant symbols among these of $\mathbb{F}$. The class of Herbrand interpretations of P consists of all total structures of the language of P - $\mathcal{L}_P$ - with domain $\mathcal{S}$, where the constant and functional symbols are interpreted in the usual way. If Q is a predicate symbol in $\mathcal{L}_P$ and I is a Herbrand interpretation for P then with I(Q) we shall denote the respective predicate on $\mathcal{S}$. An interpretation I of P is called model for P if all clauses of P are true in I.

For every natural number n, by $\underline{n}$ we shall denote the term $f_0^n(\underline{0})$. Let $\underline{N}$ denote the set $\{\underline{n} : n \in N\}$.

The following proposition is a reformulation of a well known result.

<u>Proposition</u> <u>5.</u> For every recursive enumerable (r. e.) subset W of N^k and for every k-ary predicate symbol Q, there exists a logic program P with the following properties:

(i) If $(x_1,\dots, x_k) \in W$ then $P \vdash Q(\underline{x}_1,\dots, \underline{x}_k)$;

(ii) There exists a Herbrand interpretation I of P, which is a model for P and

$I(Q)(a_1,\dots, a_k) = 0 \leftrightarrow \exists x_1\dots\exists x_k((x_1,\dots, x_k) \in W \ \&\ a_1 = \underline{x}_1 \ \&\ \dots \ \&\ a_k = \underline{x}_k)$.

Such interpretations for P we shall call standard.

If P is a logic program with the above properties then we shall say that P represents the set W with respect to the predicate symbol Q.

As usual, substitutions are called finite sets of the form $\{X_1/a_1, \dots, X_l/a_l\}$, where $X_1,\dots, X_l$ are distinct variables.

If a is a term or an atomic formula, $\varkappa = \{X_1/b_1, \dots, X_l/b_l\}$ is a substitution then with $a\varkappa$ we shall denote the term or the formula obtained from a by simultaneously replacing each occurrence of the variable X_i in a by the term b_i, $i = 1,\dots, l$.

Analogously, if $G = \{\Pi_1,\dots, \Pi_l\}$ is a finite set of atoms then with $G\varkappa$ we shall denote the set $\{\Pi_1\varkappa,\dots, \Pi_l\varkappa\}$.

Every substitution of the form $\{X_1/\underline{m}_1, \dots, X_l/\underline{m}_l\}$, $l \geq 0$, where $m_1,\dots, m_l$ are distinct elements of $N^0\backslash dom(\alpha')$, is called <u>correspondence.</u>

For every natural number x the term x^* is defined by the following inductive clauses:

(i) if $x \in Dom(\alpha')$ and $x = w_i$, $1 \leq i \leq r$, then $x^* = c_{s_i}$;

(ii) if $x \in N^0\backslash Dom(\alpha')$ then $x^* = \underline{x}$;

(iii) if $x = \langle i,y\rangle$, $1 \leq i \leq n$, then $x^* = f_i(y^*)$.

Let $K = \{c_{s_1},\dots, c_{s_r}\}$ and $\mathcal{L}_K = \mathcal{L} \cup K$.

<u>Proposition</u> <u>6.</u> For every natural x, there exists a term τ of the language $\mathcal{L}_K$ and a correspondence λ such that $\tau\lambda = x^*$.

<u>Proposition</u> <u>7.</u> Suppose that τ is a term of the language $\mathcal{L}_K$, x is a natural number, $\lambda = \{X_1/\underline{m}_1, \dots, X_l/\underline{m}_l\}$ is a correspondence and $\tau\lambda = x^*$.

Then for every normal enumeration $\langle\alpha, \mathfrak{B}\rangle$ of $\mathfrak{U}$, if $\alpha' \le \alpha$ then
$\alpha(x) \cong \tau_{\mathfrak{U}^*}(X_1/\alpha(m_1), \ldots, X_1/\alpha(m_1))$.

Proof. The proof follows from proposition 2, by induction on $|x|$, where $|x| = 0$ if $x \in N^0$ and $|\langle i,x\rangle| = |x| + 1$, $i = 1, \ldots, n$.

Lists are elements of $\mathscr{S}$, defined by the following inductive definition.

(i) nil is list;

(ii) if a is a term and ℓ is list then $g(a,\ell)$ is list.

For lists we shall use the following standard notational conventions. With $[a|\ell]$ we shall denote the term $g(a,\ell)$. For any elements $a_1, \ldots, a_k$ of $\mathscr{S}$, with $[a_1, \ldots, a_k]$ we shall denote the list $g(a_1, \ldots, g(a_k, \underline{nil})\ldots)$. In particular, with [] we shall denote the empty list - nil.

If $\lambda = \{X_1/\underline{m}_1, \ldots, X_1/\underline{m}_1\}$ is a correspondence then the list $[[X_1,\underline{m}_1], \ldots, [X_1,\underline{m}_1]]$ is called representation of λ in $\mathscr{S}$. We shall identify every correspondence with its representation in $\mathscr{S}$. So, the correspondences will be considered simultaneously as lists and as substitutions. We shall use the letters ℓ and f to denote correspondences.

If ℓ and f are correspondences then we shall write $\ell \le_1 f$ to denote that $\ell = f$ or $f = \text{append}(\ell,[X,\underline{m}])$, for some X and $\underline{m}$, i. e. if $\ell = [[X_1,\underline{m}_1], \ldots, [X_1,\underline{m}_1]]$ then $f = [[X_1,\underline{m}_1], \ldots, [X_1,\underline{m}_1],[X,\underline{m}]]$.
We shall use the sign $\le$ to denote the reflexive and transitive closure of the relation $\le_1$.

Let Nat, Neq and Cod be r.e. sets of natural numbers such that
$Nat = N^0 \setminus Dom(\alpha')$, $Neq = \{(x,y): x \ne y\}$ and $Cod = \{(x,y,z): z=\langle x,y\rangle\}$.

Let nat be a new unary predicate symbol, neq be a new binary predicate symbol and cod be a new ternary predicate symbol. Let $P_{\underline{nat}}$, $P_{\underline{neq}}$ and $P_{\underline{cod}}$ be logic programs which represent the sets Nat, Neq and Cod with respect to the predicate symbols nat, neq and cod. We shall assume that there are no common predicate symbols in $P_{\underline{nat}}$, $P_{\underline{neq}}$ and $P_{\underline{cod}}$.

The following program we shall denote with P_0. There alpha is a new predicate symbol, X, Y, Z, X_1, Y_1, Z_1 are distinct variables.

$\underline{alpha}(\underline{w}_i, c_{s_i}, X, X) \leftarrow$ $\quad i = 1, \ldots, r.$

$\underline{alpha}(X,Y,[],[[Y,X]]) \leftarrow \underline{nat}(X)$.

$\underline{alpha}(X,Y,[[Y,X]|Z],[[Y,X]|Z]) \leftarrow \underline{nat}(X).$

$\underline{alpha}(X,Y,[[Y_1,X_1]|Z_1],[[Y_1,X_1]|Z_2]) \leftarrow \underline{nat}(X),\ \underline{nat}(X_1),\ \underline{neg}(X,X_1),$
$\underline{alpha}(X,Y,Z_1,Z_2).$

$\underline{alpha}(X,f_i(Y),Z_1,Z_2) \leftarrow \underline{cod}(\underline{i},X_1,X),\ \underline{alpha}(X_1,Y,Z_1,Z_2).\quad i = 1,\ldots, n.$

$P_{\underline{nat}}$

$P_{\underline{neg}}$

$P_{\underline{cod}}$

A term a is called pseudo correspondence iff there exist a correspondence ℓ and a substitution $\varkappa$, such that $a = \ell\varkappa$. Note that a term a is pseudo correspondence iff a is of the form $[[\mathcal{b}_1, \underline{m}_1],\ldots, [\mathcal{b}_l,\underline{m}_l]]$, where $l \geq 0$ and $m_1,\ldots, m_l$ are distinct elements of $N^0\setminus Dom(\alpha')$.

Proposition 8. Let $x \in N$ and c be a pseudo correspondence. Let $\mathcal{b}$ and d be elements of $\mathcal{S}$. Then, $P_0 \vdash \underline{alpha}(\underline{x},\mathcal{b},c,d)$ if and only if there exist a substitution $\varkappa$, a term τ of $\mathcal{L}_K$ and two correspondences ℓ and f such that $\tau\varkappa = \mathcal{b}$, $\ell\varkappa = c$, $f\varkappa = d$ and

(1) $\ell \leq_1 f$, $var(\tau) \cup var(\ell) = var(f)$ and $\tau f = x^*$.

Proof. To prove the "if" part of the proposition it is sufficient to show that if $x \in N$, τ is a term of $\mathcal{L}_K$, ℓ and f are correspondences and (1) is true then $P_0 \vdash \underline{alpha}(\underline{x},\tau,\ell,f)$. This can be done by induction on $|x|$.

To prove the "only if" part, we shall define a special Herbrand interpretation for P_0. Since $P_{\underline{nat}}$, $P_{\underline{neg}}$ and $P_{\underline{cod}}$ do not have common predicate symbols, there exists a Herbrand interpretation I of the predicates which occur in $P_{\underline{nat}}$, $P_{\underline{neg}}$ and $P_{\underline{cod}}$ which is standard for them. Take this interpretation I and define the predicate I(alpha) as follows.

Let a, $\mathcal{b}$, c and d be arbitrary elements of $\mathcal{S}$. Then,

1. $I(\underline{alpha})(a,\mathcal{b},c,d) = 0$, if $a \notin \underline{N}$ or c is not a pseudo correspondence;

2. If $a = \underline{x}$ for some $x \in N$ and c is a pseudo correspondence then $I(\underline{alpha})(\underline{x},\mathcal{b},c,d) = 0$ iff there exist a term τ of $\mathcal{L}_K$, correspondences ℓ and f and a substitution $\varkappa$, such that $\mathcal{b} = \tau\varkappa$, $c = \ell\varkappa$, $d = f\varkappa$ and for x, τ, ℓ and f the condition (1) holds.

A long but straightforward proof shows that the interpretation I is a model for P_0. This fact together with the definition of I(alpha) prove the proposition.

A finite subset E of N is said to be correct if all elements of E are of the form $\langle j,x\rangle$, $1 \le j \le k+n$.

Suppose that a correct finite set E is given. Let $v \in E$ and $v = \langle j,x\rangle$. The atomic predicate $v^{\sim}$ is defined as $T_j(x^*)$, if $1 \le j \le k$, and as $T_1(f_{j-k}(x^*))$, if $k+1 \le j \le k+n$. With $E^{\sim}$ we shall denote the set $\{v^{\sim} : v \in E\}$.

<u>Proposition</u> <u>9.</u> For every correct finite set of natural numbers E and for every $x \in N$, there exists a finite set G of atoms of $\mathcal{L}_K$, a term τ of $\mathcal{L}_K$ and a correspondence ℓ such that $G\ell = E^{\sim}$ and $\tau\ell = x^*$.

<u>Proposition</u> <u>10.</u> Let E be a correct finite subset of N, let G be a finite set of atoms of $\mathcal{L}_K$, let $\ell = \{X_1/\underline{m}_1, \ldots, X_l/\underline{m}_l\}$ be a correspondence and $G\ell = E^{\sim}$. Then, for every normal enumeration $\langle \alpha, \mathfrak{B}\rangle$, if $\alpha' \le \alpha$ then $E \le R_{\mathfrak{B}} \leftrightarrow G_{\mathfrak{A}^*}(X_1/\alpha(m_1), \ldots, X_l/\alpha(m_l)) \cong 0$.

As usual, we shall identify every finite set $G = \{\Pi_1, \ldots, \Pi_l\}$ of atoms with the formula $\Pi_1\&\ldots\&\Pi_l$, if $l \ge 1$, and with the logical constant "true", if G is empty.

Consider the following logic program P_1. There <u>pi</u> is a new binary predicate symbol, X, Y, Z, X_1, Y_1, Z_1 are distinct variables.

<u>pi</u>([],Z) ←.

<u>pi</u>([X|Y],Z) ← cod(<u>j</u>,X_1,X), <u>alpha</u>(X_1,Y_1,Z,Z_1),
$T_j(Y_1)$, <u>pi</u>(Y,Z_1). $j = 1, \ldots, k$.

<u>pi</u>([X|Y],Z) ← cod(<u>j</u>,X_1,X), <u>alpha</u>(X_1,Y_1,Z,Z_1),
$T_1(f_{j-k}(Y_1))$, <u>pi</u>(Y,Z_1). $j = n+1, \ldots, n+k$.

P_0.

<u>Proposition</u> <u>11.</u> Let $E = \{v_1, \ldots, v_l\}$ be a correct finite subset of N. Let $\upsilon = [\underline{v}_1, \ldots, \underline{v}_l]$ be the list of the elements of E. Let δ be a pseudo correspondence. Then, for every finite set G of atoms of $\mathcal{L}_K$, $P_1 \vdash G \Rightarrow \underline{pi}(\upsilon, \delta)$ if and only if there exist a substitution $\varkappa$, a finite set G^0 of atoms of $\mathcal{L}_K$ and correspondences ℓ and f, such that $G = G^0\varkappa$, $\delta = \ell\varkappa$, $\ell \le f$ and $G^0 f \supseteq E^{\sim}$.

Proof. As in the proof of proposition 8, the "if" part is easy.

To prove the "only if" part, we define a class $\mathcal{K}$ of Herbrand interpretations of P_1.

A Herbrand interpretation I of P_1 belongs to $\mathcal{K}$ if the following conditions are satisfied:

(i) The predicate symbols which occur in the part P_0 of P_1 are interpreted as in the proof of proposition 8.

(ii) Let $\boldsymbol{a}$ and $\boldsymbol{b}$ be elements of $\mathcal{S}$. Then,

a) $I(\underline{pi})(\boldsymbol{a},\boldsymbol{b}) = 0$ if $\boldsymbol{a}$ is not of the form $[\underline{v}_1,\ldots, \underline{v}_l]$, $l \geq 0$, where $\{v_1,\ldots, v_l\}$ is a correct finite set of natural numbers, or $\boldsymbol{b}$ is not a pseudo correspondence.

b) If $\boldsymbol{a} = [\underline{v}_1,\ldots, \underline{v}_l]$, $l \geq 0$, is a list of elements of a correct finite set E of natural numbers and $\boldsymbol{b}$ is a pseudo correspondence then $I(\underline{pi})(\boldsymbol{a},\boldsymbol{b}) = 0$ iff there exist a finite set $G = \{T_{j_1}(\boldsymbol{d}_1),\ldots, T_{j_q}(\boldsymbol{d}_q)\}$, $q \geq 0$, such that $I(T_{j_1})(\boldsymbol{d}_1) =\ldots = I(T_{j_q})(\boldsymbol{d}_q) = 0$, a finite set G^0 of atoms of $\mathcal{L}_K$, a substitution $\varkappa$ and correspondences ℓ and f, such that

$G = G^0\varkappa$, $\boldsymbol{b} = \ell\varkappa$, $\ell \leq f$ and $G^0 f \geq E^\sim$.

Using the definition above, one can check that every I in $\mathcal{K}$ is a model for P_1.

Suppose that $G = \{T_{j_1}(\tau_1),\ldots, T_{j_q}(\tau_q)\}$ is a finite set of atoms of $\mathcal{L}_K$, $E = \{v_1,\ldots, v_l\}$ is a correct finite set of natural numbers, $\boldsymbol{b}$ is a pseudo correspondence and $P_1 \vdash G \rightarrow \underline{pi}([\underline{v}_1,\ldots, \underline{v}_l],\boldsymbol{b})$.

Let I be an element of $\mathcal{K}$ such that for each $1 \leq j \leq k$ and $\boldsymbol{d} \in \mathcal{S}$, $I(T_j)(\boldsymbol{d})=0$ iff $T_j(\boldsymbol{d}) \in G$.

Since I is a model for P_1, $I(\underline{pi})([\underline{v}_1,\ldots, \underline{v}_l],\boldsymbol{b}) = 0$, and therefore, by the definition of I, there exist a substitution $\varkappa$, a finite set of atoms G^0 of $\mathcal{L}_K$ and correspondences f and ℓ, such that

$G = G^0\varkappa$, $\boldsymbol{b} = \ell\varkappa$, $\ell \leq f$ and $G^0 f \geq E^\sim$.

Thereby the proof is completed.

Assume that an effective coding of the finite sets of natural numbers is fixed. We shall use the notation E_u for the finite set with code u.

<u>Proposition 12.</u> For every enumeration operator Γ there exists a logic program P such that the subset A, definable in $\mathfrak{U}$ by means of P and H is sufficient for α' and Γ.

Proof. Let Γ be an enumeration operator and let W be the r.e. set which determines Γ, i. e. if $R \subseteq N$ then $x \in \Gamma(R) \leftrightarrow \exists u(\langle u,x\rangle \in W \;\&\; E_u \subseteq R)$.

Let $W_1 = \{\langle u,x\rangle : \langle u,x\rangle \in W$ and E_u is correct$\}$.

Let Q be a new unary predicate symbol and let P_2 be a logic program which represents W_1 with respect to Q.

Let <u>list</u> be a new binary predicate symbol and let P_3 be a logic program which does not have common predicate symbols with P_1 and P_2 and

(i) if u is a code of the finite set $\{v_1, \ldots, v_l\}$, $l \geq 0$, then $P_3 \vdash \underline{list}(\underline{u}, [\underline{v}_1, \ldots, \underline{v}_l])$;

(ii) there exists a (standard) Herbrand interpretation I of P_3, which is a model for P_3 and if $u \in N$ and $E_u = \{v_1, \ldots, v_l\}$ then $I(\underline{list})(\underline{u}, \delta) = 0 \leftrightarrow \delta = [\underline{v}_1, \ldots, \underline{v}_l]$.

Consider the logic program P.

$H(Y) \leftarrow Q(Z)$, <u>cod</u>(U,X,Z), <u>alpha</u>$(X,Y,[\,],F)$, <u>list</u>(U,U_1), <u>pi</u>(U_1,F).

P_1

P_2

P_3

Using the same methods as in the proofs of the propositions 8 and 11, one can obtain the following characterization of P. For every finite set G of atoms of $\mathcal{L}_K$ and every term τ of $\mathcal{L}_K$, $P \vdash G \rightarrow H(\tau)$ iff there exist a substitution $\varkappa$, an ordered pair $\langle u,x\rangle \in W_1$, a finite set G^0 of atoms of $\mathcal{L}_K$, a term τ^0 of $\mathcal{L}_K$ and a correspondence ℓ, such that

$G = G^0\varkappa$, $\tau = \tau^0\varkappa$, $G^0\ell \supseteq E_u^{\sim}$ and $\tau^0\ell = x^*$.

Let A be the subset of B, definable by means of P and H, i. e.

$s \in A \leftrightarrow \exists\tau(\tau \in \mathcal{T} \;\&\; \tau_{\mathfrak{A}^*} \cong s \;\&\; \mathcal{D}(\mathfrak{A}) \cup \{P\} \vdash H(\tau))$.

We shall prove that A is sufficient for α' and Γ.

Suppose that $\langle \alpha, \mathfrak{B}\rangle$ is a normal enumeration and $\alpha' \subseteq \alpha$. Let s be an element of $\alpha(\Gamma(R_{\mathfrak{B}}))$ and let x be an element of $\Gamma(R_{\mathfrak{B}})$ such that $\alpha(x) \cong s$. Due to the definition of $R_{\mathfrak{B}}$, there exists an element $\langle u,x\rangle$ of W_1, such that $E_u \subseteq R_{\mathfrak{B}}$. Let G be a finite set of atoms of $\mathcal{L}_K$, τ be a term of $\mathcal{L}_K$ and ℓ be a correspondence, such that $G\ell = E_u^{\sim}$ and $\tau\ell = x^*$.

It is clear that $P \vdash G \Rightarrow H(\tau)$. Suppose that $\ell = \{X_1/m_1, \ldots, X_l/m_l\}$. Then, by proposition 7 and by proposition 10,

(2) $s \cong \alpha(x) \cong \tau_{\mathfrak{U}^*}(X_1/\alpha(m_1), \ldots, X_l/\alpha(m_l))$ and

(3) $G_{\mathfrak{U}^*}(X_1/\alpha(m_1), \ldots, X_l/\alpha(m_l)) \cong 0$.

It follows from (2), (3) and from the definition of A that $s \in A$.

So, the inclusion $\alpha(\Gamma(R_{\mathfrak{B}})) \subseteq A$ is proved.

Suppose that s is an element of A. Due to the reduction theorem, the theorem on constants and the fact that all of the constants of C which occur in P are among these of K, there exist a finite set G of atoms of $\mathcal{L}_K$ and a term τ of $\mathcal{L}_K$ such that $P \vdash G \Rightarrow H(\tau)$ and if $X_1, \ldots, X_l$ are the variables in the formula $G \Rightarrow H(\tau)$ then for some elements $p_1, \ldots, p_l$ of B,

(4) $G_{\mathfrak{U}^*}(X_1/p_1, \ldots, X_l/p_l) \cong 0$ and

(5) $\tau_{\mathfrak{U}^*}(X_1/p_1, \ldots, X_l/p_l) \cong s$.

By the characterization of P, there exist an ordered pair $\langle u,x\rangle \in W_1$, a substitution $\varkappa$, a finite set G^0 of atoms of $\mathcal{L}_K$, a term τ^0 of $\mathcal{L}_K$ and a correspondence ℓ, such that $G = G^0\varkappa$, $\tau = \tau^0\varkappa$, $G^0\ell \supseteq E_u^{\sim}$ and $\tau^0\ell = x^*$.

Let $Y_1, \ldots, Y_q$ be the variables in G^0 and τ^0, let $\ell = \{Y_1/m_1, \ldots, Y_q/m_q\}$ and $\varkappa = \{Y_1/\mu^1, \ldots, Y_q/\mu^q\}$ and let $\mu^i_{\mathfrak{U}^*}(X_1/p_1, \ldots, X_l/p_l) \cong t_i$, $i = 1, \ldots, q$.

Then, $G_{\mathfrak{U}^*}(X_1/p_1, \ldots, X_l/p_l) \cong G^0_{\mathfrak{U}^*}(Y_1/t_1, \ldots, Y_q/t_q)$ and $\tau_{\mathfrak{U}^*}(X_1/p_1, \ldots, X_l/p_l) \cong \tau^0_{\mathfrak{U}^*}(Y_1/t_1, \ldots, Y_q/t_q)$.

Hence, by (4) and (5),

(6) $G^0_{\mathfrak{U}^*}(Y_1/t_1, \ldots, Y_q/t_q) \cong 0$ and

(7) $\tau^0_{\mathfrak{U}^*}(Y_1/t_1, \ldots, Y_q/t_q) \cong s$.

Define the finite mapping α'' of N^0 into B, by $\alpha''(m_i) \cong t_i$, $i = 1, \ldots, q$, and $\alpha''(x) \cong \alpha'(x)$ if $x \notin \{m_1, \ldots, m_q\}$. Since the numbers $m_1, \ldots, m_l$ are distinct, the definition of α'' is correct.

Let $\langle \alpha, \mathfrak{B}\rangle$ be a normal enumeration such that $\alpha'' \subseteq \alpha$. By (6), (7), proposition 7 and proposition 10, $\alpha(x) \cong \tau^0_{\mathfrak{U}^*}(Y_1/t_1, \ldots, Y_q/t_q) \cong s$ and $E_u \subseteq R_{\mathfrak{B}}$. Hence, $x \in \Gamma(R_{\mathfrak{B}})$ and $\alpha(x) \cong s$.

Thereby the proposition is proved.

Combining the last proposition and proposition 4, we obtain and the following:

<u>Proposition 13.</u> Every $\forall$-weak admissible set is definable in $\mathfrak{A}$ by means of logic programs.

5. Search computability and definability by means of logic programs.

In this section we shall establish the relationships between search computability and definability by means of logic programs.

We begin with a definition from [3]. A subset A of B is said to be of Range type iff there exists a partial multiple-valued (p.m.v.) function ψ, which is search computable in $\mathfrak{A}$ with constants derived from B, such that $s \in A \leftrightarrow \exists t(s \in \psi(t))$.

The following theorem is proved in [3].

<u>Theorem 1.</u> A subset A of B is of Range type iff it is $\forall$-weak admissible in $\mathfrak{A}$.

A set A is said to be semi-computable in $\mathfrak{A}$ if it is domain of some search computable in $\mathfrak{A}$ p.m.v. function. If the equality relation is semi-computable in $\mathfrak{A}$ then the sets of Range type coinside with the semi-computable ones. In general the semi-computable sets are a proper subclass of the sets of Range type.

<u>Theorem 2.</u> Let A be a subset of B. Then, the following assertions are equivalent:

(i) A is definable by means of logic programs in $\mathfrak{A}$.

(ii) A is logically definable in $\mathfrak{A}$.

(iii) A is $\forall$-weak admissible.

(iv) A is of Range type.

Proof. (i) $\rightarrow$ (ii) is obvious.

(ii) $\rightarrow$ (iii) follows from proposition 1.

(iii) $\leftrightarrow$ (iv) follows from theorem 1.

(iii) $\rightarrow$ (i) follows from proposition 13.

6. Uniform definability

Suppose that a finite first order language $\mathcal{L} = \{q_1, \ldots, q_r; f_1, \ldots, f_n; T_1, \ldots, T_k\}$ is fixed. Here $q_1, \ldots, q_r$ are constants, $f_1, \ldots, f_n$ are functional symbols and $T_1, \ldots, T_k$ are predicate symbols, r, k , n $\geq$ 0.

Let $\mathfrak{A} = (B; t_1, \ldots, t_r; \theta_1, \ldots, \theta_n; \Sigma_1, \ldots, \Sigma_k)$ be an arbitrary partial structure of the language $\mathcal{L}$, let $\mathcal{F}$ be a r. e. set of closed first order formulae and H be an unary predicate symbol, which does not belong to $\{T_1, \ldots, T_k\}$. Let C be a set of constants - names for the elements of B - such that no one of the elements of C occurs in the formulae of $\mathcal{F}$. Denote with $\mathcal{T}$ the set of all ground terms of $\mathcal{L}_C = \mathcal{L} \cup C$, with $\mathfrak{A}^*$ the enrichment of $\mathfrak{A}$ to the language $\mathcal{L}_C$ and define $\partial(\mathfrak{A})$ by the equality

$$\partial(\mathfrak{A}) = \{T_j(\tau^1, \ldots, \tau^{b_j}) : 1 \leq j \leq k \ \& \ \tau^1, \ldots, \tau^{b_j} \in \mathcal{T} \ \& \ \Sigma_j(\tau^1_{\mathfrak{A}^*}, \ldots, \tau^{b_j}_{\mathfrak{A}^*}) \cong 0\}.$$

Let A be the subset of B defined by

$$s \in A \leftrightarrow \exists \tau (\tau \in \mathcal{T} \ \& \ \tau_{\mathfrak{A}^*} \cong s \ \& \ \partial(\mathfrak{A}) \cup \mathcal{F} \vdash H(\tau)).$$

Then, A is said to be <u>uniformly</u> <u>definable</u> in $\mathfrak{A}$ by means of $\mathcal{F}$ and H.

For every r.e. set $\mathcal{F}$ of closed first order formulae and every unary predicate symbol H, denote with $\mathcal{D}(\mathcal{F}, H)$ the set of all formulae of the form $G \Rightarrow H(\tau)$, where G is a finite conjunction of atoms of $\mathcal{L}$ and τ is a term in $\mathcal{L}$ and $\mathcal{F} \vdash G \Rightarrow H(\tau)$.

The following proposition shows that the notion of uniform definability does not depend on the choice of the set of names C.

<u>Proposition 14.</u> Let $\mathfrak{A}$ be a partial structure of the language $\mathcal{L}$. Let $\mathcal{F}$ be a r.e. set of closed first order formulae and H be a unary predicate symbol, $H \notin \mathcal{L}$. Let A be uniformly definable in $\mathfrak{A}$ by means of $\mathcal{F}$ and H. Then, an element s belongs to A iff there exist a formula $G \Rightarrow H(\tau)$ of $\mathcal{D}(\mathcal{F}, H)$ with variables $X_1, \ldots, X_k$, $k \geq 0$, and elements $p_1, \ldots, p_k$ of B, such that $G_{\mathfrak{A}}(X_1/p_1, \ldots, X_k/p_k) \cong 0$ and $\tau_{\mathfrak{A}}(X_1/p_1, \ldots, X_k/p_k) \cong s$.

In this section the following theorem will be proved.

<u>Theorem 3.</u> For every r.e. set of closed first order formulae $\mathcal{F}$ and every unary predicate symbol $H \notin \mathcal{L}$, there exists a logic program P such that $\mathcal{D}(\mathcal{F}, H) = \mathcal{D}(\{P\}, H)$.

<u>Corollary.</u> In every partial structure $\mathfrak{A}$, uniform definability by means of arbitrary r.e. sets of first order formulae is equivalent with uniform definability by means of logic programs.

A partial structure $\mathfrak{A} = (B; t_1, \ldots, t_{r'}; \theta_1, \ldots, \theta_n; \Sigma_1, \ldots, \Sigma_k)$ of the language $\mathcal{L}$ is called $\mathcal{L}$ - structure iff the following conditions are satisfied:

(i) The set B is denumerable;

(ii) The basic functions $\theta_1, \ldots, \theta_n$ are totally defined;

(iii) The basic predicates $\Sigma_1, \ldots, \Sigma_k$ obtain only the value 0, whenever are defined.

Suppose that a $\mathcal{L}$ - structure $\mathfrak{A} = (B; t_1,\ldots, t_r; \theta_1,\ldots, \theta_n; \Sigma_1,\ldots, \Sigma_k)$ is given. Normal enumerations of $\mathfrak{A}$ are defined as in section 3. In addition we shall assume that if $w_1,\ldots, w_r$ are the first r elements of N^0 then for every normal enumeration $\langle\alpha, \mathfrak{B}\rangle$ of $\mathfrak{A}$, $\alpha(w_1) \cong t_1,\ldots, \alpha(w_r) \cong t_r$. By N_r we shall denote the set $N^0\setminus\{w_1,\ldots,w_r\}$.

Note that, if $\langle\alpha, \mathfrak{B} = (N; \varphi_1,\ldots, \varphi_n; \sigma_1,\ldots, \sigma_k)\rangle$ is a normal enumeration of a $\mathcal{L}$-structure $\mathfrak{A}$ then α is totally defined over N and the predicates $D_1,\ldots, D_n$ are equal to $\lambda x.0$. Hence, in this case it is sufficient to define $R_{\mathfrak{B}} = \{\langle j,x\rangle : 1 \le j \le k \;\&\; \sigma_j(x) \cong 0\}$.

The following proposition is an uniform variant of proposition 1.

<u>Proposition 15.</u> For every r.e. set of closed first order formulae $\mathcal{F}$ and every predicate symbol $H \notin \mathcal{L}$, there exists an enumeration operator $\Gamma_{\mathcal{F},H}$ such that if $\mathfrak{A}$ is a $\mathcal{L}$-structure, A is the set uniformly definable in $\mathfrak{A}$ by means of $\mathcal{F}$ and H and $\langle\alpha, \mathfrak{B}\rangle$ is a normal enumeration of $\mathfrak{A}$ then $\alpha(\Gamma_{\mathcal{F},H}(R_{\mathfrak{B}})) = A$.

It is hardly true that for every enumeration operator Γ, there exists a logic program P such that $\Gamma = \Gamma_{\{P\},H}$. For our purposes the following weaker proposition is sufficient.

<u>Proposition 16.</u> For every enumeration operator Γ and every predicate symbol $H \notin \mathcal{L}$, there exists a logic program P with the following property:

If $\mathfrak{A}$ is a $\mathcal{L}$ - structure, A is the set uniformly definable in $\mathfrak{A}$ by means of {P} and H then

(i) For every normal enumeration $\langle\alpha, \mathfrak{B}\rangle$ of $\mathfrak{A}$, $\alpha(\Gamma(R_{\mathfrak{B}})) \subseteq A$ and

(ii) For every element s of A, there exists a normal enumeration $\langle\alpha, \mathfrak{B}\rangle$ of $\mathfrak{A}$, such that $s \in \alpha(\Gamma(R_{\mathfrak{B}}))$.

Proof. Since the proof is very similar to those of proposition 12, we shall give here only a sketch.

Let W be the r.e. subset which determines Γ, i.e. if $R \subseteq N$ then $x \in \Gamma(R) \Longleftrightarrow \exists u(\langle u,x\rangle \in W \;\&\; E_u \subseteq R)$. Say that a finite set of natural numbers E is correct if each $v \in E$ is of the form $\langle j,x\rangle$, $1 \le j \le k$.
Denote with W_1 the set $\{\langle u,x\rangle : \langle u,x\rangle \in W \;\&\; E_u \text{ is correct}\}$.

Using almost the same definitions as in section 4, one can define a logic program P with the following property.

For every finite set G of atoms of $\mathcal{L}$ and every term τ of $\mathcal{L}$, $P \vdash G \rightarrow H(\tau)$ iff there exist a substitution $\varkappa$, an element $\langle u,x\rangle$ of W_1 a finite set G^0 of atoms of $\mathcal{L}$, a term τ^0 of $\mathcal{L}$ and a correspondence ℓ, such that $G = G^0\varkappa$, $\tau = \tau^0\varkappa$, $G^0\ell \supseteq E^\sim$ and $\tau^0\ell = x^*$.

Suppose that $\mathfrak{A}$ is a $\mathcal{L}$ - structure and let A be the set, uniformly definable in $\mathfrak{A}$ by means of {P} and H. Now, the same proof as in the last part of the proof of proposition 12 shows that for every normal enumeration $\langle\alpha, \mathfrak{B}\rangle$ of $\mathfrak{A}$, $\alpha(\Gamma(R_{\mathfrak{B}})) \subseteq A$ and for every element s of A, there exist distinct elements $m_1, \ldots, m_q$ of N_r and elements $p_1, \ldots, p_q$ of B, such that for every normal enumeration $\langle\alpha, \mathfrak{B}\rangle$ of $\mathfrak{A}$, if $\alpha(m_i) \cong p_i$, $i = 1, \ldots, q$, then $s \in \alpha(\Gamma(R_{\mathfrak{B}}))$.

Now, we are ready to prove theorem 3.

<u>Proof of theorem 3.</u> Let $\mathcal{F}$ be a r.e. set of closed first order formulae and let H be an unary predicate symbol which does not belong to $\mathcal{L}$. Due to proposition 16, there exists a logic program P, such that if $\mathfrak{A}$ is a $\mathcal{L}$ - structure, A is the set uniformly definable in $\mathfrak{A}$ by means of {P} and H in $\mathfrak{A}$ then

(i) For every normal enumeration $\langle\alpha, \mathfrak{B}\rangle$ of $\mathfrak{A}$, $\alpha(\Gamma_{\mathcal{F},H}(R_{\mathfrak{B}})) \subseteq A$ and

(ii) For every element s of A, there exists a normal enumeration $\langle\alpha, \mathfrak{B}\rangle$ of $\mathfrak{A}$, such that $s \in \alpha(\Gamma_{\mathcal{F},H}(R_{\mathfrak{B}}))$.

We shall prove that $\mathcal{D}(\mathcal{F},H) = \mathcal{D}(\{P\},H)$.

Suppose that $G \Rightarrow H(\tau) \in \mathcal{D}(\mathcal{F},H)$. Let $X_1, \ldots, X_l$ be the variables which occur in G and τ. Let $\underline{k}_1, \ldots, \underline{k}_l$ be new constant symbols.

Define the $\mathcal{L}$ - structure $\mathfrak{A}$ as follows.

The domain B of $\mathfrak{A}$ consists of all ground terms of $\mathcal{L} \cup \{\underline{k}_1, \ldots, \underline{k}_l\}$. The i-th constant t_i of $\mathfrak{A}$ is eqaul to q_i, $i = 1, \ldots, r$. The i-th basic function θ_i is defined by $\theta_i(\tau) = f_i(\tau)$ for $\tau \in B$.

Let $G = \{T_{j_1}(\tau_1), \ldots, T_{j_q}(\tau_q)\}$ and let $G^\sim$ be the finite set of atoms of $\mathcal{L} \cup \{\underline{k}_1, \ldots, \underline{k}_l\}$, obtained from G by replacing each occurrence of X_i in G by $\underline{k}_i$, $i = 1, \ldots, l$. Define the partial predicates $\Sigma_1, \ldots, \Sigma_k$ by $\Sigma_j(\tau) \cong 0$ if $T_j(\tau) \in G^\sim$ and $\Sigma_j(\tau)$ is undefined otherwise, for $\tau \in B$, $1 \leq j \leq k$.

Let $\tau^\sim$ be the element of B, obtained from τ by replacing each occurrence of X_i in τ by $\underline{k}_i$, $i = 1, \ldots, l$.

Let $A \subseteq B$ and let A be uniformly definable in $\mathfrak{A}$ by means of $\mathcal{F}$ and H. Since $G \Rightarrow H(\tau) \in \mathcal{D}(\mathcal{F},H)$, by proposition 14 and the definition of $\mathfrak{A}$, $\tau^\sim \in A$.

Let $\langle\alpha, \mathfrak{B}\rangle$ be a normal enumeration of $\mathfrak{A}$. It is clear that $\alpha(\Gamma_{\mathcal{F},H}(R_{\mathfrak{B}})) = A$, and hence, $\tau^\sim \in \alpha(\Gamma_{\mathcal{F},H}(R_{\mathfrak{B}}))$.

Let A_P be the subset of B, uniformly definable by means of {P} and H in $\mathfrak{A}$. Then $\alpha(\Gamma_{\mathcal{F},H}(R_{\mathfrak{B}})) \subseteq A_P$, and hence, $\tau^{\sim} \in A_P$. By proposition 14, there exist an element $G^1 \to H(\tau^1)$ of $\mathcal{D}(\{P\},H)$ such that if $Y_1,\ldots, Y_q$ are the variables which occur in G^1 and τ^1 then for some elements $p_1,\ldots, p_q$ of B,

(8) $G^1_{\mathfrak{A}}(Y_1/p_1,\ldots, Y_q/p_q) \cong 0$ and

(9) $\tau^1_{\mathfrak{A}}(Y_1/p_1,\ldots, Y_q/p_q) \cong \tau^{\sim}$.

Denote by $\varkappa$ the substitution $\{Y_1/p_1,\ldots, Y_q/p_q\}$ and by G^0 the set $G^1\varkappa$.

From (8) and from the definition of $\mathfrak{A}$ it follows that $G^0 \subseteq G^{\sim}$.

By (9), $\tau^{\sim} = \tau^1\varkappa$.

Since $P \vdash G^1 \to H(\tau^1)$, $P \vdash G^0 \to H(\tau^{\sim})$, and hence, $P \vdash G^{\sim} \to H(\tau^{\sim})$. From here, by the theorem on constants, $P \vdash G \to H(\tau)$.

Let us suppose that $G \to H(\tau) \in \mathcal{D}(\{P\},H)$. Define the structure $\mathfrak{A}$ as above. Let A_P be the subset of B, uniformly definable in $\mathfrak{A}$ by means of {P} and H. By the definition of $\mathfrak{A}$, $\tau^{\sim} \in A_P$. Let $\langle\alpha, \mathfrak{B}\rangle$ be a normal enumeration of $\mathfrak{A}$, such that $\tau^{\sim} \in \alpha(\Gamma_{\mathcal{F},H}(R_{\mathfrak{B}}))$. Let A be the subset of B, uniformly definable by means of $\mathcal{F}$ and H. It is clear that $\tau^{\sim} \in A$. From here, as in the previous case, it follows that $\mathcal{F} \vdash G \to H(\tau)$, and hence, that $G \to H(\tau) \in \mathcal{D}(\mathcal{F},H)$.

Acknowledgements

The author is indebted to St. Nikolova for the valuable suggestions concerning uniform definability.

References

1. Y. N. Moschovakis, Abstract first order computability I, Trans. Amer. Math. Soc., 138 (1969), pp. 427 - 464.
2. R. Fraisse, Une notion de recursivite relative, in: "Infinistic methods", Proceedings of the symposium of mathematics, Warsaw, 1959, Pergamon Press, (1961), pp. 323 - 328.
3. I. N. Soskov, Definability via enumerations, J. Symb. Logic (to appear).
4. Y. N. Moschovakis, Abstract computability and invariant definability, J. Symb. Logic, 34:4, (1969).
5. M. C. Fitting, "Fundamentals of Generalized Recursion Theory", North-Holland, Amsterdam, (1981).
6. D. Lacombe, Recursion theoretic structures for relational systems, in: "Logic Colloquium'69", R. O. Gandy, C. M. E. Yates. eds., North-Holland, Amsterdam, (1971).
7. H. Rogers, Jr., "Theory of recursive functions and effective computability", McGraw-Hill Book Company, (1967).
8. J. R. Shoenfield, "Mathematical Logic". Addison-Wesley Pub. Comp., (1967).

SOME RELATIONS AMONG SYSTEMS FOR BOUNDED ARITHMETIC

Gaisi Takeuti

Department of Mathematics, University of Illinois

1409 West Green Street Urbana, Illinois 61801

In [1], S. Buss introduced systems S_2^i (i = 0, 1, 2,...), $\overset{\circ}{U}{}_2^1$, and $\overset{\circ}{V}{}_2^1$ for Bounded Arithmetic which are closely related to Δ_i^p (if $i > 0$) in the polynomial hierarchy, PSPACE, and EXPTIME respectively. In Bounded Arithmetic weaker systems and stronger systems are interacted each other. As we think in physics that the basic principles are the principles on elementary particles, we believe that the intrinsic nature of Bounded Arithmetic is hidden in very weak systems of Bounded Arithmetic. Therefore it seems to us very useful to define sharply bounded arithmetic and to study it. Unfortunately S_2^0 is too weak to be a good candidate for sharply bounded arithmetic since it is proved in [7] that S_2^0 does not prove $\exists x(a = 0 \vee x + 1 = a)$. On the other hand, $\mathfrak{S}_2^0$ introduced in [8] seems to us a good candidate for sharply bounded arithmetic. We shall prove that $\mathfrak{S}_2^0 = \mathfrak{S}_2^1$ implies EXPTIME = PSPACE, where $\mathfrak{S}_2^1$ is a conservative extension of S_2^1, and that if $\overset{\circ}{V}{}_2^1$ proves

$$\forall \ulcorner\varphi\urcorner \exists w(PA - Prf(w, \ulcorner\varphi\urcorner) \vee PA - Prf(w, \ulcorner\neg\varphi\urcorner))$$

then PSPACE = NP $\cap$ co-NP, where $\ulcorner\varphi\urcorner$ is a Gödel number of a sharply bounded sentence and $PA - Prf(w, \ulcorner\varphi\urcorner)$ is a formalized statement "w is a proof of $\ulcorner\varphi\urcorner$ in Peano Arithmetic".

A very interesting problem on Bounded Arithmetic is to find out what systems in Bounded Arithmetic prove the consistency of a weaker system in

Mathematical Logic
Edited by P. P. Petkov
Plenum Press, New York, 1990

Bounded Arithmetic. If we take consistency to be the oridnary consistency, then the answer is very negative since Wilkie and Paris proved in [8] that $S_2^1 + \exp$ does not prove the consistency of Robinson's arithmetic Q, where exp is the axiom $\forall x(2^x$ exists). Their result shows that the meaningful consistency for Bounded Arithmetic must be consistency which is not involved with unbounded quantifiers. Therefore the next question is about the bounded consistency of T denoted by BdCon(T) which means that there is no bounded proof (i.e. a proof which uses only bounded formulas) of contradiction form T. In [4], P. Pudlák proved that S_2 does not prove BdCon (S_2^1). In [3], S. Buss improved Pudlák's result by showing that S_2 does not prove $\mathrm{BdCon}(S_2^{-1})$ where S_2^{-1} is the induction-free fragment of S_2. Their main tool is an analysis of the notion $\gamma(\varphi)$ which is the minimal Gödel number of a bounded proof of φ. In [6], we introduced a notion "i-normal proof". In this paper we shall define a similar notion "T_2^i-normal proof". T_2^i was also introduced in [1] by S. Buss. T_2^i is a little stronger than S_2^i and S. Buss also proved in [2] that T_2^i is closely related to Δ_{i+1}^p (if $i > 0$) in the polynomial hierarchy. By NCon (T_2^i) we mean that there are no T_2^i-normal proof of contradiction. We shall show

$$S_2^1 \vdash \mathrm{Con}^{i+5}(S_2^{-1}) \longrightarrow \mathrm{N\,Con}(T_2^i),$$

where by $\mathrm{Con}^{i+5}(S_2^{-1})$ we mean that there are no S_2^{-1}-proof of contradiction in which every formula belongs to $\sum_{i+5}^b \cup \prod_{i+5}^b$. This is an improvement of Pudlák-Buss' result since $T_2^i \nvdash \mathrm{N\,Con}\ (T_2^i)$ and

$$S_2^1 \vdash \mathrm{BdCon}\ (S_2^{-1}) \longrightarrow \mathrm{Con}^{i+5}(S_2^{-1}).$$

We use the terminology of [1] and [6].

SHARPLY BOUNDED ARITHMETIC

In this section, by sharply bounded arithmetic we mean the system $\mathfrak{S}_2^0$ in [6]. $\mathfrak{S}_2^0$ is obtained from S_2^0 by adding PTIME functions $a \dot{-} b$, max $(a,b),\ldots,\beta(i,w)$ in §2.4 and §2.5 up to $\beta(i,w)$ in [1] and their finitely many defining axioms and by replacing $\sum_0^b$-PIND by the following $\sum_0^b$-LIND

$$\frac{A(a),\ \Gamma \longrightarrow \Delta,\ A(a+1)}{A(0),\ \Gamma \longrightarrow \Delta,\ A(|t|)} ,$$

where $A(a)$ is sharply bounded i.e. all quantifiers in $A(a)$ are sharply bounded.

Lemma 1. Let $A(x,a)$ and $B(y,a)$ be sharply bounded and $\breve{S}_2^0$ prove

$$\forall x A(x,a) \longleftrightarrow \exists y B(y,a).$$

Then there exists a sharply bounded formula $C(a)$ such that $\breve{S}_2^0$ proves

$$\forall x A(x,a) \longleftrightarrow C(a).$$

Proof. First we extend the language of $\breve{S}_2^0$ by introducing $\mu x \le |t|$ such that if t and $s(a)$ are terms, then $\mu x \le |t| s(x)$ is also a term. $\hat{S}_2^0$ is obtained from $\breve{S}_2^0$ by adding the following axioms.

(1) $s(b) = 0,\ b \le |t|,\ 0 = \mu x \le |t| s(x) \longrightarrow s(0) = 0$
(2) $\longrightarrow \mu x \le |t| s(x) \le |t|$
(3) $\longrightarrow \mu x \le |t| s(x) = 0,\ s(\mu x \le |t| s(x)) = 0$
(4) $s(b) = 0 \longrightarrow \mu x \le |t| s(x) \le b$
(5) $s(b) = 0 \longrightarrow 0 = \mu x \le |t| s(x)$

In $\hat{S}_2^0$, we can eliminate sharply bounded quantifiers by using $\mu x \le |t|$. Therefore there exists quantifier free formulas $A'(b,a)$ and $B'(b,a)$ such that the following formulas are provable in $\hat{S}_2^0$

$$A(b,a) \longleftrightarrow A'(b,a)$$

and

$$B(b,a) \longleftrightarrow B'(b,a).$$

Therefore $\hat{S}_2^0$ proves the following two sequents:

(1) $\qquad B(b,a) \longrightarrow A(c,a),$

where b and c are new free variables, and

(2) $\qquad \forall x A'(x,a) \longrightarrow \exists y B'(y,a).$

Since $\sum_0^b$-LIND can be eliminated by using $\mu x \leq |t|$, there exists a $\hat{S}_2^0$-proof P of (2) such that P has no LIND, no free variables except a, and no cut except inessential cuts i.e. no cut with quantifiers. Then it is easily seen that there exist terms $t_1(a),\ldots,t_m(a)$ and $s_1(a),\ldots,s_n(a)$ such that the following sequent is provable in $\hat{S}_2^0$

$$A'(t_1(a), a),\ldots,A'(t_m(a), a) \longrightarrow B'(s_1(a), a),\ldots,B'(s_n(a), a).$$

We define $C'(a)$ to be

$$A'(t_1(a),a) \wedge \cdots \wedge A'(t_m(a), a).$$

Then $C'(a) \longleftrightarrow \forall x A(x,a)$ is provable in $\hat{S}_2^0$. Now by eliminating $\mu x \leq |t|$ in $C'(a)$ by using sharply bounded quantifiers, we can find sharply bounded formula $C(a)$ without any μx such that $C'(a) \longleftrightarrow C(a)$ is provable in $\hat{S}_2^0$. Obviously $\forall x A(x,a) \longleftrightarrow C(a)$ is provable in $\tilde{S}_2^0$.

Let $\tilde{S}_2^1$ be an extension of S_2^1 obtained by introducing functions $a \dot{-} b$, max $(a,b),\ldots,\beta(i,w)$ i.e. new functions in $\tilde{S}_2^0$ and their finitely many defining axioms. By §2.4 and §2.5 in [1], $\tilde{S}_2^1$ is a conservative extension of S_2^1.

<u>Theorem 1</u>. If $\tilde{S}_2^0 = \tilde{S}_2^1$, then

EXPTIME = PSPACE.

<u>Proof</u>. Let M be a universal Turing machine. Let $\tilde{T}u(w, x, a)$ be a Δ_1^b-formula with respect to $\tilde{S}_2^1$ which expresses "w is a code expressing a Turing machine Mw, x is a code of an input of Mw, and M accepts <w.x> in the time less than $|a|$". If $\tilde{S}_2^0 = \tilde{S}_2^1$, then $\tilde{T}u(w, x,$

a) is Δ_1^b-formula with respect to $\mathfrak{S}_2^0$ and can be regarded as a sharply bounded formula by Lemma 1. As in [6], let $\tilde{e}(a)$ be $1 \# 2 \# \cdots \# 2$, where the number of 2 is $|a|$. Then $\ulcorner \tilde{e}(a) \urcorner$ is a formalized closed term in $\mathfrak{S}_2^0$ and therefore $\ulcorner \mathfrak{T}u(Iw, Ia, \tilde{e}(a)) \urcorner$ is a formalized sharply bounded sentence in $\mathfrak{S}_2^0$, where Ia is a numeral expressing a. Now let $\mathfrak{T}$ be a $\Delta_1^{1,b}$-truth definition of the sharply bounded sentences developed in §3 in [6]. Since the value of $\tilde{e}(a)$ is greater than 2^a by §1 in [6], $\mathfrak{T}(\ulcorner \mathfrak{T}u(Iw, Ia, \tilde{e}(a)) \urcorner)$ enumerates all predicates on a which can be computed in the less than the time a if w ranges over all the natural numbers. Obviously $\mathfrak{T}(\ulcorner \mathfrak{T}u(Iw, Ia, \tilde{e}(a)) \urcorner)$ is a complete predicate of EXPTIME and $\Delta_1^{1,b}$ with respect to $\overset{\circ}{U}{}_2^1$ by §3 in [6]. Therefore $\mathfrak{T}(\ulcorner \mathfrak{T}u(Iw, Ia, \tilde{e}(a)) \urcorner)$ is in PSPACE by §10.5 in [1]. Hence follows EXPTIME = PSPACE.

<u>Remark</u>. P. Pudlak has observed the folloiwng. If we do not include functions [a/b] and Rem(a,b) in $\mathfrak{S}_2^0$ and $\mathfrak{S}_2^1$ then Theorem 1 can be improved to

$$\mathfrak{S}_2^0 = \mathfrak{S}_2^1 \longrightarrow P = \text{LSPACE},$$

where LSPACE is "logarithmic space". This follows from Lemma 1 and the fact that sharply bounded formulas define predicates which are in LSPACE.

The proof of Lemma 1 gives also the following fact: for $\varphi(a,b)$ sharply bounded $\mathfrak{S}_2^0 \vdash \exists y \varphi(a,y) \Rightarrow \exists f$ log space computable such that $\mathbb{N} \models \varphi(a,f(a))$.

By PA we mean a system of Peano Arithmetic which includes all the function symbols of $\mathfrak{S}_2^0$ and their axioms i.e. which is an extension of $\mathfrak{S}_2^0$. PA may have extra functions e.g. 2^a. What we need on PA is the following

(1) Let PA - Prf $(w, \ulcorner \varphi \urcorner)$ be a formalized notion of "w is a formalized proof in PA of $\ulcorner \varphi \urcorner$". Then PA - Prf can be expressed by a bounded formula in S_2^1.

(2) If φ is provable in PA, then φ is true.

Let $SBS(\ulcorner\varphi\urcorner)$ be a $\sum_1^b$-formalized statement "$\ulcorner\varphi\urcorner$ is a Gödel number of a sharply bounded sentence in $S_2^{\aleph_0}$".

Theorem 2. Suppose that $\overset{\circ}{V}{}_2^1$ proves the following sequent:

(1) $SBS(\ulcorner\varphi\urcorner) \longrightarrow \exists w(PA - Prf(w, \ulcorner\varphi\urcorner) \vee PA - Prf(w, \ulcorner\neg\varphi\urcorner))$.

Then PSPACE = NP ∩ co-NP.

Proof. If (1) is provable in $\overset{\circ}{V}{}_2^1$, then by Parihk's theorem, there exists a term $t(a)$ in S_2 such that the following sequent is provable in $\overset{\circ}{V}{}_2^1$

(2) $SBS(\ulcorner\varphi\urcorner) \longrightarrow \exists w \leq t(\ulcorner\varphi\urcorner)(PA - Prf(w,\ulcorner\varphi\urcorner) \vee PA - Prf(w,\ulcorner\neg\varphi\urcorner))$.

Now by [5], the quantified Boolean formula probelm i.e. "$F = Q_1x_1 \cdots Q_nx_nE$ is true, where E is a Boolean expression involving the variables $x_1, x_2,\ldots,x_n$ and each Q_i is either $\exists$ or $\forall$" is PSACE-complete. We interprete propositional variable x by a number variable x with $x \leq 1$. Then $Q_1x_1 \cdots Q_nx_nE$ can be expressed by a sharply bounded formula. Therefore by (2) "$\ulcorner\varphi\urcorner = \ulcorner Q_1x_1 \cdots Q_nx_nE\urcorner$) is true" is expressed by "$\ulcorner\varphi\urcorner$ is a quantified Boolean sentence" $\wedge\ \exists w \leq t(\ulcorner\varphi\urcorner)\ PA - Prf(w, \ulcorner\varphi\urcorner)$ and by "$\ulcorner\varphi\urcorner$ is a quantified Boolean sentence" $\wedge\ \forall w \leq t(\ulcorner\varphi\urcorner)\ \neg PA - Prf(w, \ulcorner\neg\varphi\urcorner)$. Therefore our PSPACE-complete predicate is in NP and also in co-NP.

The system $\hat{S}_2^1$ is obtained from $S_2^{\aleph_1}$ by introducing a function symbol Tu and its finitely many defining axioms, where the intended meaning of $Tu(w, x, a)$ is that $Tu(w, x, a) = 1$ if w is a code expressing a Turing machine Mw and if x is a code of an input of Mw and universl Turing machine M accept $\langle w.x\rangle$ in the time less than $|a|$ and $Tu(w, x, a) = 0$ otherwise. $\hat{S}_2^1$ is a conservative extension of S_2^1. In the next theorem, PA is a system of Peano Arithmetic which is an extension of $\hat{S}_2^1$.

Let $CT(\ulcorner t \urcorner)$ be a formalized $\sum_0^b$-statement expressing "$\ulcorner t \urcorner$ is a closed term in $\hat{S}_2^1$".

Theorem 3. Suppose that $\overset{\circ}{V}{}_2^1$ proves the following sequent:

$$CT(\ulcorner t_1 \urcorner), CT(\ulcorner t_2 \urcorner) \longrightarrow \exists w(PA\text{-}Prf(w, \ulcorner t_1 \le t_2 \urcorner) \lor PA - Prf(w, \ulcorner t_2 \le t_1 \urcorner)).$$

Then EXPTIME = NP $\cap$ co-NP.

Proof goes in the same way as in Theorem 2, by using $\ulcorner Tu(Iw, Ia, \tilde{e}(a)) \urcorner$ and $\ulcorner 0 \urcorner$ as $\ulcorner t_1 \urcorner$ and $\ulcorner t_2 \urcorner$ respectively.

Remark. In Theorem 2 and Theorem 3, $\overset{\circ}{V}{}_2^1$ can be replaced by any system in which Parihk's theorem holds and we use the same function symbols as in S_2^1. Therefore we can replace $\overset{\circ}{V}{}_2^1$ e.g. by $\overset{\circ}{V}_2 = \bigcup_n \overset{\circ}{V}{}_2^n$.

NORMAL CONSISTENCY OF T_2^i.

A proof P in T_2^i is said to be T_2^i-normal if the following conditions are satisfied.

(1) P is in free variable normal form.

(2) Let $\vec{c}$ be all parameter variables in P (cf. §4.5 in [1]) and $\vec{b}$ be an enumeration of all other free variables in P satisfying the condition that if the elimination inference for b_i is below the elimination inference for b_j then $i < j$. There exists an assignment $t_i(\vec{c})$ for b_i satisfying the following conditions.

(i) $t_i(\vec{c})$ is a term in the language of S_2.

(ii) If the elimination inference of b_i is

$$\frac{A(b_i), \Gamma \longrightarrow \Delta. A(Sb_i)}{A(0), \Gamma \longrightarrow \Delta, A(t(b_1,\dots,b_{i-1}, \vec{c}))}$$

or

$$\frac{b_i \le t(b_1,\dots,b_{i-1}, \vec{c}), A(b_i), \Gamma \longrightarrow \Delta}{\exists x \le t(b_1,\dots,b_{i-1}, \vec{c}) \ A(x), \Gamma \longrightarrow \Delta}$$

or

$$\frac{b_i \leq t(b_1,\ldots,b_{i-1}, \vec{c}),\ \Gamma \longrightarrow \Delta,\ A(b_i)}{\Gamma \longrightarrow \Delta,\ \forall x \leq t(b_1,\ldots,b_{i-1}, \vec{c})\ A(x)}$$

then

$$a_1 \leq t_1(\vec{c}),\ldots,a_{i-1} \leq t_{i-1}(\vec{c}) \longrightarrow t(a_1,\ldots,a_{i-1}, \vec{c}) \leq t_i(\vec{c})$$

is provable without using logical inference, induction, or any free varriables other than $a_1,\ldots,a_{i-1}$ and $\vec{c}$. Such a proof is provided. All the information for the condition (ii) is called a supplementary proof.

Precisely P together with a supplementary proof is called a T_2^i-normal proof. Let $\Gamma_0 \longrightarrow \Delta_0$ be a sequent in which every formula belongs to $\sum_i^b \cup \prod_i^b$. Then by using free cut free proof of $\Gamma_0 \longrightarrow \Delta_0$, one can easily obtain a T_2^i-normal proof of $\Gamma_0 \longrightarrow \Delta_0$. We denote the formalized notion "w is a T_2^i-normal proof of $\ulcorner\Gamma_0 \longrightarrow \Delta_0\urcorner$" by T_2^i-NPrf(w, $\ulcorner\Gamma_0 \longrightarrow \Delta_0\urcorner$). NCon($T_2^i$) is defined to be $\forall w \neg T_2^i$-NPrf (w, $\ulcorner\longrightarrow\urcorner$). Since one can get a T_2^i-normal proof for any T_2^i-provable sequent in which every formula belongs to $\sum_i^b \cup \prod_i^b$, the standard argument shows that T_2^i does not prove NCon(T_2^i).

Remark. Though a normal proof is easily obtained from a free cut free proof, the formalized notion "a normal proof" is much more useful than the formalized notion "a free cut free proof". The reason is that the formalized notion "a free cut free proof" is very difficult to handle in Bounded Arithmetic since one needs exponential functions for the cut elimination procedure.

Relations between cut elimination procedures and exponential functions are knwon in work of R. Statman, J. Paris and A. Wilkie, and P. Pudlak. Let CET and CET^0 be a formalized statement of the cut elimination theorem of LK and of the propositional part of LK respectively. Then one can say the following as precise statements.

(A) $\quad S_2^1 \vdash \exp \longleftrightarrow CET^0$

and

(B) $S^1_2 \vdash \text{super exp} \longleftrightarrow CET^0$,

where exp is the axiom $\forall x(2^x$ exists) and superexp(a) is defined by

$$\text{superexp}(0) = 1 \qquad \text{and}$$
$$\text{superexp}(a+1) = 2^{\text{superexp}(a)}$$

and superexp is the axiom $\forall x$ (superexp (x) exists)

An outline of proofs of (A) and (B).

(A) The part $S^1_2 \vdash \exp \longrightarrow CET^0$ is obtained by analyzing Gentzen's original proof of CET. For a proof of $S^1_2 \vdash CET^0 \longrightarrow \exp$, let $n = |a|$, and $C_1, C_2, \ldots, C_{2n}, D_1, D_2, \ldots, D_{2n}$ be 4n propositional variables. Define $F_1, \ldots, F_n, A_1, \ldots, A_n, B_1, \ldots, B_n$ by the following equivalences

$$\begin{aligned}
F_i &\Leftrightarrow (C_2 \vee D_2) \wedge \cdots \wedge (C_{2i} \vee D_{2i}) \\
A_1 &\Leftrightarrow (C_1 \supset C_2) \wedge (D_1 \supset D_2) \\
B_1 &\Leftrightarrow (C_1 \supset D_2) \wedge (D_1 \supset C_2) \\
&\ \ \vdots \\
A_{i+1} &\Leftrightarrow F_i \supset ((C_{2i+1} \supset C_{2i+2}) \wedge (D_{2i+1} \supset D_{2i+2})) \\
B_{i+1} &\Leftrightarrow F_i \supset ((C_{2i+1} \supset D_{2i+2}) \wedge (D_{2i+1} \supset C_{2i+2})).
\end{aligned}$$

Let S be the sequent

$$C_1 \vee D_1,\ A_1 \vee B_1,\ C_3 \vee D_3,\ A_2 \vee B_2, \ldots, C_{2n-1} \vee D_{2n-1},\ A_n \vee B_n \longrightarrow C_{2n},\ D_{2n}.$$

Then "S is provable with cuts" can be formalized in S^1_2. To see this, first we have

$$C_1 \vee D_1,\ A_1 \vee B_1 \longrightarrow C_2 \vee D_2.$$

Then we have by induction on k

$$C_1 \vee D_1,\ A_1 \vee B_1, \ldots, C_{2i-1} \vee D_{2i-1},\ A_i \vee B_i \longrightarrow \bigwedge_{k=1}^{i} (C_{2k} \vee D_{2k}).$$

It is easily seen that

$$\bigwedge_{k=1}^{i} (C_{2k} \vee D_{2k}),\ A_{i+1} \vee B_{i+1},\ C_{2i+1} \vee D_{2i+1} \longrightarrow \bigwedge_{k=1}^{i+1} (C_{2k} \vee D_{2k}).$$

By using cuts, S is obtained from these sequents. Now suppose S were provable without cuts and let P be a cut-free proof of S. Let $E_{j0} = A_j$ and $E_{j1} = B_j$. Then for every $i_1,\ldots,i_n$ ($i_k = 0$ or 1), there is a branch of P corresponding to

$$E_{1i_1},\ldots,E_{ni_n}$$

which will come to $A_1 \vee B_1,\ldots,A_n \vee B_n$ in the antecedent of S. Therefore P has at least 2^n sequents and $\ulcorner P \urcorner \geq 2^a$.

(B) The part $S_2^1 \vdash$ super exp $\longrightarrow$ CET is also obtained by analyzing Gentzen's original proof of CET. For a proof of $S_2^1 \vdash$ CET $\longrightarrow$ super exp, it suffices to show

$$S_2^1 + \text{exp} + \text{CET} \vdash \text{super exp}.$$

Let L be the language consiting of $=$, a unary predicate $A(\ )$, constants $0,1$, and functions $+$, $\cdot$, 2^x. Let Γ_0 be finitely many simple defining axioms on 0, 1, $+$, $\cdot$, 2^x, and $=$ together with

$$\forall x \forall y\ (x = y \wedge A(x) \supset A(y)).$$

Let spexp (n) be an abbreviation of a term

$$\left.2^{2^{\cdot^{\cdot^{\cdot^{2}}}}}\right\} n.$$

Let S be sequent

$$\Gamma_0,\ A(0),\ \forall x(A(x) \supset A(x+1)) \longrightarrow P(\text{spexp}(n)).$$

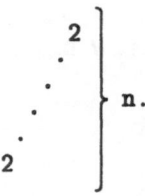

Then we can formalize a proof of S in $S_2^1 + \exp$ by using cuts. In order to see this, define $B(a)$ to be

$$\forall z(A(z) \supset A(z + 2^a)).$$

Then we have

(1) $\Gamma_0, \forall x(A(x) \supset A(x + 1)) \longrightarrow B(0)$

and

(2) $\Gamma_0 \longrightarrow \forall x(B(x) \supset B(x + 1)).$

Let $P(n)$ be a proof of

$$\Gamma_0, A(0), \forall x(A(x) \supset A(x + 1)) \longrightarrow A(\text{spexp }(n)).$$

Then replacing $A(x)$ by $B(x)$ in $P(n)$, we obtain

$$\Gamma_0, \forall x(A(x) \supset A(x + 1)) \longrightarrow \forall z(A(z) \supset A(z + \text{spexp}(n + 1))).$$

Therefore we can get

$$\Gamma_0, A(0), \forall x(A(x) \supset A(x + 1)) \longrightarrow A(\text{spexp }(n + 1)).$$

Now suppose there were a cut-free proof of S. Then such a proof must have at least superexp (n) many inferences since it must contain $A(m) \supset A(m + 1)$ for every m with $0 \leq m < \text{spexp}(n)$.

P. Pudlak has pointed out that the formulation of initial sequents of S_2^i and T_2^i in [1] is not adequate when we discuss S_2^{-1} since S_2^{-1} seems not to prove certain inequalitites we need. E.g. S_2^{-1} seems not to prove $|a| \leq a$. This is also inconvenient when we discuss S_2^i and T_2^i for $i \geq 0$. Therefore we assume that the following form of initial sequents are always included in S_2^i or T_2^i.

(1) $t \leq s \longrightarrow f(t) \leq f(s)$,

where f is S, $[\frac{1}{2}x]$, or $|x|$,

(2) $t_1 \leq s_1,\ t_2 \leq s_2 \longrightarrow f(t_1, t_2) \leq f(s_1, s_2)$,

where f is $+$, $\cdot$, or $\#$.

(3) $\longrightarrow [\frac{1}{2}s] \leq s$

and

$\longrightarrow |s| \leq s$.

<u>Definition</u>. We introduce a new unary predicate $\alpha(a)$ into S_2^{-1} and define Ind_α, J_α^1, J_α^2, and J_α by the following equivalences.

$$Ind_\alpha(a) \longleftrightarrow \forall x \leq a(\alpha(0) \wedge \forall y < x\ (\alpha(y) \supset \alpha(Sy)) \supset \alpha(x))$$
$$J_\alpha^1(a,b) \longleftrightarrow \forall x \leq b(Ind_\alpha(x) \supset Ind_\alpha(x + a))$$
$$J_\alpha^2(a,b) \longleftrightarrow \forall x \leq b(J_\alpha^1(x,b) \supset J_\alpha^1(x \cdot a, b))$$
$$J_\alpha(a,b) \longleftrightarrow \forall x \leq b(J_\alpha^2(x,b) \supset J_\alpha^2(x \# a, b)).$$

Let $A(a)$ be a formula. $Ind_A(a)$ and $J_A(a,b)$ are obtained from $Ind_\alpha(a)$ and $J_\alpha(a,b)$ respectively by substituting $\{x\}A(x)$ for α.

The following lemma is proved in a slightly different form in Pudlák [4] by using S_2^1 in the place of S_2^{-1}. Later S. Buss observed in [3] that S_2^{-1} is sufficient to prove all these sequents.

<u>Lemma 1</u>. S_2^{-1} proves the following.

(1) $J_\alpha(0,b)$.

(2) $J_\alpha(a,b) \longrightarrow J_\alpha(Sa,b),\ Ind_\alpha(b)$.

(3) $J_\alpha(a,b) \longrightarrow J_\alpha(|a|,b),\ Ind_\alpha(b)$.

(4) $J_\alpha(a,b) \longrightarrow J_\alpha([\frac{1}{2}a],b),\ Ind_\alpha(b)$.

(5) $J_\alpha(a_1,b),\ J_\alpha(a_2,b) \longrightarrow J_\alpha(a_1 + a_2,b),\ Ind_\alpha(b)$.

(6) $J_\alpha(a_1,b),\ J_\alpha(a_2,b) \longrightarrow J_\alpha(a_1 \cdot a_2,b),\ Ind_\alpha(b)$.

(7) $J_\alpha(a_1,b),\ J_\alpha(a_2,b) \longrightarrow J_\alpha(a_1 \# a_2,b),\ Ind_\alpha(b)$.

(8) $a \le b,\ Ind_\alpha(b) \longrightarrow Ind_\alpha(a)$.

(9) $J_\alpha(a,b) \longrightarrow Ind_\alpha(a)$.

In S_2^i and T_2^i, we use Gentzen type inferences. Unfortunately Gentzen type formulation of inferences is not adequate for an economical proof. Here we made the following slight modification. In original Gentzen type formulation the principal formula and the auxiliary formula are always at the end of segment e.g.

$$\frac{\Gamma \longrightarrow \Delta,\ A}{\neg A,\ \Gamma \longrightarrow \Delta}\ .$$

We generalize every inference-schema so that the principal formula and the auxiliary formula may not be at the end of the sequent. E.g. the inference $\neg$ left is now of the form

$$\frac{\Gamma,\ \Pi \longrightarrow \Delta,\ A,\ \Lambda}{\Gamma,\ \neg A,\ \Pi \longrightarrow \Delta,\ \Lambda}\ .$$

Now the following lemmas are self-explaining.

<u>Lemma 2</u>. Let $\ulcorner A \urcorner$ be the Gödel number of some expression A and $\ulcorner C \urcorner$ be the Gödel number of C which is obtained from A by substituting B for some variable in A. Then

$$\ulcorner C \urcorner \le \ulcorner A \urcorner \cdot (8\ \ulcorner B \urcorner)^{|\ulcorner A \urcorner|}\ .$$

Therefore

$$\ulcorner C \urcorner \le (\ulcorner A \urcorner)^4 \cdot (\ulcorner A \urcorner \# \ulcorner B \urcorner)\ .$$

Lemma 3. Let $\ulcorner P \urcorner$ be the Gödel number of some proof P including α and $\ulcorner A \urcorner$ be the Gödel number of a formula A. We substitute A for α in P after we change the names of bound variables in P in order to make A substitutable in P. The maximum of possible Gödel number of new bound variables is less than $15 + 2|k|$, where $k = \max(\ulcorner A \urcorner, \ulcorner P \urcorner)$. Let P' be the proof obtained from P by renaming bound variables. Then

$\ulcorner P' \urcorner \leq k^4 \cdot (k \# k)$ by Lemma 2. Let P'' be the proof obtained from P' by substituting A for α. Then by Lemma 1,

$$\ulcorner P'' \urcorner \leq (\ulcorner P' \urcorner)^4 \cdot (\ulcorner P' \urcorner \# \ulcorner A \urcorner).$$

Therefore there exists a fixed polynomial p_0 such that

$$\ulcorner A \urcorner \leq p_0(k \# k \# k).$$

Lemma 4. Let t be a closed term in S_2 and $A(a)$ be a formula. Let M be the maximum of the Gödel numbers of S_2^{-1}-proofs of (1),...,(9) in Lemma 1. Since t or $A(a)$ does not occur in (1),...,(9), M does not depend on $\ulcorner A(a) \urcorner$ or $\ulcorner t \urcorner$. Now

$$\longrightarrow J_A(t,t),\ Ind_A(t)$$

has a $O(k_1)$-length proof P in S_2^{-1}, where $k_1 = \max(\ulcorner t \urcorner, \ulcorner A(a) \urcorner)$. Therefore by Lemmas 2 and 3, there exists a fixed polynomial p_1 such that

$$\ulcorner P \urcorner \leq p_1(k_1 \# k_1 \# k_1 \# k_1).$$

Now from (9) of Lemma 1,

$$J_A(t,t) \longrightarrow Ind_A(t)$$

has some S_2^{-1}-proof Q such that

$$\ulcorner Q \urcorner \leq q(k_1 \# k_1 \# k_1)$$

where q is some fixed polynomial. Therefore the formula $Ind_A(t)$ has some S^{-1}_2-proof P_2 such that

$$\ulcorner P_2 \urcorner \leq p_2(k_1 \# k_1 \# k_1 \# k_1),$$

where p_2 is some fixed polynomial.

Using the notation defined in the introduction, we have the following theorem.

<u>Theorem</u>. $S_2^1 \vdash Con^{i+5}(S_2^{-1}) \longrightarrow N\ Con(T_2^i)$.

<u>Proof</u>. Let P be a T_2^i-normal proof to $\longrightarrow$. Notice that there are no parameter variables in P. Let $\Pi(\vec{a}) \longrightarrow \Lambda(\vec{a})$ be a sequent in P and P' be the subproof of P to $\Pi(\vec{a}) \longrightarrow N(\vec{a})$. Let $\vec{a}$ be $a_1,\ldots,a_k$ and $t_1,\ldots,t_k$ be closed terms assigned to $a_1,\ldots a_k$ respectively. Then we construct an S_2^{-1}-proof of

$$a_1 \leq t_1,\ldots,a_k \leq t_k,\ \Pi(\vec{a}) \longrightarrow \Lambda(\vec{a}),$$

in which every formula belongs to Π^b_{i+5}, We treat only the case that the last inference to $\Pi(\vec{a}) \longrightarrow \Lambda(\vec{a})$ be of the form

$$\frac{A(a_i),\ \Gamma \longrightarrow \Delta,\ A(Sa_i)}{A(0),\ \Gamma \longrightarrow \Delta,\ A(t(a_1,\ldots,a_{i-1}))}\ .$$

By the induction hypothesis, we have a S_2^{-1}-proof of

$$a_1 \leq t_1,\ldots,a_i \leq t_i,\ A(a_i),\ \Gamma \longrightarrow \Delta,\ A(Sa_i).$$

Since S_2^{-1} proves $a_1 \leq t_1,\ldots,a_{i-1} \leq t_{i-1} \longrightarrow t(a_1,\ldots,a_{i-1}) \leq t_i$, the following sequents are S_2^{-1}-provable.

$$a_1 \leq t_1,\ldots,a_{i-1} \leq t_{i-1},\ a_i \leq t(a_1,\ldots,a_{i-1}) \longrightarrow a_i \leq t_i$$

$$a_1 \leq t_1,\ldots,a_{i-1} \leq t_{i-1}, a_i \leq t(a_1,\ldots,a_{i-1}), A(a_i), \Gamma \longrightarrow \Delta,\ A(Sa_i)$$

$$a_1 \leq t_1,\ldots,a_{i-1} \leq t_{i-1}, Ind_A(t(a_1,\ldots,a_{i-1})), A(0), \Gamma \longrightarrow \Delta, A(t(a_1,\ldots,a_{i-1}))$$

and

$$\longrightarrow Ind_A(t_i).$$

Therefore the following sequent is S_2^{-1}-provable

$$a_1 \leq t_1, \ldots, a_{i-1} \leq t_{i-1}, A(0), \Gamma \longrightarrow \Delta, A(t(a_1, \ldots, a_{i-1})).$$

Obviously in this proof S_2^{-1}-proof of

$$a_1 \leq t_1, \ldots, a_{i-1} \leq t_{i-1}, A(0), \Gamma \longrightarrow \Delta, A(t(a_1, \ldots, a_{i-1}))$$

every formula belongs to Π^b_{i+5}. Therefore what is needed to show now is a bound of the so constructed proof to

$$a_1 \leq t_1, \ldots, a_{i-1} \leq t_{i-1}, A(0), \Gamma \longrightarrow \Delta, A(t(a_1, \ldots, a_{i-1})).$$

However Lemmas 1-4 immediately gives us a bound of the so constructed proof

$$p(\ulcorner P \urcorner \# \ulcorner P \urcorner \# \ulcorner P \urcorner \# \ulcorner P \urcorner)$$

where p is a fixed polynomial.

References

1. S. Buss, Bounded Arithmetic, Bibliopolis, 1986, Napoli.
2. S. Buss, Axiomatizations and Conservation Results for Fragments of Bounded Arithmetic, Manuscripts, 1987.
3. S. Buss, A letter to P. Pudlak, July, 1986.
4. P. Pudlak, $S_2 \vdash \mathrm{BDCon}(S_2^1)$, Manuscripts, June, 1986. See also "A note on bounded arithmetic" in Abstracts of the 8-th International Congress of Logic, Methodology, and Philosophy of Science, Moscow 1987, vol 1. pp. 156-160.
5. L. J. Stockmeyer and A. R. Meyer, Word problems requiring exponential time, Proc. 5th Ann. ACM Symp. on Theory of Computing, Association Computing Machinery, New York, 1-9, 1973.
6. G. Takeuti, Bounded Arithmetic and Truth Definition, Annals of Pure and Applied Logic, 39, 75-104, 1988.
7. G. Takeuti, Sharply bounded arithmetic and the function $a \dot{-} 1$, to appear in Contemporary Mathematics AMS, Proc. of Workshop in Logic and Computation.
8. A. Wilkie and J. Paris, On the scheme of induction for bounded arithmetic formulas, Logic Colloquium '84, Proc. of an ASL Conference in Manchester, England, North-Holland.

A SURVEY OF INTUITIONISTIC DESCRIPTIVE SET THEORY

Wim Veldman
Mathematisch Instituut, Katholieke Universiteit
Toernooiveld
6525 ED Nijmegen, The Netherlands

INTRODUCTION

In descriptive set theory (cf. Moschovakis 1980), a subject which was founded in the early decades of this century by French and Russian mathematicians like Baire, Borel, Lebesgue, Lusin and Suslin, one describes and studies classes of subsets of the set $\mathbb{R}$ of real numbers. Examples of such classes are: the class of *open* subsets of $\mathbb{R}$, the class of *closed* subsets of $\mathbb{R}$, the class of those subsets of $\mathbb{R}$ which are the union of a countable sequence of closed subsets of $\mathbb{R}$, and its dual: the class of those subsets of $\mathbb{R}$ which are the intersection of a countable sequence of open subsets of $\mathbb{R}, \ldots$, the class of *Borel* subsets of $\mathbb{R}$, i.e.: the least class of subsets of $\mathbb{R}$ which contains the closed subsets of $\mathbb{R}$ and the open subsets of $\mathbb{R}$ and is closed under the operations of countable union and countable intersection, the class of *analytical* subsets of $\mathbb{R}$, i.e.: the class of those subsets of $\mathbb{R}$ that result from projecting a closed subset of $\mathbb{R}^2$ on one of the coordinate-axes, the class of *co-analytical* subsets of $\mathbb{R}$, i.e.: the class of those subsets of $\mathbb{R}$ whose complement is analytical $, \ldots ,$ the class of *projective* subsets of $\mathbb{R}$, i.e.: the class of those subsets of $\mathbb{R}$ which result from a closed subset of some $\mathbb{R}^n$ by a finite number of applications of the operations of projection and complementation.

We want to study these classes from an intuitionistic point of view. As is usual in the subject, we will discuss classes of subsets of Baire space $\mathcal{N}$, rather than classes of subsets of the set $\mathbb{R}$ of real numbers. It is somewhat easier to work in Baire space and, without difficulty, one may obtain results on subsets of $\mathbb{R}$ from results on subsets of $\mathcal{N}$. For an intuitionistic mathematician, Baire space $\mathcal{N}$ coincides with the so-called *universal spread.*

We will interpret the set-theoretical operations of union, intersection, existential and universal projection in conformity with intuitionistic logic. Our further assumptions are the usual axioms for intuitionistic analysis as they have been formulated, for instance, in Kleene and Vesley 1965. We will describe them in the first section of

Mathematical Logic
Edited by P. P. Petkov
Plenum Press, New York, 1990

this paper. In the second section we describe the contents of the remaining part of the paper.

It is well-known that the early descriptive set theorists weighed their arguments carefully and that their criticisms of classical reasoning partially agree with Brouwer's. (cf. Lusin 1930). Let us try, therefore, to develop descriptive set theory intuitionistically.

1. SOME PRELIMINARIES

In this section, we will give some definitions, establish some notations and list the axioms for intuitionistic analysis that we will use.

1.1. $\mathbb{N}$ is the set of natural numbers, $\mathbb{N}^*$ is the set of finite sequences of natural numbers.
We suppose that some bijective mapping from $\mathbb{N}^*$ to $\mathbb{N}$ is given which associates to each finite sequence $(a_0, a_1, \ldots, a_{k-1})$ in $\mathbb{N}^*$ its code number $\langle a_0, a_1, \ldots, a_{k-1}\rangle$.
For each $a \in \mathbb{N}$, we may calculate $k = \text{length}(a)$ and numbers $a(0), a(1), \ldots, a(k-1)$ such that $a = \langle a(0), a(1), \ldots, a(k-1)\rangle$.
We assume that our coding function fulfils the condition:
$\forall a \in \mathbb{N} \forall j < \text{length}(a)[a(j) < a]$.
For each $a \in \mathbb{N}$, $p \in \mathbb{N}$ such that $p < \text{length}(a)$, we define
$\overline{a}(p) := \langle a(0), a(1), \ldots, a(p-1)\rangle$, the initial segment of a of length p.
$*$ is a function from $\mathbb{N}^2$ to $\mathbb{N}$, corresponding to concatenation:
for each $a = \langle a(0), a(1), \ldots, a(k-1)\rangle$ and $b = \langle b(0), b(1), \ldots, b(l-1)\rangle$ in $\mathbb{N}$:

$$a * b = \langle a(0), a(1), \ldots a(k-1), b(0), b(1), \ldots, b(l-1)\rangle.$$

We define, for $a, b \in \mathbb{N}$: $a \sqsubseteq b := \exists c \in \mathbb{N}[a * c = b]$.
$\langle\,\rangle$ is the empty sequence, the only sequence whose length is 0. Remark: $\langle\,\rangle = 0$.

1.2. $\mathcal{N}$ is the set of all functions from $\mathbb{N}$ to $\mathbb{N}$, i.e. the set of all infinite sequences of natural numbers.
Let $\alpha \in \mathcal{N}$ and $n \in \mathbb{N}$. We define: $\overline{\alpha}n := \langle \alpha(0), \alpha(1), \ldots, \alpha(n-1)\rangle$.
Let $\alpha \in \mathcal{N}$ and $a \in \mathbb{N}$. We define: $\alpha \in a := \exists n \in \mathbb{N}[\overline{\alpha}n = a]$.
We also define $a * \alpha$ in $\mathcal{N}$ by: for all $j < \text{length}(a)$: $a * \alpha(j) = a(j)$ and for all $j \geq \text{length}(a)$: $a * \alpha(j) = \alpha(j - \text{length}(a))$.
Let $\alpha \in \mathcal{N}$ and $n \in \mathbb{N}$. We define elements ${}^n\alpha$ and α^n of $\mathcal{N}$ by: for all $m \in \mathbb{N}$: ${}^n\alpha(m) = \alpha(n * m)$ and $\alpha^n(m) = \alpha(\langle n\rangle * m)$.
Let $\alpha \in \mathcal{N}$ and $m, n \in \mathbb{N}$. We define: $\alpha^{m,n} := (\alpha^m)^n$.
Let $\alpha, \beta \in \mathcal{N}$. We define: $\alpha \# \beta := \exists n \in \mathbb{N}[\alpha(n) \neq \beta(n)]$.
$\#$ is the well-known *apartness* relation on $\mathcal{N}$.
Let $n \in \mathbb{N}$. We define an element $\underline{n}$ of $\mathcal{N}$ by: for all $m \in \mathbb{N}$: $\underline{n}(m) = n$.
For each $k \in \mathbb{N}$, for each $\alpha_0, \alpha_1, \ldots, \alpha_{k-1} \in \mathcal{N}$ we define a sequence $\langle \alpha_0, \alpha_1, \ldots, \alpha_{k-1}\rangle$ in $\mathcal{N}$ by: $\forall j < k[\langle \alpha_0, \alpha_1, \ldots \alpha_{k-1}\rangle^j = \alpha_j], \forall j \geq k[\langle \alpha_0, \alpha_1, \ldots, \alpha_{k-1}\rangle^j = \underline{0}]$ and $\langle \alpha_0, \alpha_1, \ldots, \alpha_{k-1}\rangle(\langle\,\rangle) = 0$.
Let $\alpha, \beta \in \mathcal{N}$. We define $\beta \circ \alpha \in \mathcal{N}$ by: $\forall n \in \mathbb{N}[\beta \circ \alpha(n) = \beta(\alpha(n))]$.

1.3. We will use the following two *axioms of countable choice*:

1.3.1. $\mathbf{AC}_{00}$: For all subsets X of $\mathbb{N} \times \mathbb{N}$:
If $\forall m \in \mathbb{N} \exists n \in \mathbb{N}[X(m,n)]$, then $\exists \alpha \in \mathcal{N} \forall m \in \mathbb{N}[X(m, \alpha(m))]$.

1.3.2. $\mathbf{AC}_{01}$: For all subsets X of $\mathbb{N} \times \mathcal{N}$:
If $\forall m \in \mathbb{N} \exists \alpha \in \mathcal{N}[X(m,\alpha)]$, then $\exists \alpha \in \mathcal{N} \forall m \in \mathbb{N}[X(m, \alpha^m)]$.

(We use the notation "$X(m,n)$" for "$(m,n) \in X$").
In the classical theory also the contrapositives of $\mathbf{AC}_{00}$ and $\mathbf{AC}_{01}$ are important, but they are not true constructively. (Cf. Veldman 1983.)

1.4. $\mathcal{F}_0$ is the set of all (continuous) functions from $\mathcal{N}$ to $\mathbb{N}$.
We assume that, for all $f \in \mathcal{F}_0$, there exists $\gamma \in \mathbb{N}$ such that
$\forall \alpha \in \mathcal{N} \exists m \in \mathbb{N}[\gamma(\overline{\alpha}m) = f(\alpha) + 1 \wedge \forall n < m[\gamma(\overline{\alpha}n) = 0]]$;
we might say, then, that γ is a *code* for f.
$\mathcal{F}$ is the set of all (continuous) functions from $\mathcal{N}$ to $\mathcal{N}$.
If f belongs to $\mathcal{F}$ and α to $\mathcal{N}$, $f|\alpha$ is the image of α under f. We assume that, for all $f \in \mathcal{F}$, there exists $\gamma \in \mathcal{N}$ such that
$\forall p \in \mathbb{N} \forall \alpha \in \mathcal{N} \exists m \in \mathbb{N}[\gamma^p(\overline{\alpha}m) = (f|\alpha)(p) + 1 \wedge \forall n < m[\gamma^p(\overline{\alpha}n) = 0]$;
we might say, then, that γ is a *code* for f.

1.5. Let X, Y be subsets of $\mathcal{N}$. We define:

$X \preceq Y$ ("X is *reducible* to Y") $:= \exists f \in \mathcal{F} \forall \alpha \in \mathcal{N}[X(\alpha) \leftrightarrow Y(f|\alpha)]$.

This notion, classically known as *Wadge-reducibility*, will be important. One easily verifies that the relation $\preceq$ is reflexive and transitive.
We also define, for subsets X, Y of $\mathcal{N}$:

$X \prec Y$ ("X is *strictly reducible* to Y") $:= X \preceq Y \wedge \neg(Y \preceq X)$.

1.6. We will use the following three *axioms of continuity*:

1.6.1. **CP** (*continuity principle*): For all subsets X of $\mathcal{N} \times \mathbb{N}$:
If $\forall \alpha \in \mathcal{N} \exists n \in \mathbb{N}[X(\alpha, n)]$,
then $\forall \alpha \in \mathcal{N} \exists m \in \mathbb{N} \exists n \in \mathbb{N} \forall \beta \in \mathcal{N}[\overline{\alpha}m = \overline{\beta}m \rightarrow X(\beta, n)]$.

1.6.2. $\mathbf{AC}_{10}$: For all subsets X of $\mathcal{N} \times \mathbb{N}$:
If $\forall \alpha \in \mathcal{N} \exists n \in \mathbb{N}[X(\alpha, n)]$, then $\exists f \in \mathcal{F}_0 \forall \alpha \in \mathcal{N}[X(\alpha, f(\alpha))]$.

1.6.3. $\mathbf{AC}_{11}$: For all subsets X of $\mathcal{N} \times \mathcal{N}$:
If $\forall \alpha \in \mathcal{N} \exists \beta \in \mathcal{N}[X(\alpha, \beta)]$, then $\exists f \in \mathcal{F} \forall \alpha \in \mathcal{N}[X(\alpha, f|\alpha)]$.

The three axioms have been given here in order of increasing strength.
In Kleene and Vesley 1965 $\mathbf{AC}_{10}$ is called "Brouwer's principle for numbers" and $\mathbf{AC}_{11}$ is called "Brouwer's principle for functions".

1.7. Let $\sigma \in \mathcal{N}$. σ is called a *spreadlaw* if:

$$\sigma(\langle\,\rangle) = 0 \text{ and } \forall a \in \mathbb{N}[\sigma(a) = 0 \leftrightarrow \exists n \in \mathbb{N}[\sigma(a * \langle n \rangle) = 0]].$$

If σ is a spreadlaw and α belongs to $\mathcal{N}$, we say:
$\alpha \in \sigma$ ("σ *admits* α") $:= \forall n \in \mathbb{N}[\sigma(\overline{\alpha}n) = 0]$.
The set of all $\alpha \in \mathcal{N}$ such that the spreadlaw σ admits α is called a *spread*, and is a namesake of the spreadlaw itself.
The axioms 1.6.1-3 generalize to spreads; 1.6.3, for instance, extends to the following principle:

1.7.1. GAC$_{11}$: For every spread σ, and every subset X of $\sigma \times \mathcal{N}$:

If $\forall \alpha \in \sigma \exists \beta \in \mathcal{N}[X(\alpha, \beta)]$, then $\exists f \in \mathcal{F} \forall \alpha \in \sigma[X(\alpha, f|\alpha)]$.

$\mathcal{N}$ itself is a spread, sometimes $\mathcal{N}$ is called the *universal spread.*
Famous spreadlaws are σ_2 and σ_{2mon}, which are defined as follows: for all $a \in \mathbb{N}$:

$$\sigma_2(a) = 0 \leftrightarrow \forall j < \text{length}(a)[a(j) = 0 \vee a(j) = 1]$$
$$\sigma_{2mon}(a) = 0 \leftrightarrow (\sigma_2(a) = 0 \wedge \forall j \in \mathbb{N}[j + 1 < \text{length}(a) \rightarrow a(j) \leq a(j + 1)])$$

σ_2, the *binary fan*, is intuitionistic Cantor space.
σ_{2mon}, the *monotonous binary fan*, is an example of a simple but not completely trivial spread.
(A spreadlaw σ is called a *fan-law* if:
$\forall a \in \mathbb{N}[\sigma(a) = 0 \rightarrow \exists m \in \mathbb{N} \forall n \in \mathbb{N}[\sigma * \langle n \rangle) = 0 \rightarrow n < m]].$)
We will make use also of the fan τ which is defined by: for all $a \in \mathbb{N}$:

$$\tau(a) = 0 \leftrightarrow \quad (\forall i < \text{length}(a)[a(i) = 0 \vee a(i) = 1] \wedge$$
$$\forall i \in \mathbb{N} \forall j \in \mathbb{N}[(i < j < \text{length}(a) \wedge a(i) = a(j) = 1) \rightarrow i = j])$$

There exists a bijective mapping from the fan τ to the fan σ_{2mon}.

1.8. We introduce a class \$ of subsets of $\mathbb{N}$ by means of the following inductive definition:

(i) $\{\langle\ \rangle\} \in \$$
(ii) If $S_0, S_1, S_2, \ldots$ is a sequence of members of \$,
also $S := \{\langle\ \rangle\} \cup \bigcup_{n \in \mathbb{N}} \{\langle n \rangle * a \mid a \in S_n\}$ belongs to \$
(The following picture may be helpful:)

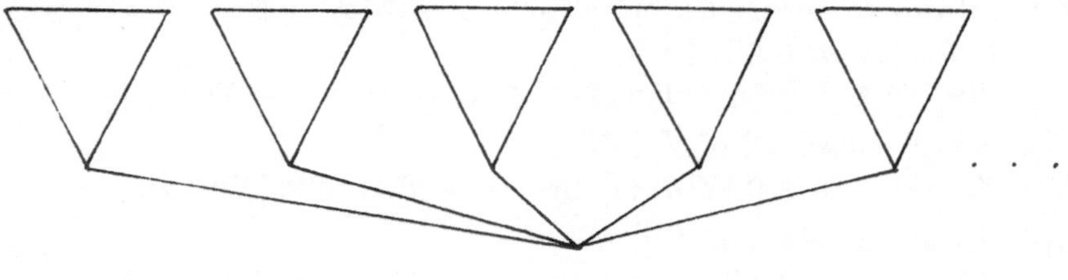

A stump and its substumps

(iii) Clauses (i) and (ii) give all members of \$.

Elements of \$ will be called *stumps.*
The rôle of stumps in the intuitionistic theory is comparable to that played by countable ordinals in the classical theory.
In particular, we will use stumps as indexes for Borel classes. One easily verifies by induction that each stump σ has the following properties:

(1) $\forall m \in \mathbb{N}[m \in \sigma \vee \neg(m \in \sigma)]$
(2) $\forall m \in \mathbb{N} \forall n \in \mathbb{N}[(m \in \sigma \wedge n \sqsubseteq m) \rightarrow n \in \sigma]$
(3) $\forall \alpha \in \mathcal{N} \exists n \in \mathbb{N}[\overline{\alpha}n \notin \sigma]$.

We define, for any $\alpha \in \mathcal{N}$, a subset B_α of $\mathbb{N}$ by:

$\forall a \in \mathbb{N}[a \in B_\alpha \leftrightarrow \forall i < \text{length}(a)[\alpha(\overline{a}(i)) \neq 0]]$.

One may prove, again by induction, using $\mathbf{AC}_{01}$, the axiom of countable choice mentioned in 1.3.2: $\forall \sigma \in \$ \exists \alpha \in \mathcal{N}[\sigma = B_\alpha]$.

We are now able to formulate the last of our axioms:

1.8.1. BT (*Brouwer's Thesis*): $\forall \alpha \in \mathcal{N}[\forall \beta \in \mathbb{N} \exists n \in \mathbb{N}[\alpha(\overline{\beta}n) = 0] \rightarrow B_\alpha \in \$]$.

BT is a consequence of Brouwers famous *bar theorem*.
In Kleene and Vesley 1965 the axiom does not occur in this form, but there is a *principle of induction on decidable bars* which is easily seen to follow from the axiom.

1.9. We use "iff" for "if and only if" and ◇ to denote the end of a proof.

2. PROSPECTUS

2.1. When building Borel subsets or projective subsets of Baire space $\mathcal{N}$, *we try to avoid the use of the operation of complementation.* As we admit closed subsets of $\mathcal{N}$ as well as open subsets of $\mathcal{N}$ as basic sets, this may seem harmless. But, intuitionistically, the complement of a closed set need not be open, the complement of a countable intersection of open sets need not be a countable union of closed sets, and so on. Negatively defined subsets of $\mathcal{N}$ will fall outside our scope.
We feel that positively defined subsets of $\mathcal{N}$ deserve of more attention than negatively defined ones.

2.2. The avoidance of the operation of complementation makes it difficult to prove the Borel hierarchy theorem, i.e.: to establish that no single Borel class contains all Borel sets. The classical argument which consists in "diagonalizing on universal elements", breaks down.

> It is true, intuitionistically as well as classically, that any Borel class $\mathcal{K}$ possesses a *universal element*, i.e.: there exists $U \in \mathcal{K}$ such that for all $X \in \mathcal{K}$ there exists $\alpha \in \mathcal{N}$ such that $\forall \beta \in \mathcal{N}[X(\beta) \leftrightarrow U(\langle \alpha, \beta \rangle)]$. So we may form a subset D of $\mathcal{N}$, by saying: $\forall \alpha \in \mathcal{N}[D(\alpha) \leftrightarrow \neg U(\langle \alpha, \alpha \rangle)]$. Then D does not belong to $\mathcal{K}$.
> *But we have no reason to believe that D is a positive Borel set.*

In sections 3-5 we give an intuitionistic argument.
In section 3 we consider the easy first level of closed and open sets,and we learn how the hierarchy theorem could be formulated intuitionistically.
In section 4 the second level is treated: together with $\mathbf{AC}_{00}$, the axiom of countable choice mentioned in 1.3.1, the continuity principle **CP** plays a key rôle in the very unclassical reasoning.
The general theorem is formulated in section 5, but not proved: a complete proof would take too much space.

2.3. In intuitionistic real analysis, the set $[0,1] \cup [1,2]$, i.e. the union of the closed segments $[0,1]$ and $[1,2]$, does not coincide with the closed segment $[0,2]$: there are real numbers x in $[0,2]$ for which we are unable to decide: $x \leq 1 \vee 1 \leq x$.
Apparently, a union of two closed subsets of the set $\mathbb{R}$ of real numbers need not be closed.

In section 6, we will see that the same is true in Baire space, and that more may be said: in general, the class of unions of $n+1$ closed sets properly contains the class of unions of n closed sets.
The class of countable unions of closed sets will be called $\mathbf{\Sigma}^0_2$. The just mentioned result implies that there exists a strictly increasing infinite sequence in the structure $<\mathbf{\Sigma}^0_2, \preceq>$.
We will prove a strong theorem, enabling us to see that the structure $<\mathbf{\Sigma}^0_2, \preceq>$ is even more complicated : to any strictly increasing sequence withing $<\mathbf{\Sigma}^0_2, \preceq>$ we may form an upper bound which itself is the first element of a new strictly increasing sequence within $<\mathbf{\Sigma}^0_2, \preceq>$.
The results of section 6 are typically intuitionistic and use the continuity principle **CP**. (cf. 1.6.1).

2.4. In section 7 we turn to (positively) analytical and (positively) co-analytical subsets of $\mathcal{N}$. We will see that all Borel subsets of $\mathcal{N}$ are analytical, but that there exists a union of two closed subsets of $\mathcal{N}$ which is not co-analytical. It is not true that every inhabited closed subset of $\mathcal{N}$ is the range of a (continuous) function from $\mathcal{N}$ to $\mathcal{N}$. Using the axiom **BT** (Brouwer's thesis, cf. 1.8.1), we are able to prove a version of a famous theorem of Suslin's, viz. that every subset of $\mathcal{N}$ which is both co-analytical and the range of a function from $\mathcal{N}$ to $\mathcal{N}$, is a Borel subset of $\mathcal{N}$.

2.5. In section 8 we will see that every positively projective subset of $\mathcal{N}$ belongs to the class $\mathbf{\Sigma}^1_2$, i.e.: to any such set X we may determine an open subset C of $\mathcal{N}$ such that $\forall\alpha \in \mathcal{N}[X(\alpha) \leftrightarrow \exists\beta \in \mathcal{N}\forall\gamma \in \mathcal{N}[C(\langle\alpha,\beta,\gamma\rangle)]]$. Thus, there is no projective hierarchy beyond $\mathbf{\Sigma}^1_2$.
This follows rather easily from the axiom $\mathbf{AC}_{11}$. (Cf. 1.6.3).

2.6. In section 9 we consider the subset S of $\mathcal{N}$ which is defined by:

$$\forall\alpha \in \mathcal{N}[S(\alpha) \leftrightarrow \exists\gamma \in \sigma_{2mon}\forall n \in \mathbb{N}[\alpha(\overline{\gamma}n) = 0]]$$

(The fan σ_{2mon} has been mentioned in 1.7).
It turns out that the set S is not a positively Borel subset of $\mathcal{N}$, although, from classical experience, one might surmise that S belongs to the class $\mathbf{\Sigma}^0_2$, or, even more, that it is a closed subset of $\mathcal{N}$. This fact confirms the intuitionist's feeling that quantification over a spread, even over such a small spread as σ_{2mon}, is completely different from quantification over a set like the set $\mathbb{N}$ of natural numbers.
We will give some indications as to how to establish that the set S cannot be obtained from closed sets and open sets by means of the operations of countable union and intersection, without giving the complete proof, for that would be too long.

2.7. We list the remaining sections of the paper:

3. Closed sets and open sets
4. Countable intersections of open sets and countable unions of closed sets
5. The Borel hierarchy theorem
6. Some remarks on the structure $<\mathbf{\Sigma}^0_2, \preceq>$
7. Analytical sets and co-analytical sets
8. The collapse of the projective hierarchy

9. A remarkable set
10. Some concluding remarks

2.8. Several theorems will only be mentioned in this paper and not proved, or not proved completely. Full proofs may be found in Veldman 1981.

3. CLOSED SETS AND OPEN SETS

3.1. We define: a subset X of $\mathcal{N}$ is *closed* iff:

there exists $\beta \in \mathcal{N}$ such that $\forall \alpha \in \mathcal{N}[X(\alpha) \leftrightarrow \forall n \in \mathbb{N}[\beta(\overline{\alpha}n) = 0]]$.

$\mathbf{\Pi}_1^0$ is the class of closed subsets of $\mathcal{N}$.
We introduce a subset A_1 of $\mathcal{N}$ by: $\forall \alpha \in \mathcal{N}[A_1(\alpha) \leftrightarrow \forall n \in \mathbb{N}[\alpha(n) = 0]]$.
The set A_1 belongs to the class $\mathbf{\Pi}_1^0$.

3.2. One verifies easily that for every subset X of $\mathcal{N}$: $X \in \mathbf{\Pi}_1^0 \leftrightarrow X \preceq A_1$.

3.3. We define: a subset X of $\mathcal{N}$ is *open* iff:

there exists $\beta \in \mathcal{N}$ such that $\forall \alpha \in \mathcal{N}[X(\alpha) \leftrightarrow \exists n \in \mathbb{N}[\beta(\overline{\alpha}n) = 0]]$.

$\mathbf{\Sigma}_1^0$ is the class of open subsets of $\mathcal{N}$.
We introduce a subset E_1 of $\mathcal{N}$ by: $\forall \alpha \in \mathcal{N}[E_1(\alpha) \leftrightarrow \exists n \in \mathbb{N}[\alpha(n) = 0]]$.
The set E_1 belongs to the class $\mathbf{\Sigma}_1^0$.

3.4. One verifies easily that for every subset X of $\mathcal{N}$: $X \in \mathbf{\Sigma}_1^0 \leftrightarrow X \preceq E_1$.

3.5. We wish to show that the class $\mathbf{\Pi}_1^0$ is not included in the class $\mathbf{\Sigma}_1^0$ and that the class $\mathbf{\Sigma}_1^0$ is not included in the class $\mathbf{\Pi}_1^0$.
In view of 3.2. and 3.4, it suffices to show that the set A_1 is not reducible to the set E_1 and that the set E_1 is not reducible to the set A_1. We will prove, in 3.7 and 3.8, two positive statements from which these assertions follow.

3.6. We introduce subsets A_1^* and E_1^* of $\mathcal{N}$ by:
$\forall \alpha \in \mathcal{N}[A_1^*(\alpha) \leftrightarrow \forall n \in \mathbb{N}[\alpha(n) \neq 0]]$ and $\forall \alpha \in \mathcal{N}[E_1^*(\alpha) \leftrightarrow \exists n \in \mathbb{N}[\alpha(n) \neq 0]]$.
Observe that $\langle A_1^*, E_1 \rangle$ is a *separate pair of subsets of* $\mathcal{N}$, i.e.:
$\forall \alpha \in \mathcal{N} \forall \beta \in \mathcal{N}[(A_1^*(\alpha) \wedge E_1(\beta)) \rightarrow \alpha \# \beta]$.
Also $\langle E_1^*, A_1 \rangle$ is a separate pair of subsets of $\mathcal{N}$.

3.7. Theorem: Let $f \in \mathcal{F}$ be such that $\forall \alpha \in \mathcal{N}[A_1(\alpha) \rightarrow E_1(f|\alpha)]$.
Then $\exists \alpha \in \mathcal{N}[E_1^*(\alpha) \wedge E_1(f|\alpha)]$.

Proof: Suppose: $f \in \mathcal{F}$ and $\forall \alpha \in \mathcal{N}[A_1(\alpha) \rightarrow E_1(f|\alpha)]$. Then $E_1(f|\underline{0})$.
Calculate $m_0 \in \mathbb{N}$ such that $(f|\underline{0})(m_0) = 0$.
Calculate $n_0 \in \mathbb{N}$ such that $\forall \alpha \in \mathcal{N}[\overline{\alpha}n_0 = \overline{\underline{0}}n_0 \rightarrow (f|\alpha)(m_0) = (f|\underline{0})(m_0)]$.
Consider $\alpha := \overline{\underline{0}}n_0 * \underline{1}$ and observe: $E_1^*(\alpha)$ and $E_1(f|\alpha)$. ◇

3.8. Theorem: Let $f \in \mathcal{F}$ be such that $\forall \alpha \in \mathcal{N}[E_1(\alpha) \rightarrow A_1(f|\alpha)]$.
Then $\exists \alpha \in \mathcal{N}[A_1^*(\alpha) \wedge A_1(f|\alpha)]$.

Proof: Suppose $f \in \mathcal{F}$ and $\forall \alpha \in \mathcal{N}[E_1(\alpha) \rightarrow A_1(f|\alpha)]$. We claim even more than the conclusion of the theorem, viz. $\forall \alpha \in \mathcal{N}[A_1(f|\alpha)]$.
Proof of this claim: let $\alpha \in \mathcal{N}$ and $n \in \mathbb{N}$. Calculate $(f|\alpha)(n)$. Calculate $m_0 \in \mathbb{N}$ such that $\forall \beta \in \mathcal{N}[\overline{\beta}m_0 = \overline{\alpha}m_0 \rightarrow (f|\beta)(n) = (f|\alpha)(n)]$. Consider $\beta := \overline{\alpha}m_0 * \underline{0}$ and observe: $E_1(\beta)$, therefore: $A_1(f|\beta)$, and $(f|\alpha)(n) = (f|\beta)(n) = 0$. ◇

4. COUNTABLE INTERSECTIONS OF OPEN SETS AND COUNTABLE UNIONS OF CLOSED SETS

4.1. We define a class $\mathbf{\Pi}_2^0$ of subsets of $\mathcal{N}$. A subset X of $\mathcal{N}$ belongs to $\mathbf{\Pi}_2^0$ iff there exists a sequence $Y_0, Y_1, \dots$ of members of $\mathbf{\Sigma}_1^0$ such that $X = \bigcap_{n \in \mathbb{N}} Y_n$.
We introduce a subset A_2 of $\mathcal{N}$ by: $\forall \alpha \in \mathcal{N}[A_2(\alpha) \leftrightarrow \forall n \in \mathbb{N}[E_1(\alpha^n)]]$.
The set A_2 belongs to the class $\mathbf{\Pi}_2^0$.

4.2. Using $\mathbf{AC}_{01}$, the axiom of countable choice introduced in 1.3.2, one may verify: for every subset X of $\mathcal{N}$: $X \in \mathbf{\Pi}_2^0 \leftrightarrow X \preceq A_2$.
Also: $\mathbf{\Pi}_1^0 \subseteq \mathbf{\Pi}_2^0$ and: $\mathbf{\Sigma}_1^0 \subseteq \mathbf{\Pi}_2^0$ and: the class $\mathbf{\Pi}_2^0$ is closed under the operation of countable intersection.

4.3. We define a class $\mathbf{\Sigma}_2^0$ of subsets of $\mathcal{N}$. A subset of $\mathcal{N}$ belongs to $\mathbf{\Sigma}_2^0$ iff there exists a sequence $Y_0, Y_1, \dots$ of members of $\mathbf{\Pi}_1^0$ such that $X = \bigcup_{n \in \mathbb{N}} Y_n$.
We introduce a subset E_2 of $\mathcal{N}$ by: $\forall \alpha \in \mathcal{N}[E_2(\alpha) \leftrightarrow \exists n \in \mathbb{N}[A_1(\alpha^n)]]$.
The set E_2 belongs to the class $\mathbf{\Sigma}_2^0$.

4.4. Using again the axiom $\mathbf{AC}_{01}$ (cf. 1.3.2), one may verify:
for every subset X of $\mathcal{N}$: $X \in \mathbf{\Sigma}_2^0 \leftrightarrow X \preceq E_2$.
Also: $\mathbf{\Pi}_1^0 \subseteq \mathbf{\Sigma}_2^0$ and: $\mathbf{\Sigma}_1^0 \subseteq \mathbf{\Sigma}_2^0$ and: the class $\mathbf{\Sigma}_2^0$ is closed under the operation of countable union.

4.5. We wish to prove that the class $\mathbf{\Sigma}_2^0$ is not included in the class $\mathbf{\Pi}_2^0$ and that the class $\mathbf{\Pi}_2^0$ is not included in the class $\mathbf{\Sigma}_2^0$. In view of 4.2 and 4.4, it suffices to show that the set E_2 is not reducible to the set A_2, and that the set E_2 is not reducible to the set A_2.
We will prove two constructive theorems from which these negatively formulated statements follow easily. The two theorems, 4.7 and 4.9, have very different proofs. Theorem 4.7 might be said to be an elementary result, whereas the proof of theorem 4.9 uses $\mathbf{AC}_{00}$, the axiom of countable choice, (cf. 1.3.1), and the typically intuitionistic continuity principle **CP** (cf. 1.6.1).

4.6. We introduce subsets A_2^* and E_2^* of $\mathcal{N}$ by:
$\forall \alpha \in \mathcal{N}[A_2^*(\alpha \leftrightarrow \forall n \in \mathbb{N}[E_1^*(\alpha^n)]]$ and $\forall \alpha \in \mathcal{N}[E_2^*(\alpha) \leftrightarrow \exists n \in \mathbb{N}[A_1^*(\alpha^n)]]$.
(The sets E_1^* and A_1^* were introduced in 3.6.)
Observe that $\langle A_2^*, E_2 \rangle$ is a *separate pair of subsets of* $\mathcal{N}$, i.e.:
$\forall \alpha \in \mathcal{N} \forall \beta \in \mathcal{N}[(A_2^*(\alpha) \wedge E_2(\beta)) \rightarrow \alpha \# \beta]$.
Also $\langle E_2^*, A_2 \rangle$ is a separate pair of subsets of $\mathcal{N}$.

4.7. Theorem: Let $f \in \mathcal{F}$ be such that $\forall \alpha \in \mathcal{N}[E_2(\alpha) \rightarrow A_2(f|\alpha)]$.
Then $\exists \alpha \in \mathcal{N}[A_2^*(\alpha) \wedge A_2(f|\alpha)]$.

Proof: Suppose: $f \in \mathcal{F}$ and $\forall \alpha \in \mathcal{N}[E_2(\alpha) \rightarrow A_2(f|\alpha)]$.
We will construct an infinite sequence $a_0, a_1, \dots$ of (finite sequences of) natural numbers such that:

(1) $\forall n \in \mathbb{N}[a_n \sqsubseteq a_{n+1} \wedge a_n \neq a_{n+1}]$
(2) $\forall n \in \mathbb{N} \forall \alpha \in \mathcal{N}[\alpha \in a_n \rightarrow (E_1^*(\alpha^n) \wedge E_1((f|\alpha)^n))]$
(3) $\forall n \in \mathbb{N} \exists \alpha \in \mathcal{N}[\alpha \in a_n \wedge \forall m > n[\alpha^m = \underline{0}]]$

If we succeed in constructing such a sequence, we form $\alpha \in \mathcal{N}$ such that

$\forall n \in \mathbb{N}[\alpha \in a_n]$. Then $A_2^*(\alpha)$ and $A_2(f|\alpha)$.
We build the sequence $a_0, a_1, \ldots$ inductively:

Suppose $n \in \mathbb{N}$ and we have defined $a_0, a_1, \ldots, a_{n-1}$ in such a way that they fulfil the requirements (1), (2) and (3).
Define $\alpha := a_{n-1} * \underline{0}$. (If $n = 0$, define $\alpha := \underline{0}$.)
Remark: $E_2(\alpha)$, therefore: $A_2(f|\alpha)$, in particular $E_1((f|\alpha)^n)$.
Calculate $n_0 \in \mathbb{N}$ such that $(f|\alpha)^n(n_0) = 0$.
Calculate $p_0 \in \mathbb{N}$ such that $\forall \beta \in \mathcal{N}[\overline{\beta}p_0 = \overline{\alpha}p_0 \rightarrow (f|\beta)^n(n_0) = (f|\alpha)^n(n_0)]$.
Calculate $q_0 \in \mathbb{N}$ such that $p_0 \leq \langle n, q_0 \rangle$ and $\text{length}(a_{n-1}) \leq \langle n, q_0 \rangle$.
Define $a_n \in \mathbb{N}$ such that $a_{n-1} \sqsubseteq a_n$ and $\text{length}(a_n) = \langle n, q_0 \rangle + 1$ and $\overline{a_n}(p_0) = \overline{\alpha}p_0$, $\forall j \in \mathbb{N}[p_0 \leq j < \langle n, q_0 \rangle \rightarrow a_n(j) = 0]$ and $a_n(\langle n, q_0 \rangle) = 1$.
Then $\forall \alpha \in \mathcal{N}[\alpha \in a_n \rightarrow (\alpha^n(q_0) = 1 \wedge (f|\alpha)^n(n_0) = 0)]$.
Observe: a_n fulfils the requirements (1), (2), (3). ◇

4.8. Let $\gamma, \alpha \in \mathcal{N}$. We define a sequence $\gamma \bowtie \alpha$ (read: "α-*as-corrected-by*-γ") in the following way: $\gamma \bowtie \alpha(\langle\,\rangle) = \alpha(\langle\,\rangle)$, for all $m \in \mathbb{N}$: $(\gamma \bowtie \alpha)^m(\gamma(m)) = 0$ and for all $m, n \in \mathbb{N}$ such that $n \neq \gamma(m)$: $(\gamma \bowtie \alpha)^m(n) = \alpha^m(n)$.
Using $\mathbf{AC_{00}}$, the axiom of countable choice mentioned in 1.3.1, one verifies easily:

$\forall \alpha \in \mathcal{N}[A_2(\alpha) \leftrightarrow \exists \gamma \in \mathcal{N}[\alpha = \gamma \bowtie \alpha]]$.

4.9. Theorem: Let $f \in \mathcal{F}$ be such $\forall \alpha \in \mathcal{N}[A_2(\alpha) \rightarrow E_2(f|\alpha)]$.
Then $\exists \alpha \in \mathcal{N}[E_2^*(\alpha) \wedge E_2(f|\alpha)]$.

Proof: Suppose: $f \in \mathcal{F}$ and $\forall \alpha \in \mathcal{N}[A_2(\alpha) \rightarrow E_2(f|\alpha)]$. Observe:
$\forall \gamma \in \mathcal{N} \forall \alpha \in \mathcal{N}[A_2(\gamma \bowtie \alpha)]$, therefore: $\forall \gamma \in \mathcal{N} \forall \alpha \in \mathcal{N}[E_2(f|(\gamma \bowtie \alpha))]$ i.e.
$\forall \gamma \in \mathcal{N} \forall \alpha \in \mathcal{N} \exists n \in \mathbb{N}[A_1((f|(\gamma \bowtie \alpha))^n)]$.
Using the continuity principle **CP** (cf. 1.6.1), we calculate $p, n \in \mathbb{N}$ such that
$\forall \gamma \in \mathcal{N} \forall \alpha \in \mathcal{N}[\overline{\gamma}p = \overline{\alpha}p = \overline{0}p \rightarrow (f|(\gamma \bowtie \alpha))^n = \underline{0}]$.
Let $\beta \in \mathcal{N}$ be such that $\forall j \in \mathbb{N}[j < p \rightarrow \beta^j = \underline{0}]$.
Observe: $\forall m \in \mathbb{N} \exists \gamma \in \mathcal{N} \exists \alpha \in \mathcal{N}[\overline{\gamma}p = \overline{\alpha}p = \overline{0}p \wedge \overline{\gamma \bowtie \alpha}m = \overline{\beta}m]$.
We claim: $(f|\beta)^n = \underline{0}$.
(Proof of this claim:

Let $q \in \mathbb{N}$. Calculate $(f|\beta)^n(q)$. Calculate $m \in \mathbb{N}$ such that
$\forall \delta \in \mathcal{N}[\overline{\delta}m = \overline{\beta}m \rightarrow (f|\delta)^n(q) = (f|\beta)^n(q)]$
Calculate $\gamma, \alpha \in \mathcal{N}$ such that $\overline{\gamma \bowtie \alpha}m = \overline{\beta}m$ and $\overline{\gamma}p = \overline{\alpha}p = \overline{0}p$.
Then $(f|\beta)^n(q) = (f|(\gamma \bowtie \alpha))^n(q) = 0$. End of proof of claim.)

Let $\alpha \in \mathcal{N}$ be such that $\forall j \in \mathbb{N}[j < p \rightarrow \alpha^j = \underline{0}]$ and $\alpha^p = \underline{1}$.
Then $E_2^*(\alpha)$ and $E_2(f|\alpha)$. ◇

5. THE BOREL HIERARCHY THEOREM

5.1. The set of stumps, \$, has been defined in 1.8.
Let $\sigma \in \$$, $\sigma \neq \{\langle\,\rangle\}$ and $n \in \mathbb{N}$. We define $\sigma^n := \{a \mid a \in \mathbb{N} \mid \langle n \rangle * a \in \sigma\}$.
σ^n itself is a stump; it is called the n-th *substump* of σ.

5.2. We define, by induction on the structure of the stumps, for every stump σ, classes $\mathbf{\Sigma}(\sigma)$ and $\mathbf{\Pi}(\sigma)$ of subsets of $\mathbb{N}$, in the following way:

(i) $\mathbf{\Sigma}(\{\langle\,\rangle\}) := \mathbf{\Sigma}_1^0$ and $\mathbf{\Pi}(\{\langle\,\rangle\}) := \mathbf{\Pi}_1^0$

(ii) Suppose $\sigma \in \$$, $\sigma \neq \{\langle\,\rangle\}$.

A subset X of $\mathcal{N}$ belongs to $\mathbf{\Sigma}(\sigma)$ iff there exists a sequence $Y_0, Y_1, \ldots$ of subsets of $\mathcal{N}$ such that $\forall n \in \mathbb{N}[Y_n \in \mathbf{\Pi}(\sigma^n)]$ and $X = \bigcup_{n \in \mathbb{N}} Y_n$

A subset X of $\mathcal{N}$ belongs to $\mathbf{\Pi}(\sigma)$ iff there exists a sequence $Y_0, Y_1, \ldots$ of subsets of $\mathcal{N}$ such that $\forall n \in \mathbb{N}[Y_n \in \mathbf{\Sigma}(\sigma^n)]$ and $X = \bigcap_{n \in \mathbb{N}} Y_n$.

A subset X of $\mathcal{N}$ is a *(positively) Borel* set iff there exists a stump $\sigma \in \$$ such that $X \in \mathbf{\Sigma}(\sigma)$ or $X \in \mathbf{\Pi}(\sigma)$

We also introduce, for each $\sigma \in \$$, subsets $E(\sigma)$ and $A(\sigma)$ of $\mathcal{N}$, as follows:

(i) $E(\{\langle\,\rangle\}) := E_1$ and $A(\{\langle\,\rangle\}) := A_1$.

(ii) Suppose $\sigma \in \$$, $\sigma \neq \{\langle\,\rangle\}$. Then:
$\forall \alpha \in \mathcal{N}[(E(\sigma))(\alpha) \leftrightarrow \exists n \in \mathbb{N}[(A(\sigma^n))(\alpha^n)]$ and
$\forall \alpha \in \mathcal{N}[(A(\sigma))(\alpha) \leftrightarrow \forall n \in \mathbb{N}[(E(\sigma^n))(\alpha^n)]$.

For each $\sigma \in \$$, the set $E(\sigma)$ belongs to the class $\mathbf{\Sigma}(\sigma)$ and the set $A(\sigma)$ belongs to the class $\mathbf{\Pi}(\sigma)$.

5.3. Using the axiom $\mathbf{AC}_{01}$ (cf. 1.3.2) one may verify: for every $\sigma \in \$$, every subset X of $\mathcal{N}$: $X \in \mathbf{\Sigma}(\sigma) \leftrightarrow X \preceq E(\sigma)$ and $X \in \mathbf{\Pi}(\sigma) \leftrightarrow X \preceq A(\sigma)$.

5.4. We also introduce, for each $\sigma \in \$$, subsets $E^*(\sigma)$ and $A^*(\sigma)$ of $\mathcal{N}$, as follows:

(i) $E^*(\{\langle\,\rangle\}) := E_1^*$ and $A^*(\{\langle\,\rangle\}) := A_1^*$.

(ii) Suppose $\sigma \in \$$, $\sigma \neq \{\langle\,\rangle\}$. Then:
$\forall \alpha \in \mathcal{N}[(E^*(\sigma))(\alpha) \leftrightarrow \exists n \in \mathbb{N}[(A^*(\sigma^n))(\alpha^n)]$ and
$\forall \alpha \in \mathcal{N}[(A^*(\sigma))(\alpha) \leftrightarrow \forall n \in \mathbb{N}[(E^*(\sigma^n))(\alpha^n)]$.

Observe, that, for each $\sigma \in \$$, $\langle A(\sigma), E^*(\sigma)\rangle$ and $\langle E(\sigma), A^*(\sigma)\rangle$ are separate pairs of subsets of $\mathcal{N}$, i.e.: $\forall \alpha \in \mathcal{N} \forall \beta \in \mathcal{N}[((A(\sigma))(\alpha) \wedge (E^*(\sigma))(\beta)) \rightarrow \alpha \# \beta]$, and similarly for the second pair.

5.5. We introduce a subclass $HI\$$ of $\$$, consisting of the so-called *hereditarily iterative stumps*, by means of the following inductive definition:

(i) $\{\langle\,\rangle\} \in HI\$$.

(ii) If $S_0, S_1, S_2, \ldots$ is a sequence of members of $HI\$$, also
$S := \{\langle\,\rangle\} \cup \bigcup_{m,n \in \mathbb{N}} \{\langle 2^m(2n+1) - 1\rangle * a \mid a \in S_n\}$ belongs to $HI\$$.

(iii) Clauses (i) and (ii) give all members of $\$$.

5.6. Borel hierarchy theorem: Let $\sigma \in HI\$$ and $f \in \mathcal{F}$.

(1) If $\forall \alpha \in \mathcal{N}[(A(\sigma))(\alpha) \rightarrow (E(\sigma))(f|\alpha)]$, then $\exists \alpha \in \mathcal{N}[(E^*(\sigma))(\alpha) \wedge (E(\sigma))(f|\alpha)]$.

(2) If $\forall \alpha \in \mathcal{N}[(E(\sigma))(\alpha) \rightarrow (A(\sigma))(f|\alpha)]$, then $\exists \alpha \in \mathcal{N}[(A^*(\sigma))(\alpha) \wedge (A(\sigma))(f|\alpha)]$.

Proof: In Veldman 1981. $\diamond$

The proof extends the methods used in the proofs of theorems 4.7 and 4.9. The only intuitionistic axioms used are $\mathbf{AC}_{01}$ and **CP** (cf. 1.3.2 and 1.6.1).

6. SOME REMARKS ON THE STRUCTURE $<\mathbf{\Sigma}_2^0, \preceq>$

The structure $<\mathbf{\Sigma}_2^0, \preceq>$ is rather complex. The fact that the union of two closed subsets of $\mathcal{N}$ is not necessarily closed, is the first one of its many unclassical properties.

6.1. Let X be a subset of $\mathcal{N}$ and $n \in \mathbb{N}$. We define a subset D^nX of $\mathcal{N}$ by:

$\forall\alpha \in \mathcal{N}[D^n X(\alpha) \leftrightarrow \exists j < n[X(\alpha^j)]$.

6.2. Consider $D^2 A_1 = \{\alpha \mid \alpha \in \mathcal{N} \mid \alpha^0 = \underline{0} \vee \alpha^1 = \underline{0}\}$ and observe:
for each subset X of $\mathcal{N}$:
$X \preceq D^2 A_1 \leftrightarrow$ there exists closed subsets Y, Z of $\mathcal{N}$ such that $X = Y \cup Z$.

6.3. Theorem: $D^2 A_1$ is not a closed subset of $\mathcal{N}$, i.e.: $\neg(D^2 A_1 \preceq A_1)$.

Proof: Suppose that $D^2 A_1$ is a closed subset of $\mathcal{N}$. Determine $\beta \in \mathcal{N}$ such that $\forall\alpha \in \mathcal{N}[D^2 A_1(\alpha) \leftrightarrow \forall n \in \mathbb{N}[\beta(\overline{\alpha}n) = 0]]$
Consider the spread τ which has been introduced in 1.7. Remark:
$\forall\alpha \in \mathcal{N}[\alpha \in \tau \leftrightarrow (\forall i \in \mathbb{N}[\alpha(i) \leq 1] \wedge \forall i \in \mathbb{N}\forall j \in \mathbb{N}[\alpha(i) = \alpha(j) = 1 \rightarrow i = j])]$.
Observe: $\forall\alpha \in \tau\forall n \in \mathbb{N}\exists\gamma \in \mathcal{N}[D^2 A_1(\gamma) \wedge \overline{\gamma}n = \overline{\alpha}n]$.
Therefore: $\forall\alpha \in \tau\forall n \in \mathbb{N}[\beta(\overline{\alpha}n) = 0]$, and: $\forall\alpha \in \tau[D^2 A_1(\alpha)]$, i.e.:
$\forall\alpha \in \tau[\alpha^0 = \underline{0} \vee \alpha^1 = \underline{0}]$.
We apply the *generalized* continuity principle **CP** (cf. 1.6.1 and 1.7) and calculate $n \in \mathbb{N}$ such that: $\forall\alpha \in \tau[\overline{\alpha}n = \overline{\underline{0}}n \rightarrow \alpha^0 = \underline{0}] \vee \forall\alpha \in \tau[\overline{\alpha}n = \overline{\underline{0}}n \rightarrow \alpha^1 = \underline{0}]$.
But, obviously, $\exists\alpha \in \tau[\overline{\alpha}n = \overline{\underline{0}}n \wedge \alpha^0 \neq 0]$ and $\exists\alpha \in \tau[\overline{\alpha}n = \overline{\underline{0}}n \wedge \alpha^1 \neq \underline{0}]$.
Contradiction. ◇

6.4. Theorem: $D^3 A_1$ is not the union of two closed subsets of $\mathcal{N}$, i.e.:
$\neg(D^3 A_1 \preceq D^2 A_1)$.

Proof: Suppose: X_0, X_1 are closed subsets of $\mathcal{N}$ and $D^3 A_1 = X_0 \cup X_1$. Consider the subsets Q_0, Q_1, Q_2 of $\mathcal{N}$ which are defined by: $\forall i < 2\forall\alpha \in \mathcal{N}[Q_i(\alpha) \leftrightarrow \alpha^i = \underline{0}]$ and observe: Q_0, Q_1 and Q_2 are spreads (cf. 1.7.) and $D^3 A_1 = Q_0 \cup Q_1 \cup Q_2$. Also: $\underline{0} \in Q_0 \cap Q_1 \cap Q_2$. We apply the (generalized) continuity principle **CP** three times and find $m_0, m_1, m_2 \in \mathbb{N}$ and $n_0, n_1, n_2 \in \{0,1\}$ such that:
$\forall i < 2\forall\alpha \in \mathcal{N}[(Q_i(\alpha) \wedge \overline{\alpha}m_i = \overline{\underline{0}}m_i) \rightarrow X_{n_i}(\alpha)]$.
Without loss of generality, we assume: $m = m_0 = m_1$ and $n_0 = n_1 = 0$. Then: $\forall\alpha \in \mathcal{N}[((Q_0(\alpha) \vee Q_1(\alpha)) \wedge \overline{\alpha}m = \overline{\underline{0}}m) \rightarrow X_0(\alpha)]$.
We consider again the spread τ which has the property:
$\forall\alpha \in \mathcal{N}[\alpha \in \tau \leftrightarrow (\forall i \in \mathbb{N}[\alpha(i) \leq 1] \wedge \forall i \in \mathbb{N}\forall j \in \mathbb{N}[\alpha(i) = \alpha(j) = 1 \rightarrow i = j])]$.
Observe: $\forall\alpha \in \tau\forall p \in \mathbb{N}\exists\gamma \in \mathcal{N}[(Q_0(\gamma) \vee Q_1(\gamma)) \wedge \overline{\gamma}p = \overline{\alpha}p]$
Determine $\beta \in \mathcal{N}$ such that $\forall\alpha \in \mathcal{N}[X_0(\alpha) \leftrightarrow \forall n \in \mathbb{N}[\beta(\overline{\alpha}n) = 0]]$
Conclude: $\forall\alpha \in \tau[\overline{\alpha}m = \overline{\underline{0}}m \rightarrow \forall n \in \mathbb{N}[\beta(\overline{\alpha}n) = 0]]$, and:
$\forall\alpha \in \tau[\overline{\alpha}m = \overline{\underline{0}}m \rightarrow X_0(\alpha)]$.
Observe: $X_0 \subseteq D^3 A_1$ and $\forall\alpha \in \tau[\overline{\alpha}m \neq \overline{\underline{0}}m \rightarrow D^3 A_1(\alpha)]$.
Therefore: $\forall\alpha \in \tau[D^3 A_1(\alpha)]$.
This leads to a contradiction, like the statement $\forall\alpha \in \tau[D^2 A_1(\alpha)]$ did in the proof of theorem 6.4. ◇

6.5. Theorem: The sequence $A_1, D^2 A_1, D^3 A_1, \ldots$ is *strictly increasing* i.e.:
$\forall n \in \mathbb{N}[D^n A_1 \prec D^{n+1} A_1]$.

Proof: A straightforward generalization of the proof of theorem 6.4. ◇

6.6. Obviously, the set E_2 is an upper bound for the sequence $A_1, D^2 A_1, \ldots$ i.e.: $\forall n \in \mathbb{N}[D^n A_1 \prec E_2]$, but, as we will see, there are many much smaller upper bounds. We first observe that there exists a *least upper bound* for the sequence $A_1, D^2 A_1, \ldots$; in fact, the structure $<\Sigma^0_2, \preceq>$ is a *countably complete upper semi-lattice.*

In the following, S denotes the successor function on $\mathbb{N}$.
If $\alpha = \alpha(0), \alpha(1), \alpha(2), \ldots$ belongs to $\mathcal{N}$, then $\alpha \circ S = \alpha(1), \alpha(2), \ldots$.

6.7. Let X, Y be subsets of $\mathcal{N}$. We define a subset $\mathcal{M}(X,Y)$ of $\mathcal{N}$, as follows:
$\forall \alpha \in \mathcal{N}(\mathcal{M}(X,Y))(\alpha) \leftrightarrow \forall n \in \mathbb{N}[(\alpha(0) = 0 \to X(\alpha \circ S)) \wedge (\alpha(0) \neq 0 \to Y(\alpha \circ S))]]$.
Let $X_0, X_1, \ldots$ be a sequence of subsets of $\mathcal{N}$. We define a subset $\mathcal{M}_{n=0}^{\infty}(X_n)$, as follows: $\forall \alpha \in \mathcal{N}[(\mathcal{M}_{n=0}^{\infty}(X_n))(\alpha) \leftrightarrow \forall n \in \mathbb{N}[\alpha(0) = n \to X_n(\alpha \circ S)]]$.

6.8. We observe the following:
For all subsets X, Y of $\mathcal{N}$:

(i) $X \preceq \mathcal{M}(X,Y)$ and $Y \preceq \mathcal{M}(X,Y)$.
(ii) For each subset Z of $\mathcal{N}$: if $X \preceq Z$ and $Y \preceq Z$, then $\mathcal{M}(X,Y) \preceq Z$.

For each sequence $X_0, X_1, \ldots$ of subsets of $\mathcal{N}$:

(i) $\forall n \in \mathbb{N}[X_n \preceq \mathcal{M}_{k=0}^{\infty}(X_k)]$
(ii) For each subset Z of $\mathbb{N}$: if $\forall n \in \mathbb{N}[X_n \preceq Z]$, then $\mathcal{M}_{k=0}^{\infty}(X_k) \preceq Z$.

The proofs are straightforward. Let us look at the argument for the second part of the second statement. Suppose $Z, X_0, X_1, \ldots$ is a sequence of subsets of $\mathcal{N}$ and $\forall n \in \mathbb{N}[X_n \preceq Z]$. Using the axiom of countable choice, $\mathbf{AC}_{01}$, (cf. 1.3.2), determine a sequence $f_0, f_1, \ldots$ of (codes of) elements of $\mathcal{F}$ such that
$\forall n \in \mathbb{N} \forall \alpha \in \mathcal{N}[X_n(\alpha) \leftrightarrow Z(f_n|\alpha)]$.
Define $f \in \mathcal{F}$ such that $\forall \alpha \in \mathcal{N}[f|\alpha = f_{\alpha(0)} \mid (\alpha \circ S)]$.
Then $\forall \alpha \in \mathcal{N}[(\mathcal{M}_{k=0}^{\infty}(X_k))(\alpha) \leftrightarrow Z(f|\alpha)]$, i.e.: $\mathcal{M}_{k=0}^{\infty}(X_k) \preceq Z$.

6.9. The class $\mathbf{\Sigma}_2^0$ is closed under the infinitary operation $\mathcal{M}$, introduced in 6.7, as, for each sequence $X_0, X_1, \ldots$ of subsets of $\mathcal{N}$:
$\forall \alpha \in \mathbb{N}[(\mathcal{M}_{n=0}^{\infty}(X_n))(\alpha) \leftrightarrow \exists n \in \mathbb{N}[\alpha(0) = n \wedge X_n(\alpha \circ S)]]$.
In particular, the set $\mathcal{M}_{n=0}^{\infty}(D^n A_1)$ belongs to $\mathbf{\Sigma}_2^0$.
For our purposes, however, it is more rewarding to consider another upper bound for the sequence $A_1, D^2 A_1, \ldots$

6.10. Let X, Y be subsets of $\mathcal{N}$. We define a subset $D(X,Y)$ of $\mathcal{N}$, called the *disjunction* of X and Y, as follows: $\forall \alpha \in \mathcal{N}[(D(X,Y))(\alpha) \leftrightarrow (X(\alpha^0) \vee Y(\alpha^1))]$.

6.11. Let $P_0, P_1, \ldots$ be a sequence of subsets of $\mathcal{N}$.
We call the sequence $P_0, P_1, \ldots$ *disjunctively closed* iff:

$\forall m \in \mathbb{N} \forall n \in \mathbb{N} \exists k \in \mathbb{N}[D(P_m, P_n) \preceq P_k]$.

We call the sequence $P_0, P_1, \ldots$ *strictly increasing* iff: $\forall m \in \mathbb{N}[P_m \prec P_{m+1}]$.
We introduce a subset $\mathcal{L}_{n=0}^{\infty}(P_n)$ of $\mathcal{N}$, called *the limit of the sequence* $P_0, P_1, \ldots$, as follows:

$$\forall \alpha \in \mathcal{N}[(\mathcal{L}_{n=0}^{\infty}(P_n))(\alpha) \leftrightarrow \exists n \in \mathbb{N}[\alpha^0(n) = 0 \wedge \forall p < n[\alpha^0(p) \neq 0] \wedge P_n(\alpha^{n+1})]].$$

(A word of comment on the latter definition.
When does $\alpha \in \mathbb{N}$ belongs to the limit set $\mathcal{L}_{n=0}^{\infty}(P_n)$?
There are two necessary and sufficient conditions:

(1) α^0, the first subsequence of α, must give a *sign*, i.e.: $\exists n \in \mathbb{N}[\alpha^0(n) = 0]$.
(2) If n is this sign, i.e.: $\alpha^0(n) = 0$ and $\forall p < n[\alpha^0(p) \neq 0]$, then α^{n+1}, the $(n+1)$-st subsequence of α, must belong to P_n.

We call α^0 the *signalling* subsequence of α.)

6.12. Theorem: Let $P_0, P_1, \ldots$ be a sequence of subsets of $\mathcal{N}$ which is strictly

increasing and disjunctively closed.
Let $R := \mathcal{L}_{n=0}^{\infty}(P_n)$. Then:
(i) $\forall n \in \mathbb{N}[P_n \preceq R]$.
(ii) $\forall n \in \mathbb{N}[D^n R \prec D^{n+1} R]$.

Proof: (i) Let $n \in \mathbb{N}$. Define $f \in \mathcal{F}$ such that
$\forall \alpha \in \mathcal{N}[(f|\alpha)^0(n) = 0 \wedge \forall q < n[(f|\alpha)^0(q) \neq 0] \wedge (f|\alpha)^{n+1} = \alpha]$.
Then $\forall \alpha \in \mathcal{N}[P_n(\alpha) \leftrightarrow R(f|\alpha)]$.

(ii) It is obvious that $\forall n \in \mathbb{N}[D^n R \preceq D^{n+1} R]$.

Assume: $n \in \mathbb{N}$ and $D^{n+1} R \preceq D^n R$.
Determine $f \in \mathcal{F}$ such that $\forall \alpha \in \mathcal{N}[D^{n+1} R(\alpha) \leftrightarrow D^n R(f|\alpha)]$.
We define, for each $\alpha \in \mathcal{N}$ and each $p \in \mathbb{N}$, natural numbers $c_p(\alpha)$ and $d_p(\alpha)$, as follows:

$$c_p(\alpha) := \#\{j \mid j < n+1 \mid \overline{\alpha^{j,0}}p = \overline{\underline{0}}p\}$$
$$d_p(\alpha) := \#\{j \mid j < n \mid \overline{(f|\alpha)^{j,0}}p = \overline{\underline{0}}p\}$$

(Here we use the symbol # to denote the cardinality of a finite set.
$c_p(\alpha)$ counts the number of signalling sequences among $\alpha^{0,0}, \alpha^{1,0}, \ldots, \alpha^{n,0}$ which, up to moment p, have given no sign.
$d_p(\alpha)$ is the number of signalling sequences among $(f|\alpha)^{0,0}, (f|\alpha)^{1,0}, \ldots, (f|\alpha)^{n-1,0}$ which, up to moment p, did not give a sign.)
We now venture the following <u>claim</u>:

$$\forall j < n \forall \alpha \in \mathcal{N}[\forall p \in \mathbb{N}[c_p(\alpha) > j] \rightarrow \forall p \in \mathbb{N}[d_p(\alpha) > j]]$$

This claim may be proved by induction, starting with $j = 0$, then for $j = 1$, and so on.
For lack of space, we do not spell out the proof.
After n steps we have the following conclusion:

$$\forall \alpha \in \mathcal{N}[\forall p \in \mathbb{N}[c_p(\alpha) \geq n] \rightarrow \forall p \in \mathbb{N}[d_p(\alpha) \geq n]]$$

Therefore: $\forall \alpha \in \mathcal{N}[\forall j < n[\alpha^{j,0} = \underline{0}] \rightarrow \forall j < n[(f|\alpha)^{j,0} = \underline{0}]]$
and: $\forall \alpha \in \mathbb{N}[\forall j < n[\alpha^{j,0} = \underline{0}] \rightarrow \neg D^n R(f|\alpha)]$.
On the other hand: $\exists \alpha \in \mathbb{N}[\forall j < n[\alpha^{j,0} = \underline{0}] \wedge D^{n+1} R(\alpha)]$.
This is an obvious contradiction. ◇

6.13. Observe that the class $\mathbf{\Sigma}_2^0$ is closed under the limit-operation $\mathcal{L}$, introduced in 6.11. Therefore, theorem 6.12 may be applied arbitrarily many times, even within $\mathbf{\Sigma}_2^0$.
The limit-operation $\mathcal{L}$ does not coincide with the infinitary operation $\mathcal{M}$ of taking least upper bounds: one may prove that $D^2 \mathcal{M}_{n=0}^{\infty}(D^n A_1) \preceq \mathcal{M}_{n=0}^{\infty}(D^n A_1)$.

7. ANALYTICAL SETS AND CO-ANALYTICAL SETS

7.1. We define: a subset X of $\mathcal{N}$ is *analytical* iff:
there exists a closed subset C of $\mathcal{N}$ such that $\forall \alpha \in \mathcal{N}[X(\alpha) \leftrightarrow \exists \beta \in \mathcal{N}[C(\langle \alpha, \beta \rangle)]]$.
$\mathbf{\Sigma}_1^1$ is the class of analytical subsets of $\mathcal{N}$.
We introduce a subset E_1^1 of $\mathcal{N}$ by: $\forall \alpha \in \mathcal{N}[E_1^1(\alpha) \leftrightarrow \exists \beta \in \mathbb{N} \forall n \in \mathbb{N}[\alpha(\overline{\beta}n) = 0]]$.
The set E_1^1 belongs to the class $\mathbf{\Sigma}_1^1$.

7.2. One may verify that for every subset X of $\mathcal{N}$: $X \in \mathbf{\Sigma}_1^1 \leftrightarrow X \preceq E_1^1$.

Using $\mathbf{AC}_{01}$, the countable axiom of choice introduced in 1.3.2, one may prove that the class $\mathbf{\Sigma}^1_1$ is closed under the operations of countable union and countable intersection. Also, every closed subset of $\mathcal{N}$ is analytical. Therefore, for every $\sigma \in \$$: $\mathbf{\Pi}(\sigma) \subseteq \mathbf{\Sigma}^1_1$ and $\mathbf{\Sigma}(\sigma) \subseteq \mathbf{\Sigma}^1_1$, i.e.: every subset of $\mathcal{N}$ which is (positively) Borel, is analytical.
It now follows from theorem 5.6 that E^1_1 is not a Borel set.

7.3. We define: a subset X of $\mathcal{N}$ is ***co-analytical*** iff:
there exists an open subset D of $\mathcal{N}$ such that $\forall\alpha \in \mathcal{N}[X(\alpha) \leftrightarrow \forall\beta \in \mathcal{N}[D(\langle\alpha,\beta\rangle)]]$.
$\mathbf{\Pi}^1_1$ is the class of co-analytical subsets of $\mathcal{N}$.
We introduce a subset A^1_1 of $\mathcal{N}$ by: $\forall\alpha \in \mathcal{N}[A^1_1(\alpha) \leftrightarrow \forall\beta \in \mathbb{N}\exists n \in \mathbb{N}[\alpha(\overline{\beta}n) = 0]]$.
The set A^1_1 belongs to the class $\mathbf{\Pi}^1_1$.

7.4. One may verify that for every subset X of $\mathcal{N}$: $X \in \mathbf{\Pi}^1_1 \leftrightarrow X \preceq A^1_1$.
Using $\mathbf{AC}_{01}$, the axiom of countable choice, one may prove that the class $\mathbf{\Pi}^1_1$ is closed under the operation of countable intersection.
The following theorem shows that the class $\mathbf{\Pi}^1_1$ is not closed under the operation of finite union.
Observe that the set A_1 belongs to the class $\mathbf{\Pi}^1_1$, i.e.: $A_1 \preceq A^1_1$.

7.5. Theorem: The set D^2A_1 is not co-analytical, i.e.: $\neg(D^2A_1 \preceq A^1_1)$.

Proof: Suppose:
$f \in \mathcal{F}$ and $\forall\alpha \in \mathcal{N}[(\alpha^0 = \underline{0} \vee \alpha^1 = \underline{0}) \leftrightarrow \forall\beta \in \mathcal{N}\exists n \in \mathbb{N}[(f|\alpha)(\overline{\beta}n) = 0]]$.
Let τ be the spread with the property:
$\forall\alpha \in \mathcal{N}[\alpha \in \tau \leftrightarrow (\forall i \in \mathbb{N}[\alpha(i) \le 1] \wedge \forall i \in \mathbb{N}\forall j \in \mathbb{N}[\alpha(i) = \alpha(j) = 1 \rightarrow i = j])]$.
We claim: $\forall\beta \in \mathcal{N}\forall\alpha \in \tau\exists n \in \mathbb{N}[(f|\alpha)(\overline{\beta}n) = 0]$.
Proof of claim: let $\beta \in \mathcal{N}$. Calculate $n_0 \in \mathbb{N}$ such that $(f|\underline{0})(\overline{\beta}n_0) = 0$.
Calculate $q_0 \in \mathcal{N}$ such that
$\forall\alpha \in \mathcal{N}[\overline{\alpha}q_0 = \overline{0}q_0 \rightarrow (f|\alpha)(\overline{\beta}n_0) = (f|\underline{0})(\overline{\beta}n_0) = 0]$.
Let $\alpha \in \tau$ and distinguish two cases:
(1) $\overline{\alpha}q_0 = \overline{0}q_0$. Then: $(f|\alpha)(\overline{\beta}n_0) = 0$.
(2) $\overline{\alpha}q_0 \neq \overline{0}q_0$. Then: $D_2A_1(\alpha)$, therefore $A^1_1(f|\alpha)$,
in particular $\exists n \in \mathbb{N}[(f(\alpha)(\overline{\beta}n) = 0]$.
In both cases, we have $\exists n \in \mathbb{N}[(f|\alpha)(\overline{\beta}n) = 0]$. End of proof of claim.
Therefore: $\forall\alpha \in \tau[A^1_1(f|\alpha)]$, and: $\forall\alpha \in \tau[D^2A_1(\alpha)]$. But this is contradictory, as we saw in the proof of theorem 6.3. ◇

7.6. Observe that the class $\mathbf{\Pi}^0_2$ is a subclass of the class $\mathbf{\Pi}^1_1$. We may conclude from theorem 7.5 that $D^2A_1 \notin \mathbf{\Pi}^0_2$. Another conclusion from this theorem is: $\neg(E^1_1 \preceq A^1_1)$, i.e.: E^1_1 is not co-analytical.
One may prove the following stronger result, which, by its wording, is reminiscent of theorem 5.6:

Let $f \in \mathcal{F}$ be such that $\forall\alpha \in \mathcal{N}[E^1_1(\alpha) \rightarrow A^1_1(f|\alpha)]$, i.e.:
$\forall\alpha \in \mathcal{N}[\exists\gamma \in \mathcal{N}\forall n \in \mathbb{N}[\alpha(\overline{\gamma}n) = 0] \rightarrow \forall\gamma \in \mathcal{N}\exists n \in \mathbb{N}[(f|\alpha)(\overline{\gamma}n) = 0]]$.
Then $\exists\alpha \in \mathcal{N}[\forall\gamma \in \mathcal{N}\exists n \in \mathbb{N}[\alpha(\overline{\gamma}n) \neq 0] \wedge \forall\gamma \in \mathcal{N}\exists n \in \mathbb{N}[(f|\alpha)(\overline{\gamma}n) = 0]]$.

We do not give the proof but we hope that the following hint suffices: given a function $f \in \mathcal{F}$ which satisfies the conditions of the theorem, construct $\alpha \in \mathcal{N}$ in such a way that, for every $c \in \mathbb{N}$: $\alpha(c) \neq 0$ if and only if $\overline{\alpha}c$ is sufficiently long for concluding: $\exists d \in \mathbb{N}[d \sqsubseteq c \wedge (f|\alpha)(d) = 0]$).

7.7. We introduce a subset *Fun* of $\mathcal{N}$ by:
$\forall\gamma \in \mathcal{N}[Fun(\gamma) \leftrightarrow \forall p \in \mathbb{N}\forall\alpha \in \mathcal{N}\exists m \in \mathbb{N}[\gamma^p(\overline{\alpha}m) \neq 0]]$.
Observe that for each $\gamma \in \mathcal{N}$, if $Fun(\gamma)$, then γ codes a function f_γ from $\mathcal{N}$ to $\mathcal{N}$ (cf. 1.4).
If $\gamma, \alpha \in \mathbb{N}$ and $Fun(\gamma)$, we define: $\gamma|\alpha := f_\gamma|\alpha$.
If $\gamma \in \mathcal{N}$ and $Fun(\gamma)$, we define: $Ran(\gamma) := Ran(f_\gamma)$, therefore:
$\forall\alpha \in \mathcal{N}[\alpha \in Ran(\gamma) \leftrightarrow \exists\beta \in \mathcal{N}[\alpha = \gamma|\beta]]$.
We call a subset X of $\mathcal{N}$ *strictly analytical* iff $\exists\gamma \in \mathcal{N}[Fun(\gamma) \wedge X = Ran(\gamma)]$. Observe that each strictly analytical subset of $\mathcal{N}$ is analytical. Of course, the empty set, $\emptyset$, is analytical and not strictly analytical.
Is it true that any subset X of $\mathcal{N}$ which is *inhabited* (i.e.: $\exists\alpha \in \mathcal{N}[X(\alpha)]$) and analytical, is strictly analytical?
Let us introduce, for each $\alpha \in \mathcal{N}$ a subset D_α of $\mathcal{N}$ by:
$\forall\beta \in \mathcal{N}[D_\alpha(\beta) \leftrightarrow (\forall n \in \mathbb{N}[\overline{\beta}n = \overline{\underline{0}}n \vee (\overline{\beta}n = \overline{\underline{1}}n \vee \overline{\alpha}n = \overline{\underline{0}}n)])]$.
Observe that, for each $\alpha \in \mathcal{N}$, D_α is a closed subset of $\mathcal{N}$, and $\forall\alpha \in \mathcal{N}[D_\alpha(\underline{0})]$.
Also, for each $\alpha \in \mathcal{N}$: $\forall\beta \in \mathcal{N}[D_\alpha(\beta) \rightarrow (\beta = \underline{0} \vee \beta = \underline{1})]$, i.e.: the set D_α has at most two members.
Assume that we should know that every set D_α is strictly analytical in the following strong sense: there exists a function $g \in \mathcal{F}$ such that
$\forall\alpha \in \mathcal{N}[Fun(g|\alpha) \wedge D_\alpha = Ran(g|\alpha)]$. Let g be such a function. Observe that $\underline{1} \in D_{\underline{0}}$ and determine $\beta \in \mathcal{N}$ such that $(g|\underline{0})|\beta = \underline{1}$.
Calculate $p_0 \in \mathbb{N}$ such that $(g|\underline{0})^0(\overline{\beta}p_0) = 2$ and $\forall q < p_0[(g|\underline{0})^0(\overline{\beta}q) = 0]$.
Calculate $m_0 \in \mathbb{N}$ such that $\forall\alpha \in \mathcal{N}[\overline{\alpha}m_0 = \overline{\underline{0}}m_0 \rightarrow \forall q \leq p[(g|\alpha)^0(\overline{\beta}q) = (g|\underline{0})^0(\overline{\beta}q)]]$.
Then $\forall\alpha \in \mathcal{N}[\overline{\alpha}m_0 = \overline{\underline{0}}m_0 \rightarrow \underline{1} \in D_\alpha]$, and this is obviously false.

7.8. We now prepare for proving a version of Suslin's theorem (theorem 7.11).
Let A, B be subsets of the set $\mathbb{N}$ of (finite sequences of) natural numbers.
We say: $A \leq B$ ("A is *embeddable* into B") iff $\exists\gamma \in \mathcal{N}(\forall a \in A[\gamma(a) \in B] \wedge$
$\forall a \in \mathbb{N}[\text{length}(a) = \text{length}(\gamma(a))] \wedge \forall a \in A\forall b \in A[a \sqsubseteq b \rightarrow \gamma(a) \sqsubseteq \gamma(b)]]$.
We define, for each $\alpha \in \mathcal{N}$, a subset B_α of $\mathbb{N}$ by:
$\forall a \in \mathbb{N}[a \in B_\alpha \leftrightarrow \forall i < \text{length}(a)[\alpha(\overline{a}(i)) \neq 0]]$.

7.9. Boundedness lemma: Let $f \in \mathcal{F}$ be such that $\forall\alpha \in \mathcal{N}[A_1^1(f|\alpha)]$.
Then there exists $\beta \in \mathcal{N}$ such that $A_1^1(\beta)$ and $\forall\alpha \in \mathcal{N}[B_{f|\alpha} \leq B_\beta]$.

Proof: Suppose: $f \in \mathcal{F}$ and $\forall\alpha \in \mathcal{N}[A_1^1(f|\alpha)]$.
Then $\forall\alpha \in \mathcal{N}\forall\gamma \in \mathcal{N}\exists n \in \mathbb{N}[(f|\alpha)(\overline{\gamma}n) = 0$. Determine $\delta \in \mathcal{N}$ such that $Fun(\delta)$ and $f = f_\delta$. (Cf. 7.7).
Observe that: $\forall\alpha \in \mathcal{N}\forall\gamma \in \mathcal{N}\exists n \in \mathbb{N}\exists m \in \mathbb{N}[\delta^{\overline{\gamma}n}(\overline{\alpha}m) = 1 \wedge \forall q < m[\delta^{\overline{\gamma}n}(\overline{\alpha}q) = 0]]$.
We now have a bar in $\mathcal{N} \times \mathcal{N}$.
We find β by translating this bar - via some pairing device - into a bar in $\mathcal{N}$.
We do not carry out this translation in detail. ◇

7.10. Observe that, for each $\sigma \in \$$, the set $\{\alpha \mid \alpha \in \mathcal{N} \mid B_\alpha \leq \sigma\}$ is a Borel set, because:

1) $\forall\alpha \in \mathcal{N}[B_\alpha \leq \{\langle\,\rangle\} \leftrightarrow \alpha(\langle\,\rangle) = 0]$

   and:

2) for each $\sigma \in \$$, $\sigma \neq \{\langle\,\rangle\}$:

$$\forall\alpha \in \mathcal{N}[B_\alpha \leq \sigma \leftrightarrow (\alpha(\langle\, \rangle) = 0 \vee \forall m \in \mathbb{N}\exists n \in \mathbb{N}[B_{\alpha^m} \leq \sigma^n])].$$

Observe also that, for each $\sigma \in \$$: $\forall\alpha \in \mathcal{N}[B_\alpha \leq \sigma \rightarrow A^1_1(\alpha)]$.
Brouwer's thesis, which we mentioned in 1.8.2, says:
$\forall\alpha \in \mathcal{N}[A^1_1(\alpha) \rightarrow \exists\sigma \in \$[B_\alpha = \sigma]]$.

7.11. Intuitionistic Suslin theorem:

Every subset of $\mathcal{N}$ which is both co-analytical and strictly analytical, is Borel.

Proof: Let X be a subset of $\mathcal{N}$ which is both co-analytical and strictly analytical. Determine $f \in \mathcal{F}$ such that $\forall\alpha \in \mathcal{N}[X(\alpha) \leftrightarrow A^1_1(f|\alpha)]$. Determine $g \in \mathcal{F}$ such that $X = Ran(g)$.
Observe: $\forall\alpha \in \mathcal{N}[A^1_1(f|(g|\alpha))]$. Apply the boundedness lemma 7.9 and determine $\beta \in \mathcal{N}$ such that $A^1_1(\beta)$ and $\forall\alpha \in \mathcal{N}[B_{f|(g|\alpha)} \leq B_\beta]$. Apply Brouwer's thesis and find $\sigma \in \$$ such that $B_\beta = \sigma$.
Observe: $\forall\alpha \in \mathcal{N}[X(\alpha) \leftrightarrow B_{f|\alpha} \leq \sigma]$, therefore (cf. 7.10), X is Borel. ◇

7.12. We mention two more theorems which follow from Brouwer's thesis, although we do not give the proofs.
The first one is an intuitionistic version of the Lusin *separation theorem*:

Let $\langle X, Y\rangle$ be a separate pair of strictly analytical subsets of $\mathcal{N}$.
There exists a separate pair $\langle A, B\rangle$ of Borel subsets of $\mathcal{N}$, such that $X \subseteq A$ and $Y \subseteq B$.

(A pair $\langle X, Y\rangle$ of subsets of $\mathcal{N}$ is called *separate* iff:
$\forall\alpha \in \mathcal{N}\forall\beta \in \mathcal{N}[(X(\alpha) \wedge Y(\beta)) \rightarrow \alpha\#\beta]$.)

The second one is an intuitionistic version of another famous result from the classical theory:

Let $f \in \mathcal{F}$ be strongly injective, i.e.: $\forall\alpha \in \mathcal{N}\forall\beta \in \mathcal{N}[\alpha\#\beta \rightarrow f|\alpha\# f|\beta]$.
Then $Ran(f)$ is a Borel subset of $\mathcal{N}$.

8. THE COLLAPSE OF THE PROJECTIVE HIERARCHY

8.1. We define a class $\mathbf{\Sigma}^1_2$ of subsets of $\mathcal{N}$:
A subset X of $\mathcal{N}$ belongs to $\mathbf{\Sigma}^1_2$ iff there exists an open subset D of $\mathcal{N}$ such that $\forall\alpha \in \mathcal{N}[X(\alpha) \leftrightarrow \exists\beta \in \mathcal{N}\forall\gamma \in \mathcal{N}[D(\langle\alpha, \beta, \gamma\rangle)]]$.

8.2. One may prove easily that the class $\mathbf{\Sigma}^1_2$ is closed under the operations of union and intersection. Using $\mathbf{AC}_{01}$, the countable axiom of choice, (cf. 1.3.2), one may establish that the class $\mathbf{\Sigma}^1_2$ is closed under the operations of countable union and countable intersection. Finally, one shows without difficulty that the class $\mathbf{\Sigma}^1_2$ is closed under the operations of taking pre-images-by-continuous functions (i.e.: for all subsets X, Y of $\mathcal{N}$, if $Y \in \mathbf{\Sigma}^1_2$ and $X \preceq Y$, then $X \in \mathbf{\Sigma}^1_2$) and existential projection (i.e.: for all subsets X, Y of $\mathcal{N}$, if $X \in \mathbf{\Sigma}^1_2$ and $\forall\alpha \in \mathcal{N}[Y(\alpha) \leftrightarrow \exists\delta \in \mathcal{N}[X(\langle\alpha, \delta\rangle)]]$, then $Y \in \mathbf{\Sigma}^1_2$).
Somewhat surprisingly, also the following holds:

8.3. Theorem: The class $\mathbf{\Sigma}_2^1$ is closed under the operation of universal projection, i.e.: for all subsets X, Y of $\mathcal{N}$:
if $X \in \mathbf{\Sigma}_2^1$ and $\forall\alpha \in \mathcal{N}[Y(\alpha) \leftrightarrow \forall\delta \in \mathcal{N}[X(\langle\alpha,\delta\rangle)]]$, then $Y \in \mathbf{\Sigma}_2^1$.

Proof: Let X, Y be subsets of $\mathcal{N}$ such that $X \in \mathbf{\Sigma}_2^1$ and
$\forall\alpha \in \mathcal{N}[Y(\alpha) \leftrightarrow \forall\delta \in \mathcal{N}[X(\langle\alpha,\delta\rangle)]]$.
We may determine a decidable subset E of the set $\mathbb{N}$ of natural numbers such that:
$\forall\alpha \in \mathcal{N}[Y(\alpha) \leftrightarrow \forall\delta \in \mathcal{N}\exists\beta \in \mathcal{N}\forall\gamma \in \mathcal{N}\exists n \in \mathbb{N}[E(\langle\overline{\alpha}n, \overline{\beta}n, \overline{\gamma}n, \overline{\delta}n\rangle)]]$.
Using the axiom $\mathbf{AC}_{11}$ (cf. 1.6.3), we observe that for each $\alpha \in \mathcal{N}$:

$$\begin{aligned} Y(\alpha) &\leftrightarrow \exists\zeta[Fun(\zeta) \wedge \forall\delta \in \mathcal{N}\forall\gamma \in \mathcal{N}\exists n \in \mathbb{N}[E(\langle\overline{\alpha}n, \overline{\zeta|\delta}n, \overline{\gamma}n, \overline{\delta}n\rangle)] \\ &\leftrightarrow \exists\zeta[Fun(\zeta) \wedge \forall\delta \in \mathcal{N}\forall\gamma \in \mathcal{N}\exists n \in \mathbb{N}\exists p \in \mathbb{N}[E(\langle\overline{\alpha}n, p, \overline{\gamma}n, \overline{\delta}n\rangle) \wedge \\ &\qquad \wedge \operatorname{length}(p) = n \wedge \forall j < n\exists m \in \mathbb{N}[\zeta^j(\overline{\delta}m) = p(j) + 1 \wedge \forall l < m[\zeta^j(\overline{\delta}l) = 0]]]] \end{aligned}$$

Using the fact that $Fun \in \mathbf{\Pi}_1^1$, and some other obvious properties of $\mathbf{\Pi}_1^1$, we conclude: $Y \in \mathbf{\Sigma}_2^1$. ◇

8.4. It is clear, now, that every subset of $\mathcal{N}$ which can be obtained from closed subsets of $\mathcal{N}$ and open subsets of $\mathcal{N}$ by means of the operations countable union, countable intersection, existential projection and universal projection, in short: every positively projective subset of $\mathcal{N}$, belongs to $\mathbf{\Sigma}_2^1$.

9. A REMARKABLE SET

9.1. We introduce a subset S of $\mathcal{N}$ by:
$\forall\alpha \in \mathcal{N}[S(\alpha) \leftrightarrow \exists\gamma \in \sigma_{2mon}\forall n \in \mathbb{N}[\alpha(\overline{\gamma}n) = 0]]$
(σ_{2mon} is the spread which has been mentioned in 1.7, i.e.:
$\forall\gamma \in \mathcal{N}[\gamma \in \sigma_{2mon} \leftrightarrow \forall n \in \mathbb{N}[\gamma(n) \leq \gamma(n+1) \leq 1]]$.)

9.2. From a classical point of view, the spread σ_{2mon} seems to be a countable set, and the following seems to be a complete enumeration of its elements:

$$\underline{0}, \qquad 0^* := \underline{1}, \qquad 1^* := \langle 0\rangle * \underline{1}, \qquad 2^* := \langle 0, 0\rangle * \underline{1}, \qquad \ldots$$

Therefore, from a classical point of view:

$$\forall\alpha \in \mathcal{N}[S(\alpha) \leftrightarrow (\forall n \in \mathbb{N}[\alpha(\overline{\underline{0}}n) = 0] \vee \exists k \in \mathbb{N}\forall n \in \mathbb{N}[\alpha(\overline{k^*}n) = 0])]$$

So, it looks as if $S \in \mathbf{\Sigma}_2^0$.
But, as we will see, this is far from true intuitionistically.

9.3 We claim that the following is an *elementary* intuitionistic theorem:

For every subset C of $\mathbb{N}$:
If $\forall\gamma \in \sigma_{2mon}\exists n \in \mathbb{N}[C(\overline{\gamma}n)]$, then $\exists N \in \mathbb{N}\forall\gamma \in \sigma_{2mon}\exists n \leq N[C(\overline{\gamma}n)]$.

(Proof of this claim: let C be a subset of $\mathbb{N}$ such that $\forall\gamma \in \sigma_{2mon}\exists n \in \mathbb{N}[C(\overline{\gamma}n)]$.
Determine $m \in \mathbb{N}$ such that $C(\overline{\underline{0}}m)$. Then consider the sequences $0^*, 1^*, \ldots, (m-1)^*$, (introduced in 9.2), which belong to σ_{2mon}.
Calculate natural numbers $n_0, n_1, \ldots, n_{m-1}$ such that $\forall j < m[C(\overline{j^*}n_j)]$.
Define $N := \max\{m, n_0, n_1, \ldots, n_{m-1}\}$.

Then $\forall\gamma \in \mathbf{\Sigma}_{2mon}\exists n \le N[C(\overline{\gamma}n)]$. End of proof.)

We call the theorem *elementary* as we do not use any axiom in its proof, only (intuitionistic) logic.

We claim that also the following is an elementary intuitionistic theorem:

For every subset C of $\mathbb{N}$:

If $\forall\gamma \in \sigma_{2mon}\neg\neg\exists n \in \mathbb{N}[C(\overline{\gamma}n)]$, then $\neg\neg\exists N \in \mathbb{N}\forall\gamma \in \sigma_{2mon}\exists n \le N[C(\overline{\gamma}n)]$.

(Proof of this second claim: let C be a subset of $\mathbb{N}$ such that
$\forall\gamma \in \sigma_{2mon}\neg\neg\exists n \in \mathbb{N}[C(\overline{\gamma}n)]$. In particular: $\neg\neg\exists n \in \mathbb{N}[C(\overline{\underline{0}}n)]$.
Make a *first* extra assumption: $\exists n \in \mathbb{N}[C(\overline{\underline{0}}n)]$. Determine $m \in \mathbb{N}$ such that $C(\overline{\underline{0}}m)$.
We then know: $\forall j < m\neg\neg\exists n \in \mathbb{N}[C(\overline{j^*}n)]$. Make m *further* extra assumptions:
$\forall j < m\exists n \in \mathbb{N}[C(\overline{j^*}n)]$. We conclude, as in the proof of the first claim:
$\exists N \in \mathbb{N}\forall\gamma \in \sigma_{2mon}\exists n \le N[C(\overline{\gamma}n)]$. Give up the *latter* m extra assumptions, and conclude:
$\neg\neg\exists N \in \mathbb{N}\forall\gamma \in \sigma_{2mon}\exists n \le N[C(\overline{\gamma}n)]$
Give up the *first* extra assumption, and conclude:
$\neg\neg\exists N \in \mathbb{N}\forall\gamma \in \sigma_{2mon}\exists n \le N[C(\overline{\gamma}n)]$.
End of proof of second claim.)

In the proof of the second claim, we repeatedly used the following intuitionistically sound principle of reasoning:

For all propositions P, Q:

If we can prove Q from P, then we can prove $\neg\neg Q$ from $\neg\neg P$.

9.4 The second claim of 9.3 justifies the following observation:
For all $\alpha \in \mathcal{N}$:

$$\begin{aligned}\neg\neg S(\alpha) &\leftrightarrow \neg\neg\exists\gamma \in \sigma_{2mon}\forall n \in \mathbb{N}[\alpha(\overline{\gamma}n) = 0]\\ &\leftrightarrow \neg\forall\gamma \in \sigma_{2mon}\neg\forall n \in \mathbb{N}[\alpha(\overline{\gamma}n) = 0]\\ &\leftrightarrow \neg\forall\gamma \in \sigma_{2mon}\neg\neg\exists n \in \mathbb{N}[\alpha(\overline{\gamma}n) \neq 0]\\ &\leftrightarrow \neg\exists N \in \mathbb{N}\forall\gamma \in \sigma_{2mon}\exists n \le N[\alpha(\overline{\gamma}n) \neq 0]\\ &\leftrightarrow \forall N \in \mathbb{N}\exists\gamma \in \sigma_{2mon}\forall n \le N[\alpha(\overline{\gamma}n) = 0]\end{aligned}$$

Remark that, for each $N \in \mathbb{N}$, the set $\{\overline{\gamma}n \mid \gamma \in \sigma_{2mon} \mid n \le N\}$ is finite. Therefore, the above shows that the subset $Neg(Neg(S))$ of $\mathcal{N}$, which is defined by:
$\forall\alpha \in \mathcal{N}[(Neg(Neg(S)))(\alpha) \leftrightarrow \neg\neg S(\alpha)]$, belongs to $\mathbf{\Pi}^0_1$.
(We may improve upon the observation in 9.2: from a classical point of view, S belongs to $\mathbf{\Pi}^0_1$, i.e.: S is a closed set.)

9.5. We introduce a class $\mathcal{H}$ of subsets of $\mathcal{N}$, which we call: *Borel-approximations to the set S*, by means of the following inductive definition:

(i) $Neg(Neg(S))$ belongs to $\mathcal{H}$.

(ii) If $P_0, P_1, \ldots$ is a sequence of members of $\mathcal{H}$ such that $\forall n \in \mathbb{N}[P_{n+1} \subseteq P_n]$, then the set $\mathcal{K}^{\infty}_{n=0}(P_n)$ belongs to S. The set $\mathcal{K}^{\infty}_{n=0}(P_n)$ is defined by:

$$\begin{aligned}\forall\alpha \in \mathcal{N}[(\mathcal{K}^{\infty}_{n=0}(P_n))(\alpha) \leftrightarrow &(\forall n \in \mathbb{N}[\exists j < n\forall k \in \mathbb{N}[\alpha(\overline{j^*}k) = 0]\vee\\ &(\forall j \le n[\alpha(\overline{\underline{0}}j) = 0] \wedge P_n({}^{\underline{0}^n}\alpha)))]\end{aligned}$$

(iii) The clauses (i) and (ii) give all members of $\mathcal{H}$.

One may prove, for each subset P of $\mathcal{N}$ which belongs to $\mathcal{H}: S \subseteq P$.

One may prove, that for each subset P of $\mathcal{N}$ which belongs to $\mathcal{H}$ there exists a function $f \in \mathcal{F}$ such that $\forall \alpha \in \mathcal{N}[P(f|\alpha)]$ and $\neg\forall \alpha \in \mathcal{N}[S(f|\alpha)]$.
Therefore, no member of the class $\mathcal{H}$ coincides with the set S.

One may prove, for each sequence $P_0, P_1, \ldots$, of members of $\mathcal{H}$ which fulfils the condition: $\forall n \in \mathbb{N}[P_{n+1} \subseteq P_n]$, that
$\forall n \in \mathbb{N}[\mathcal{K}_{j=0}^{\infty}(P_j) \subseteq P_n \wedge \mathcal{K}_{j=0}^{\infty}(P_j) \neq P_n]$.
Therefore, the class $\mathcal{H}$ contains better and better approximations to S.

One may prove, that for each Borel set C the following holds: if $S \subseteq C$, then there exists a subset P of $\mathcal{N}$ which belongs to $\mathcal{H}$, such that $S \subseteq P \subseteq C$.
Therefore, a Borel set C cannot coincide with S.

The proofs of these facts are not easy, and we do not give them here. We want to emphasize:
(1) S is not a Borel set.
(2) Uncountably many Borel sets may be intercalated in between S and $Neg(Neg(S))$.

10. SOME CONCLUDING REMARKS

10.1. We feel that the set A_1^1 (introduced in 7.3), which characterizes the class $\mathbf{\Pi}_1^1$, should be studied further. Brouwer's thesis (cf. 1.8.1) is a bold assertion on the structure of this set. We would like to see arguments showing that the set A_1^1 is not a Borel subset of $\mathcal{N}$, not a strictly analytical subset of $\mathcal{N}$, and not an analytical subset of $\mathcal{N}$.

10.2. Some annoying remaining problems can be "solved" if one adds some semi-classical assumption to the axioms of intuitionistic analysis. One such assumption is the following Generalized Markov Principle:

$$\textbf{GMP}: \quad \forall \alpha \in \mathcal{N}[\neg\neg\exists n \in \mathbb{N}[\alpha(n) = 0] \rightarrow \exists n \in \mathbb{N}[\alpha(n) = 0]]$$

Acceptance of this principle makes it possible to establish the Borel hierarchy by means of the classical diagonal argument, and also to show that the set A_1^1 is not analytical.
Of course, such "solutions" are not acceptable from a genuinely intuitionistic point of view. We even think that is is dangerous to experiment with such semi-classical assumptions as they might make one lose sight of truly constructive arguments.

10.3. We want to draw an amusing conclusion from the fact that $\mathbf{\Sigma}_2^1$ is the class of all positively projective sets. (Cf. 8.4).
It is not difficult to show, that, for any subset X of $\mathcal{N}$ which belongs to $\mathbf{\Sigma}_2^1$, we may find $\beta \in \mathcal{N}$ such that

$$\forall \alpha \in \mathcal{N}[X(\alpha) \leftrightarrow \exists \gamma \in \mathcal{N}\forall \delta \in \mathcal{N}\exists n \in \mathbb{N}[\beta(\langle \overline{\alpha}n, \overline{\gamma}n, \overline{\delta}n\rangle) = 0.]]$$

Consider the subset X_0 of $\mathcal{N}$ which is defined by

$$\forall \alpha \in \mathcal{N}[X_0(\alpha) \leftrightarrow \forall \gamma \in \mathcal{N}\exists \delta \in \mathcal{N}\forall n \in \mathbb{N}[\alpha(\langle \overline{\alpha}n, \overline{\gamma}n, \overline{\delta}n\rangle) \neq 0]].$$

Constructing the appropriate β for this particular see X_0, we find:

$$\exists\gamma\in\mathcal{N}\forall\delta\in\mathcal{N}\exists n\in\mathbb{N}[\beta(\langle\overline{\beta}n,\overline{\gamma}n,\overline{\delta}n\rangle)=0]\leftrightarrow$$
$$\forall\gamma\in\mathcal{N}\exists\delta\in\mathcal{N}\forall n\in\mathbb{N}[\beta(\langle\overline{\beta}n,\overline{\gamma}n,\overline{\delta}n\rangle)\neq 0]$$

and therefore:

$$\neg\exists\gamma\in\mathcal{N}\forall\delta\in\mathcal{N}\exists n\in\mathbb{N}[\beta(\langle\overline{\beta}n,\overline{\gamma}n,\overline{\delta}n\rangle)=0]\wedge$$
$$\neg\forall\gamma\in\mathcal{N}\exists\delta\in\mathcal{N}\forall n\in\mathbb{N}[\beta(\langle\overline{\beta}n,\overline{\gamma}n,\overline{\delta}n\rangle)\neq 0].$$

This is a rather strong refutation of de Morgan's laws.
The reader may divert himself by establishing the following simpler case of this phenomenon:

$$\neg\forall\alpha\in\mathcal{N}\exists n\in\mathbb{N}\forall m\in\mathbb{N}[\alpha(n)=0\rightarrow\alpha(m)=0]\wedge$$
$$\neg\exists\alpha\in\mathcal{N}\forall n\in\mathbb{N}\exists m\in\mathbb{N}[\alpha(n)=0\wedge\alpha(m)\neq 0].$$

Observe finally that we may produce sets which do not belong to $\mathbf{\Sigma}^1_2$, and, therefore, are not (positively) projective, simply by the classical diagonal argument: the subset X of $\mathcal{N}$ defined by:

$$\forall\alpha\in\mathcal{N}[X(\alpha)\leftrightarrow\neg\exists\gamma\in\mathcal{N}\forall\delta\in\mathcal{N}\exists n\in\mathbb{N}[\alpha(\langle\overline{\alpha}n,\overline{\gamma}n,\overline{\delta}n\rangle)=0]],$$

is such a set.

REFERENCES

S.C. Kleene and R.E. Vesley (1965), The foundations of intuitionistic mathematics, especially in relation to recursive functions. *Studies in Logic and the Foundations of Mathematics.* Amsterdam (North Holland Publ. Co.).

N. Lusin (1930), Leçons sur les ensembles analytiques et leurs applications. *Collection de monographies sur la théorie des fonctions.* Paris (Gauthier-Villars). Reprint (1972): New York (Chelsea Publ. Co.).

Y.N. Moschovakis (1980), Descriptive Set Theory. *Studies in Logic and the Foundations of Mathematics*, vol. 100. Amsterdam (North Holland Publ. Co.).

W. Veldman (1981), Investigations in intuitionistic hierarchy theory. Thesis. Katholieke Universiteit. Nijmegen, the Netherlands.

W. Veldman (1983), On the constructive contrapositions of two axioms of countable choice. In: the L.E.J. Brouwer centenary symposium, Proceedings of the conference held in Noordwijkerhout, 8-13 June, 1981, ed. A.S. Troelstra and D. van Dalen, *Studies in Logic and the Foundations of Mathematics*, vol. 110, Amsterdam (North Holland Publ. Co.), pp. 513-523.

INTERPRETABILITY LOGIC

Albert Visser

Department of Philosophy
University of Utrecht
Utrecht, The Netherlands

1 Introduction

Interpretations are much used in metamathematics. The first application that comes to mind is their use in reductive Hilbert-style programs. Think of the kind of program proposed by Simpson, Feferman or Nelson (see Simpson[1988], Feferman[1988], Nelson[1986]). Here they serve to compare the strength of theories, or better to prove conservation results within a properly weak theory. An advantage of using interpretations is that even if their use should -perhaps- be classified as a proof-theoretical method, it is often possible to employ a model-theoretical heuristics. An example is given in section 7.2 where a conservation result due to Paris & Wilkie, which is proven by a model-theoretical argument, is formalized in a weak theory. For more discussion of and perspective on the use of interpretability in reductive programs the reader is referred to Feferman[1988].

A second application is the use of an interpretation of Elementary Syntax e.g in proving Gödel's Second Incompleteness Theorem: here the interpretation is essential both for the significance of the result and for the heuristics of the argument.

The notion of relative interpretability was made explicit in Tarski, Mostowski, Robinson [1953]; it was systematically studied in the twin pioneering papers Feferman[1960] and Orey[1961]. Lattices of interpretability types were considered in much detail e.g in Montague[1958], Mycielski[1962, 1977], Svejdar[1978], Lindström[1979], Pudlák[1983a]. The interest in these lattices is clearly motivated by the view that interpretability is an adequate means for comparison of strength of theories. Characterizations of relative interpretability for various kinds of theories were obtained by Hájek applying the Orey Compactness Theorem (for essentially reflexive theories) and by Friedman and Pudlák independently (for finitely

Mathematical Logic
Edited by P. P. Petkov
Plenum Press, New York, 1990

axiomatized sequential theories; see respectively Smorynski[1985b] and Pudlák[1985]; a presentation of part of Friedman's result is given in sections 7.2, 7.3).

Both Solovay and Lindström proved that relative interpretability over essentially reflexive theories like PA or ZFC is complete Π_2 (see Lindström[1979], Solovay[?]). To be more specific for example the set of Σ_1-sentences S such that ZFC interprets ZFC+S is complete Π_2. This awesome complexity has suggested to some that the usual notion allows too many interpretations. I'm not quite convinced: nobody said we have to *use* all of them. Another response is to study restrictions of the usual notion: there is still room for a lot of experimentation here.

Modal logics for interpretability were first studied by Hájek and then by Svejdar (see Hájek [1981], Svejdar[1983]). They studied logics with modal operators for provability and interpretability and with witness comparison relations. In Svejdar's system a number of important arguments can be formulated. Moreover Svejdar provides a number of different interpretations of his system. What one seeks in a Svejdar-type approach (which is analogous to Smorynski's approach in his "Ubiquitous Fixed Point Calculation") is a system that is as weak as possible, but still codifies the relevant class of arguments, the point being unification and simplification of a number of specific arguments from the literature. There is no need for the system to be complete w.r.t. any set of interpretations.

The approach in this paper is somewhat different: the focus of interest is to find logics that are sound and complete for interpretations in a given theory (or class of theories). If we know that a logic is sound and complete for interpretations in a given theory and a modal formula ϕ is consistent with the logic, then we know that we can find an interpretation of ϕ that is consistent with the given theory. Typically this interpretation is explicitly given by the proof of the Completeness Theorem.

Solovay's Completeness Theorem for *provability logic* is remarkably general: we have the same logic, viz. Löb's Logic L, for all theories T with the following properties: (i) they have a Σ_1-provability predicate, (ii) they extend $I\Delta_0$+EXP, and (iii) they do not prove their own n-iterated inconsistency (i.e. $\Box_T{}^n\bot$) for any n. (If a theory T satisfies (i) and (ii), but not (iii) let n* be the least n such that $T\vdash\Box_T{}^n\bot$, then the provability logic of T is L+ $\Box^{n^*}\bot$. Suppose T has an $R_1{}^+$-provability predicate, extends $I\Delta_0+\Omega_1$ and has property (iii), then we know that L is sound for interpretations in T, but we do not know in general whether L is complete for interpretations in T. Specifically it is an open question what the provability logic of $I\Delta_0+\Omega_1$ is.) From one point of view the generality of Solovay's theorem is a disadvantage: one cannot expect information from it connected with specific properties of the theory considered. In this respect interpretability fares better: it turns out, for example, that properties like finiteness and essential reflexivity induce essentially different interpretability principles.

We study two kinds of questions. Let some property Ξ of theories be given: (i) which interpretability principles are valid in all theories satisfying Ξ?; and (ii) does Ξ determine the interpretability principles valid for interpretations in any given theory T satisfying Ξ? In this paper the following specific instances of questions (i) and (ii) are considered: (a) which interpretability principles are valid in all R_1^+-axiomatized theories extending $I\Delta_0+\Omega_1$?; (b) what is the interpretability logic of a given verifiably essentially reflexive theory U?; (c) what is the interpretability logic of a given finitely axiomatized sequential theory U extending $I\Delta_0+\Omega_1$? For questions (a), (b) conjectures are formulated. An answer is available for (c) in case U extends $I\Delta_0$+SUPEXP.

2 Contents

Section 5 contains the necessary preliminaries. In section 6 the systems of interpretability logic IL, ILW, ILP and ILM are introduced. We take a brief look at their consequences and discuss their Kripke semantics and arithmetical significance. In section 7 the form of Friedman's characterization of interpretability for finitely axiomatized sequential theories that is needed to prove our arithmetical completeness result is derived. It turns out that it is convenient to prove this result from a technical lemma (7.2). This lemma is the formalized version of a result of Paris & Wilkie which provides a connection between $I\Delta_0+\Omega_1$ and $I\Delta_0$+EXP. I think this lemma is of some independent interest. Finally in section 8 it is shown that ILP is a complete axiomatization of the interpretability logic of finitely axiomatized sequential theories extending $I\Delta_0$+SUPEXP

3 Acknowledgements

The research on which this paper reports is part of a project together with Dick de Jongh, Craig Smorynski and Frank Veltman. Discussions with them were very important for me. Correspondence with George Kreisel and with Franco Montagna has been invariably stimulating.

4 Prerequisites

We presuppose some knowledge of Smorynski[1985a], Paris & Wilkie[1987], Pudlák[1985, 1986].

5 Conventions, Notions & Elementary Facts.

5.1 Languages

In this paper we consider only relational languages, i.e. languages without function symbols and constants. So for example in the case of arithmetic, instead of + we have a ternary relation symbol, etc. . Of course this is a severe and unjustifiable restriction. I am convinced that the restriction can be dropped almost everywhere. My only excuse is that at some places - especially where tableaux provability is involved- the use of a language with function symbols asks for some extra work: work I have not yet done.

After this is said officially we will of course often *pretend* that we are working in a language with function symbols. Here one has to be careful: for example at a certain point we are working in $I\Delta_0+\Omega_1$ and we consider a function assigning to n the Gödelnumber of $\exists y\ y=\underline{n}$, where $\underline{n}$ is the numeral in the sense of Paris and Wilkie[1987] corresponding to n. For the functional language it is easy to see that this function is total (in $I\Delta_0+\Omega_1$). Inspection of the translation procedure into the corresponding relational language shows that the fomulas become only polynomially longer, so the function is also total for the relational language.

In our languages there are only finitely many relation symbols including identity.

5.2 Special Classes of Formulas

We refer the reader to the discussion of special classes of formulas in Paris & Wilkie[1987].

Δ_0-formulas are formulas where all quantifiers are bounded by terms in 0, S, + and . (or rather the translations of such formulas in the relational language), where the variable of quantification does not occur in the bounding term. If the theory we are working in proves that some function f with Δ_0-graph is total, we may want to consider $\Delta_0(f)$-formulas, where the bounding terms also involve f. In Gaifman & Dimitracopoulos[1982] it is shown that if f is reasonable -roughly: if it doesn't jump up an down wildly- then $I\Delta_0$+"f is total" implies $I\Delta_0(f)$. For our purposes it is sufficient to know that ω_1 and exp are reasonable; here: $\exp(x):=2^x$.

5.3 Theories and Provability

We consider only theories with identity for which a fixed formulas of their language are specified giving us a set of natural numbers, 0, successor, addition and multiplication. We assume in most cases that $I\Delta_0+\Omega_1$ is provable for these natural numbers. Variables x,y,z,u,v,... will be taken to range over the designated numbers. As variables for general objects of the theory we will use a,b,... . Syntactical notions will always be formalized in the designated natural numbers.

We consider a theory T as given by a formula $\alpha_T(x)$ having just x free plus the relevant information on what the set of natural numbers of the theory is. α_T gives the set of codes of the (non-predicate-logical) axioms of the theory. Different α different theories; same α same theory. Unless explicitely stated otherwise we will always assume that α is an R_1^+-formula.

Let $Proof_T(x,y)$ be the R_1^+-formula representing the relation: x is the Gödelnumber of a T-proof of the formula with Gödelnumber y. $Proof_T$ will be built in some standard way from α_T. The precise choice of the system on which $Proof_T$ is based is immaterial: any Hilbert style system or Natural Deduction system or Genzen style sequent system will do. If we want to stress that we are looking at the Proof-relation based on a certain specific formula β we write: $Proof_\beta$.

We assume for convenience that: $I\Delta_0+\Omega_1 \vdash \forall x \exists ! y \ Proof_T(x,y)$. Let $Prov_T(y) := \exists x Proof_T(x,y)$.

We write par abus de langage '$Proof_T(u, \phi(x_1,...,x_n))$' for: $Proof_T(u, \ulcorner\phi(\dot{x}_1,...,\dot{x}_n)\urcorner)$, here:

i) all free variables of ϕ are among those shown.
ii) $\ulcorner\phi(\dot{x}_1,...,\dot{x}_n)\urcorner$ is the "Gödelterm" for $\phi(x_1,...,x_n)$ as defined in Smoryński [1985], p43. Here we use instead of the usual numerals the efficient numerals of Paris & Wilkie[1987], so that: $I\Delta_0+\Omega_1 \vdash \forall x_1,...,x_n \exists y \ \ulcorner\phi(\dot{x}_1,...,\dot{x}_n)\urcorner = y$.

$\Box_T\phi(x_1,...,x_n)$ will stand for: $Prov_T(\ulcorner\phi(\dot{x}_1,...,\dot{x}_n)\urcorner)$.

Occurrences of terms inside $\Box_T$ should be treated with some care. Is $\Box_T(\phi[t/x])$ intended or $(\Box_T\phi(x))[t/x]$? We will always use the first, i.e. the small scope reading. In cases where: U proves that t is total and $U \vdash t=x \rightarrow \Box_V t=x$, the scope distinction may be ignored within U w.r.t. $\Box_V$. We have: $U \vdash (\Box_V\phi(x))[t/x] \leftrightarrow \Box_V(\phi[t/x])$.

We will also need normalized or cut-free provability: here we could choose Herbrand provability (as used in Pudlák[1985]) or cut-free provability in a sequent system or tableaux provability. We use tableaux provability as in Paris & Wilkie[1987]: we write $Tproof_T(x,y)$ for: Tabinconproof(U,x), where U is T plus the negation of the formula coded by y. $Tproof_T(x,y)$ is given by an R_1^+-formula. $Tprov_T(y)$ is $\exists x Tproof(x,y)$. Tcon(T) is $\forall x \neg Tproof(x, \ulcorner\bot\urcorner)$. $\Delta_T\phi(x_1,...,x_n)$ will stand for: $Tprov_T(\ulcorner\phi(\dot{x}_1,...,\dot{x}_n)\urcorner)$. Of course our remarks about scope of terms carry over to Δ.

$\Diamond_T$ will stand for: $\neg\Box_T\neg$, and ∇_T for: $\neg\Delta_T\neg$.

Let the axiom set of T be given by $\alpha(x)$ then $\Box_T\restriction y$ stands for provability in the theory whose axiom set is given by $(\alpha(x) \wedge x<y)$. $\Box_{T,x}$ will stand for restricted provability in the sense of Paris & Wilkie[1987].

For convenience we write $\Box_\Omega$ for provability in $I\Delta_0+\Omega_1$ and $\Box_{EXP}$ for provability in $I\Delta_0$+EXP.

5.4 Special Properties of Theories

A theory T, with designated natural numbers satisfying $I\Delta_0+\Omega_1$, is *sequential* if in it one can form sequences of any of its objects i.e. there is a relation $(s)_x=a$ such that T proves:

(i) $\forall s,x,a,b\ (\ ((s)_x=a \wedge (s)_x=b) \rightarrow a=b\ )$,

(ii) $\forall s \exists x \forall y\ (\ \exists b\ (s)_y=b \leftrightarrow y<x\ )$,

(iii) $\exists s \forall x,a \neg (s)_x=a$,

(iv) $\forall s,a,x(\ \forall y<x \exists b\ (s)_y=b \rightarrow \exists s' \forall b \forall y \leq x\ (\ (s')_y=b \leftrightarrow ((y<x \wedge (s)_y=b) \vee (y=x \wedge a=b))\ )\ )$

Our notion of sequentiality is only seemingly more restrictive than those in the literature: for any theory that is sequential e.g. in the sense of Pudlák[1983] one can define set of natural numbers satisfying $I\Delta_0+\Omega_1$ and a relation $(s)_x=a$ making the theory sequential in our sense. The notion of sequentiality is due to Pudlák. We will describe several important properties of sequential theories later.

A theory is *finitely axiomatized* if its axiom set is given by a disjunction of formulas of the form $x=\underline{n}$, where n codes a formula.

A theory T is *essentially reflexive* if for all fomulas $\phi(x,...)$ of its language and for all natural numbers n: $T \vdash \forall x,...(\ \Box \ulcorner \underline{n}\ \phi(x,...) \rightarrow \phi(x,...)\ )$. T is *verifiably* essentially reflexive if T is essentially reflexive and T proves the formalization of "T is essentially reflexive".

5.5 Interpretability

Interpretations are in this paper: one dimensional global relative interpretations without parameters. Consider two languages L and L'. An interpretation M of L' in L is given by (i) a function F from the relation symbols of L' to formulas of the language of L and (ii) a formula $\delta(a)$ of L having just a free. The image of a relation symbol has precisely $a_1,...,a_n$ free, where n is the arity of the relation symbol. The image of = need not be $a_1=a_2$. The function F is canonically extended in the following way: $(R(b_1,...,b_n))^M:=\phi(b_1,...,b_n)$, where $\phi=F(R)$. (To make substitution of the b's possible we rename bound variables in ϕ if necessary. In fact it would be neater to set apart bound variables for the F(R) and for δ that do not occur in the original L'.) $(.)^M$ commutes with the propositional connectives. $(\forall b\psi)^M:=\forall b(\delta(b)\rightarrow\psi^M)$. We can easily extend $(.)^M$ again to map proofs π (from assumptions) in L' to proofs π^M from the translated assumptions in L in the obvious way (for free variables b one adds $\delta(b)$ as a hypothesis). As is easily seen for a given interpretation M the lengths of the translated objects are given by a fixed polynomial in the lengths of the originals. The graphs of ψ^M (considered as a function in ψ and M) and of π^M (considered as a function in π and M) can be arithmetized by R_1^+-formulas in such a way that the recursive clauses are verifiable in $I\Delta_0+\Omega_1$. Because of the bound on the lengths of the values $I\Delta_0+\Omega_1$ proves that these functions are total.

Consider theories T (with language L) and T' (with language L'). What could it mean to say that T' is interpretable in T via M? I think the obvious interpretation is this: for every axiom ψ of T' there is a proof in T of ψ^M. (I assume in this discussion that we are dealing with sentences, in the case of formulas one should consider: $(\delta[\psi]\rightarrow \psi^M)$, where $\delta[\psi]$ is the conjunction of $\delta(b)$'s, for all free variables b of ψ.) Given the definition the next step is to show: if T' is interpretable in T via M and if T' proves χ, say by π, then there is a proof π^* in T of ψ^M. Roughly π^* is π^M with proofs of the translated T'-axioms plugged in at the relevant places. Now here is the problem: the verification of the existence of π^* requires (prima facie) $I\Sigma_1$, so in weak theories we don't have this step available. On the other hand what is the point of interpretability if we don't have the π^*?

Let us say that:

T' is *a-interpretable* via M in T if for every axiom ψ of T' there is a proof in T of ψ^M.

T' is *t-interpretable* via M in T if for every theorem χ of T' there is a proof in T of χ^M.

The proof π^* as described above could be said to *simulate* π.

T' is *s-interpretable* via M in T if for every proof π of T' there is a simulating proof π^* in T.

Clearly (in $I\Delta_0+\Omega_1$) s-interpretability implies t-interpretability which in turn implies a-interpretability. My choice to solve the problem mentioned above is simply to take t-interpretability as my notion of interpretability. One could argue that from the philosophical point of view s-interpretability would be the best choice. However t-interpretability is somewhat easier to define and somewhat easier to work with. Moreover I am not aware of any point where the difference between the notions becomes important.

Note that our problem vanishes if T' is finitely axiomatized: it is easy to see that in this case $I\Delta_0+\Omega_1$ proves that a-interpretability implies t-interpretability. A further idea is to impose a bound on the proofs of the translated axioms of T':

T' is *e-interpretable* via M in T if there is a polynomial p such that for every axiom ψ of T' there is a proof in T of ψ^M that is shorter than p of the length of ψ.

Again it is not difficult to see that $I\Delta_0+\Omega_1$ proves that e-interpretability implies t-interpretability. Moreover by applying a well known result we find: if $I\Delta_0+\Omega_1$ proves that T' is a-interpretable in T via M, then $I\Delta_0+\Omega_1$ proves that T' is e-interpretable in T via M and hence that T' is t-interpretable in T via M. So if we verify in $I\Delta_0+\Omega_1$ that M is an interpretation of T' in T we need only worry about the axioms.

We write:

$M:U\rhd V$, for the arithmetization of: V is t-interpretable in U via M.

We can arrange it so that M occurs in the arithmetization as a number, so it is possible to quantify over M in the theory. Define:

$U\rhd V \quad :\Leftrightarrow \exists M\, M: U\rhd V$

$M{:}\phi \rhd_U \psi \;:\Leftrightarrow M{:}(U+\phi) \rhd (U+\psi)$

$\phi \rhd_U \psi \;:\Leftrightarrow (U+\phi) \rhd (U+\psi)$

$U \equiv V \;:\Leftrightarrow U \rhd V \wedge V \rhd U$

$\phi \equiv_U \psi \;:\Leftrightarrow (U+\phi) \equiv (U+\psi)$

Finally let me mention an important fact (which is just a variation of the similar fact stated for Herbrand consistency, see e.g. Pudlák[1985]):

5.5.1 Fact: for T R_1^+-axiomatized: $I\Delta_0+\Omega_1 \vdash (I\Delta_0+\Omega_1+\nabla_T\top) \rhd T$.

Proof: The proof will be given in detail in Marianne Kalsbeek's Masters Thesis. It involves carefully constructing the systematic tableaux τ for T on a suitable cut I and then producing a path that is provably infinite on a cut J shortening I. □

5.6 Cuts

Consider a theory T with designated natural numbers satisfying $I\Delta_0+\Omega_1$. A T-cut is a definable set I of natural numbers such that T proves that: $\underline{0} \in I$, $((x<y \wedge y \in I) \rightarrow x \in I)$, "I is closed under $S,+,.,\omega_1$". This definition of cut is a bit stronger that usual, but because any cut in the weaker sense can be shortened to a cut in our sense the difference in definition does no harm. For an introduction to Solovay's method of shortening cuts the reader is refered to Paris & Wilkie[1987]. We collect a few facts to be used later.

5.6.1 Fact: Let I be a T-cut, then $I\Delta_0+\Omega_1 \vdash \forall x \Box_T x \in I$.

This fact is due indepently to Pudlák (see Pudlák[1985]) and Paris & Wilkie (see Paris & Wilkie[1987]). It depends crucially on the use of efficient numerals and is proved by carefully constructing the proof of $x \in I$ from x and from the proof in T that I is a cut. A slightly sharpend version (due to Paris and Wilkie) is:

5.6.2 Fact: Let I be a T-cut, then for some n $I\Delta_0+\Omega_1 \vdash \forall x \Box_{T,\underline{n}} x \in I$.

Let $exp(x) := 2^x$. Define: $itexp(x,0) := x$, $itexp(x,Sy) := exp(itexp(x,y))$, and $supexp(y) := itexp(1,y)$. One can find a Δ_0-formula representing the graph of itexp, such that the recursive clauses for itexp are verifiable in $I\Delta_0+\Omega_1$. We have:

5.6.3 Fact:
$I\Delta_0+\Omega_1 \vdash \forall y($ (exp(y) exists) $\rightarrow \exists\, I\Delta_0+\Omega_1$-cut I such that $\Box_\Omega(\forall x \in I$ itexp(x,y) exists$)\,)$.

Proof: This is an immediate consequence of the proof of lemma 2.2 of Pudlák[1986]. □

5.6.4 Consequence: $I\Delta_0+\Omega_1 \vdash \forall x,y(\text{ (exp(y) exists)} \rightarrow \Box_\Omega(\text{itexp(x,y) exists}))$.

Proof: By 5.6.2 and 5.6.3. □

5.6.5 Consequence: For T R_1^+-axiomatized: $I\Delta_0+\Omega_1 \vdash \Box_T\phi \rightarrow \Box_\Omega\Delta_T\phi$.

Proof: If a proof x is converted in a tableaux proof, the result is of order itexp(x,|x|), where |x| is the length of x in the sense of the number of symbols (as in Paris & Wilkie[1987]). So $|x| \approx \log(x)$. This estimate can be extracted from the one concerning cut-elimination on p876 of Schwichtenberg[1977], using the close connection between cut-free and tableaux proofs. We have: $I\Delta_0+\Omega_1 \vdash \Box_T\phi \rightarrow \exists x \Box_\Omega \text{Proof}_T(x,\phi)$ and $I\Delta_0+\Omega_1 \vdash \forall x \Box_\Omega(\text{itexp}(x,|x|) \text{ exists})$. So our result follows by induction inside $\Box_\Omega$ using itexp(x,|x|) as a bound. □

An important property of sequential theories is the presence of partial truthpredicates (see Pudlák[1986]). As a consequence of this a finitely axiomatized sequential theory T proves its own tableaux consistency on a T-cut I, i.e.:

5.6.6 Fact: $T \vdash \nabla_T{}^I \top$

It follows that $T \rhd (I\Delta_0+\Omega_1+\nabla_T\top)$ and hence by 5.5.1: $T \equiv (I\Delta_0+\Omega_1+\nabla_T\top)$.

At this point is is perhaps good to mention a possible source of confusion. $I\Delta_0$+EXP is infinitely axiomatized but finitely axiomatizable. In this paper we will use the results stated for finitely axiomatized sequential theories freely for $I\Delta_0$+EXP. The simplest way to justify this is simply to stipulate that by $I\Delta_0$+EXP we will understand the theory given by some fixed finite axiomatization. Another way is to check the results directly for $I\Delta_0$+EXP under its obvious axiomatization: this is possible because in $I\Delta_0$+EXP the usual truthpredicates for Σ_n-formulas are available and because of the agreeable form of the Δ_0-induction scheme. A third way is to prove in $I\Delta_0$+EXP the equivalence of tableaux provability in its finitely axiomatized form and tableaux provability in its infinitely axiomatized form. For simplicity I will opt for the first way out. Of course similar remarks hold for extensions of $I\Delta_0$+EXP with finitely many axioms and for $I\Delta_0$+SUPEXP.

5.7 Some Facts about $I\Delta_0+\Omega_1$ and $I\Delta_0$+EXP

5.7.1 Fact: For $\psi \in \Pi_2$: $I\Delta_0+\text{EXP} \vdash \forall x(\Delta_\Omega\psi(x) \rightarrow \psi(x))$.

Proof: This is the contraposition of Lemma 8.10 of Paris & Wilkie[1987] with a parameter added. The extra parameter doesn't require any significant changes in Paris & Wilkie's proof. □

5.7.2 Fact: For every $\psi(x,y)\in\Delta_0$, having only x,y free, there is an $I\Delta_0+\Omega_1$-cut I such that: $I\Delta_0+\Omega_1\vdash\forall x\in I\exists y\psi(x,y) \Leftrightarrow I\Delta_0+EXP\vdash\forall x\exists y\psi(x,y)$.

Proof:"$\Leftarrow$" This is an entirely trivial variation of corollary 8.8 of Paris & Wilkie[1987]: the extra existential quantifier rides along for free. "$\Rightarrow$" Suppose I is an $I\Delta_0+\Omega_1$-cut and $I\Delta_0+\Omega_1\vdash\ \forall x\in I\exists y\psi(x,y)$. It follows that for some m: $I\Delta_0+EXP\vdash\Box_{\Omega,\underline{m}}\forall x\in I\exists y\psi(x,y)$. On the other hand for some k: $I\Delta_0+EXP\vdash\forall x\Box_{\Omega,\underline{k}}x\in I$, so it follows that for some n: $I\Delta_0+EXP\vdash\forall x\Box_{\Omega,\underline{n}}\exists y\psi(x,y)$. By the estimate in Paris & Wilkie[1987], p293, we can prove cut-elimination for restricted provability in $I\Delta_0+EXP$, so $I\Delta_0+EXP\vdash\forall x\Delta_\Omega\exists y\psi(x,y)$. By 5.7.1 we may conclude that: $I\Delta_0+EXP\vdash\forall x\exists y\psi(x,y)$. □

5.7.3 Consequence: for S,S' in Σ_1: $I\Delta_0+EXP\vdash S\rightarrow S' \Rightarrow I\Delta_0+\Omega_1\vdash\Box_\Omega S\rightarrow\Box_\Omega S'$.

Proof: Suppose $I\Delta_0+EXP\vdash S\rightarrow S'$, then for some $I\Delta_0+\Omega_1$-cut I $I\Delta_0+\Omega_1\vdash S^I\rightarrow S'$, so $I\Delta_0+\Omega_1\vdash\Box_\Omega S^I\rightarrow\Box_\Omega S'$. On the other hand: $I\Delta_0+\Omega_1\vdash\Box_\Omega S\rightarrow\Box_\Omega S^I$. □

5.8 Π_1-cut-conservativity

Define: $T\vdash^c\phi :\Leftrightarrow$ there is a T-cut I such that $T\vdash\phi^I$.

We say that U is Π_1-cut-conservative over V if for all Π_1-sentences P: $V\vdash^c P \Rightarrow U\vdash^c P$.

We show that for sequential U: U interprets V $\Rightarrow$ U is Π_1-cut-conservative over V. The proof will be verifiable in $I\Delta_0+\Omega_1$.

Proof: (The proof is really just a proof of lemma 3.3 of Pudlák[1985].) Suppose U interprets V. We will use outline for variables ranging over the domain assigned to V in the translation, for translated constants and predicates of V. Suppose I is a V-cut, P a Π_1-sentence and $V\vdash P^I$. Reason in U:

> The idea is to try to map the numbers of U into the 'translated numbers' of V. A small complication is that translated identity need only be an equivalence relation. So the 'function' we define will be multivalued.
>
> Define for $x\in\omega$:
>
> $F(x,y) :\leftrightarrow$ there is a sequence σ of elements of $\mathbb{\omega}$ such that $(\sigma)_0=\mathbb{0}$, for u<x $(\sigma)_{u+1}=\mathbb{S}((\sigma)_u)$, $(\sigma)_x=\mathbb{y}$.
>
> Let I_0 be the set of x's such that: $\exists\mathbb{y}\in\mathbb{\omega}\ (F(x,\mathbb{y})\wedge\forall\mathbb{z}\in\mathbb{\omega}\ (F(x,\mathbb{z})\rightarrow\mathbb{y}=\mathbb{z})$. As is easily seen I_0 contains 0 and is closed under successor. Clearly F behaves like a function w.r.t. = on I_0, so we will write f(x)=y instead of F(x,y) for $x\in I_0$.

Define $I_1:=\{x\in I_0|\forall y\in I_0\ (y+x\in I_0 \wedge f(y)+f(x)=f(y+x))\}$. It is easily seen that I_1 contains 0 and is closed under sucessor and addition. $I_2:=\{x\in I_1|\forall y\in I_1\ (y.x\in I_1 \wedge f(y).f(x)=f(y.x))\}$. Again it is easily seen that I_2 contains 0 and is closed under successor, addition and multiplication. Clearly on I_2 f commutes with 0,S, + and . .

Let $I_3:=\{x\in I_2|\forall y\leq x\ y\in I_2 \wedge \forall y\leq f(x)\ \exists z\leq x\ f(z)=y\}$. I_3 contains 0 and is closed under successor. Finally let I^* be the result of shortening I_3 to a cut that is closed under S, +, . and ω_1. Let $\mathbb{I}^*$ be the image of I^* under f. Both I^* and $\mathbb{I}^*$ are initial segments of their respective natural numbers, which are isomorphic w.r.t. 0, S, +, and . . Note that $\mathbb{I}^*$ need not be definable in V: for V it is an "external cut". We find for Δ_0-formulas $\phi(x_1,...,x_n)$:

$$\forall x_1,...,x_n\in I^*\ \phi(x_1,...,x_n) \leftrightarrow \phi(f(x_1),...,f(x_n)),$$

and thus for Π_1-sentences ψ: $\psi^{I^*} \leftrightarrow \psi^{\mathbb{I}^*}$.

By assumption we had $\mathbb{P}^{\mathbb{I}}$ where $\mathbb{I}$ is a translated V-cut. Let $\mathbb{J}:=\mathbb{I}\cap\mathbb{I}^*$ and let $J:=f^{-1}(\mathbb{J})$. We find that J is an U-cut isomorphic to $\mathbb{J}$ and thus P^J. □

Suppose V is also sequential and suppose U is Π_1-cut-conservative over V. We show that in this case V is locally interpretable in U.

Proof: Consider a finite subtheory V_0 of V. We have for some V-cut I: $V\vdash Tcon^I(V_0)$. So for some U-cut J $U\vdash Tcon^J(V_0)$. Ergo U interprets V_0. □

5.8.1 Application: We show for finite sequential U and V:

$$I\Delta_0+\Omega_1\vdash U\rhd V \leftrightarrow \exists\ I\Delta_0+\Omega_1\text{-cut } I\ \Box_\Omega(Tcon(U)\rightarrow Tcon^I(V)).$$

Proof: Reason in $I\Delta_0+\Omega_1$:

First suppose $U\rhd V$. Clearly $(I\Delta_0+\Omega_1+Tcon(U))\rhd U$ and hence $(I\Delta_0+\Omega_1+Tcon(U))\rhd V$. There is a V-cut J such that $\Box_V Tcon^J(V)$, so by Π_1-cut-conservativity there is an $I\Delta_0+\Omega_1+Tcon(U)$-cut J^* such that $\Box_\Omega(Tcon(U)\rightarrow Tcon^{J^*}(V))$. Define: $x\in I :\leftrightarrow (x\in J^*\vee\neg Tcon(U))$. As is easily seen I is an $I\Delta_0+\Omega_1$-cut and $\Box_\Omega(Tcon(U)\rightarrow Tcon^I(V))$.

Suppose $\exists\ I\Delta_0+\Omega_1$-cut I $\Box_\Omega(Tcon(U)\rightarrow Tcon^I(V))$. We have:

$$U\rhd(I\Delta_0+\Omega_1+Tcon(U))\rhd(I\Delta_0+\Omega_1+Tcon(V))\rhd V.$$ □

5.8.2 Consequence: let T be a finitely axiomatized and sequential. Let $U:=T+\phi$, $V=T+\bot$, we find: $I\Delta_0+\Omega_1\vdash \Box_T\phi \leftrightarrow \Box_\Omega\Delta_T\phi$.

5.8.3 Application: Let U be finitely axiomatized and sequential and let P be a Π_1-sentence. We have: $U\rhd P \Leftrightarrow I\Delta_0+EXP+Tcon(U)\vdash P$.

Proof: As is easy to see: $U \rhd P \Leftrightarrow$ for some $I\Delta_0+\Omega_1$-cut I: $I\Delta_0+\Omega_1+Tcon(U) \vdash P^I$. Apply 5.7.2. □

6 Principles

The language of IL is the language of modal propositional logic with one extra binary operator $\rhd$. An interpretation of this language in a theory T with a designated set of natural numbers satisfying $I\Delta_0+\Omega_1$ is a function $(.)^*$ that maps the atoms of the modal language on arbitrary sentences of the language of T, commutes with the propositional connectives (including $\perp$) and satisfies: $(\Box\phi)^*=\Box_T\phi^*$ and $(\phi\rhd\psi)^*=\phi^*\rhd_T\psi^*$. Here $\Box_T$ and $\rhd_T$ are the arithmetizations in the designated set of natural numbers of respectively provability in T and interpretability over T.

6.1 IL, the basic logic

The theory IL is useful as a basic theory from the modal standpoint. From the point of view of arithmetical interpretations it is too weak: as we will see the principle W, which is not derivable in IL, is valid for interpretations in all reasonable theories. The theory IL is given as Propositional Logic plus:

L1 $\vdash\phi \Rightarrow \vdash\Box\phi$

L2 $\vdash \Box(\phi\rightarrow\psi) \rightarrow (\Box\phi\rightarrow\Box\psi)$

L3 $\vdash \Box\phi \rightarrow \Box\Box\phi$

L4 $\vdash \Box(\Box\phi\rightarrow\phi)\rightarrow\Box\phi$

J1 $\vdash \Box(\phi\rightarrow\psi) \rightarrow \phi\rhd\psi$

J2 $\vdash (\phi\rhd\psi\wedge\psi\rhd\chi) \rightarrow \phi\rhd\chi$

J3 $\vdash (\phi\rhd\chi\wedge\psi\rhd\chi) \rightarrow (\phi\vee\psi)\rhd\chi$

J4 $\vdash \phi\rhd\psi \rightarrow (\Diamond\phi\rightarrow\Diamond\psi)$

J5 $\vdash \Diamond\phi\rhd\phi$

Note that the principle L3 is doubly superfluous: it follows both from L1, L2, L4 (by a well known argument) and from L1, L2, J4, J5 (by a trivial argument).

6.1.1 Reasoning in IL

It is pleasant to get some feeling for reasoning in IL. This section aims to provide some examples.

K1 $\vdash \phi\equiv(\phi\vee\Diamond\phi)$

Proof: immediate by J1, J5, J3. □

Let $F\phi := (\phi\vee\Diamond\phi)$, $G\phi := (\phi\wedge\Box\neg\phi)$, then:

K2 $\vdash F\phi \leftrightarrow FF\phi$

$\vdash F\phi \leftrightarrow FG\phi$

$\vdash G\phi \leftrightarrow GG\phi$

$\vdash G\phi \leftrightarrow GF\phi$

Immediate consequences are:

K3 $\vdash \phi \rhd (\phi\wedge\Box\neg\phi)$

K4 $\vdash \phi \equiv (\phi\wedge\Box\neg\phi)$

Note that: K3 is an alternative for axiom J5.

K5 $\vdash \phi\rhd\bot \rightarrow \Box\neg\phi$

Proof: by J4. □

Feferman's Principle is the following:

F $\vdash \Diamond\phi \rightarrow \neg(\phi\rhd\Diamond\phi)$

A Kripke Model argument shows that F is *not* derivable in IL. However the following weakening is derivable:

K6 $\vdash \Diamond\phi\rhd\neg(\phi\rhd\Diamond\phi)$

Proof: It is sufficient to show: IL$\vdash$ $(\Diamond\phi\wedge\Box\neg\Diamond\phi) \rightarrow \neg(\phi\rhd\Diamond\phi)$. We have:

$$\begin{aligned}\vdash (\Diamond\phi\wedge\Box\neg\Diamond\phi\wedge(\phi\rhd\Diamond\phi)) &\rightarrow (\Diamond\phi\wedge\Box\Box\neg\phi\wedge(\phi\rhd\Diamond\phi))\\ &\rightarrow (\Diamond\phi\wedge\phi\rhd\bot)\\ &\rightarrow (\Diamond\phi\wedge\Box\neg\phi)\\ &\rightarrow \bot\end{aligned}$$

□

In IL one can already derive the existence of unique and explicit fixed points for modalized formulas. For a (model-theoretical) proof the reader is referred to de Jongh & Veltman[?], this volume.

6.2 The logic ILW

ILW is IL plus the principle W:

W $\vdash \phi\rhd\psi \rightarrow \phi\rhd(\psi\wedge\Box\neg\phi)$

It may amuse the reader to show that ILW can be more efficiently axiomatized using only L1, L2, J1, J2, J3, J4, W.

W characterizes the set of IL-frames such that $R \circ S_x$ is upwards wellfounded for all x in their domain (see de Jongh & Veltman[?], this volume). I conjecture that ILW is complete for this set of structures. One can show that completeness w.r.t this set of frames implies completeness w.r.t a more restricted class of frames, namely those in which there are no infinite R,S-chains, where the index of S may vary. ILW is valid for interpretations in theories T with designated natural numbers satisfying $I\Delta_0+\Omega_1$, whose axiom sets can be represented by a R_1^+-formula. I conjecture that:

ILW$\vdash\phi \Leftrightarrow$ for all T with designated natural numbers satisfying $I\Delta_0+\Omega_1$, with R_1^+ axiom sets, for all interpretations $(.)^*$ into T: $T\vdash\phi^*$

6.2.1 Consequences of W

A first consequence of W is Feferman's Principle F:

F $\quad \vdash \Diamond\phi \rightarrow \neg(\phi \rhd \Diamond\phi)$

This is immediate by substituting $\Diamond\phi$ for ψ. A second consequence is the Contraposition Principle:

KW1 $\quad \vdash \phi \rhd \Diamond\top \rightarrow \top \rhd \neg\phi$

Proof:
$$\begin{aligned} \vdash \phi \rhd \Diamond\top &\rightarrow \phi \rhd (\Diamond\top \wedge \Box\neg\phi) \\ &\rightarrow \phi \rhd \Diamond\neg\phi \\ &\rightarrow \phi \rhd \neg\phi \end{aligned}$$

Ergo by J1 and J3 we have KW1. □

Both F and KW1 characterize the same class of IL structures as W. However I do not know whether W is derivable either in ILF or in IL(KW1).

Given the arithmetical validity of ILW we have the following consequence: Paris & Wilkie show that $EXP \rhd_\Omega \Diamond_\Omega \top$, ergo by KW1: $\top \rhd_\Omega \neg EXP$, i.o.w.: $(I\Delta_0+\Omega_1) \rhd (I\Delta_0+\Omega_1+\neg EXP)$.

Autobiographical note: this proof of the interpretability of $I\Delta_0+\Omega_1+\neg EXP$ in $I\Delta_0+\Omega_1$ could have been a nice example of how the logic allows one to discover new interpretations. Alas, things did not go like that. First I sketched a proof in $I\Delta_0+EXP$ of the tableaux consistency of $I\Delta_0+\Omega_1+\neg EXP$ adapting a method from Paris & Wilkie[1987]. This gives us $(I\Delta_0+\Omega_1) \rhd (I\Delta_0+\Omega_1+\neg EXP)$. Then I constructed an interpretation of $I\Delta_0+\Omega_1+\neg EXP$ in $I\Delta_0+\Omega_1+con(I\Delta_0+\Omega_1)$ using the Henkin construction described in 7.2.2.1. This again gives us $(I\Delta_0+\Omega_1) \rhd (I\Delta_0+\Omega_1+\neg EXP)$. Then I started to wonder about the connection of this fact and $(I\Delta_0+EXP) \rhd (I\Delta_0+\Omega_1+con(I\Delta_0+\Omega_1))$. This led me to prove the arithmetical validity of KW1 directly. Then I showed that KW1 is valid in all IL-structures with RoS_x upwards wellfounded. And *finally* I gave the simple modal proof of KW1.

6.2.2 Arithmetical validity of ILW

We verify that ILW is valid for arithmetical interpretations in theories T with designated natural numbers satisfying $I\Delta_0+\Omega_1$, whose axiom sets can be represented by a R_1^+-formula.

The axioms L1-4 are verified in Paris & Wilkie[1987]. J1, J2, J3 and J4 are trivial, given the fact that we opted for t-interpretability. We turn to J5: the Interpretation Existence Lemma. Before we proceed let me answer an obvious question: J5 follows from the stronger principle $\nabla\phi \rhd \phi$, which is assumed in this paper, so why bother to prove it? The answer is (i) to fix a number of concepts that we will use later on in the paper, and (ii) because the assumptions on provability in the proof are so weak that the argument also works for alternative notions like Feferman provability. The present construction is essentially Henkin's, refined by Feferman (see Feferman[1960]), with some twists due to Pudlák and Friedman.

6.2.2.1 The Henkin Construction

Let U be any theory with designated natural numbers satisfying $I\Delta_0+\Omega_1+\mathrm{con}V$, where V is a theory whose language L is given by a $\Delta_0(\omega_1)$-formula. We assume that $V\vdash\phi \Rightarrow U\vdash \Box_V\phi$. Define an extension of L, L^+ as follows: L^+ is the smallest extension such that if ϕ is in L^+ then there are constants $c[\exists x\phi]$ and $c[\forall x\phi]$ in L^+. L^+ is again $\Delta_0(\Omega_1)$. We choose an efficient coding of 0,1-sequences, where 0 is the empty sequence. Sequences are written: 0110, etc.. |x| is the length of the sequence coded by x. $\prec$ is the 'initial sequence' ordering. Define:

$u\in T[x] :\Leftrightarrow$ u is a T^+-sentence; $(x)_u=0$ or (u =NEG(v) and $(x)_v=1$) or (there is a w of the form $\exists z\phi(z)$ such that $(x)_w=0$ and u codes $\phi(c[\exists z\phi])$) or (there is a w of the form $\forall z\phi(z)$ such that $(x)_w=1$ and u codes $\neg\phi(c[\forall z\phi])$).

Note that $u\in T[x]$ is $\Delta_0(\omega_1)$. Moreover: $U\vdash x\prec y \rightarrow T[x]\subseteq T[y]$ and $U\vdash T[0]=\emptyset$.

Define further: $x\in TREE :\Leftrightarrow Con(V+T[x])$. Clearly $U\vdash (x\prec y \wedge y\in TREE) \rightarrow x\in TREE$. Moreover: $U\vdash 0\in TREE$. We show that $U\vdash x\in TREE \rightarrow (x0\in TREE \vee x1\in TREE)$. Reason in U:

> Suppose $x\in TREE$, i.e. Con(V+T[x]). Let u:=|x|+1. In case u does not code an L^+-sentence we have: T[x0]=T[x1]=T[x], so we are done. We treat the case that u codes a sentence of the form $\exists z\phi(z)$, the other cases are analogous or easier. So suppose u codes $\exists z\phi(z)$. Then $T[x0] = T[x]+\{ \exists z\phi(z), \phi(c[\exists z\phi(z)]) \}$ (note that the existence of $\phi(c[\exists z\phi(z)])$ requires Ω_1) and $T[x1] = T[x]+\{\neg\exists z\phi(z)\}$. The constant $c[\exists z\phi(z)]$ does not occur in V+T[x] (because we used the natural Gödelnumbering), hence it is easy to

convert a proof of falsity in V+ T[x]+{$\exists z\phi(z)$, $\phi(c[\exists z\phi(z)])$} in a proof of falsity in V+T[x]+{$\exists z\phi(z)$}. Thus if both V+T[x0] and V+T[x1] were inconsistent, we could convert the proofs of inconsistency in a proof of inconsistency of V+T[x] in the usual way. (All these conversions are available in $I\Delta_0+\Omega_1$.) □

Define PATH:={x∈TREE|"there is no y in TREE to the left of x"}. As is easily seen: U⊢ x∈PATH → (x0∈PATH ∨ x1∈PATH) and U⊢0∈PATH. Also U⊢ (x∈PATH ∧ y∈PATH) → (x≺y ∨ y≺x ∨ x=y).

Let X:={x| for some y in PATH x=|y|}. By the above U proves that 0 is in X and that 0 is closed under successor. By Solovay's methods we can shorten X to a U-cut I. For purposes of presentation we will define our interpretation for L with just one unary relation symbol R. The general case is, of course precisely the same. Define:

$x\in L^0$	:⇔	x∈I and x is a code of an L-sentence.
$x\in L^1$	:⇔	x∈I and x is a code of an L^+-sentence.
$x\in F^1(y)$	:⇔	x,y∈I and y is a code of a variable, x is a code of an L^+-formula with at most the variable coded by y free.
x∈D	:⇔	$x\in L^1$ and x codes a sentence of the form $\exists u\phi(u)$ or $\forall u\phi(u)$.
K(x)	:⇔	x∈I and there is an y∈PATH with |y|≤x and x∈T[y].
$R^K(x)$	:⇔	x is in D, x codes ψ and K(⌜R(c[ψ])⌝).

We have:

(i) U⊢$\forall x\in L^0$ $Prov_V(x)$ → K(x).

Reason in U:

Suppose $x\in L^0$ and $Prov_V(x)$. Since x is in I there is a y in PATH with |y|=x. Say x codes ψ. V+T[y] is consistent, and either ψ or ¬ψ is in T[y]. Clearly ¬ψ cannot be in T[y], so ψ is. □

(ii) K 'commutes' provably in U with the logical constants on L^1.

We first show: (a) U⊢$\forall x\in L^1$ K(x)∨K(NEG(x)) and (b) U⊢$\forall x\in L^1$ ¬(K(x)∧K(NEG(x))). Reason in U:

a) Consider x in L^1. x is in I so there is an y in PATH with |y|=x. In case $(y)_x=0$ we have x∈T[y], hence K(x). In case $(y)_x=1$ we have NEG(x)∈T[y], hence K(NEG(x)).
b) Suppose K(x) and K(NEG(x)). There are y and y' in PATH with x in T[y] and NEG(x) in T[y']. We have y=y' or y≺y' or y'≺y. Let z be the ≺-maximum of y, y'. Clearly both x and NEG(x) are in T[z]. But T[z] is consistent. Contradiction. □

We treat the cases of negation, conjunction and universal quantification: we show

(c) $U\vdash\forall x\in L^1\ K(NEG(x))\leftrightarrow\neg K(x)$

(d) $U\vdash\forall x,y\in L^1\ K(CONJ(x,y))\leftrightarrow(K(x)\wedge K(y))$

(e) $U\vdash\forall y\in I\ VAR(y)\rightarrow\forall x\in F^1(y)\ (\ K(UQ(y,x))\leftrightarrow\forall z\in D\ K(SUB(z,y,x))\ )$
(Here if z codes ψ, x codes $\phi(u)$ and y codes u: $SUB(z,y,x)=\ulcorner\phi(c[\psi])\urcorner$. Note that by Ω_1 both UQ(y,x) and SUB(z,y,x) are in L^1.)

(c) is immediate from (a) and (b). For (d) and (e) reason in U:

d) Consider x, y in L^1 and suppose K(x) and K(y). Let z:=CONJ(x,y). As is easily seen z is in I and hence in L^1. There is a w in PATH with $|w|=z$. Either z or NEG(z) are in T[w]. As is easily seen x and y are in T[w], so by the consistency of T[w] z must be in T[w], so K[z]. In case e.g. $\neg K(x)$ we have K(NEG(x)) and reasoning as before we find K(NEG(CONJ(x,y))), so $\neg K(CONJ(x,y))$.

e) Consider $y\in I$ with VAR(y) and $x\in F^1(y)$. First suppose K(UQ(y,x)). Clearly UQ(y,x) is in L^1. Consider z in D. As is easily seen SUB(z,y,x) is in L^1. Let v be the maximum of UQ(y,x) and SUB(z,y,x). There is a w in PATH with $|w|=v$. We have UQ(y,x) in T[w] and either SUB(z,y,x) or NEG(SUB(z,y,x)). By the consistency of T[w] we must have SUB(z,y,x) in T[w] and hence K(SUB(z,y,x)). Suppose for the converse that $\neg K(UQ(y,x))$. Let v:=UQ(y,x) and let w be in PATH with $|w|=v$. Reasoning as before we find that $(v)_w=1$ and thus that $NEG(SUB(v,y,x))\in T[v]$. Clearly v is in D and we have $\neg K(SUB(v,y,x))$. □

We write ϕ^K for the interpretation in the language of U of sentences ϕ of L using D and the translation of the relation symbols described above. As is easily seen: $U\vdash\ \phi^K\leftrightarrow K(\phi)$. So we have by (i): $U\vdash\ \Box_V\phi\rightarrow\phi^K$. Conclude: $V\vdash\phi\Rightarrow U\vdash\phi^K$.

NOTE:

I) Clearly the above reasoning can be verified in any theory T extending $I\Delta_0+\Omega_1$ such that $T\vdash\Box_V\phi\rightarrow\Box_U\Box_V\phi$.

II) We didn't make any assumption on the complexity of the formula defining the axiom set of V. So we can use the result for Feferman style predicates.

III) If provability in V is representable by a Σ_1-predicate then by a result of Wilkie $I\Delta_0+\Omega_1+$ con(V) is interpretable on a cut in Q+con(V). So in this case we can reduce our assumption that $I\Delta_0+\Omega_1+$con(V) is contained in U to the assumption that Q+conV is contained in U. In fact we may assume that U contains Q and proves con(V) on a cut (simply take as the natural numbers of U the elements of this cut).

6.2.2.2 The principle W

Let U and V be theories axiomatized by R_1^+-formulas extending $I\Delta_0+\Omega_1$. Suppose V is interpretable in U. We show that $V+\Box_U\bot$ is interpretable in U. The argument below is designed to be verifiable in $I\Delta_0+\Omega_1$.

The argument uses a trick that is due to Feferman. Let M be the interpretation of V in U. M is given by a finite amount of information and the associated translation of formulas is R_1^+-definable in $I\Delta_0+\Omega_1$. Define: $Prov_{V^*}(x) :\Leftrightarrow Prov_V(x) \wedge Prov_U(x^M)$. Trivially V^* is extensionally equal to V. So Id: $(V^*+\Box_U\bot)\rhd(V+\Box_U\bot)$. Also: $\Box_{V^*}(\Box_{V^*}\bot\to\Box_U\bot)$. Clearly the principles of IL can be verified for $\Box_{V^*}$ and $\rhd_{V^*}$ (using the fact that $Prov_{V^*}(x)$ can be written as an R_1^+-formula preceded by existential quantifiers). By K3: $V^*\rhd(V^*+\Box_{V^*}\bot)$ and hence $V^*\rhd(V^*+\Box_U\bot)$. Conclude: $U\rhd V\rhd V^*\rhd(V^*+\Box_U\bot)\rhd(V+\Box_U\bot)$. $\Box$

6.3 The Logic ILP

ILP is IL+P, where P is the Persistence Principle:

P $\quad \vdash \phi\rhd\psi\to\Box(\phi\rhd\psi)$

ILP is arithmetically valid for interpretations in finitely axiomatized theories with designated natural numbers satisfying $I\Delta_0+\Omega_1$. The verification of the arithmetical validity of P is trivial. We will show that ILP is complete for interpretations in finitely axiomatized sequential theories with designated natural numbers satisfying $I\Delta_0$+SUPEXP that do not prove their iterated inconsistency for any finite number of iterations. These include ACA_0 and GB.

ILP is also arithmetically sound and complete for a different interpretation, namely when we interpret $\Box$ as provability in PA and $\phi\rhd\psi$ as: for some primitive recursive term tx with just x free PA$\vdash \forall x\ Proof_{PA+\psi}(x,\ulcorner\bot\urcorner)\to Proof_{PA+\phi}(tx,\ulcorner\bot\urcorner)$. This strong notion of relative consistency is studied in Christian Bennet's Thesis (Bennet[1986a]). More on the alternative interpretation in section 8.3.

P characterizes IL structures with the following property: $yRzS_xu \Rightarrow yRu$. De Jongh & Veltman show the completeness of ILP w.r.t. (finite) IL structures satisfying this property (see de Jongh & Veltman[?], this volume).

We show that ILP extends ILW:

$$
\begin{aligned}
\vdash \phi\rhd\psi &\to \Box(\phi\rhd\psi)\\
&\to \Box(\Diamond\phi\to\Diamond\psi)\\
&\to \Box(\Box\neg\psi\to\Box\neg\phi)\\
&\to \Box((\psi\wedge\Box\neg\psi)\to(\psi\wedge\Box\neg\phi))\\
&\to (\psi\wedge\Box\neg\psi)\rhd(\psi\wedge\Box\neg\phi)
\end{aligned}
$$

$\rightarrow \psi \rhd (\psi \wedge \Box \neg \phi)$

The desired result is immediate. □

6.4 The logic ILM

ILM is IL plus Montagna's Principle M:

M $\vdash \phi \rhd \psi \rightarrow (\phi \wedge \Box \chi) \rhd (\psi \wedge \Box \chi)$

Fact: Montagna's Principle is arithmetically valid in verifiably essentially reflexive Δ_1-axiomatized theories with designated natural numbers satisfying $I\Delta_0+\Omega_1$.

Before we prove this fact first a few useful observations:

Observation 1: Suppose U has designated natural numbers satisfying $I\Delta_0+\Omega_1$. Let Q* be (Q+the axioms for linear ordering for the usual ordering on the natural numbers) extended to the language of U. Suppose U proves the Uniform Reflection Principle for Q*. Then U proves full Induction.

Proof of observation 1: Consider any formula $\phi(x)$ of the language of U. Let X:= $\{x \mid ((\phi(\underline{0}) \wedge \forall y(\phi(y) \rightarrow \phi(Sy))) \rightarrow \phi(x))\}$. We shorten X to a Q*-cut I and find: $U \vdash \forall x \Box_{Q^*} x \in I$. Ergo by URP for Q*: $U \vdash \forall x\ x \in I$. □

Observation 2: Let U be sequential, satisfying full Induction. Then U is essentially reflexive.

Proof of observation 2: This is just by the usual argument for the essential reflexiveness of PA, using the existence of partial truth-predicates in sequential theories. □

Proof of claim: Let T be an essentially reflexive Δ_1-axiomatized theory with designated natural numbers satisfying $I\Delta_0+\Omega_1$. We prove the slightly stronger principle: for S a Σ_1-sentence:

$\wedge\Sigma$ $T \vdash \phi \rhd \psi \rightarrow (\phi \wedge S) \rhd (\psi \wedge S)$

By observation 1 T satsfies full induction. So the Orey-Hájek theorem is verifiable: $T \vdash \chi \rhd_T \rho \leftrightarrow \forall x \Box_T(\chi \rightarrow \Diamond_T \ulcorner x \rho)$. Reason in T:

Let S be a Σ_1-sentence. Suppose $\chi \rhd_T \rho$ so $\forall x \Box_T(\chi \rightarrow \Diamond_T \ulcorner x \rho)$. Let q be so big that for all $x>q$ $\Box_T(S \rightarrow \Box_T \ulcorner x(S \leftrightarrow T))$. It follows that: $\Box_T(S \rightarrow (\Diamond_T \ulcorner x \rho \leftrightarrow \Diamond_T \ulcorner x(\rho \wedge S)))$, ergo: $\forall x \Box_T((\chi \wedge S) \rightarrow \Diamond_T \ulcorner x(\rho \wedge S))$ and thus $(\chi \wedge S) \rhd_T (\rho \wedge S)$. □

If T is sequential the following proof can be used: Reason in T:

Let S be a Σ_1-sentence. Suppose $M:\chi \rhd \rho$. The natural numbers of T+χ are on a T+χ-cut isomorphic to the natural numbers of the interpretation on a suitable 'external' cut. T+χ satisfies full induction, so this means that the natural numbers of T+χ are isomorphic to the natural numbers of the interpretation on a suitable 'external' cut, say I*. We have $\Box_T(\chi \rightarrow (S \rightarrow (S^{I^*})^M))$, hence by upwards persistence of Σ_1-sentences: $\Box_T(\chi \rightarrow (S \rightarrow S^M))$. □

Question: Is some strengthened version of $\wedge\Sigma$ equivalent to essential reflexiveness?

I conjecture that ILM is complete for arithmetical interpretations in verifiably essentially reflexive Δ_1-axiomatized theories with designated natural numbers satisfying $I\Delta_0+\Omega_1$.

M characterizes IL-frames with the following property: $yS_xzRu \Rightarrow yRu$. De Jongh & Veltman show that ILM is complete w.r.t. (finite) IL-models satisfying this property.

6.4.1 Consequences of M

We leave the simple verification that W is derivable in ILM to the reader. Two important consequences of M are:

KM1 $\vdash \phi \rhd \Diamond\psi \rightarrow \Box(\phi \rightarrow \Diamond\psi)$

KM2 $\vdash \phi \rhd \psi \rightarrow (\Box(\psi \rightarrow \Diamond\chi) \rightarrow \Box(\phi \rightarrow \Diamond\chi))$

Clearly these principles show us what is 'visible' of the Π_1-conservativety of essentially reflexive theories over theories interpreted in them. First we prove KM1:

Proof: $\vdash \phi \rhd \Diamond\psi \rightarrow (\phi \wedge \Box\neg\psi) \rhd (\Diamond\psi \wedge \Box\neg\psi)$

$\rightarrow (\phi \wedge \Box\neg\psi) \rhd \bot$

$\rightarrow \Box\neg(\phi \wedge \Box\neg\psi)$

$\rightarrow \Box(\phi \rightarrow \Diamond\psi)$ □

Next we show derive KM2 from KM1:

Proof: $\vdash \phi \rhd \psi \rightarrow (\Box(\psi \rightarrow \Diamond\chi) \rightarrow (\psi \rhd \Diamond\chi)$

$\rightarrow (\phi \rhd \Diamond\chi)$

$\rightarrow \Box(\phi \rightarrow \Diamond\chi))$ □

Both KM1 and KM2 characterize IL-frames satisfying $yS_xzRu \Rightarrow yRu$. But it is unknown whether any of them implies M over IL.

7 $I\Delta_0+\Omega_1$, $I\Delta_0+EXP$ & Friedman's Characterization

7.1 Tableaux provability in $I\Delta_0+EXP$

Consider any R_1^+-axiomatized theory T. Transforming an ordinary T-proof into a T-tableaux-proof is a superexponential process. To be precise it is of order: itexp($|x|,\rho(x)$), where x is our original proof and where $\rho(x)$ is the cut rank of x, i.e. the supremum of the lengths of the cut formulas in x. So in general $I\Delta_0+EXP$ will not prove: $\Box_T\phi \rightarrow \Delta_T\phi$. On the other hand using the above estimate as a bound one can show for *sentences* ϕ and ψ:

$$I\Delta_0+EXP \vdash \Delta_T(\phi\rightarrow\psi) \rightarrow (\Delta_T\phi \rightarrow \Delta_T\psi).$$

The point is of course that the cut-rank involved is standard and thus the rate of growth is just multi-exponential. It would be very pleasant if we had this fact also for formulas (under our convention for free varables within the scope of Δ). It seems to me that the more general fact should hold, but I do not really know it. Another familiar principle is:

$$I\Delta_0+EXP \vdash \Delta_T\phi \rightarrow \Delta_T\Delta_T\phi .$$

In fact I conjecture that this principle is already verifiable in $I\Delta_0+\Omega_1$. (To prove this one would have to inspect how much cuts are involved in Paris & Wilkie's procedure to produce a proof of a R_1^+-formula ϕ given ϕ.)

The above observations imply that we have the usual provability logic for Δ_T with T extending $I\Delta_0+EXP$. One can verify Solovay's completeness proof in $I\Delta_0+EXP$, so it follows that Löb's Logic L is precisely the logic of such Δ_T. (The fact that we talk about tableaux proofs does not matter at all.)

So surprisingly $\Box_{EXP}$ and Δ_{EXP} satisfy the same provability principles without being provably the same over $I\Delta_0+EXP$. The next section's result will imply that one extra principle characterizes the logic of $\Box_{EXP}$ and Δ_{EXP} together.

7.2 Formalizing a result of Paris & Wilkie

We want to formalize 5.7.2: i.e.: for every $\psi(x,y)\in \Delta_0$ with only x,y free, there is an $I\Delta_0+\Omega_1$-cut I such that: $I\Delta_0+\Omega_1\vdash\forall x\in I\exists y\psi(x,y) \Leftrightarrow I\Delta_0+EXP\vdash\forall x\exists y\psi(x,y)$. An obvious first guess at the correct formulation of the formalization is e.g.: for every $\psi(x,y)\in \Delta_0$ with only x,y free: $I\Delta_0+EXP\vdash \Box_{EXP}\forall x\exists y\ \psi(x,y) \leftrightarrow \exists\ I\Delta_0+\Omega_1$-cut $I\Box_\Omega\forall x\in I\ \exists y\ \psi(x,y)$. But this cannot be right. Taking $\psi:=\perp$ we would get: $I\Delta_0+EXP\vdash \neg\Box_\Omega\perp \rightarrow \neg\Box_{EXP}\perp$, contradicting theorem 8.19 of Paris & Wilkie[1987].

(A somewhat simplified proof of this theorem is as follows: suppose $I\Delta_0+EXP\vdash \neg\Box_\Omega\perp \rightarrow \neg\Box_{EXP}\perp$, then by 5.7.2 for some $I\Delta_0+\Omega_1$-cut I: $I\Delta_0+\Omega_1\vdash \neg\Box_\Omega\perp \rightarrow \neg\Box^I_{EXP}\perp$. Let J be an $I\Delta_0+EXP$-cut such that $I\Delta_0+EXP\vdash \forall x\in J$ supexp(x) exists. $I\Delta_0+EXP\vdash \neg\Delta_\Omega\perp$ by 5.7.1 so by cut-elimination: $I\Delta_0+EXP\vdash \neg\Box^J{}_\Omega\perp$. Because $I\Delta_0+\Omega_1$ is verifiable on J we find by composing cuts that for some $I\Delta_0$+EXP-cut J*: $I\Delta_0+EXP\vdash \neg\Box^{J*}_{EXP}\bot$. This contradicts Pudlák's sharpening of the second incompleteness theorem (see Pudlák[1985]) (or alternatively: it contradicts Feferman's Principle F (see section 6.2.1)).)

The correct form for the formalization turns out to be this: for every $\psi(x,y)\in\Delta_0$ with only x,y free:

$$I\Delta_0+EXP\vdash \Delta_{EXP}\forall x\exists y\ \psi(x,y) \leftrightarrow \exists\ I\Delta_0+\Omega_1\text{-cut } I\Box_\Omega\forall x\in I\ \exists y\ \psi(x,y).$$

Proof: For the "→"-direction I briefly sketch how this can be shown by transforming proofs and then give a more elaborate simulation of the model-theoretical argument of Paris & Wilkie. Reason in $I\Delta_0$+EXP:

Let z be an EXP-tableaux-proof of $\ulcorner\forall x\exists y\ \psi(x,y)\urcorner$. The tableaux will move once from $\ulcorner\neg\forall x\exists y\ \psi(x,y)\urcorner$ to $\ulcorner\neg\exists y\ \psi(c,y)\urcorner$ to obtain a contradiction from this last formula plus the axioms of $I\Delta_0$+EXP. The only principles used in the rest of the proof that are not Π_1 or negated Σ_1 are the axioms for S, +, . and EXP. So the only "growing constants" introduced are due to these axioms and their maximal rate of growth is due to EXP. Our tableaux system is assumed to be relational, so in every step the growth is only caused by one application of EXP. So the biggest constant in the proof will be something like: exp(exp(...c)...), where the exp is iterated |z| times. Using an estimate of Pudlák we can find for any u an $I\Delta_0+\Omega_1$-cut $I_u\leq exp(u)$ such that $\Box_\Omega(\forall v\in I_u\ exp(v)\in I_{u-1})$. Choose $I:=I_{|z|}$.

Now we transform z into an $I\Delta_0+\Omega_1$-proof z* of $\ulcorner\forall x\in I\ \exists y\ \psi(x,y)\urcorner$ as follows. We may start from the axioms of $I\Delta_0+\Omega_1$ plus $\ulcorner c\in I\urcorner$ and $\ulcorner\neg\exists y\ \psi(c,y)\urcorner$. We follow z, but add on proofs for any constant e introduced that e≤exp(exp(...c)...) say for u≤|z| iterations of exp, plus proofs that: $exp(exp(...c)...)\in I_{|z|-u}$. Application of EXP to e can then be replaced by a use of the fact that e is in $I_{|z|-u}$.

We turn to the alternative argument: by contraposition it is sufficient to show that for χ in Δ_0 with only x,y free:

$$I\Delta_0+EXP\vdash\ \forall\ I\Delta_0+\Omega_1\text{-cuts } I\Diamond_\Omega\exists x\in I\ \forall y\ \chi(x,y) \rightarrow \nabla_{EXP}\exists x\forall y\ \chi(x,y),$$

hence by 5.7.1 it is sufficient to show that: for some $I\Delta_0+\Omega_1$-cut J:

$$I\Delta_0+\Omega_1\vdash\ \forall\ I\Delta_0+\Omega_1\text{-cuts } I\Diamond_\Omega\exists x\in I\ \forall y\ \chi(x,y) \rightarrow \nabla^J_{EXP}\exists x\forall y\ \chi(x,y).$$

The above in its turn is immediate by Π_1-cut-conservativity from:

$$(\forall\ I\Delta_0+\Omega_1\text{-cuts } I\Diamond_\Omega\exists x\in I\ \forall y\ \chi(x,y))\rhd_\Omega\nabla_{EXP}\exists x\forall y\ \chi(x,y),$$

Because $\nabla_{EXP}\exists x\forall y\ \chi(x,y)\equiv(EXP\wedge\exists x\forall y\ \chi(x,y))$ (see section 5.6) this last statement follows from:

$$(\forall\ I\Delta_0+\Omega_1\text{-cuts } I\Diamond_\Omega\exists x\in I\ \forall y\ \chi(x,y))\rhd_\Omega(EXP\wedge\exists x\forall y\ \chi(x,y)).$$

Reason in $I\Delta_0+\Omega_1$:

Suppose that for every $I\Delta_0+\Omega_1$-cut I: $\Diamond_\Omega\exists x\in I\,\forall y\,\chi(x,y)$. By 5.6.4 we can find a (standard) $I\Delta_0+\Omega_1$-cut J such that $\forall u\in J\,\exists\, I\Delta_0+\Omega_1$-cut I $\Box_\Omega(\forall v\in I$ itexp(v,u) exists). It follows that: $\forall u\in J\,\Diamond_\Omega\exists x$ (itexp(x,u) exists $\wedge\,\forall y\,\chi(x,y)$). Let c be a new constant and let $V:=I\Delta_0+\Omega_1+\forall y\,\chi(c,y)+\{$itexp(c,u) exists $\mid u\in J\}$. As is easily seen V is consistent. Let I, K, D, f be as in 6.2.2.1 and let $c^*:=\ulcorner\exists x\; x=c\urcorner$. Define:

$$x\in D^* :\Leftrightarrow x\in D\wedge\exists y\in I\; x\leq^K \mathrm{itexp}^K(c^*,f(y)).$$

Let K* be the interpretation based on I, K and D*. As is easily seen D* is closed under $\exp^K$ and thus under $\exp^{K^*}$. Moreover $(\forall y\,\chi(c,y))^{K^*}$, and thus $(\exists x\forall y\,\chi(x,y))^{K^*}$. The (standard) instances ρ of Δ_0-induction have Π_1 form, moreover we have ρ^K, so: ρ^{K^*}.

"$\leftarrow$" Let $\mathfrak{I}$ be an $I\Delta_0$+EXP-cut such that $I\Delta_0+\mathrm{EXP}\vdash\forall u\in\mathfrak{I}\,\forall v$ itexp(v,u) exists. We first show:

$I\Delta_0+\mathrm{EXP}\vdash\forall I\in\mathfrak{I}(\,\Box_\Omega{}^{\mathfrak{I}}$"I is a cut" $\rightarrow(\,\exists z\in\mathfrak{I}\;\Box_{\Omega,z}\forall x\in I\,\exists y\,\psi(x,y)\rightarrow\forall x\,\exists y\,\psi(x,y)\,)\,)$.

Reason in $I\Delta_0$+EXP:

Suppose $I\in\mathfrak{I}$, $\Box_\Omega{}^{\mathfrak{I}}$"I is a cut", $z\in\mathfrak{I}$ and $\Box_{\Omega,z}\forall x\in I\,\exists y\,\psi(x,y)$. Inspecting the argument for 5.6.2 we find that for some $u\in\mathfrak{I}$ and for all v $\Box_{\Omega,u}v\in I$. It follows that for some $w\in\mathfrak{I}$: $\forall x\Box_{\Omega,w}\exists y\,\psi(x,y)$. Using the estimate on cut-elimination in Paris and Wilkie[1978], p293 we may conclude: $\forall x\Delta_\Omega\exists y\,\psi(x,y)$. By 5.7.1: $\forall x\exists y\,\psi(x,y)$.

To prove our theorem reason again in $I\Delta_0$+EXP:

Suppose that for some $I\Delta_0+\Omega_1$-cut I $\Box_\Omega\forall x\in I\,\exists y\,\psi(x,y)$. By the sharp version of $R_1{}^+$-completeness we find that for some standard $\underline{m}$ and for some u and z: $\Box_{\mathrm{EXP},\underline{m}}\mathrm{Proof}_\Omega$(u,"I is a cut") and $\Box_{\mathrm{EXP},\underline{m}}\Box_{\Omega,z}\forall x\in I\,\exists y\,\psi(x,y)$. By 5.7.1 for some standard $\underline{k}$: $\Box_{\mathrm{EXP},\underline{k}}I\in\mathfrak{I}$, $\Box_{\mathrm{EXP},\underline{k}}u\in\mathfrak{I}$, $\Box_{\mathrm{EXP},\underline{k}}z\in\mathfrak{I}$. By our auxiliary result for some standard $\underline{n}$:

$\Box_{\mathrm{EXP},\underline{n}}\forall I\in\mathfrak{I}(\,\Box_\Omega{}^{\mathfrak{I}}$"I is a cut" $\rightarrow(\,\exists z\in\mathfrak{I}\;\Box_{\Omega,z}\forall x\in I\,\exists y\,\psi(x,y)\rightarrow\forall x\,\exists y\,\psi(x,y)\,)\,)$.

Conclude that for some standard $\underline{p}$: $\Box_{\mathrm{EXP},\underline{p}}\forall x\,\exists y\,\psi(x,y)$. By applying cut elimination we find: $\Delta_{\mathrm{EXP}}\forall x\,\exists y\,\psi(x,y)$. □

A variant of our theorem can be easily obtained as follows: by 5.7.3 we find: for every $\psi(x,y)\in\Delta_0$ with only x,y free:

$$I\Delta_0+\Omega_1\vdash\Box_\Omega\Delta_{\mathrm{EXP}}\forall x\exists y\,\psi(x,y)\leftrightarrow\Box_\Omega\exists\, I\Delta_0+\Omega_1\text{-cut } I\Box_\Omega\forall x\in I\,\exists y\,\psi(x,y).$$

Combining this with 5.8.2 we get:

$$I\Delta_0+\Omega_1\vdash\Box_{\mathrm{EXP}}\forall x\exists y\,\psi(x,y)\leftrightarrow\Box_\Omega\exists\, I\Delta_0+\Omega_1\text{-cut } I\Box_\Omega\forall x\in I\,\exists y\,\psi(x,y).$$

7.3 Some Consequences

1 $(\forall\, I\Delta_0+\Omega_1\text{-cut } I\Diamond_\Omega\exists x\in I\,\forall y\,\chi(x,y))\equiv_\Omega(\mathrm{EXP}\wedge\exists x\forall y\,\chi(x,y))$.

2 $\Diamond_\Omega\top\equiv_\Omega\mathrm{EXP}$.

3 For S in Σ_1 we have: $I\Delta_0+EXP\vdash \Delta_{EXP}S \leftrightarrow \Box_{\Omega}S$.

4 For S in Σ_1 we have: $I\Delta_0+\Omega_1\vdash \Box_{EXP}S \leftrightarrow \Box_{\Omega}\Box_{\Omega}S$.

5 For S and S' in Σ_1 we have: $I\Delta_0+\Omega_1\vdash \Box_{EXP}(S \rightarrow S') \rightarrow \Box_{\Omega}(\Box_{\Omega}S \rightarrow \Box_{\Omega}S')$.

Proof:
$$\begin{aligned} I\Delta_0+\Omega_1\vdash \Box_{EXP}(S \rightarrow S') &\rightarrow \Box_{\Omega}\exists\, I\Delta_0+\Omega_1\text{-cut } I\ \Box_{\Omega}(S^I \rightarrow S') \\ &\rightarrow \Box_{\Omega}\exists\, I\Delta_0+\Omega_1\text{-cut } I\ (\Box_{\Omega}S^I \rightarrow \Box_{\Omega}S') \\ &\rightarrow \Box_{\Omega}(\Box_{\Omega}S \rightarrow \Box_{\Omega}S') \end{aligned}$$
□

6 For ψ in Π_2 we have: $I\Delta_0+\Omega_1\vdash \Box_{\Omega}\Box_{\Omega}\psi \rightarrow \Box_{EXP}\psi$.

7 $I\Delta_0$+SUPEXP proves Π_2-reflection for $I\Delta_0$+EXP. Let $\phi(x,y)$ be Δ_0, having only x,y free:

Proof:
$$\begin{aligned} I\Delta_0+SUPEXP\vdash \Box_{EXP}\forall x\exists y\ \phi(x,y) &\rightarrow \Delta_{EXP}\forall x\exists y\ \phi(x,y) \\ &\rightarrow \exists\ I\Delta_0+\Omega_1\text{-cut } I\Box_{\Omega}\forall x\in I\ \exists y\ \phi(x,y) \\ &\rightarrow \forall x\Box_{\Omega}\exists y\ \phi(x,y) \\ &\rightarrow \forall x\Delta_{\Omega}\exists y\ \phi(x,y) \\ &\rightarrow \forall x\exists y\ \phi(x,y) \end{aligned}$$
□

Combining 5.8.2 and 7.3.3 we get the desired missing principle for the combined provability logic of $\Box_{EXP}$ and Δ_{EXP}:

8 $I\Delta_0+EXP\vdash \Box_{EXP}\psi \leftrightarrow \Delta_{EXP}\Delta_{EXP}\psi$

It is immediately clear that this last principle together with the complete set of principles for Δ_{EXP} fully describes the mixed logic. Note that we have this variation of Löb's Principle:

9 $I\Delta_0+\Omega_1\vdash \Box_{EXP}(\Box_{EXP}\psi \rightarrow \Delta_{EXP}\psi) \rightarrow \Box_{EXP}\Delta_{EXP}\psi$.

This principle "says" in some sense that $I\Delta_0$+EXP does not prove cut-elimination.

7.4 Friedman's Characterization

Let U and V be finitely axiomatized sequential theories. Combining 5.8.1 with 7.2 we find:

$$I\Delta_0+EXP\vdash U\rhd V \leftrightarrow \Delta_{EXP}(\mathrm{Tcon}(U)\rightarrow \mathrm{Tcon}(V)).$$

A variant is:

$$I\Delta_0+\Omega_1\vdash \Box_{\Omega}U\rhd V \leftrightarrow \Box_{EXP}(\mathrm{Tcon}(U)\rightarrow \mathrm{Tcon}(V)).$$

Note that for interpretability over $I\Delta_0$+EXP this implies:

$$I\Delta_0+EXP \vdash \phi \rhd_{EXP} \psi \leftrightarrow \Delta_{EXP}(\nabla_{EXP}\phi \rightarrow \nabla_{EXP}\psi).$$

Combining this with the fact that we know the complete provability logic of Δ_{EXP} we get a *Kripke model characterization* of the interpretability logic of $I\Delta_0$+EXP. It is unknown what modal theory corresponds precisely with this Kripke semantics.

8 Arithmetical Completeness for Interpretations in Finitely Axiomatized, Sequential Theories extending $I\Delta_0$+SUPEXP

In this section we prove: ILP is complete for arithmetical interpretations in any finitely axiomatized sequential theory with designated natural numbers that satisfy $I\Delta_0$+SUPEXP (plus a minimal extra condition). It is convenient to use a slightly different Kripke semantics in the proof of the arithmetical completeness theorem. Because this semantics is strongly suggested by Friedman's characterization of interpretability, I propose to call it Friedman Semantics.

8.1 Friedman semantics

A *Friedman structure* is a tuple <K,b,P,Q>, where:

i) K is a non-empty set,
ii) $b \in K$ and for every $x \in K$ bWQx,
iii) R is a binary relation on K,
iv) Q is a transitive, irreflexive, upwards wellfounded, binary relation,
v) $P \subseteq Q$,
vi) $xQyPz \Rightarrow xPz$.

Note that (v) and (vi) imply that P is transitive. Let $R := Q \circ P$, i.e. $xRy \Leftrightarrow$ for some z xQzPy.

A relation $\Vdash$ between K and *L* is a *forcing relation* if it satisfies the usual clauses, with R as the accessibility relation for the $\Box$, plus:

$$x \Vdash \phi \rhd \psi \Leftrightarrow \forall u\, (xQu \Rightarrow (\exists y\, (uPy \wedge y \Vdash \phi) \Rightarrow \exists z\, (uPz \wedge z \Vdash \psi)))$$

If F is a Friedman structure and $\Vdash$ is a forcing relation on F, then <F,$\Vdash$> is a *Friedman model*. It is easy to check that ILP is valid in any Friedman model.

Consider an IL-model W=<K,R,S,$\Vdash$> and a Friedman model G'. β is a *bisimulation* between W and G' if:

i) β is a relation between K and K'.
ii) $b\beta b'$.
iii) Let x,y,... range over K, let x',y',... range over K': we have for any x,x' with $x\beta x'$:
$\forall y(xRy \Rightarrow \exists u',y'(y\beta y' \wedge x'Q'u'P'y' \wedge \forall z'(u'P'z' \Rightarrow \exists z(z\beta z' \wedge yS_xz))))$.

iv) We have for any x,x' with $x\beta x'$:

$$\forall u',y'(x'Q'u'P'y' \Rightarrow \exists y(y\beta y' \wedge xRy \wedge \forall z(yS_x z \Rightarrow \exists z'(z\beta z' \wedge u'P'z'))))$$.

v) $x\beta x' \Rightarrow (x \Vdash p \Leftrightarrow x' \Vdash' p)$, for all atoms p.

Note that for every $x \in K$ there is an $x' \in K'$ with $x\beta x'$, but that possibly there are $u' \in K'$ such that for no $u \in K$ $u\beta u'$. We do have: for all $x' \in \{b'\} \cup \text{range } R'$ there is an $x \in K$ such that $x\beta x'$.

We have: $x\beta x' \Rightarrow$ for all ϕ $x \Vdash \phi \Leftrightarrow x' \Vdash' \phi$. The proof is by a trivial induction on ϕ.

To prove completeness for ILP w.r.t finite Friedman models it is clearly sufficient to show that every finite IL-model W satisfying: $xRyRzS_x u \Rightarrow zS_y u$, can be bisimulated by a finite Friedman model G'.

Construction: Let W be a finite IL-model for ILP. We construct a bisimulating Friedman model G':

$K' := \{\langle x_1,\ldots,x_n\rangle \mid n \geq 1,\ x_1 = b,\ x_{2i-1}Rx_{2i}$ (for $1 \leq i$ and $2i \leq n$) and $x_{2i}S[x_{2i-1}]x_{2i+1}$ for $1 \leq i$ and $2i < n\}$,

$b' = \langle b\rangle$,

$\langle x_1,\ldots,x_n\rangle\ Q'\ \langle y_1,\ldots,y_m\rangle :\Leftrightarrow\ m > n$; for all $i \leq n$ $x_i = y_i$,

$\langle x_1,\ldots,x_n\rangle\ P'\ \langle y_1,\ldots,y_m\rangle :\Leftrightarrow \langle x_1,\ldots,x_n\rangle\ Q'\ \langle y_1,\ldots,y_m\rangle$ and m is odd,

$\langle x_1,\ldots,x_n\rangle \Vdash p :\Leftrightarrow x_n \Vdash p$.

It is easy to see that the model constructed is a finite Friedman model. Note that Q' is a tree and that G' is really a Carlson model, i.e. there is a set $X' \subseteq K'$ such that $x'R'y' \Leftrightarrow x'Q'y'$ and $y' \in X'$. We always take $X' := \{\langle y_1,\ldots,y_m\rangle \mid m \text{ is odd}\}$, so that $b' \in X'$. The nodes x' of X' also satisfy the additional property: if x'P'y' then x'R'y'. This property will be needed in our arithmetical completeness proof for reasons to be explained later.

Define: $x\beta x' :\Leftrightarrow x' = \langle x_1,\ldots,x_{2m-1}\rangle$ and $x_{2m-1} = x$. We show that β is a bisimulation between W and G'.

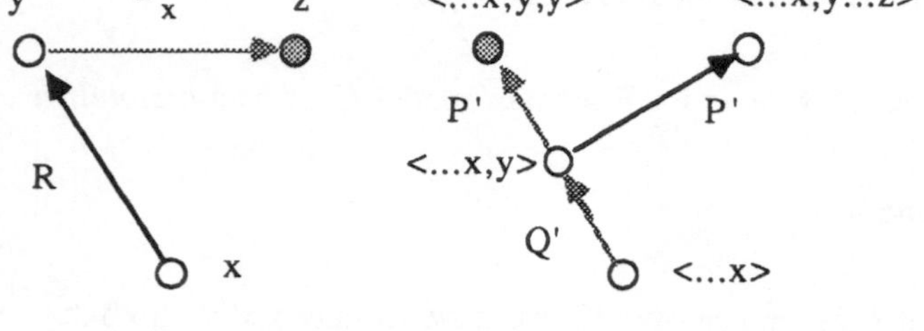

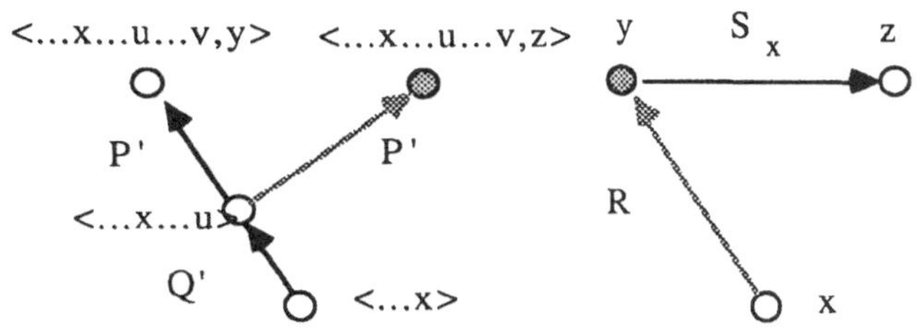

The conventions for interpreting these pictures are as follows: the black arrows and open nodes are 'given', 'universal'; the grey arrows and grey nodes are 'produced', 'existential'. If a 'given' arrow (node) in the left (right) half of the picture corresponds to a 'produced' arrow (node) in the left (right) half of the picture, then the 'produced' arrow (node) is produced, given the corresponding 'given' arrow (node). Corresponding nodes bisimulate under the given bisimulation β.

Let's first discuss the upmost picture. Here it is to be shown that indeed yS_xz. We have: either <...x,y...z>=<...x,y,z> or <...x,y...z>=<...x,y,u,v...z>. In the first case we have trivially yS_xz. In the second case one easily shows uRz. So we have: yS_xuRz, hence yS_xz. Secondly we look at the second picture. We only have to show that for the unique w with <...x...u...v,y>= <...x...w,v,y> we have: vS_wz. It is possible that x=w: this case is easy. So suppose $x\neq w$. As is easily seen: xRw, so $xRwRvS_wyS_xz$, and thus $xRwRyS_xz$. Hence yS_wz and thus vS_wz. □

8.2 A Solovay-style Completeness Proof

Let U be a finite sequential extension of $I\Delta_0$+SUPEXP. We only need to assume that U in fact extends $I\Delta_0$+SUPEXP (as an infinitely axiomatized theory), we do not need to stipulate that $I\Delta_0+EXP\vdash \Delta_{SUPEXP}\psi \rightarrow \Delta_U\psi$. This because we will only need actual theorems of $I\Delta_0$+SUPEXP. We will need however the following lemma:

Lemma: For sentences $\psi\in\Pi_2$: $I\Delta_0+EXP\vdash \Delta_{EXP}\psi \rightarrow \Delta_U\psi$.

Proof: We have: $I\Delta_0+EXP\vdash \Delta_{EXP}\psi\rightarrow\Delta_U\Delta_{EXP}\psi$ and $I\Delta_0+EXP\vdash \Delta_U(\Delta_{EXP}\psi\rightarrow\psi)$. Hence: $I\Delta_0+EXP\vdash \Delta_{EXP}\psi \rightarrow \Delta_U\psi$. □

(Note that even a stronger version of the lemma is provable with $\Box_{EXP}$ instead of Δ_{EXP}.)

We assume that for no k $U\vdash\Box^k{}_U\bot$. (If $U\vdash\Box^k{}_U\bot$ for some k, let k* be the smallest such k. The corresponding logic will then be $ILP+\Box^{k^*}\bot$, as is seen by an easy adaptation of the argument below.)

Arithmetical Completeness Theorem

$ILP\vdash\phi \Leftrightarrow$ for all * $U\vdash\phi^*$.

Proof: the proof is a refinement of an earlier proof by Smorynski and the author for the case that U=GB or $U=ACA_0$. Its basic idea is close to that behind the completeness proof for PRL_{ZF} due to Carlson (see Smorynski[1985], p205-214).

The "$\Rightarrow$" side is clear. For the "$\Leftarrow$" part suppose $ILP\nvdash\phi$. Let G_0 be a finite Friedman countermodel to ϕ with Q_0 upwards wellfounded. We may assume that K_0 is $\{1,...,N\}$, that 1 is the bottom element and $1\nVdash_0\phi$, and that our model satisfies:

- (i) Q is a tree,
- (ii) P is given "Carlson-style" by a set X: so $xPy \Leftrightarrow xQy$ and $y\in X$; $1\in X$.
- (iii) if $x\in X$ and xPy, then xRy.

Arithmetically speaking the nodes in X correspond with the 'point of view' of U. P considered as an accessibility relation corresponds to Δ_U, Q similarly corresponds to Δ_{EXP}. Property (iii) corresponds to the fact that U proves cut-elimination and thus proves the equivalence of Δ_U and $\Box_U$.

G is the result of hanging a new node 0 under G_0. Formally: $K:=K_0\cup\{0\}$, $xQy :\Leftrightarrow (x=0$ and $y\neq0)$ or xQ_0y, $X:=X_0\cup\{0\}$, $xPy :\Leftrightarrow xQy$ and $x\in X$, $x\Vdash p :\Leftrightarrow x\neq0$ and $x\Vdash_0 p$. Clearly $1\nVdash\phi$.

Define by the recursion theorem:

$h(0):=0$,
$h(x+1):=y$ if $h(x)Py$ and $Tproof_U(x,L\neq y)$ or if $h(x)Qy$ and $Tproof_{EXP}(x,L\neq y)$,
$h(x+1):=h(x)$ otherwise,
L:= the unique x such that $\exists y\forall z>y\ h(z)=x$.

The definition of h can be given in such a way that '$h(x)=y$'is a $\Delta_0(exp)$-formula and its elementary properties are verifiable in $I\Delta_0+EXP$ (in fact we can do much better, but that is not relevant here). We have e.g. $I\Delta_0+EXP\vdash x<y\rightarrow h(x)WQh(y)$ and $I\Delta_0+EXP\vdash$ "L exists".

Note that: $I\Delta_0+EXP\vdash L=x \leftrightarrow \exists y\ h(y)=x \wedge \forall u,v((h(u)=x\wedge v>u)\rightarrow h(v)=x)$, so $L=x$ is in fact the conjunction of a Σ_1- and a Π_1-formula.

Define for atoms p: $p^*:=\bigvee\{L=\underline{i} \mid i\Vdash p\}$.

Lemma: $U \vdash L \in X$.

Proof of lemma: Reason in U:

Suppose $L=\underline{i} \notin X$. By the definition of h: $\Delta_{EXP} L \neq \underline{i}$. By Π_2-reflection: $L \neq \underline{i}$. Contradiction, so $L \neq \underline{i}$. □

We show by induction on ψ for $1 \leq i \leq N$:

$i \Vdash \psi \;\Rightarrow I\Delta_0 + EXP \vdash L=\underline{i} \rightarrow \psi^*$,

$i \nVdash \psi \;\Rightarrow I\Delta_0 + EXP \vdash L=\underline{i} \rightarrow \neg\psi^*$.

The cases of the atoms and the propositional connectives are trivial. The case of □ follows immediately from the case of $\rhd$. Assume the IH for χ. We show for $i \neq 0$:

for all j with iPj $j \Vdash \chi \quad \Rightarrow I\Delta_0 + EXP \vdash L=\underline{i} \rightarrow \Delta_U \chi^*$,

for some j iPj and $j \nVdash \chi \quad \Rightarrow I\Delta_0 + EXP \vdash L=\underline{i} \rightarrow \neg\Delta_U \chi^*$.

First suppose that $i \neq 0$ and for all j with iPj $j \Vdash \chi$. Reason in $I\Delta_0$+EXP:

Suppose $L=\underline{i}$. By the definition of h we have $\Delta_{EXP} L \neq \underline{i}$ or $\Delta_U L \neq \underline{i}$. In both cases $\Delta_U L \neq \underline{i}$. For some x $h(x)=\underline{i}$, so $\Delta_U \exists x\, h(x)=\underline{i}$ and thus by an easy argument: $\Delta_U \underline{i}QL$. By the lemma: $\Delta_U \underline{i}PL$. Conclude $\Delta_U \bigvee \{L=\underline{j} \mid iPj\}$. By the Induction Hypothesis: $\Delta_U \chi^*$.

Assume that $i \neq 0$ and for some j iPj and $j \nVdash \chi$. Reason in $I\Delta_0$+EXP:

Suppose $L=\underline{i}$ and $\Delta_U \chi^*$. By the Induction Hypothesis: $\Delta_U(L=\underline{j} \rightarrow \neg\chi^*)$. Hence $\Delta_U L \neq \underline{j}$. Suppose $\mathrm{Tproof}_U(z, L \neq \underline{j})$. Because $L=\underline{i}$ clearly for every y $h(y)P\underline{j}$, hence $h(z+1)=\underline{j}$. Contradiction. Conclude $\neg\Delta_U \chi^*$.

Suppose $\psi = \chi \rhd \rho$. First we assume $i \neq 0$ and $i \Vdash \psi$. By the above: for every j with iQj: $I\Delta_0 + EXP \vdash L=\underline{j} \rightarrow (\nabla_U \chi^* \rightarrow \nabla_U \rho^*)$. Moreover if $i \in X$ by the special property (iii) of our model: $I\Delta_0 + EXP \vdash L=\underline{i} \rightarrow (\nabla_U \chi^* \rightarrow \nabla_U \rho^*)$. Reason in $I\Delta_0$+EXP:

Suppose $L=\underline{i}$. For some x $h(x)=\underline{i}$, so $\Delta_{EXP} \exists x\, h(x)=\underline{i}$. Conclude: $\Delta_{EXP} \bigvee \{L=\underline{j} \mid iWQj\}$. In case $i \in X$, we have immediately by the above: $\Delta_{EXP}(\nabla_U \chi^* \rightarrow \nabla_U \rho^*)$, i.e. $\chi^* \rhd_U \rho^*$. If $i \notin X$, it follows that $\Delta_{EXP} L \neq \underline{i}$, and thus $\Delta_{EXP} \bigvee \{L=\underline{j} \mid iQj\}$. So again: $\Delta_{EXP}(\nabla_U \chi^* \rightarrow \nabla_U \rho^*)$, i.e. $\chi^* \rhd_U \rho^*$.

Secondly assume $i \neq 0$ and $i \nVdash \psi$. So for some j with iQj for some k with jPk: $k \Vdash \chi$, and for all k' with jPk': $k' \nVdash \rho$. Ergo $I\Delta_0 + EXP \vdash L=\underline{j} \rightarrow \nabla_U \chi^*$ and $I\Delta_0 + EXP \vdash L=\underline{j} \rightarrow \Delta_U \neg\rho^*$. Reason in $I\Delta_0$+EXP:

Suppose $L=\underline{i}$ and $\Delta_{EXP}(\nabla_U \chi^* \rightarrow \nabla_U \rho^*)$. We have: $\Delta_{EXP}(L=\underline{j} \rightarrow \neg(\nabla_U \chi^* \rightarrow \nabla_U \rho^*))$, so $\Delta_{EXP} L \neq \underline{j}$. Suppose $\mathrm{Tproof}_{EXP}(z, L \neq \underline{j})$. From $L=\underline{i}$, we have: for all y $h(y)Q\underline{j}$, so $h(z+1)=\underline{j}$. Contradiction. Conclude: $\neg\Delta_{EXP}(\nabla_U \chi^* \rightarrow \nabla_U \rho^*)$.

Finally: suppose $U\vdash\phi^*$. By the above we find: $U\vdash L\neq\underline{1}$. So by the definition of h and the fact that $1\in X$: $U\vdash L\neq\underline{0}$. Thus $U\vdash\bigvee\{L=\underline{i}|1<i\leq N\}$. Clearly for some k and for all i with $1<i\leq N$: $i\Vdash\Box^k\bot$. So: $U\vdash\Box^k{}_U\bot$. Quod non. □

8.3 Another Interpretation

Christian Bennet studies in his thesis (see Bennet[1986a]) the following notion of strong relative consistency over Peano Arithmetic: for ϕ,ψ sentences of the language of Peano:

$$\phi\rhd^{SRC}\psi :\Leftrightarrow \text{ there is a primitive recursive term t such that } PA\vdash\forall x(Proof_{PA+\psi}(x,\bot)\rightarrow Proof_{PA+\phi}(tx,\bot))$$

(Actually Bennet defines his notion for arbitrary theories which are verifiably in PA RE extensions of PA; we scaled his notion down to fit our framework.)

Let PRA be Primitive Recursive Arithmetic, a theory which is for our purposes equal to $I\Sigma_1$. By a remark of Kreisel we have: $\phi\rhd^{SRC}\psi \Leftrightarrow PRA\vdash\Diamond_{PA}\phi\rightarrow\Diamond_{PA}\psi$. Formalizing this in PA we get:

$$PA\vdash\ \phi\rhd^{SRC}\psi\ \leftrightarrow\Box_{PRA}(\Diamond_{PA}\phi\rightarrow\Diamond_{PA}\psi)$$

Comparing this characterization with Friedman's characterization of $\rhd_U$, for finitely axiomatized sequential U, it is easy to adapt the above proof to show completeness of ILP for arithmetical interpretations where $\Box$ is interpreted as $\Box_{PA}$ and $\rhd$ is interpreted as $\rhd^{SRC}$.

References

Bennet, C., 1986a, *On some orderings of extensions of arithmetic,* Thesis, Department of Philosophy, University of Göteborg, Göteborg.

Bennet, C., 1986b, *On a problem by D. Guaspari,* in: Furberg, M. &al eds., *Logic and Abstraction,* Acta Philosophica Gothoburgensia 1, Göteborg, 61-69.

Bennet, J.H., 1962, *On spectra,* Thesis, Princeton University, Princeton.

Bezboruah, A., Shepherdson, J.C., 1976, *Gödel's second incompleteness theorem for Q,* JSL 41, 503-512.

Buss, S., 1985, *Bounded Arithmetic,* Thesis, Princeton University, Princeton. Reprinted: 1986, Bibliopolis, Napoli.

Diaconescu, R., Kirby, L.A.S., 1987, *Models of arithmetic and categories with finiteness conditions,* Annals of pure and applied logic 35, 123-148.

Dimitricopoulos, C., 1980, *Matijasevic's theorem and fragments of arithmetic,* Thesis, Univ. of Manchester, Manchester.

Ehrenfeucht, A., Mycielski, J., ?, *Theorems and problems of the lattice of local interpretability.*

Feferman, S., 1960, *Arithmetization of metamathematics in a general setting,* Fund. Math. 49, 33-92.

Feferman, S., Kreisel G., Orey, S., 1960, *1-consistency and faithful interpretations,* Archiv für Mathematische Logik und Grundlagen der Mathematik 6, 52-63.

Feferman, S., 1988, *Hilbert's program relativized: proof-theoretical and foundational reductions,* JSL 53, 364-384.

Friedman, H., ?, *Translatability and relative consistency II.*

Gaifman, H., Dimitracopoulos, C., 1982, *Fragments of Peano's arithmetic and the MRDP theorem,* in: in: *Logic and Algorithmic,* Monography 30 de l'Enseignement Mathematique, Genève, 187-206.

Guaspari D., 1979, *Partially conservative extensions of arithmetic,* Transactions of the AMS 254, 47-68.

Hájek, P., 1971, *On interpretability in set theories I,* Commentationes Mathematicae Universitatis Carolinae 12, 73-79.

Hájek, P., 1972, *On interpretability in set theories II,* Commentationes Mathematicae Universitatis Carolinae 13, 445-455.

Hájek, P., 1979, *On partially conservative extensions of arithmetic,* in: Barwise, J. &al eds.,*Logic Colloquium '78,* North Holland, Amsterdam, 225-234.

Hájek, P., 1981, *Interpretability in theories containing arithmetic II,* Commentationes Mathematicae Universitatis Carolinae 22, 667-688.

Hájek, P., ?, *On partially conservative extensions of arithmetic II.*

Hájek, P., ?, *Positive results on fragments of arithmetic.*

Hájek, P., 1984, *On a new notion of partial conservativity,* in: Richter, M.M. &al eds, *Computation and Proof Theory,* Logic Colloquium '83, Lecture Notes in Mathematics 1104, Springer Verlag, Berlin, 217-232.

Hájek, P., Kucera, A., ?, *On recursion theory in* IS_1.

Hájek, P., ?, *On logic in fragments of arithmetic*

Hájková, M., 1971a, *The lattice of binumerations of arithmetic I,* Commentationes Mathematicae Universitatis Carolinae 12, 81-104.

Hájková, M., 1971b, *The lattice of binumerations of arithmetic II,* Commentationes Mathematicae Universitatis Carolinae 12, 281-306.

Hájková, M. & Hájek, P., 1972, *On interpretability in theories containing arithmetic,* Fundamenta Mathematica 76, 131-137.

Harrow, K., 1987, *The bounded arithmetic hierarchy,* Information and Control 36.

Jongh, D.H.J. de & Veltman F., ?, *Provability logics for relative interpretability,* this volume.

Krajícek, J., 1985, *Some theorems on the lattice of local interpretability types,* Zeitschrift für Mathematische Logik und Grundlagen der Mathematik 31, 449-460.

Krajícek, J., ?, *A note on proofs of falsehood.*

Lindström, P., 1979, *Some results on interpretability,* in: Proceedings of the 5th Scandinavian Logic Symposium, Aalborg, 329-361.

Lindström, P., 1980, *Notes on partially conservative sentences and interpretability,* Philosophical Communications, Red Series no 13, Göteborg.

Lindström, P., 1981, *Remarks on provability and interpretability,* Philosophical Communications, Red Series no 15, Göteborg.

Lindström, P., 1982, *More on partially conservative sentences snd interpretability,* Philosophical Communications, Red Series no 17, Göteborg.

Lindström, P., 1984a, *On faithful interpretability, in:* Richter, M.M. &al eds, *Computation and Proof Theory,* Logic Colloquium '83, Lecture Notes in Mathematics 1104, Springer Verlag, Berlin, 279-288.

Lindström, P., 1984b, *On certain lattices of degrees of interpretability,* Notre Dame Journal of Formal Logic 25, 127-140.

Lindström, P., 1984c, *On partially conservative sentences and interpretability,* Proceedings of the AMS 91, 436-443.

Lindström, P., 1984d, *Provability and interpretability in theories containing arithmetic,* Atti degli incontri di logica matematica 2, 431-451.

Macintyre, A., 1987, *On the strength of weak systems,* from: *Logic, Philosophy of Science and Epistemology,* Hölder-Piçhler-Tempsky, 43-59.

Minc, G., 1972, *Quantifier-free and one-quantifier induction,* Journal of Soviet Mathematics, volume 1, 71-84.

Misercque, D., 1983, *Answer to a problem by D. Guaspari,* in: Guzicki, W. &al eds., *Open days in Model Theory and Set Theory,* Proceedings of a conference held in September 1981 in Jadwisin, Poland, Leeds University, 181-183.

Misercque, D., 1985, *Sur le treillis des formules fermées universelles de l' arithmétique de Peano,* Thesis, Université Libre de Bruxelles.

Montagna, F., 1987, *Provability in finite subtheories of PA and relative interpretability: a modal investigation,* JSL 52, 494-511.

Montague, R., 1958, *The continuum of relative interpretability types,* JSL 23, 460.

Montague, R., 1962, *Theories incomparable with respect to relative interpretability,* JSL 27, 195-211

Mycielski, J., 1962, *A lattice connected with relative interpretability of theories,* Notices of the AMS 9, 407-408. [*errata,* 1971,ibidem, 18, 984].

Mycielski, J., 1977, *A lattice of interpretability types of theories,* JSL 42, 297-305.

Mycielski, J., ?, *Finististic consistency proofs*

Nelson E., 1986, *Predicative Arithmetic,* Math Notes 32, Princeton University Press, Princeton.

Orey, S., 1961, *Relative Interpretations,* Zeitschrift für Mathematische Logik und Grundlagen der Mathematik 7, 146-153.

Palúch, S., 1973, *The lattices of numerations of theories containing Peano's Arithmetic,* Commentationes Mathematicae Universitatis Carolinae 14, 339-359.

Parikh, R., 1971, *Existence and feasibility in arithmetic,* JSL 36, 494-508.

Paris, J.B., Dimitracopoulos, C., 1982, *Truth definitions for Δ_0-formulae,* in: *Logic and Algorithmic,* Monography 30 de l'Enseignement Mathematique Genève, 319-329.

Paris, J.B., Dimitracopoulos, C., 1983, *A note on the undefinability of cuts,* JSL 48, 564-569.

Paris, J., Kirby, L.A.S., 1978, *S_n-collection schema's in arithmetic,* in: Logic Colloquium '77, North Holland, Amsterdam.

Paris, J., Wilkie, A., 1981, *Models of arithmetic and the rudimentary sets,* Bulletin Soc. Math. Belg. 33, 157-169.

Paris, J., Wilkie A., 1983, *Δ_0-sets and induction,* in: Guzicki, W. &al eds., *Open days in Model Theory and Set Theory,* Proceedings of a conference held in September 1981 in Jadwisin, Poland, Leeds University, 237-248.

Paris, J., Wilkie, A., 1987, *On the scheme of induction for bounded arithmetic formulas,* Annals for Pure and Applied Logic 35, 261-302.

Pudlák, P., 1983a, *Some prime elements in the lattice of interpretability types,* Transactions of the AMS 280, 255-275.

Pudlák, P., 1983b, *A definition of exponentiation by a bounded arithmetical formula,* Commentationes Mathematicae Universitatis Carolinae 24, 667-671.

Pudlák, P., 1985, *Cuts, consistency statements and interpretability,* JSL 50, 423-441.

Pudlák, P., 1986, *On the length of proofs of finitistic consistency statements in finitistic theories,* in: Paris, J.B. &al eds., *Logic Colloquium '84,* North Holland, 165-196.

Pudlák, P., ?, *A note on bounded arithmetic.*

Quincy, J., 1981, *Sets of S_k-conservative sentences are P_2 complete,* JSL 46, [abstract], 442.

Schwichtenberg, H., *Proof Theory: Some Applications of Cut-Elimination,* in Barwise, J. ed., *Handbook of Mathematical Logic,* North Holland, 867-895.

Simpson, S.G., 1988, *Partial realizations of Hilbert's Program,* JSL 53, 349-363.

Smorynski, C., 1985a, *Self-Reference and Modal Logic,* Springer Verlag.

Smorynski, C., 1985b, *Nonstandard models and related developments in the work of Harvey Friedman,* in: Harrington, L.A. &alii eds., *Harvey Friedman's Research on the Foundations of Mathematics,* North Holland, 212-229.

Solovay, R., ?, *On interpretability in set theories.*

Svejdar, V., 1978, *Degrees of interpretability,* Commentationes Mathematicae Universitatis Carolinae 19, 789-813.

Svejdar, V., 1981, *A sentence that is difficult to interpret,* Commentationes Mathematicae Universitatis Carolinae 22, 661-666.

Svejdar, V., 1983, *Modal analysis of generalized Rosser sentences,* JSL 48, 986-999.

Takeuti, G., 1988, *Bounded arithmetic and truth definition,* Annals of Pure & Applied Logic 36, 75-104.

Tarski, A., Mostowski, A., Robinson, R.M., 1953, *Undecidable theories,* North Holland, Amsterdam.

Visser, A., 1986, *Peano's Smart Children, a provability logical study of systems with built-in consistency,* Logic Group Preprint Series nr 14, Dept. of Philosophy, University of Utrecht, Heidelberglaan 2, 3584CS Utrecht. To appear in the Notre Dame Journal of

Visser, A., 1988, *Preliminary Notes on Interpretability Logic,* Logic Group Preprint Series nr 29, Dept. of Philosophy, University of Utrecht, Heidelberglaan 2, 3584CS Utrecht.

Wilkie, A., 1980a, *Some results and problems on weak systems of arithmetic,* in: MacIntyre, A. &al eds., *Logic Colloquium '77,* North Holland, Amsterdam, 285-296.

Wilkie, A., 1980b, *Applications of complexity theory to S_0-definability problems in arithmetic,* in: *Model theory of algebra and arithmetic,* Lecture Notes in Mathematics 834, Springer, Berlin, 363-369.

Wilkie, A., 1986, *On sentences interpretable in systems of arithmetic,* in: Paris, J.B. &al eds., *Logic Colloquium '84,* North Holland, 329-342.

Woods, A.R., 1981, *Some problems in logic and number theory, and their connections,* Thesis, Manchester University.

Addendum

i) The credits in section 5.8 are heavily understated: the equivalence of Π_1-cut-conservativity and local interpretability (for sequential theories) is due to Pudlák (see Pudlák[1985]).

ii) During the conference Dick de Jongh and the author discovered an example showing that the interpretability logic of $I\Delta_0$+EXP is not ILP, but strictly extends it. Consider the following sentences: A: $\Box(\Box\bot\to(p\vee q))$, B: $\Box(\Diamond\Diamond p\leftrightarrow\Diamond\Diamond q)$, C: $\Diamond(\Box\Box\Box\bot\wedge\Diamond\Diamond p\wedge\Diamond p \rhd \Diamond q)$, D: $\Diamond\Diamond(\Diamond p\wedge\Diamond q)$, E: $\Diamond(\Box\Box\bot\wedge\Diamond p\wedge\Diamond q)$, F: $((A\wedge B\wedge C)\to(D\vee E))$. ILP does not prove F but F is arithmetically valid in $I\Delta_0$+EXP. We leave it to the reader to construct the six point ILP-countermodel to F. We prove the $I\Delta_0$+EXP-validity of F.

From our characterization of interpretability over $I\Delta_0$+EXP and from Solovay's theorem for tableaux or cut-free provability in $I\Delta_0$+EXP it is immediate that the principles valid in $I\Delta_0$+EXP are precisely those valid in all Kripke models of the following kind: they have accessibility relation P; they are finite, transitive, irreflexive; they are trees with a bottom element b; their forcing relation is given by:

$$x\Vdash\Box\phi :\Leftrightarrow (\forall y,z\ xPyPz \Rightarrow z\Vdash\phi),$$

$$x\Vdash\phi\rhd\psi :\Leftrightarrow (\forall y,z\ xPyPz \Rightarrow (z\Vdash\phi \Rightarrow \exists u\ (yPu \text{ and } u\Vdash\psi))).$$

We show that there is no countermodel to F. Suppose for a reductio we have such a model satisfying A,B,C,$\neg$D,$\neg$E at b. We write $\circ$ for relation-composition. Define $d(u):=\sup\{1+d(v)\mid uPv\}$. (So if u is a top-node $d(u)=0$.) By C there is an x with $bP\circ Px$ and $x\Vdash(\Box\Box\Box\bot\wedge\Diamond\Diamond p\wedge\Diamond p\rhd\Diamond q)$. It is easily seen that $d(x)=4$ or $d(x)=5$. Suppose $d(x)=4$. There are y and z such that $xPyP\circ P\circ Pz$ and $d(y)=3$ and $z\Vdash p$. Hence there is an u with $xPyP\circ P\circ Pu$ and $u\Vdash q$. So $y\Vdash\Box\Box\bot\wedge\Diamond p\wedge\Diamond q$, contradicting our assumption that b forces $\neg$E.

So $d(x)$ must be 5. There are y and z with $xPyP\circ P\circ P\circ Pz$. By A in z either p or q is forced; in case q is forced, we have $y\Vdash\Diamond\Diamond q$, so by B $y\Vdash\Diamond\Diamond p$. So in either case there are u,v,w such that $xPyPuPvP\circ Pw$ and $w\Vdash p$. Because $x\Vdash\Diamond p\rhd\Diamond q$ and $xPuPv$ and $v\Vdash\Diamond p$, we have for some e,f: $xPyPuPeP\circ Pf$ and $f\Vdash q$. So $u\Vdash\Diamond p\wedge\Diamond q$. Moreover $bP\circ PxPyPu$, so $b\Vdash\Diamond\Diamond(\Diamond p\wedge\Diamond q)$. Contradiction. □

Hierarchies of Provably Computable Functions

S. S. WAINER

Department of Pure Mathematics
University of Leeds
Leeds, England

Introduction

A computable number theoretic function is (total) recursive if some algorithm for it terminates on all inputs. From the algorithm we can unravel a *tree of subcomputations* such that termination is equivalent to well-foundedness with respect to the 'computation-branches'. If the tree is well-founded we can *prove* termination by induction over it. Linearisation of the tree's ordering then leads to a countable well-ordering whose size and structure reflects the *complexity* of the function, in terms of the inductive proof needed to verify its termination. We then say that the function is *provably recursive* -by induction over the given well-ordering. For example, the Ackermann Function is provably recursive, by induction over the lexicographical well-ordering of pairs of numbers; so the order-type of this well-ordering viz. the *ordinal* ω^2, in some sense measures its complexity. We further say that a function is *provably recursive in* a given formal theory (containing basic arithmetic) if the $\forall\exists$-sentence asserting its totality is a theorem of that theory, or alternatively if it is provably recursive by induction over a provable well-ordering of the theory.

The aim here is, in §1 to study certain natural hierarchies of number-theoretic functions whose ordinal levels reflect complexity in the above sense, and in §2 to isolate and characterise the large class of recursive functions which can be generated hierarchically 'from below' according to the following principle: *generate a new function only if it is provably recursive over a well-ordering which has already been coded at a previous stage*. We shall call such recursive functions *accessible*. The problem however is this: how should a well-ordering be *coded* by a number-theoretic object? The appropriate answer requires some basic category theoretic notions motivated by Girard [1].

1 The Slow and Fast Growing Hierarchies

1.1. The *Slow Growing Hierarchy* $\{G_\alpha : \alpha \in \Omega\}$ and *Fast Growing (extended Grzegorczyk) Hierarchy* $\{F_\alpha : \alpha \in \Omega\}$ of number-theoretic functions are defined as follows, by recursions over the set Ω of so-called countable *tree ordinals* $\alpha, \beta, \ldots, \lambda, \ldots$ where each 'limit' λ has a uniquely specified 'fundamental sequence' $\lambda_0, \lambda_1, \lambda_2, \ldots$:

$$G_0(x) = 0 \qquad G_{\alpha+1}(x) = G_\alpha(x) + 1 \qquad G_\lambda(x) = G_{\lambda_x}(x)$$

$$F_0(x) = x + 1 \qquad F_{\alpha+1}(x) = F_\alpha^{1+x}(x) \qquad F_\lambda(x) = F_{\lambda_x}(x)$$

where the superscript $1 + x$ denotes iteration $1 + x$ times.

DEFINITION. The set Ω is defined by the induction: $\alpha \in \Omega$ if $\alpha = 0$ or $\alpha = \beta + 1 = \beta \cup \{\beta\}$ for some $\beta \in \Omega$ or α is a function from N into Ω.

Mathematical Logic
Edited by P. P. Petkov
Plenum Press, New York, 1990

If α is a function from N into Ω we denote its value at x by α_x and write $\alpha = \sup\alpha_x$. Often, λ will be used to denote such a limit tree i.e. $\lambda = \sup\lambda_x$. Proofs and definitions will usually be by induction over the well-founded 'sub-tree' partial ordering on Ω which is denoted $\prec$.

1.2. In order (a) to ensure that the $\prec$-predecessors of α are linearly and hence *well*-ordered, and (b) to develop basic majorisation properties of the functions G_α, F_α; we need to restrict attention to tree-ordinals α possessing additional structure.

DEFINITION. The set Ω^S of *structured tree ordinals* consists of those $\alpha \in \Omega$ satisfying

$$\forall\lambda \preccurlyeq \alpha\ \forall x\ \forall y\ (x < y \to \lambda_x \in \lambda[y])$$

where for each $\alpha \in \Omega$ and each $z \in N$, $\alpha[z]$ is the finite linearly ordered set of subtrees of α given by

$$0[z] = \emptyset \qquad \beta + 1[z] = \beta[z] \cup \{\beta\} \qquad \lambda[z] = \lambda_z[z].$$

REMARK. Ketonen-Solovay [2] define the relation $\alpha \to_z \beta$ to mean that there is a sequence $\{\gamma_i : i \leq k\}$ where $\gamma_0 = \alpha$, $\gamma_k = \beta$ and for each $i < k$, $\gamma_{i+1} = \gamma_i - 1$ or $\gamma_{i+1} = (\gamma_i)_z$ according as γ_i is a successor or a limit. It is easy to see that $\alpha[z] = \{\delta : \alpha \to_z \delta + 1\}$ and hence that $\delta[z]$ is properly contained in $\alpha[z]$ whenever $\delta \in \alpha[z]$.

LEMMA. *For each $\alpha \in \Omega^S$ we have*

$$1) \quad x < y \to \alpha[x] \subset \alpha[y]$$

$$2) \quad \gamma \prec \alpha \to \exists x\ \forall z\ (x \leq z \to \gamma \in \alpha[z])$$

PROOF: by induction on α. The cases $\alpha = 0$ and $\alpha = \beta + 1$ are both trivial so suppose $\alpha = \sup\alpha_x$. For part 1), if $x < y$ then $\alpha[x] = \alpha_x[x] \subset \alpha_x[y]$ by the induction hypothesis, and $\alpha_x \in \alpha[y]$ since α is structured. Therefore $\alpha[x] \subset \alpha_x[y] \subset \alpha[y]$ by the previous remark. For part 2) suppose $\gamma \prec \alpha$. Then for some $y \in N$ we have $\gamma \prec \alpha_y$ or $\gamma = \alpha_y$, so by the induction hypothesis there is an $x > y$ such that $\gamma \in \alpha_y[z] \cup \{\alpha_y\}$ for every $z \geq x$. But then for each $z \geq x$ we have $\alpha_y \in \alpha[z]$ since α is structured, and hence $\gamma \in \alpha_y[z] \cup \{\alpha_y\} \subset \alpha[z]$ again by the previous remark.

THEOREM. *For each $\alpha \in \Omega^S$ the set $\{\gamma : \gamma \prec \alpha\}$ is linearly and hence well-ordered by the subtree relation $\prec$. Furthermore if $\gamma \prec \alpha$ then $\gamma + 1 \prec \alpha + 1$.*

PROOF: If $\gamma \prec \alpha$ and $\delta \prec \alpha$ choose any z such that $\gamma \in \alpha[z]$ and $\delta \in \alpha[z]$. Then since $\alpha[z]$ is linearly ordered by $\prec$ we must have $\gamma \prec \delta$ or $\gamma = \delta$ or $\delta \prec \gamma$. Furthermore since $\gamma \in \alpha[z]$ we have $\alpha \to_z \gamma + 1$ and so either $\gamma + 1 \prec \alpha$ or $\gamma + 1 = \alpha$.

LEMMA. *For each $\alpha \in \Omega^S$ we have*

$$1) \quad G_\alpha(x) = \text{cardinality of } \alpha[x],$$

$$2) \quad x < y \to G_\alpha(x) \leq G_\alpha(y),$$

$$3) \quad \gamma \prec \alpha \to G_\gamma(x) < G_\alpha(x) \text{ whenever } \gamma \in \alpha[x].$$

PROOF: 1) follows by a trivial induction on α. 2) then follows from 1) and the previous Lemma, since if $x < y$ then $\alpha[x] \subset \alpha[y]$. 3) also follows from 1) since if $\gamma \in \alpha[x]$ then $\gamma[x]$ is properly contained in $\alpha[x]$.

LEMMA. *For each $\alpha \in \Omega^S$ we have for all $x \in N$*

$$1) \quad x < F_\alpha(x),$$

$$2) \quad x < y \to F_\alpha(x) < F_\alpha(y),$$

$$3) \quad \gamma \prec \alpha \to F_\gamma(x) < F_\alpha(x) \text{ whenever } \gamma \in \alpha[x] \text{ and } x > 0.$$

PROOF: 1) is again by a trivial induction on α. 2) and 3) are proved simultaneously by induction on α. The case $\alpha = 0$ is trivial. If $\alpha = \beta+1$ then by 1) and the induction hypothesis with $x < y$,

$$F_\alpha(x) = F_\beta^{1+x}(x) < F_\beta^{1+x}(y) < F_\beta^{1+y}(y) = F_\alpha(y)$$

and if $\gamma \in \alpha[x]$ then $\gamma \in \beta[x] \cup \{\beta\}$ so with $x > 0$

$$F_\gamma(x) \leq F_\beta(x) < F_\beta^{1+x}(x) = F_\alpha(x).$$

If $\alpha = \sup \alpha_x$ then for $x < y$ we have $\alpha_x \in \alpha[y] = \alpha_y[y]$ so by the induction hypothesis,

$$F_\alpha(x) = F_{\alpha_x}(x) < F_{\alpha_x}(y) < F_{\alpha_y}(y) = F_\alpha(y)$$

and for $\gamma \in \alpha[x] = \alpha_x[x]$ with $x > 0$

$$F_\gamma(x) < F_{\alpha_x}(x) = F_\alpha(x).$$

1.3 The Combinatorial Significance of G. If we were to try to measure the 'size' of $\alpha \in \Omega$ in terms of an increasing number theoretic function rank_α we could proceed as follows:

Call $f : N \to N$ an *upper bound* of the sequence $\{f_n : n \in N\}$ if

$$n \leq x \to f_n(x) \leq f(x).$$

and note that if the sequence has the property

$$* \quad m \leq n \leq x \to f_m(x) \leq f_n(x)$$

then the *diagonal function* $f(x) = f_x(x)$ is the *least* upper bound.

Now define $\text{rank}_\alpha : N \to N$ by induction over $\alpha \in \Omega$:

$$\text{rank}_0 = 0, \quad \text{rank}_{\beta+1} = \text{rank}_\beta + 1, \quad \text{rank}_\lambda = \text{lub}\{\text{rank}_{\lambda_n} : n \in N\}$$

THEOREM. *For each $\alpha \in \Omega^S$,*

$$\text{rank}_\alpha = G_\alpha.$$

PROOF: by a trivial induction on α, noting that the above property * holds when $f_n = G_{\lambda_n}$ for each n.

1.4 The Combinatorial Significance of F. If we were to try to define hierarchies of combinatorially 'big' numbers we might proceed as follows:

Call a number k *α- big for x* if (i) $\alpha = 0$ and $k > x$ or (ii) $\alpha = \beta + 1$ and there are at least $x+1$ numbers $k_0, k_1, \ldots, k_x \leq k$ such that k_0 is β-big for x and each k_{i+1} is β-big for k_i or (iii) $\alpha = \sup \alpha_x$ and k is α_x-big for x.

THEOREM. *For each $x \in N$ and $\alpha \in \Omega^S$*

$$k \text{ is } \alpha\text{-big for } x \quad \text{if and only if} \quad k \geq F_\alpha(x).$$

PROOF: by induction on α. The cases $\alpha = 0$ and $\alpha = \sup \alpha_x$ are trivial, so suppose $\alpha = \beta + 1$. Then by the induction hypothesis, k is α-big for x if and only if there are numbers $k_0 < k_1 < \ldots < k_x \leq k$ such that $k_0 \geq F_\beta(x)$ and for each $i < x$, $k_{i+1} \geq F_\beta(k_i)$. But by the majorisation properties of F_β, this is then equivalent to

$$k \geq k_x \geq F_\beta(k_{x-1}) \geq F_\beta \circ F_\beta(k_{x-2}) \geq \ldots \geq F_\beta^{1+x}(x) = F_\alpha(x).$$

REMARK. The notion of 'big' number is just a reformulation of 'large' interval as in Ketonen-Solovay [2]. In fact k is α-big for x just in case the interval $\{x, x+1, x+2, \ldots, k-1\}$ is ω^α-large. For example,

$$k \text{ is 2-big for } x \quad \text{iff} \quad k \geq 2^{x+1}.(x+1)-1$$

$$k \text{ is } \omega\text{-big for } x \quad \text{iff} \quad k \geq \text{Ackermann}(x).$$

DEFINITION. Call a number theoretic function α- *big* if for each $x \in N$ its value at x is α-big for x.

COROLLARY. *For each $\alpha \in \Omega^S$, F_α is the least α-big function.*

1.5 Subrecursive Inaccessibility. The basic functions of 'subrecursion theory' are the Kalmar *elementary* ones: a number theoretic function is *elementary in* $f_0, f_1, f_2, \ldots$ if it can be defined explicitly from the zero, successor, addition, subtraction and projection functions, and the functions $f_0, f_1, f_2, \ldots$, using bounded summation and bounded multiplication. The elementary functions supply all the usual Gödel numbering apparatus for finite sequences etc.

FACT. If f is of at least exponential growth then every function elementary in f is majorised by some fixed iterate $f^k = f \circ f \circ \ldots \circ f$ and hence is eventually dominated by the function $\text{It}(f)(x) = f^{1+x}(x)$. If furthermore, f is *honest* in the sense that it is Turing-machine computable within tape-space bounded by some fixed iterate f^k of itself, then

$$g \text{ is elementary in } f \quad \text{iff} \quad g \in \cup\{f^n\text{-SPACE} : n \in N\}.$$

Thus the iteration operator used in defining $F_{\beta+1}$ from F_β is, in this context, a natural analogue of the jump operator in classical recursion theory. Furthermore the Fast Growing Hierarchy $\{F_\alpha : \alpha \in \Omega^S\}$ can be considered as a (very) rough function-analogue of the hierarchy of regular cardinals, since if $2 \prec \alpha$ then F_α eventually dominates every function elementary in $\{F_\beta : \beta \prec \alpha\}$. We carry this analogy a stage further in the following definition, recalling that κ is a (weakly) inaccessible cardinal if it is the κ-th. regular cardinal.

DEFINITION. Call $\alpha \in \Omega^S$ *subrecursively inaccessible* or just *s-inaccessible* for short, if its rank function rank_α (which is just G_α by 1.3) 'captures' F_α in the sense that for all $x \in N$,

$$F_\alpha(x) \leq G_\alpha(x+1).$$

REMARK. Every strictly increasing function $g : N \to N$ is a rank function rank_α for some $\alpha \in \Omega^S$, e.g. choose $\alpha = \sup g(x)$ so that the set theoretic ordinal height of α is ω and by 1.3

$$\text{rank}_\alpha(x) = G_\alpha(x) = G_{g(x)}(x) = g(x).$$

Thus the most obvious analogue of inaccessibility in our present context would be the requirement

$$\text{rank}_\alpha = F_\alpha \quad \text{i.e.} \quad G_\alpha = F_\alpha.$$

However this is impossible because, by an easy induction we have, for all $\alpha \in \Omega$ and all $x > 0$,

$$G_\alpha(x) < F_\alpha(x).$$

The definition of s-inaccessibility given above therefore just requires 'the next best thing'.

NOTE. Every s-inaccessible *must be a limit* $\alpha = \sup \alpha_x$, for it clearly cannot be 0 and for any $\beta \in \Omega^S$ and $x > 1$,

$$G_{\beta+1}(x+1) = G_\beta(x+1) + 1 \leq F_\beta(x+1) \leq F_\beta \circ F_\beta(x) < F_{\beta+1}(x).$$

LEMMA. *If $\alpha \in \Omega^S$ is s-inaccessible then G_α eventually dominates every F_β with $\beta \prec \alpha$.*

PROOF: If $\beta \prec \alpha$ then, since α is a limit, $\beta + 1 \prec \alpha$ also. Therefore for some k, $\beta + 1 \in \alpha[k]$, so for every $x > k$ we have $\beta + 1 \in \alpha[x]$ and hence $F_{\beta+1}(x) < F_\alpha(x)$. But $F_\beta(x+1) \leq F_\beta \circ F_\beta(x) \leq F_{\beta+1}(x)$ and so for every $x > k$, $F_\beta(x+1) < F_\alpha(x) \leq G_\alpha(x+1)$.

THEOREM. *If $\alpha = \sup \alpha_x \in \Omega^S$ is such that*

$$G_{\alpha_{n+1}} = F_{\alpha_n} \qquad \text{for every } n \in N$$

then α is s-inaccessible and, if α_0 is finite, then no $\beta \prec \alpha$ is s-inaccessible.

PROOF: If $G_{\alpha_{n+1}} = F_{\alpha_n}$ for each n then

$$F_\alpha(x) = F_{\alpha_x}(x) = G_{\alpha_{x+1}}(x) \leq G_{\alpha_{x+1}}(x+1) = G_\alpha(x+1)$$

and so α is s-inaccessible. If also α_0 is finite and $\beta \prec \alpha$ were s-inaccessible then $\alpha_o \prec \beta$ since β must be a limit and so for some n, $\alpha_n \prec \beta \prec \alpha_{n+1} + 1$. But then by the above Lemma, for sufficiently large x we have $G_{\alpha_{n+1}}(x) = F_{\alpha_n}(x) < G_\beta(x)$. This is a contradiction, since $\beta \prec \alpha_{n+1} + 1$ and therefore $G_\beta(x) \leq G_{\alpha_{n+1}}(x)$ for sufficiently large x.

This simple Theorem suggests a method for constructing a *minimal s-inaccessible*, which we shall henceforth denote

$$\tau = \sup \tau_x.$$

First choose τ_0 to be the 'first' structured tree ordinal α for which F_α eventually dominates all functions elementary in $\{F_\beta : \beta \prec \alpha\}$, namely $\tau_0 = 3$. Then if $\tau_0, \tau_1, ..., \tau_n$ have already been defined, choose $\tau_{n+1} \in \Omega^S$ so that $G_{\tau_{n+1}} = F_{\tau_n}$.

EXISTENCE PROBLEM. Although it is clearly possible, by earlier remarks, to choose $\tau_{n+1} \in \Omega^S$ so that $G_{\tau_{n+1}} = F_{\tau_n}$, it is not clear that we can make the choice sufficiently *uniform* to ensure that τ itself is structured. This requires

$$x \leq n \rightarrow \tau_x \in \tau_{n+1}[n+1].$$

Such a $\tau \in \Omega^S$ will be computed in §2.

UNIQUENESS PROBLEM. There may be many different choices of $\tau_{n+1} \in \Omega^S$ satisfying

$$G_{\tau_{n+1}} = F_{\tau_n} \qquad \text{and} \qquad x \leq n \rightarrow \tau_x \in \tau_{n+1}[n+1]$$

and hence many different τ's constructed as above.However if the identity $G_{\tau_{n+1}} = F_{\tau_n}$ is interpreted as an identity between *functors* rather than merely functions, then any two such τ's will be *isomorphic*.

1.6 G and F as number theoretic Functors. We can consider Ω^S as a category, where the morphisms $h : \alpha \to \beta$ are $\prec$-preserving functions from $\{\gamma : \gamma \prec \alpha\}$ into $\{\delta : \delta \prec \beta\}$. Then by identifying $n \in N$ with the tree ordinal $0+1+1...+1$, N becomes a subcategory of Ω^S. We let 0 denote the 'empty' morphism $\mathrm{id}_0 : 0 \to 0$ and if $x \le y$, $i_{xy} : x \to y$ denotes the identity function from x into y. If $p : x \to y$ is any morphism of N then $p+1 : x+1 \to y+1$ is the morphism such that $p+1(n) = p(n)$ if $n < x$ and $p+1(x) = y$. Note that we can thus define functors from N into N by primitive recursion. For example, given any functor $F : N \to N$ we can define the bifunctor $F^p(q)$ of iterations of F as follows, where $p : x \to y$ and $q : m \to n$

$$F^0(q) = q \qquad F^{p+1}(q) = F(F^p(q)) \qquad F^{i \circ p}(q) = i_{uv} \circ F^p(q)$$

where $i = i_{yz}$, $u = F^y(n)$ and $v = F^z(n)$. Thus for example, if F were the successor functor $F(q) = q+1$ then iteration of it defines the *addition functor* $F^p(q) = q+p$.

We can now extend G_α and F_α, for $\alpha \in \Omega^S$, to *functors* from ω into N as follows, where ω is the subcategory $\{0 \to 1 \to 2 \to \ \ldots \ \to n \to \ \ldots \}$ whose only morphisms are the i_{xy}'s. Let $i = i_{xy}$.

$$G_0(i) = \mathrm{id}_0 \qquad G_{\beta+1}(i) = G_\beta(i) + 1 \qquad G_\lambda(i) = i_{uv} \circ G_{\lambda_x}(i)$$

$$F_0(i) = i+1 \qquad F_{\beta+1}(i) = F_\beta^{1+i}(i) \qquad F_\lambda(i) = i_{uv} \circ F_{\lambda_x}(i)$$

where, in the first line $u = G_{\lambda_x}(y)$ and $v = G_{\lambda_y}(y) = G_\lambda(y)$, and in the second line $u = F_{\lambda_x}(y)$ and $v = F_\lambda(y)$.

THEOREM. *For each $\alpha \in \Omega^S$ and each $x \in N$, let $g_\alpha(x) : G_\alpha(x) \to \alpha$ be the embedding of $\alpha[x]$ in α. Then g_α is a natural transformation from G_α to the constant-α functor and any other natural transformation from G_α to a constant- α' functor must factor uniquely through g_α. In other words α is the direct limit of G_α via the natural transformation g_α and we write*

$$\alpha = \varinjlim G_\alpha \qquad \mathrm{via}\, g_\alpha.$$

PROOF: This is simply a category theoretic restatement of the first two Lemmas of 1.2.

THEOREM. *If $\alpha \in \Omega^S$ and $\alpha' \in \Omega^S$ are such that their functors G_α and $G_{\alpha'}$ are identical, then α and α' are isomorphic and hence the functors F_α and $F_{\alpha'}$ are identical also.*

PROOF: From the preceeding Theorem we have $\alpha = \varinjlim G_\alpha$ *via* g_α and $\alpha' = \varinjlim G_\alpha$ *via* $g_{\alpha'}$, so there is an *isomorphism* $h : \alpha \to \alpha'$ in Ω^S such that $h \circ g_\alpha(x) = g_{\alpha'}(x)$ for every $x \in N$. Now if $\beta \in \alpha[x] = \mathrm{range}\, g_\alpha(x)$ and $p : x \to y$ in N then (with F_β extended in the obvious way to a functor from N into N)

$$F_\beta(p) = F_{h\beta}(p).$$

This is easily proved by induction over $\beta \prec \alpha$. For suppose γ is the immediate predecessor of β in $\alpha[x]$ and δ the immediate predecessor of β in $\alpha[y]$. Then it is easy to see that

$$F_\beta(p) = i_{uv} \circ F_\gamma^{1+p}(p)$$

where $u = F_\gamma^{1+y}(y)$ and $v = F_\delta^{1+y}(y)$. Then by the induction hypothesis we have $u = F_{h\gamma}^{1+y}(y)$, $v = F_{h\delta}^{1+y}(y)$ and $F_\gamma^{1+p}(p) = F_{h\gamma}^{1+p}(p)$. But $h\gamma$ and $h\delta$ are the immediate predecessors of $h\beta$ in $\alpha'[x]$ and $\alpha'[y]$ respectively, and therefore

$$F_\beta(p) = i_{uv} \circ F_{h\gamma}^{1+p}(p) = F_{h\beta}(p)$$

as required. From this it follows that the functors F_α and $F_{\alpha'}$ are identical, for if $p : x \to y$ in N, choose β to be the $\prec$-greatest element of $\alpha[x]$. Then $h\beta$ is the $\prec$-greatest element of $\alpha'[x]$ and for appropriate u and v we have, as above

$$F_\alpha(p) = i_{uv} \circ F_\beta^{1+p}(p) = i_{uv} \circ F_{h\beta}^{1+p}(p) = F_{\alpha'}(p).$$

COROLLARY. *Suppose $\tau = \sup \tau_x$ and $\tau' = \sup \tau'_x$ are two minimal s-inaccessibles constructed so that $\tau_0 = \tau'_0 = 3$ and $G_{\tau_{n+1}} = F_{\tau_n}$, $G_{\tau'_{n+1}} = F_{\tau'_n}$ for every n, where the identities between G's and F's are identities as functors from ω into N. Then τ and τ' are isomorphic and for each n,*

$$F_{\tau_n} = F_{\tau'_n}.$$

Thus, up to isomorphism, $\tau = \sup \tau_x$ is characterised by the induction:

$$\tau_0 = 3 \qquad \tau_{n+1} = \varinjlim F_{\tau_n}.$$

2. Accessible Recursive Functions

2.1. In this section we survey some of the main results pertaining to the existence of τ and to its proof theoretic significance. The principle fact (which motivates much of our work) is

THEOREM. *(Girard [1]) τ is the ordinal of the theory $\mathrm{ID}_{<\omega}$ of arbitrary finite iterations of an inductive definition.*

However this result is embedded in the rather heavy category theoretic framework of Girard's Π^1_2-Logic and it is not easy to relate the characterisation of τ given there, with the more conventional treatments of proof theoretic ordinal assignments for ID-theories, as developed by Buchholz and Pohlers in [3]. In what follows we outline a more direct treatment of the above Theorem. Detailed proofs will appear in the author's [4] and [5].

2.2. Recursion theoretic hierarchies are generated according to the following inductive principle: *one is allowed to generate up to a new level α only if α is already coded in some earlier level β*.The problem is to decide what it should mean for an ordinal α to be *coded* in level β.

One possible meaning (since we are considering countable ordinals and hierarchies of number theoretic objects) would be that at level β of the hierarchy there occurs the characteristic function of a well ordering of N with order type α. This notion is perfectly adequate for the generation of hierarchies in generalized recursion theory (e.g. hierarchies for 1-sections of normal type two functionals etc.). But it is a useless notion if one is interested in the generation of subrecursive hierarchies. The reason is that every recursive function can be defined by a simple primitive recursion over an elementary well ordering of order type ω. Thus the resulting hierarchy would 'collapse' in the sense that all recursive functions would 'suddenly' appear at the very first limit level.

In this case therefore, we require a more delicate notion of coding. But we have already developed one in 1.6: we shall say that α is *coded in level β* if at level β there is a *functor* $F : \omega \to N$ such that

$$\alpha = \varinjlim F \qquad \text{via} g_\alpha$$

i.e. since $\alpha = \varinjlim G_\alpha$ we regard the functor G_α as a natural *code* for α.

DEFINITION. The *accessible part* of the Fast Growing Hierarchy is the segment generated by the sequence of functors $F_{\tau_0}, F_{\tau_1}, F_{\tau_2}, \ldots$ where $\tau_0 = 3$ and for each n, $\tau_{n+1} = \varinjlim F_{\tau_n}$. The *accessible recursive functions* are those which are elementary in the number theoretic functions $\{F_{\tau_n} : n \in N\}$.

THEOREM. *A recursive function is accessible if and only if it is elementary in $\{F_\alpha : \alpha \prec \tau\}$.*

PROOF: By the fact that F_α is elementary in F_β whenever $\alpha \prec \beta \prec \tau$.

2.3 Naming Big Tree Ordinals. Just as the Fast Growing Hierarchy uses countable tree ordinals α to 'name' big number theoretic functions F_α, so we can use *uncountable* tree ordinals α to name 'big' ordinal functions $\varphi_\alpha : \Omega \to \Omega$. These in turn can then be used to name even bigger number theoretic functions $F_{\varphi_\alpha(\beta)}$ etcetera. This idea originates in Bachmann's construction of hierarchies of 'large' normal functions on ordinals, and in our case leads analogously to a collection of 'higher level' Fast Growing Hierarchies $\varphi_\alpha : \Omega_n \to \Omega_n$ where α now ranges over the 'next higher' tree class Ω_{n+1}.

DEFINITION. The classes $\Omega_0, \Omega_1, \Omega_2, \ldots$ of higher level tree ordinals are generated by the iterated induction: $\alpha \in \Omega_n$ if $\alpha = 0$ or $\alpha = \beta + 1$ where $\beta \in \Omega_n$ or α is a function from Ω_k into Ω_n for some $k < n$.

Thus $\Omega_0 = N$, $\Omega_1 = \Omega$ as in 1.1 and $\Omega_k \subset \Omega_n$ whenever $k < n$. If α is a function from Ω_k into Ω_n we denote its value at $\gamma \in \Omega_k$ by α_γ and write $\alpha = \sup\{\alpha_\gamma : \gamma \in \Omega_k\}$ or simply $\alpha = \sup \alpha_\gamma$ where $\mathrm{dom}(\alpha) = \Omega_k$. As before, we will often use $\lambda = \sup \lambda_\gamma$ to denote such a limit tree. In particular for each $k < n$ we define $\omega_k \in \Omega_n$ by

$$\omega_k = \sup\ \{\gamma : \gamma \in \Omega_k\}.$$

DEFINITION. For each n the *level n Fast Growing Hierarchy* of functions $\varphi_\alpha : \Omega_n \to \Omega_n$, indexed by tree ordinals $\alpha \in \Omega_{n+1}$, is given by

$$\varphi_0(\beta) = \beta + 1, \quad \varphi_{\alpha+1}(\beta) = \varphi_\alpha^{1+\beta}(\beta), \quad \varphi_\lambda(\beta) = \sup \varphi_{\lambda_\gamma}(\beta), \quad \varphi_\Lambda(\beta) = \varphi_{\Lambda_\beta}(\beta)$$

where $\mathrm{dom}(\lambda) = \Omega_k$ for some $k < n$, $\mathrm{dom}(\Lambda) = \Omega_n$, and $\varphi^{1+\beta}(\beta) = \varphi^\beta(\varphi(\beta))$ where the iterates φ^β of φ are defined in the obvious way: $\varphi^0() = ()$, $\varphi^{\beta+1}() = \varphi(\varphi^\beta())$, $\varphi^\lambda() = \sup \varphi^{\lambda_\gamma}()$.

NOTE. At level 0 we have $\varphi_\alpha = F_\alpha : N \to N$.

DEFINITION. The subsets $T_n \subset \Omega_n$ *named* by the Fast Growing φ's, are defined inductively by: $0, 1, \omega_0, \omega_1, \ldots, \omega_{n-1} \in T_n$ and $\varphi_\alpha^\gamma(\beta) \in T_n$ whenever $\alpha \in T_{n+1}$ and $\beta, \gamma \in T_n$.

NOTE. Each T_n is closed under addition of tree ordinals, since

$$\varphi_0^\gamma(\beta) = \beta + \gamma.$$

THEOREM. *Every tree ordinal in T_1 is structured.*

PROOF: See Wainer [5].

2.4 Computing the Minimal s-Inaccessible

DEFINITION. For each $x \in N$ the function $c = c_x$ which 'collapses' each T_{n+1} to T_n, is defined by

$$c0 = 0 \qquad c\omega_0 = x \qquad c\omega_{k+1} = \omega_k \qquad c\varphi_\alpha^\gamma(\beta) = \varphi_{c\alpha}^{c\gamma}(c\beta)$$

where the second application of φ is 'one level down' from the first.

NOTE. If α is inductively generated in T_{n+1} *without any reference to* ω_0 then $c\alpha$ is independent of x.

COLLAPSING THEOREM. *For each* $x \in N$, $\alpha \in T_2$, $\beta \in T_1$,

$$G_{\varphi_\alpha(\beta)}(i_{xy}) = F_{c\alpha}(G_\beta(i_{xy})).$$

In particular, if α *is generated in* T_2 *without reference to* ω_0 *then, as* G_{ω_0} *is the identity,*

$$G_{\varphi_\alpha(\omega_0)} = F_{c\alpha}.$$

PROOF: Wainer [4] is devoted to a proof of this result, as far as an identity between number theoretic functions, and it appears there as a Corollary to a more general Collapsing Theorem. Extension to an identity between functors is fairly straightforward and will appear in [5]. The proof requires a uniform generalisation of G to each of the higher tree classes Ω_n and involves further technical machinery not discussed here. However the underlying idea is quite simple, so we give a rough outline. Proceeding by induction on α with $i = i_{xy}$ fixed: 1) If $\alpha = 0$ then $\varphi_\alpha(\beta) = \beta + 1$ and

$$G_{\beta+1}(i) = G_\beta(i) + 1 = F_0(G_\beta(i)) = F_{c\alpha}(G_\beta(i)).$$

2) For the successor case $\alpha + 1$ we have $\varphi_{\alpha+1}(\beta) = \varphi_\alpha^{1+\beta}(\beta)$ so by iterated application of the induction hypothesis, and since $G_{\gamma+1}(i) = G_\gamma(i) + 1$,

$$G_{\varphi_{\alpha+1}(\beta)}(i) = F_{c\alpha}^{1+G_\beta(i)}(G_\beta(i)) = F_{c\alpha+1}(G_\beta(i)).$$

Then the result follows because $c\alpha + 1 = c(\alpha + 1)$. 3) If $\alpha = \sup \alpha_x$ with $\mathrm{dom}(\alpha) = \Omega_0$ then setting $\lambda = \varphi_\alpha(\beta)$, we have $\lambda = \sup \lambda_x$ where $\lambda_x = \varphi_{\alpha_x}(\beta)$ and hence for appropriate u, v the induction hypothesis gives

$$G_{\varphi_\alpha(\beta)}(i) = i_{uv} \circ G_{\lambda_x}(i) = i_{uv} \circ F_{c\alpha_x}(G_\beta(i)).$$

The required result now follows since in this case, $c\alpha_x = c\alpha$. 4) If $\alpha = \sup \alpha_\gamma$ with $\mathrm{dom}(\alpha) = \Omega_1$ then $\varphi_\alpha(\beta) = \varphi_{\alpha_\beta}(\beta)$ so again using the induction hypothesis,

$$G_{\varphi_\alpha(\beta)}(i) = F_{c\alpha_\beta}(G_\beta(i)).$$

But now $c\alpha$ is a limit in Ω_1 with the following crucial property:

$$c\alpha_\beta = (c\alpha)_{G_\beta(i)}$$

and therefore, setting $z = G_\beta(i)$,

$$G_{\varphi_\alpha(\beta)}(i) = F_{(c\alpha)_z}(z) = F_{c\alpha}(z) = F_{c\alpha}(G_\beta(i)).$$

DEFINITION. We can now describe the minimal s-inaccessible $\tau = \sup \tau_x$ by setting $\tau_0 = 3$ and for each $n \in N$,

$$\tau_{n+1} = \varphi_{\varphi_{\cdot_{\cdot_{\cdot_{\varphi_3(\omega_n)}}}\cdot\cdot\cdot(\omega_1)}}(\omega_0)$$

where each successive φ is 'one level up' from its predecessor.

COROLLARY. *For each* $n \in N$, $\tau_n \in \Omega^S$; *and we have*

$$G_{\tau_{n+1}} = F_{\tau_n} \quad \text{and hence} \quad \tau_{n+1} = \underrightarrow{\lim}\, F_{\tau_n}.$$

2.5 Proof Theoretic Characterizations. The φ-hierarchies provide a systematic way of naming well known proof theoretic ordinals. For example, letting $|\alpha|$ denote the (set theoretic) ordinal rank of the tree ordinal α, we have

$$|\tau_1| = |\varphi_3(\omega_0)| = \varepsilon_0$$

$$|\varphi_{\omega_1+1}(\omega_0)| = \Gamma_0$$

$$|\tau_2| = \left|\varphi_{\varphi_3(\omega_1)}(\omega_0)\right| = \textit{Howard Ordinal}.$$

Thus $|\tau_1|$ is the ordinal of Peano Arithmetic and $|\tau_2|$ the ordinal of ID_1, the formal theory of one inductive definition.

Buchholz [6] provides a detailed (yet succinct) ordinal analysis for the theories ID_n of n-times iterated inductive definitions. He uses a different hierarchy of ordinal functions, specifically tailored to the task. However his results can be reworked using our Fast Growing φ's (see [5]) to obtain:

THEOREM. *For each* n, $|\tau_{n+1}|$ *is the ordinal of* ID_n *and the functions provably recursive in* ID_n *are exactly those elementary in* $\{F_\alpha : \alpha \prec \tau_{n+1}\}$.

THEOREM. *The accessible recursive functions are exactly those which are provably recursive in* $\mathrm{ID}_{<\omega}$ *and hence in* $(\Pi^1_1 - \mathrm{CA})_0$.

References

[1] J.-Y. Girard, Π^1_2- *Logic*, Annals of Math. Logic 21 (1981), 75–219.

[2] J. Ketonen and R. Solovay, *Rapidly Growing Ramsey Functions*, Annals of Math. 113 (1981), 267–314.

[3] W. Buchholz, S. Feferman, W. Pohlers, and W. Sieg, *Iterated Inductive Definitions and Subsystems of Analysis*, Springer Lect. Notes in Math. 897, Springer-Verlag, Berlin-Heidelberg-New York (1981).

[4] S. S. Wainer, *Slow Growing versus Fast Growing*, Jour. Symb. Logic 54 (1989), to appear.

[5] S. S. Wainer, *Accessible Recursive Functions*, in preparation.

[6] W. Buchholz, *An Independence Result for* $(\Pi^1_1 - CA) + BI$, Annals of Pure and App. Logic 33 (1987), 131–155.

C O N F E R E N C E

(contributed papers)

SEQUENT CALCULUS FOR INTUITIONISTIC LINEAR PROPOSITIONAL LOGIC

V.Michele Abrusci

Dipartimento di Scienze Filosofiche

University of Bari, Italy

INTRODUCTION

Classical linear logic and its *phase semantics* have been introduced in [GIRARD,1987], with the proof that the sequent calculus for classical linear propositional logic is complete and sound w.r. to the validity in every *topolinear space.* [GIRARD-LAFONT,1987] gives a formulation, but not the semantics, of the sequent calculus for the *intuitionistic* linear propositional logic.

In the first section we introduce a more general formulation of the sequent calculus for intuitionistic linear propositional logic, called IL. In the second section we define and investigate a class of abstract structures, called *intuitionistic linear structures*; among these we select so called *intuitionistic topophase spaces*, a generalization of the GIRARD's topolinear structures. The third section shows how the *provability* in IL leads us to an intuitionistic topophase space. In the fourth section we define the *semantics of IL w.r. to the intuitionistic linear structures*, and we prove the main result of the present paper: IL is complete and sound w.r. to the validity in every intuitionistic linear structure, moreover w.r. to the validity in every intuitionistic topophase space.

1. FORMULATION OF THE SEQUENT CALCULUS IL

1.1 Definition

The language $\mathcal{L}$(IL) of the sequent calculus for intuitionistic linear propositional logic is defined as follows.

(i) The alphabeth of $\mathcal{L}$(IL) consists of the following symbols: denumerably many propositional variables, denoted by *a,b,c*... ; propositional constants 1 T; unary connectives $\neg$! ? ; binary connectives $\otimes$ & $\oplus$ $\multimap$; sequent symbol $\Rightarrow$; usual auxiliary symbols .

(ii) The formulas of $\mathcal{L}$(IL) are defined inductively as follows:
- each propositional variable is a formula,
- 1 and T are formulas,
- if A is a formula, then $\neg$A, !A and ?A are formulas,
- if A and B are formulas , then A$\otimes$B , A$\oplus$B , A&B and A$\multimap$B are formulas,
- nothing else is a formula.

(iii) Finite multisets of formulas of $\mathcal{L}$(IL) are denoted by Γ,Δ,Θ,... and are defined as usual. The symbol $\bullet$ will denote the empty multiset, and the symbol $*$ will denote the "multiset-union", so that for every Γ,Δ,Θ :

Mathematical Logic
Edited by P. P. Petkov
Plenum Press, New York, 1990

$$\Gamma * \bullet = \Gamma \qquad \Gamma * \Delta = \Delta * \Gamma \qquad \Gamma * (\Delta * \Theta) = (\Gamma * \Delta) * \Theta \ .$$

If Γ is a finite multiset of formulas and A is a formula, then $\Gamma * A$ will denote the "multiset-union" of Γ with the one-element multiset consisting of the formula A.

(iv) The sequents of $\mathcal{L}$(IL) are defined as follows: if Γ is a finite multiset of formulas of $\mathcal{L}$(IL) and A is a formula of $\mathcal{L}$(IL) , then

$$\Gamma \Rightarrow A \quad \text{and} \quad \Gamma \Rightarrow$$

are sequents of $\mathcal{L}$(IL). - When we say that $\Gamma \Rightarrow \Delta$ is a sequent of $\mathcal{L}$(IL), we mean that Γ is a finite multiset of formulas of $\mathcal{L}$(IL) and Δ is a formula of $\mathcal{L}$(IL) or the empty place after ⇒ (the "empty right place").

The language $\mathcal{L}$(*IL) is obtained from $\mathcal{L}$(IL) by removing the symbols ! and ?.

1.2 Definition

The sequent calculus IL for intuitionistic linear propositional logic is given by the following rules concerning the sequents of $\mathcal{L}$(IL) .

(i) Basic rules: $\dfrac{}{A \Rightarrow A}(\text{id}) \qquad \dfrac{\Gamma_1 \Rightarrow A \qquad \Gamma_2 * A \Rightarrow \Delta}{\Gamma_1 * \Gamma_2 \Rightarrow \Delta}(\text{cut})$

(ii) ¬-rules: $\dfrac{\Gamma * A \Rightarrow}{\Gamma \Rightarrow \neg A}(\neg,R) \qquad \dfrac{\Gamma \Rightarrow A}{\Gamma * \neg A \Rightarrow}(\neg,L)$

(iii) 1 - rules: $\dfrac{}{\Rightarrow 1}(1,R) \qquad \dfrac{\Gamma \Rightarrow \Delta}{\Gamma * 1 \Rightarrow \Delta}(1,L)$

(iv) T -rule: $\dfrac{}{\Gamma \Rightarrow T}(T)$

(v) ⊗ -rules: $\dfrac{\Gamma_1 \Rightarrow A \qquad \Gamma_2 \Rightarrow B}{\Gamma_1 * \Gamma_2 \Rightarrow A \otimes B}(\otimes,R) \qquad \dfrac{\Gamma * A * B \Rightarrow \Delta}{\Gamma * (A \otimes B) \Rightarrow \Delta}(\otimes,L)$

(vi) &-rules: $\dfrac{\Gamma \Rightarrow A \qquad \Gamma \Rightarrow B}{\Gamma \Rightarrow A \& B}(\&,R) \qquad \dfrac{\Gamma * A \Rightarrow \Delta}{\Gamma * (A \& B) \Rightarrow \Delta}(\&,L1) \qquad \dfrac{\Gamma * B \Rightarrow \Delta}{\Gamma * (A \& B) \Rightarrow \Delta}(\&,L2)$

(vii) ⊕ -rules: $\dfrac{\Gamma \Rightarrow A}{\Gamma \Rightarrow A \oplus B}(\oplus,R1) \qquad \dfrac{\Gamma \Rightarrow B}{\Gamma \Rightarrow A \oplus B}(\oplus,R2) \qquad \dfrac{\Gamma * A \Rightarrow \Delta \qquad \Gamma * B \Rightarrow \Delta}{\Gamma * (A \oplus B) \Rightarrow \Delta}(\oplus,L)$

(viii) ⊸ -rules: $\dfrac{\Gamma * A \Rightarrow B}{\Gamma \Rightarrow A \multimap B}(\multimap,R) \qquad \dfrac{\Gamma_1 \Rightarrow A \qquad \Gamma_2 * B \Rightarrow \Delta}{\Gamma_1 * \Gamma_2 * (A \multimap B) \Rightarrow \Delta}(\multimap,L)$

(ix) ! -rules: $\dfrac{!\Gamma \Rightarrow A}{!\Gamma \Rightarrow !A}(!,R) \qquad \dfrac{\Gamma * A \Rightarrow \Delta}{\Gamma * !A \Rightarrow \Delta}(!,L) \qquad \dfrac{\Gamma \Rightarrow \Delta}{\Gamma * !A \Rightarrow \Delta}(!,W) \qquad \dfrac{\Gamma * !A * !A \Rightarrow \Delta}{\Gamma * !A \Rightarrow \Delta}(!,C)$

(x) ? -rules: $\dfrac{\Gamma \Rightarrow A}{\Gamma \Rightarrow ?A}(?,R) \qquad \dfrac{!\Gamma * A \Rightarrow ?\Delta}{!\Gamma * ?A \Rightarrow ?\Delta}(?,L) \qquad \dfrac{\Gamma \Rightarrow}{\Gamma \Rightarrow ?A}(?,W)$

(If Γ is $A_1 * \ldots * A_m$ then !Γ denotes $!A_1 * \ldots * !A_m$, if Δ is A then ?Δ denotes ?A , if Δ is the right empty place then ?Δ denotes the right empty place too).

A proof of a sequent $\Gamma \Rightarrow \Delta$ in IL is defined as usual.

A sequent $\Gamma \Rightarrow \Delta$ is provable in IL iff there is a proof of $\Gamma \Rightarrow \Delta$ in IL .

The sequent calculus *IL is obtained from IL by using the sequents of $\mathcal{L}$(*IL) and by removing !-rules and ?-rules .

1.3 Remark

(i) IL and *IL enjoy cut-elimination.

(ii) It is easy to verify the following property of IL, *IL and usual intuitionistic logic (if we replace ⊸ by the usual intuitionistic implication →):

$\Gamma \Rightarrow A$ is provable iff $\Gamma \Rightarrow (A \multimap B) \multimap B$ is provable for every formula B .

This property is a weak form of the classical property:

$$\Gamma \Rightarrow A \text{ is provable iff } \Gamma \Rightarrow \neg\neg A \text{ is provable.}$$

Moreover, for the intuitionistic linear propositional logic another weak form of that classical property holds:

$$\Rightarrow A \text{ is provable iff } \Rightarrow \neg\neg A \text{ is provable .}$$

This fact depends on the elimination of weakening rules and contraction rules: if we have a cut-free proof of a sequent

$$B\Rightarrow \quad \text{or} \quad \Rightarrow B \quad \text{(where B is a formula)}$$

and the proof does not contain weakening rules and contraction rules, then the last rule used in the proof is necessarily the (left, right) rule concerning the outermost connective in the formula B.

(ii) The rules (!,W) and (?,W) are the sole ways to do *weakenings* in IL, and the rule (!,C) is the sole way to do *contractions* in IL: the other rules for ! and ? allow to express weakening rules and contraction rules as *logical* rules .

(iii) Relationships between *IL and *logics without contraction rules* (cfr. [GRISHIN,1982] and [ONO-KOMORI,1985]) are discussed in [ABRUSCI,1988].

<u>1.4 Definition</u>

(i) The language $\mathcal{L}$(IL0) is obtained from $\mathcal{L}$(IL) as follows:

- alphabeth: remove $\neg$ and add the propositional constant **0** ;
- formulas: remove formulas $\neg A$ and add the formula **0** ;
- sequents are defined as follows : if Γ is a finite multiset of formulas of $\mathcal{L}$(IL0), and A is a formula of $\mathcal{L}$(IL0), then $\Gamma \Rightarrow A$ is a sequent of $\mathcal{L}$(IL0).

(ii) The sequent calculus IL0 is given by the same rules of IL, concerning the sequents of IL0, with the following changes:

- remove $\neg$-rules and the rule (?,W)
- add the 0-rule $\quad \dfrac{}{\Gamma * 0 \Rightarrow A}(0)$

(iii) A sequent $\Gamma \Rightarrow \Delta$ is provable in IL0 iff there is a proof of $\Gamma \Rightarrow \Delta$ in IL0 .

(iv) IL0 is the formulation of the intuitionistic linear propositional calculus given in [GIRARD-LAFONT,1987]. - The relationships between IL, IL0 and classical linear propositional logic are considered in [ABRUSCI,1988] .

2. INTUITIONISTIC LINEAR STRUCTURES

<u>2.1 Definition</u>

$\langle X, \leq, 1, \bot, T, \otimes, \&, \oplus, \multimap \rangle$ is an <u>intuitionistic</u> <u>*-linear</u> <u>structure</u> iff :

(i) X is a set

(ii) $\leq$ is a partial order relation on X

(iii) $\forall x \in X \ \ x \leq T$

(iv) $\otimes, \&, \oplus, \multimap$ are binary functions from X to X

(v) $\forall x \in X \forall y \in X : x \otimes y = y \otimes x$

(vi) $\forall x \in X \forall y \in X \forall z \in X : x \otimes (y \otimes z) = (x \otimes y) \otimes z$

(vii) $\forall x_1 \in X \forall x_2 \in X \forall y_1 \in X \forall y_2 \in X$: if $x_1 \leq y_1$ and $x_2 \leq y_2$ then $x_1 \otimes x_2 \leq y_1 \otimes y_2$

(viii) $\forall x \in X : x \otimes 1 = x$

(ix) $\forall x \in X \forall y \in X : x \otimes (x \multimap y) \leq y$

(x) $\forall x \in X \forall y \in X \forall z \in X$: if $x \otimes y \leq z$ then $x \leq y \multimap z$

(xi) $\forall x \in X \forall y \in X : x \& y \leq x$

(xii) $\forall x \in X \forall y \in X : x \& y \leq y$

(xiii) $\forall x \in X \forall y \in X \forall z \in X$: if $z \leq x$ and $z \leq y$ then $z \leq x \& y$

(xiv) $\forall x \in X \forall y \in X : x \leq x \oplus y$

(xvi) $\forall x \in X \forall y \in X : y \leq x \oplus y$

(xvi) $\forall x \in X \forall y \in X \forall z \in X$: if $x \leq z$ and $y \leq z$ then $x \oplus y \leq z$

(xvii) $\forall x \in X \forall y \in X \forall z \in X : z \otimes (x \oplus y) = (z \otimes x) \oplus (z \otimes y)$.

If $\langle X, \leq, 1, \perp, T, \otimes, \&, \oplus, \multimap \rangle$ is an intuitionistic *-linear structure, then we define the following unary operation from X to X :

$$\forall x \in X \; x^{\perp} = x \multimap \perp .$$

$\langle X, \leq, 1, \perp, T, \otimes, \&, \oplus, \multimap \rangle$ is a <u>0-intuitionistic</u> <u>*-linear</u> <u>structure</u> iff it is an intuitionistic *-linear structure and moreover the following condition is satisfied:

(xviii) $\forall x \in X : \perp \leq x$.

<u>2.2 Lemma</u>

Let $\langle X, \leq, 1, \perp, T, \otimes, \&, \oplus, \multimap \rangle$ be an intuitionistic *-linear structure.

(i) $\forall x \in X \forall y \in X \forall z \in X \forall w \in X$: if $x \leq w$ and $y \otimes w \leq z$ then $x \otimes y \leq z$

(ii) $\forall x \in X \forall y \in X \forall z \in X \forall w_1 \in X \forall w_2 \in X$: if $w_1 \leq x$ and $w_2 \otimes y \leq z$ then $w_1 \otimes w_2 \otimes (x \multimap y) \leq z$

(iii) $\forall x \in X \forall y \in X \forall z \in X$: if $x \leq y \multimap z$ then $x \otimes y \leq z$

(iv) $\forall x \in X \forall y \in X : x \leq y$ iff $1 \leq x \multimap y$

(v) $\forall x \in X : x \leq x^{\perp\perp}$

(vi) $\forall x \in X \; \forall y \in X$: if $x \leq y$ then $y^{\perp} \leq x^{\perp}$

(vii) $\forall x \in X : x^{\perp} = x^{\perp\perp\perp}$

(viii) $\forall x \in X \forall y \in X$: if $x \leq y^{\perp\perp}$ then $x^{\perp\perp} \leq y^{\perp\perp}$

(ix) $\forall x \in X$: if $1 \leq x$ then $x^{\perp} \leq \perp$

(xi) $\forall x \in X$: if $x \leq \perp$ then $1 \leq x^{\perp}$.

(xii) $1 \leq \perp^{\perp}$

(xiii) $\perp = 1^{\perp}$

(xiv) $\forall x \in X \forall y \in X \forall z \in X \forall w \in X$: if $w \otimes x \leq z$ then $w \otimes (x \& y) \leq z$

(xv) $\forall x \in X \forall y \in X \forall z \in X \forall w \in X$: if $w \otimes y \leq z$ then $w \otimes (x \& y) \leq z$

(xvi) $\forall x \in X \forall y \in X \forall z \in X \forall w \in X$: if $w \otimes x \leq z$ and $w \otimes y \leq z$ then $w \otimes (x \oplus y) \leq z$.

□ Exercise □

<u>2.3 Remark</u>

If we remove $1, \perp, T, \&, \oplus$ from an intuitionistic *-linear structure, then we get a *residuate commutative semigroup* (cfr. [FUCHS, 1963]), since $\langle X, \otimes \rangle$ is a commutative semigroup by def. 2.1 (v)-(vi), $\leq$ is a partial order on X s.t. condition 2.1(vii) is satisfied, and by def. 2.1(x) and lemma 2.2(iii) $\multimap$ is an operation satisfying the condition : $x \leq y \multimap z$ iff $x \otimes y \leq z$.

<u>2.4 Lemma</u>

Let $\langle X, \leq, 1, \perp, T, \otimes, \&, \oplus, \multimap \rangle$ be an intuitionistic *-linear structure. Let ϕ be a function from X to X such that:

$\forall x \in X : x \leq \phi(x)$ $\quad$ $\forall x \in X \forall y \in X$: if $x \leq y$ then $\phi(x) \leq \phi(y)$ $\quad$ $\forall x \in X : \phi(\phi(x)) \leq \phi(x)$

$\phi(\perp) = \perp$ $\quad$ $\phi(T) = T$

$\forall x \in X \forall y \in X : \phi(x) \otimes \phi(y) \leq \phi(x \otimes y)$ and $\phi(x) \oplus \phi(y) \leq \phi(x \oplus y)$

$\forall x \in X \forall y \in X : \phi(\phi(x) \& \phi(y)) = \phi(x) \& \phi(y)$ and $\phi(\phi(x) \multimap \phi(y)) = \phi(x) \multimap \phi(y)$.

(Such a function will be called a <u>closure</u> <u>operation</u> on $\langle X, \leq, 1, \perp, T, \otimes, \&, \oplus, \multimap \rangle$).

Define the following binary functions from X to X :

$$\forall x \in X \forall y \in X : x \otimes_{\phi} y = \phi(x \otimes y)$$
$$\forall x \in X \forall y \in X : x \oplus_{\phi} y = \phi(x \oplus y)$$

Let $\Im(\phi)$ be the range of the function ϕ .
Then $\langle \Im(\phi), \leq, \phi(1), \bot, T, \otimes_{\phi}, \&, \oplus_{\phi}, \multimap \rangle$ is an intuitionistic *-linear structure.

□ The crucial point of the proof is to state:

$\forall x \in X \forall y \in X : \phi(\phi(x) \otimes \phi(y)) = \phi(x \otimes y)$ $\quad \forall x \in X \forall y \in X : \phi(\phi(x) \oplus \phi(y)) = \phi(x \oplus y)$
(Indeed, by the hypotheses on ϕ, we have $\phi(x) \otimes \phi(y) \leq \phi(x \otimes y)$ and then we get $\phi(\phi(x) \otimes \phi(y)) \leq \phi(\phi(x \otimes y)) = \phi(x \otimes y)$; conversely, because by the hypotheses on ϕ $x \leq \phi(x)$ and $y \leq \phi(y)$, we get by 2.1(vii) $x \otimes y \leq \phi(x) \otimes \phi(y)$, so that by the hypotheses on ϕ we have $\phi(x \otimes y) \leq \phi(\phi(x) \otimes \phi(y))$. The proof of the case $\oplus$ is analogous).□

2.5 Definition

$\langle X, \leq, 1, \bot, T, \otimes, \&, \oplus, \multimap, !, ? \rangle$ is an intuitionistic linear structure (a 0-intuitionistic linear structure , resp.) iff:

(i) $\langle X, \leq, 1, \bot, T, \otimes, \&, \oplus, \multimap \rangle$ is an intuitionistic *-linear structure (a 0-intuitionistic *-linear structure, resp.);

(ii) ! and ? are functions from X to X

(iii) $\forall x \in X : x \leq ?x$

(iv) $\forall x \forall y \in X$: if $x \leq ?y$ then $?x \leq ?y$

(v) $\bot = ?\bot$ and $\forall x \in X$ $\bot \leq ?x$

(vi) $\forall x \in X$ $!x \leq x$

(vii) $\forall x e X$ $\forall y \in X$ if $!y \leq x$ then $!y \leq !x$

(viii) $1 = !1$ and $\forall x \in X$ $!x \leq 1$

(ix) $\forall x \in X$ $\forall y \in X$ $!x \otimes !y = !(x \& y)$

(x) $\forall x \in X \forall y \in X$ $!x \otimes ?y \leq ?(!x \otimes y)$

2.6 Remark

In the following part of this section we introduce particular intuitionistic linear structures, constructed from an arbitrary commutative monoid with unity.

2.7 Definition

$\langle M, i, \cdot \rangle$ is a commutative monoid with unity iff :

(i) M is a set

(ii) $i \in M$

(iii) $\cdot$ is an unary function from M to M

(iv) $\forall x \in M \forall y \in M : x \cdot y = y \cdot x$

(v) $\forall x \in M \forall y \in M \forall z \in M : x \cdot (y \cdot z) = (x \cdot y) \cdot z)$

(vi) $\forall x \in M : x \cdot i = x$

2.8 Definition

Let $\langle M, i, \cdot \rangle$ be a commutative monoid with unity.

(i) We define : $\mathcal{P}(M) = \{F \mid F \subseteq M\}$

(ii) We define a binary function from $\mathcal{P}(M)$ to $\mathcal{P}(M)$: if $F \subseteq M$ and $G \subseteq M$, then
$$F \multimap G = \{ x \mid \forall y \in F \; x \cdot y \in G \} .$$

2.9 Lemma

Let $\langle M,i,\cdot\rangle$ be a commutative monoid with unity.

(i) $\forall F\in\mathcal{P}(M)\forall G\in\mathcal{P}(M)$: $F\subseteq G$ iff $i\in F\multimap G$

(ii) $\forall F\in\mathcal{P}(M)\forall G\in\mathcal{P}(M)$: $F\subseteq(F\multimap G)\multimap G$

(iii) $\forall F\in\mathcal{P}(M)\forall G\in\mathcal{P}(M)$: if $F\subseteq G$ then $(F\multimap G)\multimap G\subseteq G$ (so that $F=(F\multimap F)\multimap F$)

(iv) $\forall F\in\mathcal{P}(M)\forall G\in\mathcal{P}(M)\forall H\in\mathcal{P}(M)$: if $F\subseteq G$ then $G\multimap H\subseteq F\multimap H$

(v) $\forall F\in\mathcal{P}(M)$: $\{i\}\multimap F=F$ and $\emptyset\multimap F=M$

□ Exercise □

2.10 Definition

Let $\langle M,i,\cdot\rangle$ be a commutative monoid with unity. We define the following binary function from $\mathcal{P}(M)$ to $\mathcal{P}(M)$: if $F\in\mathcal{P}(M)$ and $G\in\mathcal{P}(M)$, then:

$$F\cdot G = \{ x\cdot y \ / \ x\in F \text{ and } y\in G \}$$

2.11 Lemma

Let $\langle M,i,\cdot\rangle$ be a commutative monoid with unity.

(i) $\forall F\in\mathcal{P}(M)\forall G\in\mathcal{P}(M)$: $F\cdot G = G\cdot F$

(ii) $\forall F\in\mathcal{P}(M)\forall G\in\mathcal{P}(M)\forall H\in\mathcal{P}(M)$: $F\cdot(G\cdot H) = (F\cdot G)\cdot H$

(iii) $\forall F_1\ F_2\ G_1\ G_2 \in\mathcal{P}(M)$: if $F_1\subseteq G_1$ and $F_2\subseteq G_2$ then $F_1\cdot F_2 \subseteq G_1\cdot G_2$

(iv) $\forall F\in\mathcal{P}(M)$: $F\cdot\{i\} = F$

(v) $\forall F\in\mathcal{P}(M)\forall G\in\mathcal{P}(M)\forall H\in\mathcal{P}(M)$: $F\cdot G \subseteq H$ iff $F \subseteq G\multimap H$

(vi) $\forall F\in\mathcal{P}(M)\forall G\in\mathcal{P}(M)\forall H\in\mathcal{P}(M)$: $F\cdot(G\cup H) = (F\cdot G) \cup (F\cdot H)$

(vii) $\forall F\in\mathcal{P}(M)\forall G\in\mathcal{P}(M)\forall H\in\mathcal{P}(M)$: $F\cdot(G\cap H) \subseteq (F\cdot G) \cap (F\cdot H)$

□ Exercise □

2.12 Proposition

Let $\langle M,i,\cdot\rangle$ be a commutative monoid with unity. Let $\bot \in \mathcal{P}(M)$.

(i) $\langle \mathcal{P}(M), \subseteq, \{i\}, \bot, M, \cdot, \cap, \cup, \multimap\rangle$ is an intuitionistic *-linear structure (a 0-intuitionistic *-linear structure, if $\bot=\emptyset$) .

(ii) If ϕ is a closure operation on $\langle \mathcal{P}(M), \subseteq, \{i\}, \bot, M, \cdot, \cap, \cup, \multimap\rangle$, then $\langle \Im(\phi), \subseteq, \phi(\{i\}), \bot, M, \cdot_\phi, \cap, \cup_\phi, \multimap\rangle$ is an intuitionistic *-linear structure.

(iii) If ϕ is a closure operation on $\langle \mathcal{P}(M), \subseteq, \{i\}, \bot, M, \cdot, \cap, \cup, \multimap\rangle$ and $\bot=\phi(\emptyset)$, then $\langle \Im(\phi), \subseteq, \phi(\{i\}), \bot, M, \cdot_\phi, \cap, \cup_\phi, \multimap\rangle$ is a 0-intuitionistic *-linear structure.

□ (ii) follows from (i) by lemma 2.4. The proof of (i) follows from lemma 2.9 and lemma 2.11. (iii) follows from (ii) and from the fact that $\phi(\emptyset)$ is the minimum element (w.r. to the order relation $\subseteq$) in $\Im(\phi)$. □

2.13 Definition

Let $\langle M,i,\cdot\rangle$ be a commutative monoid with unity. Let $\bot \in \mathcal{P}(M)$.

Let ϕ be a closure operation on $\langle \mathcal{P}(M), \subseteq, \{i\}, \bot, M, \cdot, \cap, \cup, \multimap\rangle$. Then $\langle \Im(\phi), \subseteq, \phi(\{i\}), \bot, M, \cdot_\phi, \cap, \cup_\phi, \multimap\rangle$ is an <u>intuitionistic phase space</u> (a <u>0-intuitionistic phase space</u> , if $\bot=\phi(\emptyset)$).

The structure $\langle \mathcal{P}(M), \subseteq, \{i\}, \bot, M, \cdot, \cap, \cup, \multimap\rangle$ is a <u>simple</u> intuitionistic phase space (in this case, ϕ is the identity).

2.14 Remark

(i) Consider an intuitionistic phase space

$$\mathcal{I}=\left\langle \Im(\phi),\ \subseteq,\ \phi(\{i\}),\ \perp,\ M,\ \cdot_{\phi},\ \cap,\ \cup_{\phi},\ \multimap \right\rangle$$

where for every $F\in\mathcal{P}(M)$ and $G\in\mathcal{P}(M)$ $F\cdot_{\phi}G = F\cdot G$. Then: $\phi(\{i\}) = \{i\}$ (because $\phi(\{i\}\cdot\{i\}) = \{i\}\cdot\{i\} = \{i\}$) and for every $F\in\mathcal{P}(M)$ $\phi(F) = F$ (because $\phi(F) = \phi(F\cdot\{i\})=F\cdot\{i\}=F$). I.e. $\mathcal{I}$ is a simple intuitionistic phase space.

(ii) Let $\langle M,i,\cdot\rangle$ be a commutative monoid with unity. Let $\perp \in \mathcal{P}(M)$. Define ϕ as follows:

$$\text{if } F\in\mathcal{P}(M), \text{ then } \phi(F) = (F\multimap\perp)\multimap\perp = F^{\perp\perp} .$$

The elements of $\Im(\phi)$ are called *facts* in [GIRARD,1987] . As we show below, ϕ is a closure operation on $\langle \mathcal{P}(M),\ \subseteq,\ \{i\},\perp,\emptyset,M,\cdot,\cap,\cup,\multimap\rangle$; the structure

$$\left\langle \Im(\phi),\ \subseteq,\ \phi(\{i\}),\ \perp,\ \phi(\emptyset),\ M,\ \cdot_{\phi},\ \cap,\ \cup_{\phi},\ \multimap,\ \wp \right\rangle$$

is called *phase space* in [GIRARD,1987], where $\wp$ is the binary function

$$F\wp G = (F^{\perp}\cdot G^{\perp})^{\perp}$$

this function maps $\Im(\phi)$ into $\Im(\phi)$, i.e. maps facts into facts, because

$$F\in\Im(\phi) \text{ iff } F=G^{\perp} \text{ for some } G\in\mathcal{P}(M) .$$

Abstract structures related to phase spaces are investigated in [AVRON,1987].

2.15 Definition

Let $\langle M,i,\cdot\rangle$ be a commutative monoid with unity, $\perp \in \mathcal{P}(M)$, ϕ a closure operation on the intuitionistic phase space

$$\left\langle \mathcal{P}(M),\ \subseteq,\ \{i\},\ \perp,\ M,\ \cdot,\ \cap,\ \cup,\ \multimap\right\rangle.$$

Let $\mathbb{Q} \subseteq \mathcal{P}(M)$ and $\mathbb{R} \subseteq \mathcal{P}(M)$. $\langle \mathbb{Q},\mathbb{R}\rangle$ is a toposystem over the intuitionistic phase space

$$\left\langle \Im(\phi),\ \subseteq,\ \phi(\{i\}),\ \perp,\ M,\ \cdot_{\phi},\ \cap,\ \cup_{\phi},\ \multimap\right\rangle$$

iff :

(i) $\mathbb{Q} \subseteq \Im(\phi)$ and $\mathbb{R} \subseteq \Im(\phi)$

(ii) $\forall F\in\Im(\phi)$ there is $G \in \mathbb{Q}$ such that $F \subseteq G$

(iii) $\perp \in \mathbb{Q}$ and $\forall F\in\mathbb{Q}$ $\perp \subseteq F$

(iv) if for every i $G_i \in \mathbb{Q}$, then $\bigcap_i G_i \in \mathbb{Q}$

(v) $\forall F\in\Im(\phi)$ there is $G\in\mathbb{R}$ such that $G \subseteq F$

(vi) $\phi(\{i\})\in\mathbb{R}$ and $\forall F\in\mathbb{R}$ $F \subseteq \phi(\{i\})$

(vii) if for every i $G_i \in \mathbb{R}$, then $\phi(\bigcup_i G_i) \in \mathbb{R}$

(viii) $\forall F\in\mathbb{R}$ $\forall G\in\mathbb{R}$ $F\cdot_{\phi}G \in \mathbb{R}$ and $F\cdot_{\phi}F = F$

(ix) $\forall F\in\mathbb{R}$ $\forall G\in\mathbb{Q}$ $F\multimap G \in \mathbb{Q}$.

If $\langle \mathbb{Q},\mathbb{R}\rangle$ is a toposystem over an intuitionistic phase space

$$\left\langle \Im(\phi),\ \subseteq,\ \phi(\{i\}),\ \perp,\ M,\ \cdot_{\phi},\ \cap,\ \cup_{\phi},\ \multimap\right\rangle,$$

then we define two binary functions from $\Im(\phi)$ to $\Im(\phi)$: if $F \in \Im(\phi)$, then

$$?_{\mathbb{Q}}(F) = \bigcap \{ G \ / \ F\subseteq G \text{ and } G\in\mathbb{Q} \}$$
$$!_{\mathbb{R}}(F) = \phi(\bigcup \{ G \ / \ G\subseteq F \text{ and } G\in\mathbb{R} \}) .$$

2.16 Proposition

Let $\langle M,i,\cdot\rangle$ be a commutative monoid with unity, $\perp \in \mathcal{P}(M)$, ϕ a closure operation on $\langle \mathcal{P}(M),\ \subseteq,\ \{i\},\ \perp,\ M,\ \cdot,\ \cap,\ \cup,\ \multimap\rangle$.

Let $\langle \mathbb{Q},\mathbb{R}\rangle$ be a toposystem over the intuitionistic phase space

$$\left\langle \Im(\phi),\ \subseteq,\ \phi(\{i\}),\ \perp,\ M,\ \cdot_{\phi},\ \cap,\ \cup_{\phi},\ \multimap\right\rangle.$$

Then

$$\left\langle \Im(\phi),\ \subseteq,\ \phi(\{i\}),\ \perp,\ M,\ \cdot_{\phi},\ \cap,\ \cup_{\phi},\ \multimap,\ ?_{\mathbb{Q}},\ !_{\mathbb{R}}\right\rangle$$

is an intuitionistic linear structure (a 0-intuitionistic linear structure, if $\perp=\phi(\emptyset)$).

□ We show that the conditions 2.5(i)-(x) are satisfied.

(i) is proved in prop. 2.12 .
(ii) $?_{\mathbf{Q}}$ and $!_{\mathbb{R}}$ are functions from $\Im(\phi)$ to $\Im(\phi)$, by def. 2.15. Moreover, if $F\in\Im(\phi)$, then by def. 2.15(iv) and (vii) $?_{\mathbf{Q}}(F) \in \mathbf{Q}$ and $!_{\mathbb{R}}(F) \in \mathbb{R}$.
(iii) Let $F\in\Im(\phi)$. Trivially we have

$$F \subseteq \bigcap\{ G / F\subseteq G \text{ and } G\in\mathbf{Q}\} = ?_{\mathbf{Q}}(F)$$

(iv) Let $F\subseteq ?_{\mathbf{Q}}(G)$. Since $?_{\mathbf{Q}}(G) \in \mathbf{Q}$, we have $?_{\mathbf{Q}}(G) \in \{ H / F\subseteq H \text{ and } H\in\mathbf{Q}\}$ and so

$$?_{\mathbf{Q}}(F) = \bigcap\{ H / F\subseteq H \text{ and } H\in\mathbf{Q}\} \subseteq ?_{\mathbf{Q}}(G).$$

(v) By def. 2.15(iii) $\bot\subseteq G$ for every $G\in\mathbf{Q}$, so that $\bot \subseteq ?_{\mathbf{Q}}(F)$ for every $F\in\Im(\phi)$ because $?_{\mathbf{Q}}(F)\in\mathbf{Q}$. Moreover, by def. 2.15(iii), $\bot \in \mathbf{Q}$ so that

$$?_{\mathbf{Q}}(\bot) = \bigcap\{ G / \bot\subseteq G \text{ and } G\in\mathbf{Q}\} = \bot .$$

(vi) Let $F\in\Im(\phi)$. Trivially $\bigcup\{ G / G\subseteq F \text{ and } G\in\mathbb{R}\} \subseteq F$, so that (because ϕ is a closure operation and $F\in\Im(\phi)$) we have

$$!_{\mathbb{R}}(F) = \phi(\bigcup\{G/ G\subseteq F \text{ and } G\in\mathbb{R}\}) \subseteq \phi(F) = F .$$

(vii) Let $!_{\mathbb{R}}(G) \subseteq F$, with $F\in\Im(\phi)$ and $G\in\Im(\phi)$. Because $!_{\mathbb{R}}(G) \in \mathbb{R}$ we have

$$!_{\mathbb{R}}(G) \subseteq \bigcup\{H / H\subseteq F \text{ and } H\in\mathbb{R}\}$$

so that, because ϕ is a closure operation and $!_{\mathbb{R}}(G) \in \Im(\phi)$,

$$!_{\mathbb{R}}(G) = \phi(!_{\mathbb{R}}(G)) \subseteq \phi(\bigcup\{ H / H\subseteq F \text{ and } H\in\mathbb{R}\}) = !_{\mathbb{R}}(F) .$$

(viii) Let $F\in\Im(\phi)$. $!_{\mathbb{R}}(F) \subseteq \phi(\{i\})$ by def. 2.15(vi), because $!_{\mathbb{R}}(F)\in\mathbb{R}$. Moreover, by 2.15(vi) $\phi(\{i\}) \in \mathbb{R}$, so that

$$!_{\mathbb{R}}(\{i\}) = \phi(\bigcup\{G / G\subseteq\phi(\{i\}) \text{ and } G\in\mathbb{R}\}) = \phi(\phi(\{i\})) = \phi(\{i\}) .$$

(ix) By (viii) above, $!_{\mathbb{R}}(F) \subseteq \phi(\{i\})$ and $!_{\mathbb{R}}(G) \subseteq \phi(\{i\})$, so that from def.2.1(v),(vii),(viii), and by (vi) above,

$$!_{\mathbb{R}}(F) \cdot_{\phi} !_{\mathbb{R}}(G) \subseteq !_{\mathbb{R}}(F) \cdot_{\phi} \phi(\{i\}) = !_{\mathbb{R}}(F) \subseteq F$$
$$!_{\mathbb{R}}(F) \cdot_{\phi} !_{\mathbb{R}}(G) \subseteq \phi(\{i\}) \cdot_{\phi} !_{\mathbb{R}}(G) = !_{\mathbb{R}}(G) \subseteq G$$

and thus by def. 2.1(xiii)

$$!_{\mathbb{R}}(F) \cdot_{\phi} !_{\mathbb{R}}(G) \subseteq F\cap G .$$

But $!_{\mathbb{R}}(F) \cdot_{\phi} !_{\mathbb{R}}(G) \in \mathbb{R}$, by def. 2.15(viii), because $!_{\mathbb{R}}(F)\in\mathbb{R}$ and $!_{\mathbb{R}}(G)\in\mathbb{R}$, so that by (vii) above we get

$$!_{\mathbb{R}}(F) \cdot_{\phi} !_{\mathbb{R}}(G) \subseteq !_{\mathbb{R}}(F\cap G) .$$

By (vi) above from $F\cap G \subseteq F$ we get $!_{\mathbb{R}}(F\cap G) \subseteq !_{\mathbb{R}}(F)$ and from $F\cap G \subseteq G$ we get $!_{\mathbb{R}}(F\cap G) \subseteq !_{\mathbb{R}}(G)$; but $!_{\mathbb{R}}(F\cap G) \in \mathbb{R}$, so that by def. 2.15(viii) $!_{\mathbb{R}}(F\cap G) \cdot_{\phi} !_{\mathbb{R}}(F\cap G) = !_{\mathbb{R}}(F\cap G)$; therefore we get

$$!_{\mathbb{R}}(F\cap G) = !_{\mathbb{R}}(F\cap G) \cdot_{\phi} !_{\mathbb{R}}(F\cap G) \subseteq !_{\mathbb{R}}(F) \cdot_{\phi} !_{\mathbb{R}}(G) .$$

(x) By (iii) above, $!_{\mathbb{R}}(F) \cdot_{\phi} G \subseteq ?_{\mathbf{Q}}(!_{\mathbb{R}}(F) \cdot_{\phi} G)$; then, from def. 2.1(v),(ix) , we get

$$G \subseteq !_{\mathbb{R}}(F)\multimap ?_{\mathbf{Q}}(!_{\mathbb{R}}(F) \cdot_{\phi} G) ,$$

but by def. 2.15(x) $!_{\mathbb{R}}(F) \multimap ?_{\mathbf{Q}}(!_{\mathbb{R}}(F) \cdot_{\phi} G) \in \mathbf{Q}$ because $!_{\mathbb{R}}(F)\in\mathbb{R}$ and $?_{\mathbf{Q}}(!_{\mathbb{R}}(F) \cdot_{\phi} G) \in \mathbf{Q}$, so that by (iv) above

$$?_{\mathbf{Q}}(G) \subseteq !_{\mathbb{R}}(F)\multimap ?_{\mathbf{Q}}(!_{\mathbb{R}}(F) \cdot_{\phi} G)$$

and then from 2.1(v) and 2.2(iii) we obtain

$$!_{\mathbb{R}}F \cdot_{\phi} ?_{\mathbf{Q}}(G) \subseteq ?_{\mathbf{Q}}(!_{\mathbb{R}}(F) \cdot_{\phi} G) \ .\square$$

<u>2.17 Definition</u>

Let $\langle M,i,\cdot\rangle$ be a commutative monoid with unity, $\bot \in \wp(M)$, ϕ a closure operation on $\langle \wp(M), \subseteq, \{i\}, \bot, M, \cdot, \cap, \cup, \multimap\rangle$.

If $\langle \mathbf{Q},\mathbb{R}\rangle$ is a toposystem over the intuitionistic phase space

$$\langle \Im(\phi), \subseteq, \phi(\{i\}), \bot, M, \cdot_{\phi}, \cap, \cup_{\phi}, \multimap\rangle,$$

then the structure

$$\langle \Im(\phi), \subseteq, \phi(\{i\}), \bot, M, \cdot_{\phi}, \cap, \cup_{\phi}, \multimap, ?_{\mathbf{Q}}, !_{\mathbb{R}}\rangle$$

is called an <u>intuitionistic topophase space</u> (a <u>0-intuitionistic topophase space</u>, if $\bot=\phi(\emptyset)$).

<u>2.18 Remark</u>

(i) Let us consider what happens when, in the prop. 2.16, ϕ is the identity: then $\langle \mathbf{Q},\mathbb{R}\rangle$ is a toposystem over the simple intuitionistic phase space

$$\langle \wp(M), \subseteq, \{i\}, \bot, M, \cdot, \cap, \cup, \multimap\rangle .$$

and so

$$\mathbf{R} = \{ \emptyset , \{i\} \}$$

because by def. 2.15(v) if $F\in\mathbf{R}$ then $F\subseteq\{i\}$. Therefore, for every F and $G \in \mathcal{P}(M)$

$!_{\mathbf{R}}(F) = \emptyset$ if not $i\in F$ $\quad$ $!_{\mathbf{R}}(F) = \{i\}$ if $i\in F$

$!_{\mathbf{R}}(F)\cdot !_{\mathbf{R}}(G) = \emptyset$ if not $i\in F\cap G$ $\quad$ $!_{\mathbf{R}}(F)\cdot !_{\mathbf{R}}(G) = \{i\}$ if $i\in F\cap G$

$!_{\mathbf{R}}(F)\cdot ?_{\mathbf{Q}}(G) = \emptyset$ if not $i\in F$ $\quad$ $!_{\mathbf{R}}(F)\cdot ?_{\mathbf{Q}}(G) = ?_{\mathbf{Q}}(G)$ if $i\in F$.

Moreover, $\bot$ is forced to be $\emptyset$.

(ii) Now, we introduce particular closure operations on the simple intuitionistic phase spaces. Our construction is related to the remark 1.3(ii).

2.19 Definition

Let $\langle M,i,\cdot\rangle$ be a commutative monoid with unity. Let $\mathbf{P}\subseteq\mathcal{P}(M)$.

We define the following unary function from $\mathcal{P}(M)$ to $\mathcal{P}(M)$: if $F\in\mathcal{P}(M)$, then

$$CL_{\mathbf{P}}(F) = \{ x \mid \forall H\in\mathbf{P}\ x\in(F\multimap H)\multimap H \}$$

Moreover, we define

$$\mathbf{CL}_{\mathbf{P}} = \Im(CL_{\mathbf{P}}) = \{ F \mid F\in\mathcal{P}(M) \text{ and } F=CL_{\mathbf{P}}(G) \text{ for some } G\in\mathcal{P}(M) \}.$$

$$F\otimes_{\mathbf{P}}G = CL_{\mathbf{P}}(F\cdot G) = F \cdot_{CL_{\mathbf{P}}} G \qquad F\oplus_{\mathbf{P}}G = CL_{\mathbf{P}}(F\cup G) = F \cup_{CL_{\mathbf{P}}} G.$$

2.20 Lemma

Let $\langle M,i,\cdot\rangle$ be a commutative monoid with unity. Let $\mathbf{P}\subseteq\mathcal{P}(M)$. Let $\bot\in\mathbf{P}$. Then $CL_{\mathbf{P}}$ is a closure operation on the structure

$$\langle \mathcal{P}(M), \subseteq, \{i\}, \bot, M, \cdot, \cap, \cup, \multimap \rangle$$

□

(i) By lemma 2.9(ii) , for every $F\in\mathcal{P}(M)$

$$F\subseteq CL_{\mathbf{P}}(F) .$$

(ii) Now we show that for every $F\in\mathcal{P}(M)$,

$$CL_{\mathbf{P}}(CL_{\mathbf{P}}(F)) \subseteq CL_{\mathbf{P}}(F) .$$

Let $x\in CL_{\mathbf{P}}(CL_{\mathbf{P}}(F))$. Suppose $H\in\mathbf{P}$ and $y\in F\multimap H$. Then $y\in CL_{\mathbf{P}}(F)\multimap H$: indeed, if $z\in CL_{\mathbf{P}}(F)$, then $z\in(F\multimap H)\multimap H$, so that from $y\in F\multimap H$ we get $z\cdot y \in H$. Therefore, since $x \in (CL_{\mathbf{P}}(F)\multimap H)\multimap H$, we get $x\cdot y \in H$. I.e. we have $x \in CL_{\mathbf{P}}$.

(iii) Now we prove that for every $F,G \in \mathcal{P}(M)$,

$$\text{if } F \subseteq G \text{ then } CL_{\mathbf{P}}(F) \subseteq CL_{\mathbf{P}}(G) .$$

Let $F\subseteq G$ and $x\in CL_{\mathbf{P}}(F)$. Suppose $H\in\mathbf{P}$ and $y\in G\multimap H$. Because $F\subseteq G$, by lemma 2.9(iv) we get $G\multimap H \subseteq F\multimap H$, so that $y\in F\multimap H$. But $x\in(F\multimap H)\multimap H$, so that $x\cdot y \in H$: i.e. $x\in CL_{\mathbf{P}}(G)$.

(iv) Trivially $CL_{\mathbf{P}}(M) = M$, by (i) above. $CL_{\mathbf{P}}(\bot) = \bot$ follows from the hypothesis that $\bot\in\mathbf{P}$ and from the following general property:

$$\forall F\in\mathbf{P} \quad F= CL_{\mathbf{P}}(F) .$$

Indeed, $F\subseteq CL_{\mathbf{P}}(F)$, by (i) above; conversely, if $x\in CL_{\mathbf{P}}(F)$ then in particular $x\in(F\multimap F)\multimap F$ (because $F\in\mathbf{P}$), but by lemma 2.9(iii) $(F\multimap F)\multimap F=F$, so $x\in F$.

(v) We prove that for every $F,G \in \mathcal{P}(M)$

$$CL_{\mathbf{P}}(F)\cdot CL_{\mathbf{P}}(G) \subseteq CL_{\mathbf{P}}(F\cdot G) .$$

Let $x\in CL_{\mathbf{P}}(F)$ and $y\in CL_{\mathbf{P}}(G)$. Suppose $H\in\mathbf{P}$ and $z \in F\cdot G\multimap H$: thus for every $w_1\in F$ and $w_2\in G$ $z\cdot w_1\cdot w_2 \in H$, i.e. for every $w_1\in F$ $z\cdot w_1 \in G\multimap H$. But $y\in(G\multimap H)\multimap H$, so that $y\cdot z\cdot w_1 \in H$ for every $w_1\in F$, i.e. $y\cdot z \in F\multimap H$. But $x\in(F\multimap H)\multimap H$, so that $y\cdot z\cdot x = x\cdot y\cdot z \in H$.

(vi) We prove that for every $F,G \in \mathcal{P}(M)$

$$CL_{\mathbf{P}}(F)\cup CL_{\mathbf{P}}(G) \subseteq CL_{\mathbf{P}}(F\cup G) .$$

Let $x\in CL_{\mathbf{P}}(F)\cup CL_{\mathbf{P}}(G)$. Suppose $H\in\mathbf{P}$ and $y\in F\cup G\multimap H$. If $x\in CL_{\mathbf{P}}(F)$, then in particular $x \in (F\multimap H)\multimap H$: but $y\in F\multimap H$ (because $F\subseteq F\cup G$, so that by lemma 2.9(iv) $F\cup G\multimap H \subseteq F\multimap H$), so that $x\cdot y \in H$. Analogously, if $x\in CL_{\mathbf{P}}(G)$, then $y\in G\multimap H$ and so $x\cdot y \in H$.

(vii) We prove that for every $F,G \in \mathcal{P}(M)$

$$CL_{\mathbf{P}}(CL_{\mathbf{P}}(F)\cap CL_{\mathbf{P}}(G)) = CL_{\mathbf{P}}(F)\cap CL_{\mathbf{P}}(G) .$$

$CL_{\mathbf{P}}(F)\cap CL_{\mathbf{P}}(G) \subseteq CL_{\mathbf{P}}(CL_{\mathbf{P}}(F)\cap CL_{\mathbf{P}}(G))$, by (i) above. - Conversely, let $x \in CL_{\mathbf{P}}(CL_{\mathbf{P}}(F)\cap CL_{\mathbf{P}}(G))$. First, we show that $x\in CL_{\mathbf{P}}(F)$. Suppose $H\in\mathbf{P}$ and $y\in F\multimap H$: thus $y \in CL_{\mathbf{P}}(F)\multimap H$, because if $w\in CL_{\mathbf{P}}(F)$ then $w\in(F\multimap H)\multimap H$ so that $w\cdot y \in H$. Because $CL_{\mathbf{P}}(F)\cap CL_{\mathbf{P}}(G)\subseteq CL_{\mathbf{P}}(F)$, by lemma 2.9(iv) we have $CL_{\mathbf{P}}(F)\multimap H \subseteq CL_{\mathbf{P}}(F)\cap CL_{\mathbf{P}}(G)\multimap H$. But $x\in(CL_{\mathbf{P}}(F)\cap CL_{\mathbf{P}}(G)\multimap H)\multimap H$, so that $x\cdot y \in H$. Similarly we obtain that $x\in CL_{\mathbf{P}}(G)$.

Thus, $x \in CL_{\mathbb{P}}(F) \cap CL_{\mathbb{P}}(G)$.

(viii) We prove that for every $F,G \in \mathcal{P}(M)$

$$CL_{\mathbb{P}}(CL_{\mathbb{P}}(F) \multimap CL_{\mathbb{P}}(G)) = CL_{\mathbb{P}}(F) \multimap CL_{\mathbb{P}}(G) .$$

$CL_{\mathbb{P}}(F) \multimap CL_{\mathbb{P}}(G) \subseteq CL_{\mathbb{P}}(CL_{\mathbb{P}}(F) \multimap CL_{\mathbb{P}}(G))$ by (i) above. - Conversely, let $x \in CL_{\mathbb{P}}(CL_{\mathbb{P}}(F) \multimap CL_{\mathbb{P}}(G))$. Suppose $y \in CL_{\mathbb{P}}(F)$: we show that $x \cdot y \in CL_{\mathbb{P}}(G)$, so that $x \in CL_{\mathbb{P}}(F) \multimap CL_{\mathbb{P}}(G)$. Indeed, let $H \in \mathbb{P}$ and $z \in G \multimap H$: so, $y \cdot z \in (CL_{\mathbb{P}}(F) \multimap CL_{\mathbb{P}}(G)) \multimap H$, (because, if $w \in CL_{\mathbb{P}}(F) \multimap CL_{\mathbb{P}}(G)$, then $w \cdot y \in CL_{\mathbb{P}}(G)$ and in particular $w \cdot y \in (G \multimap H) \multimap H$, so that $w \cdot y \cdot z \in H$). But $x \in (CL_{\mathbb{P}}(F) \multimap CL_{\mathbb{P}}(G)) \multimap H$, so that $x \cdot y \cdot z \in H$.

(ix) Finally, remark that

$$\forall F \in \mathcal{P}(M)\ \forall G \in \mathbb{P} \quad F \multimap G = CL_{\mathbb{P}}(F \multimap G) .$$

Indeed, $F \multimap G \subseteq CL_{\mathbb{P}}(F \multimap G)$ by (i) above; conversely, let $x \in CL_{\mathbb{P}}(F \multimap G)$ and $y \in F$: then in particular $x \in ((F \multimap G) \multimap G) \multimap G$ (because $G \in \mathbb{P}$), and $y \in (F \multimap G) \multimap G$ (because by lemma 2.9(ii) $F \subseteq (F \multimap G) \multimap G$), so that $x \cdot y \in G$, i.e. $x \in F \multimap G$.□

<u>2.21 Proposition</u>

Let $\langle M,i,\cdot \rangle$ be a commutative monoid with unity. Let $\mathbb{P} \subseteq \mathcal{P}(M)$. Let $\perp \in \mathbb{P}$. Then

$$\langle \mathbb{CL}_{\mathbb{P}}, \subseteq, CL_{\mathbb{P}}(\{i\}), \perp, M, \otimes_{\mathbb{P}}, \cap, \oplus_{\mathbb{P}}, \multimap \rangle$$

is an intuitionistic phase space (a 0-intuitionistic phase space, if $\perp = CL_{\mathbb{P}}(\emptyset)$).

□ Immediate from propositions 2.20 and 2.12 , and def. 2.13 .□

<u>2.22 Remark</u>

(i) Let us consider the case when in def. 2.19 $\mathbb{P} = \{\perp\}$; then we have

$$CL_{\mathbb{P}}(F) = \{ x \ / \ x \in (F \multimap \perp) \multimap \perp \} = F^{\perp\perp} .$$

So, $CL_{\mathbb{P}}$ is the closure operation ϕ considered in 2.14(ii) and $\mathbb{CL}_{\mathbb{P}}$ is the set of all the *facts* of M. - Therefore our construction is a generalization of the way in [GIRARD,1987] *facts* are obtained from a commutative monoid with unity.

(ii) Another special case is when $\perp = CL_{\mathbb{P}}(\emptyset)$. - Remark that this case does not imply $CL_{\mathbb{P}}(\{i\}) = M$; moreover, we have that for every $F \in \mathbb{CL}_{\mathbb{P}}$

$$\text{if } CL_{\mathbb{P}}(\{i\}) \subseteq F \text{ then } F^{\perp} = CL_{\mathbb{P}}(\emptyset) \quad , \quad CL_{\mathbb{P}}(\{i\}) \subseteq CL_{\mathbb{P}}(\emptyset)^{\perp} = M .$$

(iii) Let $\langle M,i,\cdot \rangle$ be a commutative monoid with unity. Let $\mathbb{P} \subseteq \mathcal{P}(M)$. Then, by using lemma 2.9(v), we can easily verify that:

$$CL_{\mathbb{P}}(\{i\}) = \{ x \ / \ \forall H \in \mathbb{P}\ x \in H \multimap H \} \qquad CL_{\mathbb{P}}(\emptyset) = \{ x \ / \ \forall H \in \mathbb{P}\ x \in M \multimap H \} .$$

3. PROVABILITY IN THE SEQUENT CALCULUS IL

The study of the *provability* in IL is a tool for obtaining the completeness theorems stated in the next section, but it is important in itself too.

For each formula A of $\mathcal{L}$(IL) we consider the set *PR*(A) of all the finite multisets Γ of formulas such that Γ⇒A is provable in IL , i.e. we consider all the finite multisets whose provability in IL leads us to get a proof of ⇒A in IL. Clearly, ⇒A is provable in IL iff the empty multiset • belongs to *PR*(A).

Our goal is to know :

- what is the set *PR*(1) and what is the set *PR*(T) ;
- what function gives *PR*(¬A), *PR*(!A), *PR*(?A) resp., in terms of *PR*(A) ;
- what function gives *PR*(A∘B) in terms of *PR*(A) and *PR*(B) , for each binary connective ∘ .

The provability in IL leads us to an intuitionistic topophase space.

<u>3.1 Lemma</u>

Let $\mathbb{T}$ be the set of all the finite multisets of formulas of $\mathcal{L}$(IL), • the empty multiset, ∗ the multiset-union. Then $\langle \mathbb{T}, \bullet, * \rangle$ is a commutative monoid with unity.

□ Immediate □

3.2 Definition

(i) We define some particular elements of $\mathbb{T}$ of lemma 3.1 :
for every formula A of $\mathcal{L}$(IL), $PR(A) = \{ \Gamma \ / \ \Gamma \Rightarrow A$ is provable in IL $\}$
$\bot = \{\Gamma \ / \ \Gamma \Rightarrow$ is provable in IL $\}$

(ii) $\mathbb{PR} = \{ F \ / \ F = \bot$ or $F = PR(A)$ for some formula A of $\mathcal{L}$(IL) $\}$.

3.3 Lemma

(i) For every formula A of $\mathcal{L}$(IL), $A \in PR(A)$.

(ii) The sequent $A \Rightarrow B$ is provable in IL iff $PR(A) \subseteq PR(B)$.

□ The proof uses the rules (id) and (cut).
(i) For every formula A we have

$$\overline{A \Rightarrow A}^{\,(id)}$$

(ii) If $A \Rightarrow B$ is provable in IL and $\Gamma \in PR(A)$, then $\Gamma \in PR(B)$ because

$$\frac{\overset{\vdots}{\Gamma \Rightarrow A} \quad \overset{\vdots}{A \Rightarrow B}}{\Gamma \Rightarrow B}(cut)$$

Conversely, let $PR(A) \subseteq PR(B)$: since $A \in PR(A)$ by (i) above, then $A \in PR(B)$, i.e. $A \Rightarrow B$ is provable in IL . □

3.4 Proposition

Let $\mathbb{T}$, $\bullet$,$*$, $\bot$ and $\mathbb{PR}$ as in lemma 3.1 and definition 3.2. Define $CL_{\mathbb{PR}}$, $\mathbb{CL}_{\mathbb{PR}}$, $\otimes_{\mathbb{PR}}$ and $\oplus_{\mathbb{PR}}$ as in def.2.19.

(i) $\left\langle \mathbb{CL}_{\mathbb{PR}}, \subseteq, CL_{\mathbb{PR}}(\{\bullet\}), \bot, \mathbb{T}, \otimes_{\mathbb{PR}}, \cap, \oplus_{\mathbb{PR}}, \multimap \right\rangle$ is an intuitionistic phase space.

(ii) $PR(1) = CL_{\mathbb{P}}(\{\bullet\})$

(iii) $PR(\mathrm{T}) = \mathbb{T}$

(iv) $PR(\neg A) = PR(A)^{\bot}$, for every formula A of $\mathcal{L}$(IL)

(v) $PR(A \otimes B) = PR(A) \otimes_{\mathbb{PR}} PR(B)$, for every formula A and B of $\mathcal{L}$(IL)

(vi) $PR(A \& B) = PR(A) \cap PR(B)$, for every formula A and B of $\mathcal{L}$(IL)

(vii) $PR(A \oplus B) = PR(A) \oplus_{\mathbb{PR}} PR(B)$, for every formula A and B of $\mathcal{L}$(IL)

(viii) $PR(A \multimap B) = PR(A) \multimap PR(B)$, for every formula A and B of $\mathcal{L}$(IL)

□ (i) is given by prop. 2.21. The proof of (ii)-(viii) is a close inspection of the rules of the rules concerning the logical symbols.
(ii) Let $\Gamma \in PR(1)$. Suppose $H = PR(\Psi)$ (where Ψ is a formula or the empty right place) and $\Delta \in PR(\Psi)$: then $\Gamma * \Delta \in PR(1)$ because

$$\frac{\overset{\vdots}{\Gamma \Rightarrow 1} \quad \dfrac{\overset{\vdots}{\Delta \Rightarrow \Psi}}{\Delta * 1 \Rightarrow \Psi}(1,L)}{\Gamma * \Delta \Rightarrow \Psi}(cut)$$

Conversely, let $\Gamma \in CL_{\mathbb{PR}}(\{\bullet\})$: so, in particular, $\Gamma \in PR(1) \multimap PR(1)$ (by remark 2.22(iii)); but $\bullet \in PR(1)$ because

$$\overline{\Rightarrow 1}^{\,(1,R)}$$

so that $\Gamma * \bullet = \Gamma \in PR(1)$.
(iii) Immediate, because for every $\Gamma \in \mathbb{T}$

$$\overline{\Gamma \Rightarrow \mathbb{T}}^{\,(\mathbb{T})} \ .$$

(iv) Let $\Gamma \in PR(\neg A)$. Suppoose $\Delta \in PR(A)$: then $\Gamma * \Delta \in \bot$, because

$$\frac{\overset{\vdots}{\Gamma \Rightarrow \neg A} \quad \dfrac{\overset{\vdots}{\Delta \Rightarrow A}}{\Delta * \neg A \Rightarrow}(\neg,L)}{\Gamma * \Delta \Rightarrow}(cut)$$

Conversely, let $\Gamma\in PR(A)^{\perp}$. Since $A\in PR(A)$ by lemma 3.3(i), we have $\Gamma*A \in \perp$, so that we get $\Gamma\in PR(\neg A)$ because

$$\dfrac{\dfrac{\vdots}{\Gamma*A\Rightarrow}}{\Gamma\Rightarrow\neg A}(\neg,R)$$

(v) Let $\Gamma\in PR(A\otimes B)$. Suppose $H=PR(\Psi)$ (where Ψ is a formula or the empty right place) and $\Delta \in PR(A)\cdot PR(B) \multimap PR(\Psi)$. But by 3.3(i) $A\in PR(A)$ and $B\in PR(B)$, so that $\Delta*A*B \in PR(\Psi)$. Then $\Gamma*\Delta \in PR(\Psi)$ because

$$\dfrac{\dfrac{\vdots}{\Gamma\Rightarrow A\otimes B}\qquad \dfrac{\dfrac{\vdots}{\Delta*A*B\Rightarrow\Psi}}{\Delta*(A\otimes B)\Rightarrow\Psi}(\otimes,L)}{\Gamma*\Delta\Rightarrow\Psi}(cut)$$

i.e. $\Gamma \in (PR(A)\cdot PR(B) \multimap PR(\Psi))\multimap PR(\Psi)$. Thus, $\Gamma \in PR(A) \otimes_{\mathbb{PR}} PR(B)$.
Conversely, let $\Gamma \in PR(A) \otimes_{\mathbb{PR}} PR(B)$. Because $PR(A\otimes B) \in \mathbb{PR}$, we have that $\Gamma \in (PR(A)\cdot PR(B) \multimap PR(A\otimes B)) \multimap PR(A\otimes B)$. Now $\bullet \in PR(A)\cdot PR(B)\multimap PR(A\otimes B)$, because if $\Delta_1\in PR(A)$ and $\Delta_2 \in PR(B)$ then $\Delta_1*\Delta_2 \in PR(A\otimes B)$

$$\dfrac{\dfrac{\vdots}{\Delta_1\Rightarrow A}\qquad \dfrac{\vdots}{\Delta_2\Rightarrow B}}{\Delta_1*\Delta_2\Rightarrow A\otimes B}(\otimes,R)$$

So, $\Gamma*\bullet = \Gamma \in PR(A\otimes B)$.
(vi) Let $\Gamma \in PR(A\&B)$. Then $\Gamma \in PR(A)\cap PR(B)$ because we have

$$\dfrac{\dfrac{\vdots}{\Gamma\Rightarrow A\&B}\qquad \dfrac{\dfrac{}{A\Rightarrow A}(id)}{A\&B\Rightarrow A}(\&,L1)}{\Gamma\Rightarrow A}(cut)\qquad\qquad \dfrac{\dfrac{\vdots}{\Gamma\Rightarrow A\otimes B}\qquad \dfrac{\dfrac{}{B\Rightarrow B}(id)}{A\&B\Rightarrow B}(\&,L2)}{\Gamma\Rightarrow B}(cut)$$

Conversely, let $\Gamma\in PR(A)\cap PR(B)$: then $\Gamma \in PR(A\&B)$ because

$$\dfrac{\dfrac{\vdots}{\Gamma\Rightarrow A}\qquad \dfrac{\vdots}{\Gamma\Rightarrow B}}{\Gamma\Rightarrow A\&B}(\&,R)$$

(vii) Let $\Gamma \in PR(A\oplus B)$. Suppose $H=PR(\Psi)$ (where Ψ is a formula or the empty right place) and $\Delta\in PR(A)\cup PR(B) \multimap PR(\Psi)$. But by lemma 3.3(i) $A\in PR(A)\cup PR(B)$ and $B\in PR(A)\cup PR(B)$, so that $\Delta*A \in PR(\Psi)$ and $\Delta*B \in PR(\Psi)$. Then $\Gamma*\Delta \in PR(\Psi)$ because

$$\dfrac{\dfrac{\vdots}{\Gamma\Rightarrow A\oplus B}\qquad \dfrac{\dfrac{\vdots}{\Delta*A\Rightarrow\Psi}\quad \dfrac{\vdots}{\Delta*B\Rightarrow\Psi}}{\Delta*(A\oplus B)\Rightarrow\Psi}(\oplus,L)}{\Gamma*\Delta\Rightarrow\Psi}(cut)$$

i.e. $\Gamma \in (PR(A)\cup PR(B)\multimap PR(\Psi))\multimap PR(\Psi)$. Thus, $\Gamma \in PR(A)\oplus_{\mathbb{PR}} PR(B)$.
Conversely, let $\Gamma \in PR(A)\oplus_{\mathbb{PR}} PR(B)$. Because $PR(A\oplus B) \in \mathbb{PR}$, we have that $\Gamma \in (PR(A)\cup PR(B)\multimap PR(A\oplus B))\multimap PR(A\oplus B)$. But $\bullet \in PR(A)\cup PR(B)\multimap PR(A\oplus B)$, because if $\Delta \in PR(A)\cup PR(B)$ then $\Delta \in PR(A\oplus B)$:

$$\dfrac{\dfrac{\vdots}{\Delta\Rightarrow A}}{\Delta\Rightarrow A\oplus B}(\oplus,R1)\qquad\qquad \dfrac{\dfrac{\vdots}{\Delta\Rightarrow B}}{\Delta\Rightarrow A\oplus B}(\oplus,R2)$$

Therefore, $\Gamma*\bullet = \Gamma \in PR(A\oplus B)$.
(viii) Let $\Gamma \in PR(A\multimap B)$. Suppose $\Delta\in PR(A)$: then $\Gamma*\Delta \in PR(B)$ because

$$\dfrac{\dfrac{\vdots}{\Gamma\Rightarrow A\multimap B}\qquad \dfrac{\dfrac{\vdots}{\Delta\Rightarrow A}\qquad \dfrac{\dfrac{}{A\Rightarrow A}(id)\quad \dfrac{}{B\Rightarrow B}(id)}{A*(A\multimap B)\Rightarrow B}(\multimap,L)}{\Delta*(A\multimap B)\Rightarrow B}(cut)}{\Gamma*\Delta\Rightarrow B}(cut)$$

i.e. $\Gamma \in PR(A)\multimap PR(B)$.
Conversely, let $\Gamma \in PR(A)\multimap PR(B)$. Because $A\in PR(A)$ by 3.3(i), we have that $\Gamma*A \in PR(B)$, so that

$$\frac{\Gamma * A \Rightarrow B}{\Gamma \Rightarrow A \multimap B}(\multimap,R)$$

i.e. $\Gamma \in PR(A \multimap B)$. □

3.5 Definition

We define particular subsets of $\mathcal{P}(\mathbb{T})$:

$\mathbb{PR}? = \{ F \,/\, F = \bigcap_i PR(?A_i)$ where A_i 's are formulas of $\mathcal{L}(IL)$, or $F = \bot \}$

$\mathbb{PR}! = \{ F \,/\, F = CL_{\mathbb{PR}}(\bigcup_i PR(!A_i)$ where A_i's are formulas of $\mathcal{L}(IL)$, or $F = CL_{\mathbb{PR}}(\emptyset) \}$

3.6 Proposition

Let $\mathbb{T}$, $\bullet$, $*$, $\bot$, $\mathbb{PR}$, $CL_{\mathbb{PR}}$, $\mathbb{CL}_{\mathbb{PR}}$, $\otimes_{\mathbb{PR}}$, $\oplus_{\mathbb{PR}}$, $\mathbb{PR}?$, $\mathbb{PR}!$ as in definitions and propositions above. Then :

(i) $\langle \mathbb{PR}?\,, \mathbb{PR}! \rangle$ is a toposystem over $\langle \mathbb{CL}_{\mathbb{PR}}, \subseteq, CL_{\mathbb{PR}}(\{\bullet\}), \bot, \mathbb{T}, \otimes_{\mathbb{PR}}, \cap, \oplus_{\mathbb{PR}}, \multimap \rangle$;

(ii) $\langle \mathbb{CL}_{\mathbb{PR}}, \subseteq, CL_{\mathbb{PR}}(\{\bullet\}), \bot, \mathbb{T}, \otimes_{\mathbb{PR}}, \cap, \oplus_{\mathbb{PR}}, \multimap, ?_{\mathbb{PR}?}\,, !_{\mathbb{PR}!} \rangle$ is an intuitionistic topophase space.

(iii) For every formula A of $\mathcal{L}(IL)$,

$$PR(?A) = ?_{\mathbb{PR}?}(PR(A)) = \bigcap \{ PR(?B) \,/\, PR(A) \subseteq PR(?B) \}$$

(iv) For every formula A of $\mathcal{L}(IL)$,

$$PR(!A) = !_{\mathbb{PR}!}(PR(A)) = \bigcup \{ PR(!B) \,/\, PR(!B) \subseteq PR(A) \}$$

□ (ii) follows from (i), by prop. 2.16 and def. 2.17 . The proof of (i), (iii), (iv) consists of a close inspection of the rules concerning ? and ! .

Proof of (i). We show that the conditions 2.15(i)-(ix) are satisfied.

(i) By definition, $\mathbb{PR}! \subseteq \mathbb{CL}_{\mathbb{PR}}$. In order to prove that $\mathbb{PR}? \subseteq \mathbb{CL}_{\mathbb{PR}}$ it is enough to show that $\bigcap_i CL_{\mathbb{PR}}(F_i) = CL_{\mathbb{PR}}(\bigcap_i F_i)$. This follows from the lemma 3.7 .

(ii) We show that $\mathbb{T} \in \mathbb{PR}?$. But $\mathbb{T} = PR(?T)$, because if $\Gamma \in \mathbb{T}$ then $\Gamma \in PR(?T)$:

$$\frac{\overline{\Gamma \Rightarrow T}(T)}{\Gamma \Rightarrow ?T}(?,R)$$

(iii) $\bot \in \mathbb{PR}?$ by definition. Let $\Gamma \in \bot$: then $\Gamma \in PR(?A)$ for every formula A because

$$\frac{\overset{\vdots}{\Gamma \Rightarrow}}{\Gamma \Rightarrow ?A}(?,W)$$

(iv) - (v) Trivial, by definitions .

(vi) By proposition 3.4 (ii), $CL_{\mathbb{PR}}(\{\bullet\}) = PR(1)$; but $PR(!1) = PR(1)$ because the sequents $!1 \Rightarrow 1$ and $1 \Rightarrow !1$ are provable in IL :

$$\frac{\overline{\Rightarrow 1}(1,R)}{!1 \Rightarrow 1}(!,W) \qquad \frac{\dfrac{\overline{\Rightarrow 1}(1,R)}{\Rightarrow !1}(!,R)}{1 \Rightarrow !1}(1,L)$$

So, $CL_{\mathbb{PR}}(\{\bullet\}) \in \mathbb{PR}!$. - Now we prove that if $F \in \mathbb{PR}!$ then $F \subseteq CL_{\mathbb{PR}}(\{\bullet\}) = PR(1)$. If $F = CL_{\mathbb{PR}}(\emptyset)$, this fact is trivial. Let $F = CL_{\mathbb{PR}}(\bigcup_i PR(!B_i))$: $\bigcup_i PR(!B_i) \subseteq PR(1)$ because for every formula A the sequent $!A \Rightarrow 1$ is provable in IL :

$$\frac{\overline{\Rightarrow 1}(1,R)}{!A \Rightarrow 1}(!,W)$$

so that $PR(!A) \subseteq PR(1)$; and then $F = CL_{\mathbb{PR}}(\bigcup_i PR(!B_i)) \subseteq CL_{\mathbb{PR}}(PR(1)) = PR(1)$.

(vii) Trivial, by definition.

(viii) Let $F \in \mathbb{PR}!$ and $G \in \mathbb{PR}!$. First, suppose $F = CL_{\mathbb{PR}}(\bigcup_i PR(!A_i))$ and $G = CL_{\mathbb{PR}}(\bigcup_j PR(!B_j))$. Then by definitions and lemma 2.16:

$$F \otimes_{\mathbb{PR}} G = CL_{\mathbb{PR}}(CL_{\mathbb{PR}}(\bigcup_i PR(!A_i))\cdot CL_{\mathbb{PR}}(\bigcup_j PR(!B_j))\ ) = CL_{\mathbb{PR}}(\bigcup_i PR(!A_i)\cdot \bigcup_j PR(!B_j)\ ) .$$

Now we prove that

$$\bigcup_i PR(!A_i)\cdot \bigcup_j PR(!B_j) \subseteq \bigcup_{i,j} PR(!(A_i \ \&\ B_j)) .$$

Let $\Gamma_1 \in PR(!A_i)$ for some i, and $\Gamma_2 \in PR(!B_j)$ for some j ; then $\Gamma_1 * \Gamma_2 \in PR(!A_i)\cdot PR(!B_j)$ and thus $\Gamma_1 * \Gamma_2 \in PR(!A_i) \otimes_{\mathbb{PR}} PR(!B_j)$, and by prop.3.4 (v) $\Gamma_1 * \Gamma_2 \in PR(!A_i \otimes !B_j)$; but $PR(!A \otimes !B) = PR(!(A\&B))$ for every formula A and B, because

$$\dfrac{\dfrac{\dfrac{\dfrac{\dfrac{\dfrac{}{A\Rightarrow A}(id)}{!A\Rightarrow A}(!,L)}{!A * !B \Rightarrow A}(!,W) \qquad \dfrac{\dfrac{\dfrac{}{B\Rightarrow B}(id)}{!B\Rightarrow B}(!,L)}{!A * !B \Rightarrow B}(!,W)}{!A * !B \Rightarrow A\&B}(\&,R)}{!A * !B \Rightarrow !(A\&B)}(!,R)}{!A\otimes !B \Rightarrow !(A\&B)}(\otimes,L)$$

$$\dfrac{\dfrac{\dfrac{\dfrac{\dfrac{\dfrac{}{A\Rightarrow A}(id)}{A\&B\Rightarrow A}(\&,L1)}{!(A\&B)\Rightarrow A}(!,L)}{!(A\&B)\Rightarrow !A}(!,R) \qquad \dfrac{\dfrac{\dfrac{\dfrac{}{B\Rightarrow B}(id)}{A\&B\Rightarrow B}(\&,L)}{!(A\&B)\Rightarrow B}(!,L)}{!(A\&B)\Rightarrow !B}(!,R)}{!(A\&B) * !(A\&B) \Rightarrow !A\otimes !B}(\otimes,R)}{!(A\&B) \Rightarrow !A\otimes !B}(!,C)$$

Thus, $\Gamma_1 * \Gamma_2 \in PR(!(A_i\&B_j))$ and so $\Gamma_1 * \Gamma_2 \in \bigcup_{i,j} PR(!(A_i\&B_j))$. Therefore, we have $F\otimes_{\mathbb{PR}} G \subseteq CL_{\mathbb{PR}}(\bigcup_{i,j} PR(!(A_i\&B_j)))$. But

$$\bigcup_{i,j} PR(!(A_i\&B_j)) \subseteq !_{\mathbb{PR}!}(F) \otimes_{\mathbb{PR}} !_{\mathbb{PR}!}(G) ,$$

because, if $\Gamma \in PR(!(A_i\&B_j))$ for some i and j, then (by the proofs given above) $\Gamma \in PR(!A_i \otimes !B_j)$, i.e. $\Gamma \in PR(!A_i) \otimes_{\mathbb{PR}} PR(!B_j)$; but $PR(!A_i) \subseteq \bigcup_i PR(!(A_i)$ and $PR(!B_j) \subseteq \bigcup_j PR(!B_j)$, so that $PR(!A_i) \otimes_{\mathbb{PR}} PR(!B_j) \subseteq \bigcup_i PR(!(A_i) \otimes_{\mathbb{PR}} \bigcup_j PR(!B_j) = F \otimes_{\mathbb{PR}} G$, and $\Gamma \in F \otimes_{\mathbb{PR}} G$. Thus we have

$$CL_{\mathbb{PR}}(\bigcup_{i,j} PR(!(A_i\&B_j))) \subseteq F \otimes_{\mathbb{PR}} G$$

and finally

$$F \otimes_{\mathbb{PR}} G = CL_{\mathbb{PR}}(\bigcup_{i,j} PR(!(A_i\&B_j))) \in \mathbb{PR}! .$$

It is trivial that

$$F \otimes_{\mathbb{PR}} F = CL_{\mathbb{PR}}(\bigcup_{i,j} PR(!(A_i\&A_j))) = \quad CL_{\mathbb{PR}}(\bigcup_i PR(!A_i) = \ F$$

because $!(A_i\&A_j) \Rightarrow !A_i$, $!(A_i\&A_j) \Rightarrow !A_j$ and $!A_i \Rightarrow !(A_i\&A_j)$ are provable.
If $F = CL_{\mathbb{PR}}(\emptyset)$, then we have :

$$F \otimes_{\mathbb{PR}} G = CL_{\mathbb{PR}}(CL_{\mathbb{PR}}(\emptyset)\cdot G) = CL_{\mathbb{PR}}(\emptyset \cdot G) = CL_{\mathbb{PR}}(\emptyset) \in \mathbb{PR}! .$$

If $G = CL_{\mathbb{PR}}(\emptyset)$, the proof is analogous.

(ix) Let $F \in \mathbb{PR}!$ and $G \in \mathbb{PR}?$. Suppose $F = CL_{\mathbb{PR}}(\bigcup_i PR(!A_i))$ and $G = \bigcap_j PR(?B_j))$: then we prove

$$F \multimap G = CL_{\mathbb{PR}}(\bigcup_i PR(!A_i)) \multimap \bigcap_j PR(?B_j)) = \bigcap_j(CL_{\mathbb{PR}}(\bigcup_i PR(!A_i)) \multimap PR(?B_j)) \in \mathbb{PR}? .$$

Indeed, if $\Gamma \in F\multimap G$ and $\Delta \in F$ then for every j $\Gamma * \Delta \in PR(?B_j)$, and if $\Gamma \in F\multimap PR(?B_j)$ for every j and $\Delta \in F$ then $\Gamma * \Delta \in PR(?B_j)$ for every j , i.e. $\Gamma * \Delta \in G$. - If $F = CL_{\mathbb{PR}}(\emptyset)$ or $G = \bot$, then the proof that $F\multimap G \in \mathbb{PR}?$ is trivial.

Proof of (iii) . First we prove

$$PR(?A) = \bigcap \{\ PR(?B)\ /\ PR(A) \subseteq PR(?B)\ \}$$

Let $\Gamma \in PR(?A)$ and take a formula B of $\mathcal{L}(IL)$ such that $PR(A) \subseteq PR(?B)$; then $A \Rightarrow ?B$ is provable in IL, so that $\Gamma \in PR(?B)$ because

$$\dfrac{\dfrac{\vdots}{\Gamma\Rightarrow ?A} \qquad \dfrac{\dfrac{\vdots}{A\Rightarrow ?B}}{?A\Rightarrow ?B}(?,L)}{\Gamma\Rightarrow ?B}(cut)$$

- Conversely, let $\Gamma \in \bigcap \{PR(?B) \; / \; PR(A) \subseteq PR(?B)\}$; $A\Rightarrow ?A$ is provable in IL :

$$\dfrac{\dfrac{}{A\Rightarrow A}(id)}{A\Rightarrow ?A}(?,R)$$

so that $PR(A) \subseteq PR(?A)$, and therefore $\Gamma \in PR(?A)$.
Now, we prove

$$?_{\mathbb{PR}?}(PR(A)) = \bigcap \{ PR(?B) \; / \; PR(A) \subseteq PR(?B) \}$$

Let $\Gamma \in ?_{\mathbb{PR}?}(PR(A))$: in particular, for every formula B such that $PR(A) \subseteq PR(?B)$ $\Gamma \in PR(?B)$, i.e. $\Gamma \in \bigcap \{ PR(?B) \; / \; PR(A) \subseteq PR(?B)\}$. - Conversely, let $\Gamma \in \bigcap \{ PR(?B) \; / \; PR(A) \subseteq PR(?B)\}$ and take $F \in \mathbb{PR}?$ such that $PR(A) \subseteq F$. If $F = \bot$, then $A\Rightarrow$ is provable in IL (because $PR(A) \subseteq \bot$) and $\Gamma \in PR(?A)$ (because $A\Rightarrow ?A$ is provable), so that

$$\dfrac{\dfrac{\vdots}{\Gamma\Rightarrow ?A}(?,R) \qquad \dfrac{\dfrac{\vdots}{A\Rightarrow}}{?A\Rightarrow}(?,L)}{\Gamma\Rightarrow}(cut)$$

i.e. $\Gamma \in \bot = F$; if $F = \bigcap_i PR(?B_i)$, then for every i $PR(A) \subseteq PR(?B_i)$, so that $\Gamma \in PR(?B_i)$ for every i, i.e. $\Gamma \in \bigcap_i PR(?B_i) = F$.

<u>Proof of (iv)</u> . First, we prove

$$PR(!A) = \bigcup \{ PR(!B) \; / \; PR(!B) \subseteq PR(A) \}$$

Let $\Gamma \in PR(!A)$: because $!A\Rightarrow A$ is provable in IL

$$\dfrac{\dfrac{}{A\Rightarrow A}(id)}{!A\Rightarrow A}(!,L)$$

we get $PR(!A) \subseteq PR(A)$, so that $\Gamma \in \bigcup \{PR(!B) \; / \; PR(!B) \subseteq PR(A) \}$. - Conversely, let $\Gamma \in \bigcup\{PR(!B) \; / \; PR(!B) \subseteq PR(A) \}$ and take a formula B such that $PR(!B) \subseteq PR(A)$ and $\Gamma \in PR(!B)$; then $\Gamma \in PR(!A)$ because

$$\dfrac{\dfrac{\vdots}{\Gamma\Rightarrow !B} \qquad \dfrac{\dfrac{\vdots}{!B\Rightarrow A}}{!B\Rightarrow !A}(!,R)}{\Gamma\Rightarrow !A}(cut)$$

Now we prove

$$!_{\mathbb{PR}!}(PR(A)) = PR(!A) .$$

Let $F \subseteq PR(A)$ and $F \in \mathbb{PR}!$. If $F = CL_{\mathbb{PR}}(\emptyset)$, then $F \subseteq PR(!A)$ because $CL_{\mathbb{PR}}(\emptyset)$ is the minimum element w.r. to $\subseteq$ in $\mathbb{PR}$. If $F = CL_{\mathbb{PR}}(\bigcup_i PR(!B_i))$, then $\bigcup_i PR(!B_i) \subseteq PR(A)$ (because $\bigcup_i PR(B_i) \subseteq CL_{\mathbb{PR}}(\bigcup_i PR(B_i))$) and thus for every i $PR(!B_i) \subseteq PR(A)$, i.e. by definition $PR(!B_i) \subseteq PR(!A)$, so that $\bigcup_i PR(!B_i) \subseteq PR(!A)$ and $F = CL_{\mathbb{PR}}(\bigcup_i PR(!B_i)) \subseteq CL_{\mathbb{PR}}(PR(!A)) = PR(!A)$. Therefore $\bigcup\{ F/F \subseteq PR(A)$ and $F \in \mathbb{PR}! \} \subseteq PR(!A)$, and finally $!_{\mathbb{PR}!}(PR(A)) \subseteq CL_{\mathbb{PR}}(PR(!A)) = PR(!A)$. - Conversely, $PR(!A) \in \mathbb{PR}!$ and $PR(!A) \subseteq PR(A)$ because $!A\Rightarrow A$ is provable; therefore $PR(!A) \subseteq \bigcup\{ F/F \subseteq PR(A)$ and $F \in \mathbb{PR}! \} = !_{\mathbb{PR}!}(PR(A))$.□

<u>3.7 Lemma</u>

$$CL_{\mathbb{PR}}(\bigcap_i CL_{\mathbb{PR}}(F_i)) = \bigcap_i CL_{\mathbb{PR}}(F_i), \text{ if for every i } F_i \in \mathcal{P}(\mathbb{T}) .$$

□ Because $CL_{\mathbb{PR}}$ is a closure operation, $\bigcap_i CL_{\mathbb{PR}}(F_i) \subseteq CL_{\mathbb{PR}}(\bigcap_i CL_{\mathbb{PR}}(F_i))$. -

Conversely, let $\Gamma \in CL_{\mathbf{PR}}(\bigcap_i CL_{\mathbf{PR}}(F_i))$. Take $H \in \mathbf{PR}$ and $\Delta \in F_i \multimap H$ for some i . Thus $\Delta \in CL_{\mathbf{PR}}(F_i) \multimap H$ (because if $\Theta \in CL_{\mathbf{PR}}(F_i)$ then $\Theta \in (F_i \multimap H) \multimap H$, so that $\Theta * \Delta = \Delta * \Theta \in H$). Therefore $\Delta \in \bigcap_i CL_{\mathbf{PR}}(F_i) \multimap H$, because $\bigcap_i CL_{\mathbf{PR}}(F_i) \subseteq CL_{\mathbf{PR}}(F_i)$ so that $CL_{\mathbf{PR}}(F_i) \multimap H \subseteq \bigcap_i CL_{\mathbf{PR}}(F_i) \multimap H$. But $\Gamma \in (\bigcap_i CL_{\mathbf{PR}}(F_i) \multimap H) \multimap H$, so that $\Gamma * \Delta \in H$.□

3.8 Remark

(i) The proof of lemma 3.7 can be used to prove that, if $\langle M, i, \cdot \rangle$ is a commutative monoid with unity, and $\mathbf{P} \subseteq \mathcal{P}(M)$, and for every i $F_i \in \mathcal{P}(M)$, then

$$CL_{\mathbf{P}}(\bigcap_i CL_{\mathbf{P}}(F_i)) = \bigcap_i CL_{\mathbf{P}}(F_i) .$$

(ii) If we consider the provability in the sequent calculus *IL , then we obtain in an analogous way an intuitionistic phase space.

3.9 Remark

Let us consider the provability in IL0 .

We take the commutative monoid with unity $\langle \mathbb{T}0, \bullet, * \rangle$, where $\mathbb{T}0$ is the set of all the finite multisets of formulas of $\mathcal{L}$(IL0); moreover, we define

$$PR^0(A) = \{ \Gamma \ / \ \Gamma \Rightarrow A \text{ provable in IL0} \} \text{, for every formula A of } \mathcal{L}(\text{IL0})$$

$$\mathbf{PR0} = \{ F \ / \ F = PR^0(A) \text{ for some formula A of } \mathcal{L}(\text{IL0}) \}.$$

As in prop. 3.4 , we may prove :

$PR^0(1) = CL_{\mathbf{PR0}}(\{\bullet\})$

$PR^0(T) = \mathbb{T}0$

$PR^0(A \otimes B) = PR^0(A) \otimes_{\mathbf{PR0}} PR^0(B)$

$PR^0(A \& B) = PR^0(A) \cap PR^0(B)$

$PR^0(A \oplus B) = PR^0(A) \oplus_{\mathbf{PR0}} PR^0(B)$

$PR^0(A \multimap B) = PR^0(A) \multimap PR^0(B)$.

Further, as we show in the lemma 2.3 below, $PR^0(0) = CL_{\mathbf{PR0}}(\emptyset)$.

Thus the provability in *IL0 leads us to the 0-intuitionistic phase space

$$\langle \mathbf{CL}_{\mathbf{PR0}}, \subseteq, CL_{\mathbf{PR0}}(\{\bullet\}), CL_{\mathbf{PR0}}(\emptyset), \mathbb{T}0, \otimes_{\mathbf{PR0}}, \cap, \oplus_{\mathbf{PR0}}, \multimap \rangle$$

(remark that $CL_{\mathbf{PR0}}(\emptyset)$ is the minimum element in this structure), and the provability in IL0 leads us to a 0-intuitionistic topophase space (we can do the analogous of 3.5-3.7).

3.10 Lemma

$$PR^0(\emptyset) = CL_{\mathbf{PR0}}(\emptyset)$$

□ Let $\Gamma \in PR^0(\emptyset)$. Suppose that $H = PR^0(A)$ for some formula A, and $\Delta \in \mathbb{T}^0$. Then $\Gamma * \Delta \in PR^0(A)$, because

$$\dfrac{\dfrac{\vdots}{\Gamma \Rightarrow 0} \qquad \overline{\Delta * 0 \Rightarrow A}^{(0)}}{\Gamma * \Delta \Rightarrow A}(\text{cut})$$

But $\Gamma \in \mathbb{T}^0 \multimap PR^0(A)$, so that $\Gamma \in CL_{\mathbf{PR0}}(\emptyset)$.

Conversely, let $\Gamma \in CL_{\mathbf{PR0}}(\emptyset)$. In particular, because $PR^0(0) \in \mathbf{PR}^0$, we have that $\Gamma \in \mathbb{T}^0 \multimap PR^0(0)$, so that $\Gamma * \bullet = \Gamma \in PR^0(0)$. □

4. SEMANTICS AND COMPLETENESS THEOREM FOR IL

4.1 Definition

Let $\mathcal{S} = \langle X,\leq,1,\perp,T,\otimes,\&,\oplus,-\!\circ,!,?\rangle$ be an intuitionistic linear structure and σ an interpretation in $\mathcal{S}$ (i.e. a function from the variables of IL into X).

(i) For every formula A of $\mathcal{L}(IL)$, we define $A^{IL,\mathcal{S},\sigma} \in X$:

$a^{IL,\mathcal{S},\sigma} = \sigma(a)$, for every variable a

$1^{IL,\mathcal{S},\sigma} = 1$ $\qquad T^{IL,\mathcal{S},\sigma} = T$

$(\neg A)^{IL,\mathcal{S},\sigma} = (A^{IL,\mathcal{S},\sigma})^{\perp}$

$(A\otimes B)^{IL,\mathcal{S},\sigma} = A^{IL,\mathcal{S},\sigma} \otimes B^{IL,\mathcal{S},\sigma}$ $\qquad (A\&B)^{IL,\mathcal{S},\sigma} = A^{IL,\mathcal{S},\sigma} \& B^{IL,\mathcal{S},\sigma}$

$(A\oplus B)^{IL,\mathcal{S},\sigma} = A^{IL,\mathcal{S},\sigma} \oplus B^{IL,\mathcal{S},\sigma}$ $\qquad (A-\!\circ B)^{IL,\mathcal{S},\sigma} = A^{IL,\mathcal{S},\sigma} -\!\circ B^{IL,\mathcal{S},\sigma}$

$(?A)^{IL,\mathcal{S},\sigma} = ?(A^{IL,\mathcal{S},\sigma})$ $\qquad (!A)^{IL,\mathcal{S},\sigma} = !(A^{IL,\mathcal{S},\sigma})$.

(ii) For every finite multiset Γ of formulas of $\mathcal{L}(IL)$, we define $\Gamma^{IL,\mathcal{S},\sigma} \in X$:

$\bullet^{IL,\mathcal{S},\sigma} = 1$ $\qquad A^{IL,\mathcal{S},\sigma}$ is defined in (i) above

$(\Gamma*\Delta)^{IL,\mathcal{S},\sigma} = \Gamma^{IL,\mathcal{S},\sigma} \otimes \Delta^{IL,\mathcal{S},\sigma}$

(iii) For every sequent $\Gamma\Rightarrow\Delta$ of $\mathcal{L}(IL)$, we define $(\Gamma\Rightarrow\Delta)^{IL,\mathcal{S},\sigma}$:

$(\Gamma\Rightarrow\ )^{IL,\mathcal{S},\sigma}$ is $\Gamma^{IL,\mathcal{S},\sigma} \leq \perp$ $\qquad (\Gamma\Rightarrow A)^{IL,\mathcal{S},\sigma}$ is $\Gamma^{IL,\mathcal{S},\sigma} \leq A^{IL,\mathcal{S},\sigma}$

4.2 Definition

Let $\Gamma\Rightarrow\Delta$ be a sequent of $\mathcal{L}(IL)$. Let $\mathcal{S}$ be an intuitionistic linear structure. $\Gamma\Rightarrow\Delta$ is IL-valid in $\mathcal{S}$ iff $(\Gamma\Rightarrow\Delta)^{IL,\mathcal{S},\sigma}$ holds for every interpretation σ in $\mathcal{S}$.

4.3 Lemma

Let $\mathcal{S} = \langle X,\leq,1,\perp,T,\otimes,\&,\oplus,-\!\circ,!,?\rangle$ be an intuitionistic linear structure and σ an interpretation in $\mathcal{S}$.

(i) For every finite multiset Γ of formulas of $\mathcal{L}(IL)$, there is $x\in X$ such that

$$(!\Gamma)^{IL,\mathcal{S},\sigma} = !x$$

(ii) For every sequent $\Gamma\Rightarrow\Delta$ of $\mathcal{L}(IL)$, there is $x\in X$ such that

$$(\Gamma\Rightarrow?\Delta)^{IL,\mathcal{S},\sigma} \text{ holds iff } \Gamma^{IL,\mathcal{S},\sigma}\leq ?x \text{ holds .}$$

□

(i) Let Γ be $\bullet$: then $(!\Gamma)^{IL,\mathcal{S},\sigma} = 1$, but $1=!1$ by def. 2.5(viii). - Let Γ be a formula A : then $(!\Gamma)^{IL,\mathcal{S},\sigma} = (!A)^{IL,\mathcal{S},\sigma} = !(A^{IL,\mathcal{S},\sigma})$ by def. 4.1. - Let $\Gamma = \Gamma_1*\Gamma_2$: Then $(!\Gamma)^{IL,\mathcal{S},\sigma} = (!\Gamma_1*!\Gamma_2)^{IL,\mathcal{S},\sigma} = (!\Gamma_1)^{IL,\mathcal{S},\sigma}\otimes(!\Gamma_2)^{IL,\mathcal{S},\sigma}$ by def. 4.1, and thus by induction hypothesis there are $x_1\in X$ and $x_2\in X$ s.t. $(!\Gamma_1)^{IL,\mathcal{S},\sigma}=!x_1$ and $(!\Gamma_2)=x_2$, so that by def. 2.5(ix) $(!\Gamma)^{IL,\mathcal{S},\sigma} = !x_1 \otimes !x_2 = !(x_1\&x_2)$.

(ii) Let Δ be a formula A : then $(\Gamma\Rightarrow?A)^{IL,\mathcal{S},\sigma}$ is $\Gamma^{IL,\mathcal{S},\sigma}\leq(?A)^{IL,\mathcal{S},\sigma}$, but by definition 4.1 $(?A)^{IL,\mathcal{S},\sigma} = ?(A^{IL,\mathcal{S},\sigma})$. - Let Δ be the empty right place : then $(\Gamma\Rightarrow?\Delta)^{IL,\mathcal{S},\sigma}$ is $\Gamma^{IL,\mathcal{S},\sigma} \leq\perp$, but by definition 2.5 $\perp=?\perp$.□

4.4 Theorem (Soundness of IL)

Let $\Gamma\Rightarrow\Delta$ be a sequent of $\mathcal{L}(IL)$. If $\Gamma\Rightarrow\Delta$ is provable in IL, then $\Gamma\Rightarrow\Delta$ is IL-valid in every intuitionistic linear structure.

□ By induction on the proof of the sequent in IL. - We use all the results stated in the section 2 on intuitionistic linear structures (def. 2.1, lemma 2.2, def. 2.5) and lemma 4.3. □

4.5 Theorem (Completeness)

Let A be a formula of $\mathcal{L}$(IL). If $\Rightarrow$A is IL-valid in every intuitionistic linear structure, then $\Rightarrow$A is provable in IL.

□ Suppose $\Rightarrow$A be IL-valid in every intuitionistic linear structure; then, in particular, $\Rightarrow$A is IL-valid in the intuitionistic topophase space

$$\mathcal{S} = \left\langle \mathbb{CL}_{\mathbb{PR}}, \subseteq, CL_{\mathbb{PR}}(\{\bullet\}), \bot, \mathbb{T}, \otimes_{\mathbb{PR}}, \cap, \oplus_{\mathbb{PR}}, \multimap, ?_{\mathbb{PR}?}, !_{\mathbb{PR}!} \right\rangle$$

considered in the proposition 3.6 .

We define the following interpretation σ in $\mathcal{S}$: for every variable a, $\sigma(a) = PR(a)$.

We show that, for every formula B of $\mathcal{L}$(IL) , $B^{IL,\mathcal{S},\sigma} = PR(B)$; by induction:

(i) $a^{IL,\mathcal{S},\sigma} = \sigma(a)$, by definition, for every variable a ;

(ii) $1^{IL,\mathcal{S},\sigma} = CL_{\mathbb{PR}}(\{\bullet\}) = PR(1)$, by def. 4.1 and prop. 3.4 (ii)

(iii) $T^{IL,\mathcal{S},\sigma} = \mathbb{T} = PR(T)$, by def. 4.1 and prop. 3.4 (iii)

(iv) $(\neg B)^{IL,\mathcal{S},\sigma} = (B^{IL,\mathcal{S},\sigma})^{\bot} = (PR(B))^{\bot} = PR(\neg B)$, by def.4.1 , ind. hyp. and prop. 3.4(iv)

(v) $(B\otimes C)^{IL,\mathcal{S},\sigma} = B^{IL,\mathcal{S},\sigma} \otimes_{\mathbb{PR}} C^{IL,\mathcal{S},\sigma} = PR(B) \otimes_{\mathbb{PR}} PR(C) = PR(B\otimes C)$,

by def.4.1, ind. hyp. and prop. 3.4 (v)

(vi) $(B\&C)^{IL,\mathcal{S},\sigma} = B^{IL,\mathcal{S},\sigma} \cap C^{IL,\mathcal{S},\sigma} = PR(B) \cap PR(C) = PR(B\&C)$,

by def. 4.1, ind. hyp. and prop. 3.4 (vi)

(vii) $(B\oplus C)^{IL,\mathcal{S},\sigma} = B^{IL,\mathcal{S},\sigma} \oplus_{\mathbb{PR}} C^{IL,\mathcal{S},\sigma} = PR(B) \oplus_{\mathbb{PR}} PR(C) = PR(B\oplus C)$,

by def. 4.1, ind. hyp. and prop. 3.4 (vii)

(viii) $(B\multimap C)^{IL,\mathcal{S},\sigma} = B^{IL,\mathcal{S},\sigma} \multimap C^{IL,\mathcal{S},\sigma} = PR(B) \multimap PR(C) = PR(B\multimap C)$,

by def. 4.1, ind. hyp. and prop. 3.4 (viii)

(ix) $(?B)^{IL,\mathcal{S},\sigma} = ?_{\mathbb{PR}?}(B^{IL,\mathcal{S},\sigma}) = ?_{\mathbb{PR}?}(PR(B)) = PR(?B)$,

by def. 4.1, ind. hyp. and prop. 3.6(iii)

(x) $(!B)^{IL,\mathcal{S},\sigma} = !_{\mathbb{PR}!}(B^{IL,\mathcal{S},\sigma}) = !_{\mathbb{PR}!}(PR(B)) = PR(!B)$,

by def. 4.1, ind. hyp. and prop. 3.6(iv) .

Now, since $\Rightarrow$A is IL-valid in $\mathcal{S}$, by deff. 4.1-4.2 $CL_{\mathbb{PR}}(\{\bullet\}) \subseteq A^{IL,\mathcal{S},\sigma} = PR(A)$, so that $\bullet \in PR(A)$, i.e. $\Rightarrow$A is provable in IL. □

4.6 Theorem

Let A be a formula of $\mathcal{L}$(IL).

$\Rightarrow$A is provable in IL iff $\Rightarrow$A is IL-valid in every intuitionistic linear structure
iff $\Rightarrow$A is valid in every intuitionistic topophase space.

□ From the proofs of 4.4 and 4.5, since in the proof of 4.5 we can use the weaker hypothesis that $\Rightarrow$A is IL-valid in every intuitionistic topophase space. □

4.7 Definition

Let $\mathcal{S}$ be an intuitionistic *-linear structure and σ an interpretation in $\mathcal{S}$. We define, as in 4.1 : (i) for every formula A of $\mathcal{L}$(*IL), $A^{*IL,\mathcal{S},\sigma}$ (we remove the cases of the exponential connectives ! and ?) ; (ii) for every finite multiset Γ of formulas of $\mathcal{L}$(*IL) , $\Gamma^{*IL,\mathcal{S},\sigma}$; (iii) for every sequent Γ$\Rightarrow$A of $\mathcal{L}$(*IL) , $(\Gamma\Rightarrow A)^{*IL\_,\mathcal{S},\sigma}$.

4.8 Definition

Let Γ$\Rightarrow$A be a sequent of $\mathcal{L}$(*IL). Let $\mathcal{S}$ be an intuitionistic *-linear structure. Γ$\Rightarrow$A is <u>*IL-valid in $\mathcal{S}$</u> iff for every interpretation σ in $\mathcal{S}$ $(\Gamma\Rightarrow A)^{*IL,\mathcal{S},\sigma}$ holds.

4.9 Theorem (Soundness and completeness of *IL)

(i) If $\Gamma \Rightarrow A$ is a sequent provable in *IL, then $\Gamma \Rightarrow A$ is *IL-valid in every intuitionistic *-linear structure.

(ii) If $\Rightarrow A$ is valid in every intuitionistic *-linear structure, then $\Rightarrow A$ is provable in *IL .

(iii) Let A be a formula of $\mathcal{L}$(*IL) . $\Rightarrow A$ is provable in *IL iff $\Rightarrow A$ is valid in every intuitionistic *-linear structure iff $\Rightarrow A$ is valid in every intuitionistic phase space.

□ Analogous to the proofs of theorems 4.4 , 4.5 and 4.6 □

4.10 Definition

Let $\mathcal{S}$ be a 0-intuitionistic linear structure and σ an interpretation in $\mathcal{S}$. We define, as in def. 4.1 :

(i) for every formula A of $\mathcal{L}$(IL0), $A^{IL0,\mathcal{S},\sigma}$; we remove the case of $\neg$ and we add the clause $0^{IL0,\mathcal{S},\sigma} = \bot$;

(ii) for every finite multiset Γ of formulas of $\mathcal{L}$(IL0) , $\Gamma^{IL0,\mathcal{S},\sigma}$;

(iii) for every sequent $\Gamma \Rightarrow A$ of $\mathcal{L}$(IL0) , $(\Gamma \Rightarrow A)^{IL0,\mathcal{S},\sigma}$.

4.11 Definition

Let $\Gamma \Rightarrow A$ be a sequent of $\mathcal{L}$(IL0). Let $\mathcal{S}$ be a 0-intuitionistic linear structure. $\Gamma \Rightarrow A$ is IL0-valid in $\mathcal{S}$ iff for every interpretation σ in $\mathcal{S}$ $(\Gamma \Rightarrow A)^{IL0,\mathcal{S},\sigma}$ holds.

4.12 Theorem (Soundness and completeness of IL0)

(i) If $\Gamma \Rightarrow A$ is provable in IL0, then $\Gamma \Rightarrow A$ is valid in every 0-intuitionistic linear structure.

(ii) Let A be a formula of $\mathcal{L}$(IL0). If $\Rightarrow A$ is valid in every 0-intuitionistic linear structure, then $\Rightarrow A$ is provable in IL0 .

(iii) Let A be a formula of $\mathcal{L}$(IL0) . $\Rightarrow A$ is provable in IL0 iff $\Rightarrow A$ is valid in every 0-intuitionistic linear structure iff $\Rightarrow A$ is valid in every 0-intuitionistic topophase space.

□ The proof of (i) is obtained by means of some modifications from the proof of theorem 4.4 (we remove the cases concerning $\neg$, and we add the case where the last rule is (0)). The proof of (ii) is analogous to the proof of theorem 4.5 , by using the results on the provability in IL0 stated in the remark 3.9. The proof of (iii) is analogous to the proof of theorem 4.6 □

4.13 Remark

By theorem 4.4 (theorem 4.9(i)) , the following holds for every formula A of $\mathcal{L}$(IL) (of $\mathcal{L}$(*IL), resp.) : if $\Rightarrow A$ is provable in IL (in *IL, resp.), then $\Rightarrow A$ is IL-valid in every simple intuitionistic topophase space (*IL-valid in every simple intuitionistic phase space, resp.).

We show that the converse does not hold . i.e. we show that *IL is not complete w.r. to the validity in every simple intuitionistic phase space and IL is not complete w.r. to the validitity in every intuitionistic topophase space.

4.14 Proposition

(i) $\Rightarrow a \& (b \oplus c) \multimap (a \& b) \oplus (a \& c)$ is *IL-valid in every simple intuitionistic phase space and IL-valid in every simple intuitionistic topophase space.

(ii) $\Rightarrow a \& (b \oplus c) \multimap (a \& b) \oplus (a \& c)$ is not provable in *IL and in IL.

□

(i) Let $\mathcal{S} = \langle \mathcal{P}(M), \subseteq, \{i\}, \bot, M, \cdot, \cap, \cup, \multimap \rangle$ be a simple intuitionistic phase space and σ an interpretation in $\mathcal{S}$. Let $\sigma(a) = F$, $\sigma(b) = G$, $\sigma(c) = H$. By

definition 4.1 and 4.7 , we have
$(\Rightarrow a\&(b\oplus c)\multimap(a\&b)\oplus(a\&c))^{*IL,\mathcal{I},\sigma}$ holds iff $\{i\} \subseteq F\cap(G\cup H) \multimap (F\cap G)\cup(F\cap H)$
i.e. iff $F\cap(G\cup H) \subseteq (F\cap G) \cup (F\cap H)$.
But in $\mathcal{I}$ $F\cap(G\cup H) \subseteq (F\cap G) \cup (F\cap H)$ holds : if $x \in F\cap(G\cup H)$ then $x\in F$ and $x\in G\cup H$, so that $x\in F$ and $x\in G$, or $x\in F$ and $x\in H$, i.e. $x \in (F\cap G) \cup (F\cap H)$.

(ii) Suppose $\Rightarrow a\&(b\oplus c)\multimap(a\&b)\oplus(a\&c)$ be provable in *IL . Then , by remark 1.3(i), the sequent is cut-free provable , and by 1.3(ii) the sequent

$$a\&(b\oplus c) \Rightarrow (a\&b)\oplus(a\&c)$$

is cut-free provable in *IL and therefore one of the sequents below is cut-free provable in *IL :

1) $a \Rightarrow (a\&b)\oplus(a\&c)$ 2) $(b\oplus c) \Rightarrow (a\&b)\oplus(a\&c)$
3) $a\&(b\oplus c) \Rightarrow (a\&b)$ 4) $a\&(b\oplus c) \Rightarrow (a\&c)$.

Thus, by theorem 4.9, at least one of the above sequents is *IL-valid in every simple intuitionistic phase space. But all the above sequents are not *IL-valid in every simple intuitionistic phase space, as we show now.

1) Take $\sigma(a) = F\neq\emptyset$, $\sigma(b) = \emptyset = \sigma(c)$: then $F\subseteq(F\cap\emptyset)\cup(F\cap\emptyset) = \emptyset$ does not hold.
2) Take $\sigma(a) = \emptyset$, $\sigma(b) = G\neq\emptyset$, $\sigma(c) = H\neq\emptyset$: then $G\cup H \subseteq (\emptyset\cap G) \cup (\emptyset\cap H) = \emptyset$ does not hold.
3) Take $\sigma(a) = \{x,y\}$, $\sigma(b) = \{x\}$, $\sigma(c) = \{y\}$: then $\{x,y\} \cap (\{x\}\cup\{y\}) = \{x,y\}$, but $\{x,y\} \cap \{x\} = \{x\}$.
4) Take $\sigma(a) = \{x,y\}$, $\sigma(b) = \{x\}$, $\sigma(c) = \{y\}$: then $\{x,y\} \cap (\{x\}\cup\{y\}) = \{x,y\}$, but $\{x,y\} \cap \{y\} = \{y\}$. □

Acknowledgements

Research supported by: European Economic Community, Ministero della Pubblica Istruzione of the Italian Republic, University of Firenze (Italy).

REFERENCES

ABRUSCI V.M., *Additional results on intuitionistic linear propositional logic*, (in preparation).

AVRON A., *The Semantics and Proof Theory of Linear Logic*, LFCS , Preprint, Edinburgh, April 1987 .(Published in Theoretical Computer Science, 1988).

FUCHS L., *Partially Ordered Algebraic Systems*, Pergamon Press, Oxford, 1963.

GIRARD J.-Y. , *Linear Logic* , Theoretical Computer Science, 50 ,1987 .

GIRARD J:-Y. , LAFONT Y. , *Linear logic and lazy computation*, Proceedings of the TAPSOFT, Pisa, SLNCS, 250, 1987

GRISHIN V.N., *Predicate and set-theoretic calculi based on logic without contractions*, Math. USSR Izvestija, Vol.18(1982), No.1, pp. 44-59.

ONO H.- KOMORI Y., *Logics without contraction rules*, Journal of Symbolic Logic, 50(1985),169.

ORDER ISOMORPHISMS – A CONSTRUCTIVE MEASURE–THEORETIC VIEW

Douglas S. Bridges

University of Buckingham
Buckingham MK18 1EG
England

Although it can be argued that, as far as the needs of economists are concerned, the problem of representing a preference relation by a real–valued utility function was solved by the work of Debreu [9,10] and others, the years since the publication of Debreu's classic work "Theory of Value" have seen several significant new approaches to, and generalisations of, that problem [11–14].

From a pure mathematical point of view, the problem is one of finding order homomorphisms from a set X with a strict weak order to the real line $\mathbb{R}$ with the usual strict weak order $>$. Two earlier papers of mine [4,5] dealt, in considerable detail, with a solution of that problem that is constructive in Bishop's sense and is based on the classical approach of Arrow and Hahn [1]. In this paper, I describe an alternative constructive development, based on the classical measure theoretic one of Neuefeind [12]. Classically, Neuefeind's representation theorem is more general than, and perhaps supersedes, that of Arrow and Hahn; constructively, since the hypotheses of the two theorems are not easily comparable, each theorem stands of interest on its own.

I shall assume familiarity with the metric space theory and measure theory in Chapters 4 and 6 of [3]; the reader may also find it helpful to consult the introduction to [4]. I shall use dmn*(f)* to denote the domain of a function *f*, and $\vee,\wedge$ to denote the union and intersection of complemented sets.

Let X be a nonvoid set, and $\succ$ a binary relation on X. We require relations to be strongly extensional; so if $x \succ y$, $x = x'$, and $y = y'$, then $x' \succ y'$. Define an associated binary relation $\succeq$ as follows:

$$x \succeq y \text{ if and only if } \forall z \in X\,(y \succ z \Rightarrow x \succ z).$$

Then $x \succeq x$ for all x in X, and $\succeq$ is transitive. For each x in X define the corresponding

Mathematical Logic
Edited by P. P. Petkov
Plenum Press, New York, 1990

	upper contour set:	$C(x) \equiv \{y \in X : y \succeq x\}$,
	strict upper contour set:	$C^{+}(x) \equiv \{y \in X : y \succ x\}$,
and	***strict lower contour set:***	$C^{-}(x) \equiv \{y \in X : x \succ y\}$.

We say that $\succ$ is a ***preference relation***, or a ***strict weak order***, with associated ***preference–indifference relation*** $\succeq$, if the following conditions are satisfied:

P1 *If $x \succ y$, then $y \succ x$ is contradictory.*

P2 *If $x \succ y$, then for each z in X either $x \succ z$ or $z \succ y$.*

In that case, $\succ$ is transitive; $x \succ x$ is contradictory; $x \succ y$ entails $x \succeq y$; and $x \succeq y$ if and only if $y \succ x$ is contradictory.

Let $\succ$ be a preference relation on X. By a ***utility function*** for $\succ$ we mean a map $u: X \to \mathbb{R}$ such that $x \succ y$ if and only if $u(x) > u(y)$; such a map is also called an ***order isomorphism*** from $(X,\succ)$ to $(\mathbb{R},>)$, and is said to ***represent*** $\succ$.

Now let $\succ$ be a preference relation on a metric space (X,ρ). We say that $\succ$ is ***continuous*** if for each x in X, $C^{+}(x)$ and $C^{-}(x)$ are open in X; in which case $C(x)$ and $\{y \in X : x \succeq y\}$ are both closed subsets of X. Clearly, if $\succ$ is represented by a continuous utility function, then $\succ$ is continuous.

If X is locally compact and $\succ$ is represented by a continuous utility function, then $\succ$ satisfies the following stronger condition of ***uniform continuity on compact sets***:

UC *If $a,b \in X$, $a \succ b$, and $K \subset X$ is compact, then there exists $r > 0$ such that for all x,y in K with $\rho(x,y) < r$, either $a \succ x$ or $y \succ b$* [5, Propn. 1].

This condition can be regarded as a strengthening of P2. Note that if X is locally compact and $\succ$ satisfies UC, then $\succ$ is continuous. Classically, if $\succ$ is continuous, then it satisfies UC; however, there is a recursive example of a continuous preference relation on [0,1] that fails to satisfy UC [5, Example 1].

In order to use measure theory in the construction of a utility function representing a preference relation on a locally compact space X, we identify the contour sets $C^{+}(x)$ and $C(x)$ with the complemented sets $(\{y \in X : y \succ x\},\{y \in X : x \succeq y\})$ and $(\{y \in X : y \succeq x\},\{y \in X : x \succ y\})$ respectively. (These *are* complemented sets, since we require that a preference relation have the property of strong extensionality.) If $a,b \in X$, then we identify $\{x \in X : a \succeq x \succeq b\}$ with the complemented set $S \equiv (X - C^{+}(a)) \wedge C(b)$; so if $C^{+}(a)$ and $C(b)$ are measurable relative to a positive measure ν on X, then for all x in the full set $\mathrm{dmn}(\chi_{C^{+}(a)}) \cap \mathrm{dmn}(\chi_{C(b)})$, $\chi_S(x) = 1$ if and only if $a \succeq x \succeq b$, and $\chi_S(x) = 0$ if and only if either $x \succ a$ or $b \succ x$.

Of particular interest is the ***indifference class*** of a, $\{x \in X : x \sim a\}$, which is identified with the complemented set $(X - C^+(a)) \wedge C(a)$. Other identifications, such as that for $\{x \in X : a \succeq x \succ b\}$, are defined in the obvious way.

Before stating any of the results of this section, we recall that a nonvoid located (in particular, a locally compact) convex subset of $\mathbb{R}^N$ is Lebesgue measurable [7, Theorem 1].

Lemma 1. *If X is a locally compact, convex subset of $\mathbb{R}^N$ with positive Lebesgue measure, then the interior of X is dense X.*

Proof. Inspection of the proof of Theorem 2 of [7] uncovers a proof of the desired result. □

Theorem 1. *Let X be a locally compact, convex subset of $\mathbb{R}^N$ with positive Lebesgue measure; let ν be the measure $\mu(X)^{-1}\mu$ on X, where μ is the standard normal probability density on $\mathbb{R}^N$; and let $\succ$ be a preference relation on X that is uniformly continuous on compact sets. Suppose there exists a dense subset A of X such that*

(i) *$C^+(x)$ and $C(x)$ are Lebesgue measurable for each x in A;*
(ii) *for each compact subset K of X, and each $\epsilon > 0$, there exists $\delta > 0$ such that if $a,b \in K \cap A$ and $\rho(a,b) < \delta$, then $\nu(\{x \in X : a \succeq x \succeq b\}) < \epsilon$.*

Then

$$u(x) \equiv 1 - \nu(C^+(x))$$

defines on A a function that extends to a continuous utility function $u : X \to [0,1]$ representing $\succ$.

Proof. Consider $a,b \in X$ with $a \succ b$. We first prove that there exists $c \in X$ such that $a \succ c \succ b$. To this end, use condition UC to compute $r > 0$ such that if x,y belong to the segment joining a and b, and $\|x - y\| < r$, then either $a \succ x$ or $y \succ b$. By Lemma 5 of [7], there exist points ξ,η of that segment such that $\|\xi - \eta\| < r$, $a \succ \xi$, and $\eta \succ b$. Then either $a \succ \eta$, in which case we set $c \equiv \eta$; or else $\xi \succ b$, when we set $c \equiv \xi$. Note that, as A is dense in X and $\succ$ is continuous, we can, if necessary, replace c by a neighbouring point in A, to arrange that $a \succ c \succ b$ and $c \in A$; we assume that this has been done.

As $\succ$ is continuous and the interior of X is dense in X, we can find a ball B of positive radius in $\mathbb{R}^N$ such that $a \succ x \succ b$ for all x in B. Since $C^+(a) \subset C^+(b)$, it follows that if $a,b \in A$, then

$$\begin{aligned} \nu(C^+(b)) &= \nu(C^+(a)) + \nu(\{x \in X : a \succeq x \succ b\}) \\ &\geq \nu(C^+(a)) + \nu(B) > \nu(C^+(a)); \end{aligned}$$

whence $u(a) > u(b)$.

Conversely, if $a,b \in A$ and $u(a) > u(b)$, then, by the foregoing, $a \succsim b$; whence $C^+(a) \subset C^+(b)$,

$$\nu(C^+(b)-C^+(a)) = \nu(C^+(b)) - \nu(C^+(a)) > 0,$$

and there exists ξ in $C^+(b)-C^+(a)$. Then $a \succsim \xi \succ b$, so $a \succ b$. This completes the proof that u represents $\succ$ on A.

We now prove that if $K \subset X$ is compact, then u is uniformly continuous on $K \cap A$. Given $\epsilon > 0$, choose δ as in (ii), and consider a,b in $K \cap A$ such that $u(a) > u(b) + \epsilon$. Then $a \succ b$ and $\nu(C^+(b)) > \nu(C^+(a)) + \epsilon$. So

$$\begin{aligned}\nu(\{x \in X : a \succsim x \succsim b\}) &\geq \nu(x \in X : a \succsim x \succ b\}) \\ &= \nu(C^+(b)-C^+(a)) \\ &= \nu(C^+(b)) - \nu(C^+(a)) > \epsilon;\end{aligned}$$

whence $\|a - b\| \geq \delta$, by our choice of δ. Thus if $a,b \in K \cap A$ and $\|a - b\| < \delta$, then $u(a) \leq u(b) + \epsilon$; from which it follows that u is uniformly continuous on $K \cap A$.

As A is dense in X, u extends to a continuous function from X into $[0,1]$. Consider $x,y \in X$ with $x \succ y$. If $u(x) < u(y)$, then (as $\succ$ is continuous, A is dense in X, and u is continuous) we can find x',y' in A such that $x' \succ y'$ and $u(x') < u(y')$, thereby contradicting the fact that u represents $\succ$ on A; so $u(x) \geq u(y)$. Noting the first part of the proof, choose x',y' in A such that $x \succ x' \succ y' \succ y$; then $u(x) \geq u(x') > u(y') \geq u(y)$, so that $u(x) > u(y)$. On the other hand, if $u(x) > u(y)$, then (as X is convex, u is continuous, and A is dense in X) we can find x',y' in A such that $u(x) > u(x') > u(y') > u(y)$; whence $x' \succ y'$, as u represents $\succ$ on A. If $x' \succ x$, then $u(x') \geq u(x)$, a contradiction; so $x \succsim x'$ and similarly $y' \succsim y$. Thus $x \succsim x' \succ y' \succsim y$, and therefore $x \succ y$. This completes the proof of the theorem. □

In the notation of Theorem 1, it is worth pointing out the interpretation of $u(x) \equiv 1 - \nu(C^+(x))$, where $x \in A$: $u(x)$ is the probability (relative to the density ν) that a randomly chosen element of X is not strictly preferred to x.

Corollary 1. *Under the conditions of Theorem 1, $\succ$ is* ***locally nonconstant****, in the sense that for each x in X and each $\epsilon > 0$, there exists x' in X such that $\|x - x'\| < \epsilon$ and either $x' \succ x$ or $x \succ x'$.*

Proof. We first consider the case where $x \in A$. Since the interior of X is dense in X (by Lemma 1), there exists a ball B in $\mathbb{R}^N$, of positive radius, such that $B \subset X$ and $\|y - x\| < \epsilon$ for each y in B. As $\nu(B) > 0$ and (by condition (ii) of Theorem 1), $\nu(\{y \in X :$

$y \sim x\}) = 0$,

$$B \cap \{y \in X : \text{either } y \succ x \text{ or } x \succ y\}$$

is nonvoid, and we need only choose x' in B.

In the general case, we can find a in A such that $\|x - a\| < \epsilon/2$. By the foregoing, there exists $a' \in A$ such that $\|a - a'\| < \epsilon/2$ and either $a \succ a'$ or $a' \succ a$. By P2, we have either $x' \succ x$ or $x \succ x'$ for some $x' \in \{a,a'\}$; moreover, $\|x - x'\| < \epsilon$. □

Corollary 2. *Under the hypotheses of Theorem 1, for all x in a dense subset of X, $C(x)$ is locally compact and $C^+(x)$ is locally totally bounded.*

Proof. The proof consists of an extension of the argument in the proof of Proposition 2 of [5]. □

Our next aim is to prove the following extension of Theorem 1.

Theorem 2. *Under the hypotheses of Theorem 1, if also A is a full set, then for each x in X,*

(i) *$C(x)$ and $C^+(x)$ are integrable and have the same measure (with respect to ν);*

(ii) $u(x) = 1 - \nu(C^+(x))$.

Moreover,

(iii) *for each compact $K \subset X$ and each $\epsilon > 0$, there exists $\delta > 0$ such that if $a,b \in K$ and $\|a - b\| < \delta$, then $\nu(\{x : a \succeq x \succeq b\}) < \epsilon$.*

The proof of this theorem depends on several lemmas, the first of which requires two definitions.

A mapping u from a metric space X into $\mathbb{R}$ is said to be ***locally nonconstant*** if for each x in X and each neighbourhood U of x, there exists y in U such that $u(x) \neq u(y)$.

Let $\succ$ be a preference relation on a metric space X, and ξ a point of X. We say that ξ is an ***order interior point*** of X (relative to $\succ$) if there exist $a,b \in X$ such that $a \succ \xi \succ b$.

Lemma 2. *Let X be a locally compact, convex subset of $\mathbb{R}^N$, $\succ$ a preference relation on X that is represented by a continuous, locally nonconstant utility function u, and ξ an order interior point of X relative to $\succ$. Then there exists $x \sim \xi$ such that each neighbourhood of x contains points s,t with $s \succ x \succ t$..*

Proof. Choose $a,b \in X$ such that $a \succ \xi \succ b$; then $u(a) > y \equiv u(\xi) > u(b)$. The first half of the proof of the theorem in [6] contains the construction of a point x of X with $u(x) = y$ – whence $x \sim \xi$ – such that each neighbourhood of x contains points s,t with $u(s) > y > u(t)$. □

Classically, we can prove the following extension of Lemma 2:

If X is a connected topological space, $\succ$ a continuous, locally nonconstant preference relation on X, and ξ an order interior point of X relative to $\succ$, then there exists $x \sim \xi$ such that each neighbourhood of x contains points s,t with $s \succ x \succ t$.

Indeed, suppose the hypotheses of this statement hold, but the conclusion is false. Then there exists an order interior point ξ of X with the following property: for each $x \sim \xi$ there exists a neighbourhood U of x such that either $y \succeq \xi$ for all y in U, or else $\xi \succeq y$ for all y in U. It follows from the continuity of $\succ$ that X is the union of the two sets

$$A \equiv \{x \in X : \text{there exists a neighbourhood } U \text{ of } x \text{ in } X \text{ such that } y \succeq \xi \text{ for all } y \text{ in } U\}$$

and

$$B \equiv \{x \in X : \text{there exists a neighbourhood } U \text{ of } x \text{ in } X \text{ such that } \xi \succeq y \text{ for all } y \text{ in } U\}.$$

Clearly, both A and B are open; whence, by the connectedness of X, there exists $x \in A \cap B$. For all y in some neighbourhood V of x in X, we have both $y \succeq \xi$ and $\xi \succeq y$; whence $y \sim \xi$. This contradicts the assumption that $\succ$ is locally nonconstant.

Returning to the main business at hand, we now prove

Lemma 3. *Under the hypotheses of Theorem 1, if ξ is an order interior point of X, then $C(\xi)$ and $C^+(\xi)$ are integrable (with respect to ν) and have the same measure; moreover, $u(\xi) = 1 - \nu(C^+(\xi))$.*

Proof. In view of Lemma 2, we may assume that each neighbourhood of ξ contains points x,x' such that $x \succ \xi \succ x'$. As $\succ$ is continuous and A is dense in X, we can therefore construct sequences (a_n) and (b_n) in A such that for each n, $a_n \succ a_{n+1} \succ \xi \succ b_{n+1} \succ b_n$, $\rho(\xi,a_n) < 1/n$, and $\rho(\xi,b_n) < 1/n$. Then $C^+(a_1) \subset C^+(a_2) \subset \ldots$; $C^+(b_1) \supset C^+(b_2) \supset \ldots$; and for all m and n, $C^+(a_m) \subset C^+(b_n)$. Since u is continuous and $\rho(a_n,b_n) \to 0$ as $n \to \infty$, we see that for $m \geq n$,

$$\begin{aligned} 0 &\leq \nu(C^+(a_m)) - \nu(C^+(a_n)) \leq \nu(C^+(b_n) - C^+(a_n)) \\ &= u(a_n) - u(b_n) \to 0 \text{ as } n \to \infty. \end{aligned}$$

Hence $(\nu(C^+(a_n))$, and similarly $(\nu(C^+(b_n))$, is a Cauchy sequence in $\mathbb{R}$; moreover, these Cauchy sequences converge to the same limit λ as $n \to \infty$. So $\vee_n C^+(a_n)$ and $\wedge_n C^+(b_n)$ are integrable and have the same measure λ [3, Ch. 6, (3.9)]. But the characteristic functions of $C^+(\xi)$ and $\vee_n C^+(a_n)$ are equal on a full set – namely, the domain of the characteristic function of the latter complemented set; similarly, the characteristic functions of $C(\xi)$ and $\wedge_n C^+(b_n)$ are equal on a full set. This completes the proof of the first part of the lemma.

As u is continuous and each a_n belongs to A,

$$\begin{aligned} u(\xi) &= \lim_{n\to\infty} u(a_n) = \lim_{n\to\infty}(1 - \nu(C^+(a_n)) \\ &= 1 - \lim_{k\to\infty} \nu(\vee_{n=1}^{k} C^+(a_n)) \\ &= 1 - \nu(\vee_n C^+(a_n)) = 1 - \nu(C^+(\xi)). \quad \square \end{aligned}$$

If $\succ$ is a preference relation on a metric space X, and $C^+(\xi)$ is nonvoid, then ξ is said to be a ***nonsatiation point*** of X (relative to $\succ$).

Lemma 4. *Under the hypotheses of Theorem 1, for each nonsatiation point ξ of X, $C^+(\xi)$ is integrable, and $u(\xi) = 1 - \nu(C^+(\xi))$.*

Proof. Construct a sequence (a_n) in A converging to ξ, such that $a_1 \succ \xi$ and such that for each n,

(1) either $a_n \succ \xi$ or $\xi \succ a_n$;

(2) $|u(\xi) - u(a_{n+1})| < \frac{1}{2}|u(\xi) - u(a_n)|$.

If $a_k \succ \xi$ for $k = 1,\dots,n$, let C_n be the integrable set $C^+(a_n)$. If $\xi \succ a_k$ for some $k \leq n$, then ξ is an order interior point of X; so, by Lemma 3, $C^+(\xi)$ is integrable; in that case, let $C_n \equiv C^+(\xi)$. By (2) above, $C_1 \subset C_2 \subset \dots$. Also, $(\nu(C_n))$ is a Cauchy sequence. To see this, let K be a compact neighbourhood of ξ in X containing each a_n. Given $\epsilon > 0$, choose $\delta > 0$ as in condition (ii) of Theorem 1, and compute N_1 such that $\|a_m - a_n\| < \delta$ for all $m,n \geq N_1$. We prove that if $m > n \geq N_1$, then $|\nu(C_m) - \nu(C_n)| < \epsilon$. If $a_k \succ \xi$ for all $k \leq m$, then

$$0 \leq \nu(C_m) - \nu(C_n) = \nu(C^+(a_m) - C^+(a_n)) = \nu(C(a_m) - C^+(a_n)) < \epsilon,$$

since $\|a_m - a_n\| < \delta$. If there exists k such that $n < k \leq m$, $a_j \succ \xi$ for $1 \leq j < k$, and $\xi \succ a_k$, then, as $\|a_n - a_k\| < \delta$, we have

$$\begin{aligned} 0 \leq \nu(C_m) - \nu(C_n) &= \nu(C^+(\xi)) - \nu(C^+(a_n)) \\ &\leq \nu(C^+(a_k)) - \nu(C^+(a_n)) \\ &\leq \nu(C(a_k) - C^+(a_n)) < \epsilon. \end{aligned}$$

If there exists $k < n$ such that $\xi \succ a_k$, then $C_m = C_n = C^+(\xi)$, and $\nu(C_m) - \nu(C_n) = 0$. Thus, in all cases, $|\nu(C_m) - \nu(C_n)| < \epsilon$.

It now follows that $\lambda \equiv \lim_{n\to\infty} \nu(C_n)$ exists; whence, by [3, Ch. 6, (3.9)], the complemented set $S \equiv \vee_n C_n$ is integrable. It is straightforward to show that the characteristic functions of $C^+(\xi)$ and S are equal on the full set $\mathrm{dmn}(\chi_S)$. So $C^+(\xi)$ is integrable, with measure λ. This completes the proof of the first part of the lemma.

To prove the second part, we note Lemma 3 and observe that, in the notation of the first part of this proof, $C_n = C^+(x_n)$ for some x_n such that $u(x_n) = 1 - \nu(C^+(x_n))$; also, $\nu(C^+(x_n)) \to \nu(C^+(\xi))$ and $u(x_n) \to u(\xi)$ as $n \to \infty$. Hence $u(\xi) = 1 - \nu(C^+(\xi))$. □

Lemma 5. *Under the hypotheses of Theorem* 1, *if A is a full set and ξ is a nonsatiation point, then $C(\xi)$ is integrable and $\nu(C(\xi)) = \nu(C^+(\xi))$.*

Proof. By Lemma 4, $C^+(\xi)$ is integrable. By Lemma 8 of [7], it will suffice to show that for each $\epsilon > 0$ there exists an integrable set S such that $C^+(\xi) \subset C(\xi) \subset S$ and $\nu(S - C^+(\xi)) < \epsilon$. Either $\nu(X - C^+(\xi)) < \epsilon$, in which case we can take $S \equiv X$; or, as we assume, $\nu(X - C^+(\xi)) > 0$. As A is a full set, there exists $a \in A \cap (X - C^+(\xi))$ [3, Ch. 6, (3.4)]; so $\xi \succeq a$. Then $C(a)$ and $C^+(a)$ are integrable and have the same measure. Either $\nu(C(a) - C^+(\xi)) < \epsilon$, in which case we can take $S \equiv C(a)$; or, as we assume,

$$\nu(C^+(a) - C^+(\xi)) = \nu(C(a) - C^+(\xi)) > 0.$$

Then there exists b such that $\xi \succeq b \succ a$, so $\xi \succ a$. Hence ξ is an order interior point, and, by Lemma 3, we can take $S \equiv C(\xi)$. This completes the proof. □

We now arrive at the

Proof of Theorem 2. We first prove that $C(\xi)$ is integrable for each ξ in X. To this end, construct a sequence (a_n) in A such that for each n,

(1) either $\xi \succ a_n$ or $a_n \succ \xi$;

(2) $|u(\xi) - u(a_{n+1})| < \frac{1}{2}|u(\xi) - u(a_n)|$.

In view of Lemma 5, we may assume that $\xi \succ a_1$. If $\xi \succ a_k$ for all $k \leq n$, let C_n be the integrable set $C^+(a_n)$; if $a_k \succ \xi$ for some $k \leq n$, let C_n be the integrable set $C(\xi)$. Then $C_1 \supset C_2 \supset \ldots$, and, as in the proof of Lemma 4, $(\nu(C_n))$ is a Cauchy sequence. By [3, Ch.6, (3.9)], $S \equiv \wedge_n C_n$ is integrable, with measure $\lambda \equiv \lim_{n\to\infty} \nu(C_n)$. It is straightforward to show that for each x in $\mathrm{dmn}(\chi_S)$, $\chi_S(x) = 0$ if and only if $\xi \succ x$. It follows that the characteristic functions of $C(\xi)$ and S are equal on the full set $\mathrm{dmn}(\chi_S)$; whence $C(\xi)$ is integrable, with measure λ.

Note that, in view of Lemma 3, $C_n = C^+(x_n)$ for some x_n such that $u(x_n) = 1 - \nu(C^+(x_n))$; and that $\nu(C^+(x_n)) \to \nu(C(\xi))$ and $u(x_n) \to u(\xi)$ as $n \to \infty$. Hence

$$u(\xi) = \lim_{n\to\infty}(1 - \nu(C^+(x_n))) = 1 - \nu(C(\xi)).$$

An argument similar to that of the first paragraph shows that $X-C^+(\xi)$ is integrable; whence $C^+(\xi)$ is integrable. Suppose that $\nu(C(\xi)) > \nu(C^+(\xi))$. Then as A is a full set, there exists $a \in A \cap (C(\xi)-C^+(\xi))$; so $a \sim \xi$. Hence

$$\nu(C(\xi)) = \nu(C(a)) = \nu(C^+(a)) = \nu(C^+(\xi)),$$

a contradiction. Thus we must have $\nu(C(\xi)) = \nu(C^+(\xi))$. This completes the proof of (i) and (ii) of the theorem.

To prove (iii), let $K \subset X$ be compact and $\epsilon > 0$. Choose $\delta > 0$ such that if $a,b \in K$ and $\|a - b\| < \delta$, then $|u(a) - u(b)| < \epsilon/2$. Consider such a and b, and suppose that

$$\nu(\{x : a \succeq x \succeq b\}) = \nu(C(b)\text{-}C^+(a)) > \epsilon/2.$$

Then $a \succeq b$, so

$$\begin{aligned} u(a) - u(b) &= \nu(C^+(b)) - \nu(C^+(a)) \\ &= \nu(C^+(b)-C^+(a)) = \nu(C(b)-C^+(a)) > \epsilon/2, \end{aligned}$$

a contradiction. Hence $\nu(\{x : a \succeq x \succeq b\}) < \epsilon$. □

It is not known whether, in Theorem 2, the hypothesis that A is a full subset of X can be weakened to one that A is merely dense in X.

Condition (ii) of Theorem 1 invites further comment. For example, would it suffice for our purposes to replace it by the simpler one that for each a in A, $\{x : x \sim a\}$ has measure 0? Classically, the answer is "yes", for we can prove the following result:

> *Let ν be a positive measure on a locally compact space X, and $\succ$ a continuous preference relation on X such that for all a,b in X, $\{x \in X : a \succeq x \succeq b\}$ is measurable and $\nu(\{x : x \sim a\}) = 0$. Then conclusion* (iii) *of Theorem 2 holds.*

Indeed, under the hypotheses of this statement, suppose that the conclusion fails to hold. Then there exist a compact set $K \subset X$, a positive number α, and sequences $(a_n),(b_n)$ in K, such that for each n, $\|a_n - b_n\| < 1/n$ and $\nu(\{x \in X : a_n \succeq x \succeq b_n\}) \geq \alpha$. As K is compact, we may assume that (a_n) converges to a limit a in K; clearly, (b_n) then also converges to a. Suppose, to begin with, that $a \succeq a_n$ for infinitely many n; we may assume that $a \succeq a_n$ for all n. Then $\nu(\{x : a \succeq x \succeq b_n\}) \geq \alpha$ for each n, so that

$$\begin{aligned} \nu(\{x : x \sim a\}) &= \nu(\wedge_{n=1}^{\infty} \vee_{k\geq n}\{x : a \succeq x \succeq b_k\}) \\ &= \lim_{n\to\infty} \nu(\vee_{k\geq n}\{x : a \succeq x \succeq b_k\}) \geq \alpha, \end{aligned}$$

a contradiction. Similarly, the assumption that $b_n \succsim a$ for infinitely many n leads to a contradiction. Hence we may assume that $a_n \succ a \succ b_n$ for each n. Since $\succ$ is continuous, and $a_n, b_n \to a$ as $n \to \infty$, for each n there exists $m > n$ such that $a_n \succ a_m \succ a \succ b_m \succ b_n$. So we may further assume that $a_1 \succ a_2 \succ \ldots \succ a \succ \ldots \succ b_2 \succ b_1$. Then

$$\begin{aligned} \nu(\{x : x \sim a\}) &= \nu(\wedge_{n=1}^{\infty}\{x : a_n \succsim x \succsim b_n\}) \\ &= \lim_{k\to\infty}\nu(\wedge_{n=1}^{k}\{x : a_n \succsim x \succsim b_n\}) \\ &= \lim_{k\to\infty}\nu(\{x : a_k \succsim x \succsim b_k\}) \geq \alpha, \end{aligned}$$

a final contradiction.

We shall give an example to show that the above classical result is essentially nonconstructive. First, we define a ***covering*** of [0,1] to be a sequence $([a_n,b_n])$ of nonoverlapping proper compact subintervals of [0,1] with the following property: for each x in (0,1) there exist m,n such that $b_m = a_n$ and $a_m < x < b_n$. Such a covering is said to be ***regular*** if $\Sigma_{n=1}^{\infty}(b_n - a_n) = 1$; and ***$\epsilon$–singular***, where $0 < \epsilon < 1$, if $\Sigma_{n=1}^{m}(b_n - a_n) < \epsilon$ for each m.

Recursive example 1: *A continuous preference relation on* X ≡ [0,1], *represented by a pointwise continuous utility function u, such that conclusions* (i) *and* (ii) *of Theorem 2 obtain, the indifference class of x has Lebesgue measure* 0 *for all x in* X, *and condition* (iii) *of Theorem* 2 *fails to hold.*

Assuming Church's thesis, let (I_n) be a regular covering of X [15, Theorems 4.2 and 4.3]. For each n let u_n be the uniformly continuous function on X that vanishes on the metric complement of I_n, takes the value 1 at the midpoint of I_n, and is linear on each half of I_n. Then the function $u \equiv \Sigma_{n=1}^{\infty} u_n$ is defined and pointwise (but not uniformly) continuous on X. Clearly, u represents a continuous preference relation $\succ$ on X.

Given $\epsilon > 0$, choose y in X such that $0 < u(y) < \epsilon$. For each x in X we have either $u(x) > 0$ or $u(x) < u(y)$. In the first case, there exists m such that $a_m < x < b_m$; a simple argument using similar triangles now shows that $C(x)$ and $C^+(x)$ are Lebesgue integrable, and that

$$\nu(C(x)) = \nu(C^+(x)) = \Sigma_{n=1}^{\infty}(1 - u(x))|I_n| = 1 - u(x).$$

In the second case, $C(y) \subset C^+(x) \subset C(x) \subset X$, $C(y)$ is Lebesgue integrable (by the case just considered), and $\nu(X) - \nu(C(y)) < \epsilon$. So, in either case, there exist Lebesgue integrable sets P, Q such that $P \subset C^+(x) \subset C(x) \subset Q$ and $\nu(P) - \nu(Q) < \epsilon$. Hence $C^+(x)$ and $C(x)$ are Lebesgue measurable, and $\nu(C^+(x)) = \nu(C(x))$ (cf. [7, Lemma 8]). Moreover, if $u(x) > 0$, then $0 \leq |u(x) - (1 - \nu(C(x)))| < \epsilon$; while if $0 \leq u(x) < u(y) <$

ϵ, then

$$0 \leq 1 - \nu(C(x)) < 1 - \nu(C(y)) < \epsilon,$$

and so $0 \leq |u(x) - (1-\nu(C(x)))| < \epsilon$. Since ϵ is arbitrary, we have $u(x) = \nu(C(x))$. This completes the proof that conditions (i) and (ii) of Theorem 2 obtain.

It follows immediately that for each x in X, the indifference class $\{y \in X : x \sim y\}$ has Lebesgue measure 0. Suppose there exists $\delta > 0$ such that if $a,b \in X$ and $\|a - b\| < \delta$, then $\nu(\{x \in X : a \succeq x \succeq b\}) < 1/2$. Choose m such that $b_m - a_m < \delta$, where $I_m = [a_m, b_m]$, and let $c_m \equiv \frac{1}{2}(a_m + b_m)$. Then $\|a_m - c_m\| < \delta$ and so

$$1/2 > \nu(\{x : c_m \succeq x \succeq a_m\}) = \nu(\{x : 1 \geq u(x) \geq 0\}) = \nu(X) = 1,$$

a contradiction. □

In the above example, the preference relation $\succ$ is not uniformly continuous on compact sets: this follows from the fact that the corresponding utility function u is not uniformly continuous on [0,1] (cf. [5, Example 1]). Also, the utility function u is not Lebesgue measurable [8].)

The last part of the proof of Theorem 2 shows that if, under the hypotheses of Theorem 1, conclusions (i) and (ii) of Theorem 2 hold, then the utility function $x \to 1-\nu(C^+(x))$ on X (which is continuous – that is, uniformly continuous on bounded sets – by Theorem 1) represents a preference relation that satisfies condition (iii) of that theorem. The following example shows that the existence of a uniformly continuous utility function for a preference relation $\succ$ on X is not enough to ensure condition (iii) of Theorem 2.

Recursive example 2: *A preference relation on $X \equiv [0,1]$ that is represented by a uniformly continuous utility function (and is therefore uniformly continuous on compact sets), such that for each x in X condition* (i) *of Theorem* 2 *holds and the indifference class of x has Lebesgue measure* 0, *and such that for each positive integer n there exist $a,b \in X$ with $a \succ b$, $\|a - b\| < 1/n$, and $\nu(\{x \in X : a \succeq x \succeq b\}) > 1/2$.*

Assuming Church's thesis, let (I_n) be a 1/2–singular covering of X [2, p. 69, 6.1]. For each n let u_n be the uniformly continuous function on X that vanishes on the metric complement of I_n, takes the value 2^{-n} at the midpoint of I_n, and is linear on each half of I_n. Then the function $u \equiv \Sigma_{n=1}^{\infty} u_n$ is defined and uniformly continuous on X, and so represents a preference relation $\succ$ that is uniformly continuous on compact sets. We leave the reader to prove that for each x in X, $C(x)$ and $C^+(x)$ are Lebesgue measurable and have the same measure, and hence that the indifference class of measure 0.

Given a positive integer n, choose k such that $|I_k| < 1/n$. Let a be the midpoint, and b an endpoint, of I_k ; then $a \succ b$. Choose a positive integer j such that $2^{-j-1} < u(a)$. Then $C^+(a) \subset \cup_{n=1}^{j} I_n$; whence

$$\nu(\{x \in X : a \succeq x \succeq b\}) = 1 - \nu(C^+(a)) \geq 1 - \Sigma_{n=1}^{j} |I_n| > 1/2. \quad \square$$

There are several interesting problems relating to the above work and remaining to be discussed. For example, if $P(X)$ denotes the set of preference relations on X that satisfy the hypotheses of Theorem 1, and for all $\succ$ in $P(X)$ and x in X, $C^+(\succ,x)$ is the strict upper contour set of x relative to $\succ$, then there is the question of the continuity, with respect to a suitable topology on the Cartesian product $P(X) \times X$, of the map that carries the pair $(\succ,x)$ to $1 - \nu(C^+(\succ,x))$ (cf. Section 5 of [5]).

Acknowledgement: I am grateful to the referee who suggested both Recursive example 2 and an improvement in my original version of Recursive example 1.

REFERENCES

[1] K.J. Arrow and F.H. Hahn, *General Competitive Analysis*, Oliver and Boyd, Edinburgh, 1971.

[2] M.J. Beeson, *Foundations of Constructive Mathematics*, Ergebnisse der Math. und Ihrer Grenzgebiete, Folge 3, Bd. 6, Springer Verlag, Berlin.

[3] Errett Bishop and Douglas Bridges, *Constructive Analysis*, Grundlehren der math. Wissenschaften, Bd. 279, Springer Verlag, Berlin 1985.

[4] D.S. Bridges, *Preference and utility, a constructive development*, J. Math. Econ. 9 (1982), 165–185.

[5] D.S. Bridges, *The constructive theory of preference relations on a locally compact space*, to appear.

[6] D.S. Bridges, *A general constructive intermediate value theorem*, to appear.

[7] D.S. Bridges, *Locatedness, convexity, and Lebesgue measurability*, to appear in Quarterly J. Math.

[8] D.S. Bridges and O. Demuth, *Lebesgue measurability in constructive analysis*, to appear.

[9] G. Debreu, *Theory of Value*, John Wiley, New York, 1959.

[10] G. Debreu, *Continuity properties of Paretian utility*, Int. Econ. Review 5 (1964), 285–293.

[11] G. Mehta, *Recent developments in utility theory*, Indian Econ. J. 30 (1983), 103–124.

[12] W. Neuefeind, *On continuous utility*, J. Econ. Theory 5 (1972), 174–176.

[13] M.K. Richter, *Continuous and semi-continuous utility*, Int. Econ. Review 21 (1980), 293–299.

[14] D. Sondermann, *Utility representations for partial orders*, J. Econ. Theory 23 (1980), 183–188.

[15] I.D. Zaslavskii, *Some properties of constructive real numbers and constructive functions*, Trudy Math. Inst. Steklov 67 (1962), 385–457; English translation in AMS Transl., Series 2, 57 (1966), 1–84.

1-GENERIC ENUMERATION DEGREES BELOW $\mathbf{0}'_e$

C. S. COPESTAKE

Department of Pure Mathematics
University of Leeds
Leeds, England

1. INTRODUCTION

Enumeration reducibility is the formalisation of the natural concept of relative enumerability between sets of natural numbers. A set A is said to be enumeration reducible to a set B iff there is some effective procedure which gives an enumeration of A from any enumeration of B. This can be shown to be equivalent to the following definition:

Definition 1.1 A set of natural numbers A is *enumeration reducible* (e-reducible, $\leq_e$) to a set of natural numbers B iff there is an i such that for all x

$$x \in A \Leftrightarrow \exists z[\langle x, z\rangle \in W_i \ \& \ D_z \subset B],$$

where W_i and D_z are, respectively, the ith recursively enumerable set and the zth finite set in appropriate standard listings of such sets.

Each i in the above definition corresponds to an *enumeration operator* Ψ_i which is a total mapping from 2^ω to 2^ω. We write $\Psi_i(B) = A$ iff $A \leq_e B$ via i.

Enumeration reducibility gives rise to a reducibility between partial functions, which agrees with Turing reducibility on the total functions, by identifying a partial function with its graph. The *enumeration degrees* are the equivalence classes of sets generated by e-reducibility and the *partial degrees*, which can be identified with the e-degrees, are the corresponding equivalence classes between functions.

The structure of the enumeration degrees has been studied by MEDVEDEV (1955), CASE (1971), GUTTERIDGE (see COOPER, 1982), COOPER (1984; with MCEVOY, 1985; 1987), MCEVOY (1985) and COPESTAKE (1988; with COOPER, 1988), amongst others.

Clearly, from the definition, the bottom e-degree $\mathbf{0}_e$ is the degree of the r.e. sets. GUTTERIDGE (see COOPER, 1982) showed that there is no minimal e-degree, MEDVEDEV (1955), however, showed the existence of quasi-minimal degrees, defined as follows:

Definition 1.2 An e-degree is *total* if it contains a total function. An e-degree is *quasi-minimal* if it is not total and contains no total predecessors other than $\mathbf{0}_e$.

A related notion is that of minimal-like degree:

Definition 1.3 An e-degree $\mathbf{a}$ is *minimal-like* if all functions with degree $\mathbf{b}$ such that $\mathbf{b} <_e \mathbf{a}$ have partial recursive extensions and $\mathbf{a}$ is not $\mathbf{0}_e$. We say an e-degree is *strongly minimal-like* if it is both minimal-like and non-total.

COOPER (1984) defined a jump operator on the e-degrees that agrees with the Turing degree jump on total degrees:

Mathematical Logic
Edited by P. P. Petkov
Plenum Press, New York, 1990

Definition 1.4 Define $J(A) = \chi_{\{x \mid x \in \Psi_x(A)\}}$. Then if $A \in \mathbf{a}$, $\mathbf{a}' = deg_e J(A)$. We call $\mathbf{a}'$ the *jump* of $\mathbf{a}$. The nth *jump* of $\mathbf{a}$ is $\mathbf{a}^{(n)} = (\mathbf{a}^{(n-1)})'$.

$0'_e$ is the jump of 0_e and is the enumeration degree of $\bar{K}$. The degrees below $0'_e$ are the degrees of the $\sum_2$ sets, which are the subject of most of the literature. As in the Turing degrees, a degree is said to be *low* if its jump is $0'_e$. McEvoy (1985) has characterised the low degrees as being those containing only Δ_2 sets.

Definition 1.5 An e-degree is *properly-*Σ_2 if it contains a Σ_2 set but does not contain any Δ_2 sets.

Cooper and Copestake (1988) showed the existence of properly-Σ_2 degrees.

The concept of genericity was first defined by Cohen (1963). Since then the structure of the Turing degrees of generic and particularly 1-generic sets has been studied extensively by, amongst others, Jockusch, Chong, Haught and Downey. In this work the string formulation of 1-generic set (see definition below) due to Posner (1977), has proved the most useful.

Case (1971) considered genericity in the e-degrees as did Moore (1974) and the e-degrees of 1-generic sets and functions have been studied by Copestake (1988). A brief résumé of the main definitions and results follows:

Definition 1.6 $\mathbb{N}^* = \mathbb{N} \cup \{\omega\}$, where the intended interpretation of "$\varphi(m) = \omega$" is "φ is undefined on argument m". $(\mathbb{N}^*)^{\mathbb{N}}$ is then the set of partial functions.

Definition 1.7 A *2-valued string* (α, β, γ etc.) is a mapping from a finite initial segment of $\mathbb{N}$ into $\{0,1\}$. An $\mathbb{N}^*$*-valued string* (τ, ρ, σ etc.) is a mapping from a finite initial segment of $\mathbb{N}$ into $\mathbb{N}^*$. The *length* of string τ is $lh(\tau) = \mu x[\tau(x) \uparrow]$. We say τ *strongly extends* $\sigma(\tau \tilde{\supset} \sigma)$ if $\forall x < lh(\sigma)[\tau(x) = \sigma(x)]$. We say τ *extends* σ $(\tau \supset \sigma)$ if $\forall x < lh(\sigma)[\tau(x) = \sigma(x)$ or $\sigma(x) = \omega]$. σ is a *beginning* of φ if $\sigma \tilde{\subset} \varphi$ and $\varphi \upharpoonright x$ is the beginning of φ with length x , similarly, α is a beginning of A means $\alpha \subset \chi_A$ and $A \upharpoonright x$ is the beginning of A with length x.

Definition 1.8 A set A of natural numbers is *1-generic* if for every r.e. set S of 2-valued strings there is some beginning α of A such that either $\alpha \in S$ or $\forall \beta \supset \alpha, \beta \notin S$.

Definition 1.9 A partial function φ is *1-generic* if for every r.e. set S of $\mathbb{N}^*$-valued strings there is some beginning σ of φ such that either $\sigma \in S$ or $\forall \tau \tilde{\supset} \sigma, \tau \notin S$.

Case (1971) showed that generic functions have quasi-minimal degree and Copestake (1988) extended this result to 1-generic sets. The degrees of 1-generic sets are, in fact, strongly minimal-like, but there are no minimal-like 1-generic function degrees (Copestake, 1988). The 1-generic sets are actually the domains of the 1-generic functions and thus all 1-generic function degrees have 1-generic set degree predecessors.

In this paper we consider the structure within the 1-generic degrees and construct a non-low 1-generic set degree below $0'_e$, contrasting with the situation in the Turing degrees, where no such degree exists. In fact, we prove a stronger result by making the degree properly-Σ_2.

2. TERSE SETS AND s-DEGREES WITHIN e-DEGREES

In (1987) Beigel, Gasarch and Owings define a terse set A as one with the property that for every n, the values of the characteristic function of A on some n arguments cannot be computed from the values of the characteristic function on any $(n+1)$ arguments. We give here a similar definition for the e-degrees:

Definition 2.1 A set of natural numbers A is said to be *terse* with respect to e-reducibility, if, for all i and n, if $\langle A \rangle^n = \Psi_i^A$ then there exists $x \in \Psi_i^A$ such that $\forall D \subset A(x \in \Psi_i^D \Rightarrow \mid D \mid \geq n)$, where $\langle A \rangle^n = \{\langle y_1, \ldots y_i, \ldots y_n \rangle \mid \forall i \leq n, y_i \in A\}$.

Beigel, Gasarch and Owings state that 1-generic sets are terse, with respect to Turing reducibility. We show the equivalent result for e-reducibility. First we need a definition:

Definition 2.2 α^+ is the finite set encoded by α, i.e. $\alpha^+ = \{x \mid \alpha(x) = 1\}$. $\bar{\alpha}^+$ is its finite complement, i.e. $\bar{\alpha}^+ = \{x \mid \alpha(x) = 0\}$.

Propositon 2.3 If A is 1-generic then A is terse.

Proof Suppose A is 1-generic and A is not terse. Then $\exists i \exists n[\langle A\rangle^n = \Psi_i^A$ & $(x \in \Psi_i^A \Rightarrow \exists D \subset A, x \in \Psi_i^D$ & $|D| < n)]$. Define an r.e. set S of 2-valued strings by: $S = \{\alpha \mid \exists x \exists D \exists s(x \in \Psi_{i,s}(D)$ & $D \subset \alpha^+$ & $x = \langle y_1, ... y_n\rangle$ & $\exists i \leq n(y_i < lh(\alpha)$ & $y_i \notin \alpha^+))\}$. There cannot be a beginning of A in S, so by the genericity of A $\exists \alpha \subset \chi_A \forall \beta \supset \alpha, \beta \notin S$. Now since Ψ_i^A must be infinite, as A is 1-generic (see COPESTAKE, 1988), we can pick $x \in \Psi_i^A$ such that $x = \langle y_1, ... y_n\rangle$ and for all $i \leq n$, $y_i > lh(\alpha)$. Then by the assumption, $x \in \Psi_i^D$, where $|D| < n$ and $D \subset A$. Pick $\beta \supset \alpha$ such that $D \subset \beta^+$. Then as D has at most $n-1$ elements there must be some y_i, $i \leq n$, such that $y_i \notin D$. Define $\beta(y_i) = 0$. Then $\beta \in S$, which is a contradiction. So all 1-generic sets are terse. □

A similar proof shows that the graphs of 1-generic functions are also terse. These results have interesting consequences in relation to the *s-degrees*, which are the degrees formed by considering a restriction of e-reducibility in which singletons take the place of the usual finite sets:

Definition 2.4 $A \leq_s B$ (A is *s-reducible* to B) if there is some i such that for all x:

$$x \in A \Leftrightarrow \exists u[\langle x, u\rangle \in W_i \ \& \ u \in B].$$

Singleton reducibility was first defined by FRIEDBERG and ROGERS in (1959). The s-degrees are the equivalence classes formed under s-reducibility.

Corollary2.3 Every e-degree containing a 1-generic function or set contains an infinite ascending sequence of s-degrees.

Proof If A is terse then $\langle A\rangle^{n+1} \not\leq_s \langle A\rangle^n$. Since $\langle A\rangle^n \leq_s \langle A\rangle^{n+1}$ we have:

$$A = \langle A\rangle^1 <_s \langle A\rangle^2 <_s ... \langle A\rangle^n <_s ... <_s \cup_{n>0}\langle A\rangle^n.$$

For every n, $A \equiv_e \langle A\rangle^n$. The result then follows. □

3. 1-GENERIC SETS AND LOWNESS

The Turing degrees of 1-generic sets below $0'$ are low. In the e-degrees this gives us:

Proposition 3.1 If A is 1-generic then A is Δ_2 iff A is low.

Proof All low sets are Δ_2 (see MCEVOY, 1985), so it remains to show that Δ_2 1-generic sets are low. As A is Δ_2 we have $A \leq_T 0'$. Then, as it is known that such 1-generic sets are low in the Turing degrees (see, for example, LERMAN, 1983), we have that A is low in the e-degrees. □

This shows us that no 1-generic set is low in the e-degrees without being low in the Turing degrees. It indicates that a 1-generic set in degree $\mathbf{a}$ with $\mathbf{a} \leq_e 0'_e$, is, in a sense, the worst behaved member of $\mathbf{a}$.

On the other hand, there are 1-generic sets below $0'_e$ that are not low:

Theorem 3.2 There is a properly-Σ_2 1-generic set degree.

Proof We construct a Σ_2 set A by means of a Σ_2 approximation $\{A^s\}_{s\geq 0}$ where we define

$$x \in A \Leftrightarrow \exists t \forall s > t(x \in A^s)$$

We use a standard listing $\{B_i\}_{i\geq 0}$ of the Σ_2 sets in which the Δ_2 sets appear with Δ_2 approximations $\{B_i^s\}$. The requirements to be satisfied are:

$$S_i : \exists \alpha \subset \chi_A, \alpha \in W_i \text{ or } \exists \alpha \subset \chi_A, \forall \beta \supset \alpha, \beta \notin W_i;$$

$$R_i : A = \Psi_i(B_i) \ \& \ B_i = \Theta_i(A) \Rightarrow \exists x \lim_s B_i^s(x) \uparrow;$$

where Ψ_i and Θ_i are enumeration operators.

We adopt the convention of using quotation marks to notate the first use of an expression that has not previously been defined and *italics* when we actually define the expression.

We satisfy the R_i requirements by looking for a 'follower' (x, D, E) such that $x \in \Psi_i(D)$ and $D \subset \Theta_i(E)$. If some such triple is found we fix $E - \{x\}$ in A and also put x in A. We then wait till $D \subset B_i^s$. If this happens we extract x from A and insert it again only if $D \not\subset B_i^t$ at a later stage t. We repeat the extraction and insertion of x so that some element of D is forced in and out of B_i, thus preventing it from being Δ_2.

We satisfy S_i by looking for a string in $W_{i,s}$ that is compatible with action taken on higher priority requirements, and making it a beginning of A.

For elegance of argument we combine the requirements using a tree construction. The tree will be an ordering of strings which will encode the information needed about outcomes on higher priority requirements. At the S_i nodes we will have branchings to represent the two possibilities; either there is a string in W_i that is compatible with the information on the tree, or, there is no such string. At the R_i nodes we will also have two branchings for the two outcomes: there is a follower (x, D, E) for the ith requirement such that $x \notin A$ or there is no such follower. The strings that comprise the tree will be denoted by σ, ρ, τ etc., despite the fact they are 2-valued, to distinguish them from the strings α, β, γ etc that make up the r.e. sets W_i and which will be considered as potential beginnings of A.

Definition 1 We define $y = \mu x(\sigma(x) \neq \tau(x))$ if τ and σ are *incompatible* that is, if some such y exists. We write $\tau \subset \sigma$ (or $\sigma \supset \tau$) for compatible strings σ and τ. An *ordering* is then defined on the strings σ, τ etc. by:

$\sigma \underline{<<} \tau \Leftrightarrow \sigma \subset \tau$ or $\sigma(y) < \tau(y)$ and σ and τ are incompatible.

We write $\sigma << \tau$ if $\sigma \underline{<<} \tau$ and $\sigma \neq \tau$. We say σ is *to the left* of τ (τ is *to the right* of σ) if $\sigma << \tau$ and $\sigma \not\subset \tau$.

The 'true path' will be defined during the construction, but we note here that the empty string $\emptyset$ will be considered to be on the true path at all stages.

Similarly, the notion of a string α being σ-'compatible' for some string σ on the tree will be defined during the construction, but all strings α will be considered $\emptyset$-compatible.

The Construction

Stage 0 $A_0 = \emptyset$.

Stage s+1 Assume we have 'taken action' on all τ such that τ is on the true path at stage $s+1$ with $lh(\tau) \leq n$, say such that τ 'requires attention' at stage $s+1$.

We say σ such that $lh(\sigma) = n+1$ is on the *true path* at stage $s+1$ if $\sigma \upharpoonright (n-1)$ $(=\sigma^-$, say) is on the true path at stage $s+1$ and $n < s$ and:

I. $n = 2i$ and

(i) $\sigma(n) = 0$ and there is a σ^--'compatible' α at stage $s+1$, $\alpha \in W_{i,s}$ or

(ii). $\sigma(n) = 1$ and otherwise, or

II. $n = 2i+1$ and

(i) $\sigma(n) = 0$ and there is a 'follower' (x, D, E) of σ^- such that $D \subset B_i^s$ or

(ii) $\sigma(n) = 1$ and otherwise.

We call a stage such that σ is on the true path a σ-*true* stage.

If σ is on the true path through clauses I(i), II(i) or II(ii) of the definition of true path at stage $s+1$ we say that σ *requires attention* at stage $s+1$.

If σ requires attention we *take action* on it according as which clause applies:

I(i): Pick least such α, call it σ's *following string* at stage $s+1$. σ-*fix* all $x \in \alpha^+$ in A. σ-*extract* all $x \in \bar{\alpha}^+$ from A. *Cancel* followers of all strings τ to the right of σ. (That is, such a string no longer has a follower at any stage $t+1 > s+1$ until a follower is assigned to it). If σ's following string at stage $s+1$ is different from its following string at the last σ^--true stage, *cancel* all followers of strings $\tau >> \sigma$.

II(i): σ-*extract* x from A, provided it is not τ-fixed, $\tau << \sigma$. *Cancel* followers of all τ to the right of σ.

II(ii): There are three cases:

(a) There is already a follower (x, D, E) for σ^- at stage $s+1$. In this case we σ-*fix* x in A at stage $s+1$, provided x is not τ-extracted from A at stage $s+1$ by any $\tau \subset \sigma$. Otherwise, cancel σ^-'s follower.

(b) There exists a triple (x, D, E) such that $x \in \Psi_i^D[s]$ and $D \subset \Theta_i^E[s]$ and x is neither τ-fixed in A, nor τ's 'follower element', and neither x nor $E - \{x\}$ is τ-extracted from A at stage $s+1$ for any $\tau \underline{<<} \sigma^-$. In this case we appoint one such (x, D, E) as *follower* for σ^-. We call x σ^-'s *follower element*. We *permanently* σ^--*fix* all members of $E - \{x\}$ in A at stage $s+1$, and we σ-*fix* x in A at stage $s+1$. *Cancel* followers of all strings $\tau >> \sigma^-$.

(c) Otherwise. Pick the least x such that x is neither τ-fixed in A, nor τ's follower element, nor τ-extracted from A at stage $s+1$ for any $\tau \underline{<<} \sigma^-$. σ-*fix* x in A at stage $s+1$.

We say α is σ-*compatible* at stage $s+1$ if $\alpha^+ \cap E_\sigma(s) = \emptyset$ and $\bar{\alpha}^+ \cap F_\sigma(s) = \emptyset$, where $E_\sigma(s) = \{x \mid x \text{ is } \tau\text{-extracted at stage } s+1,\ \tau << \sigma\}$, and $F_\sigma(s) = \{x \mid x \text{ is } \tau\text{-fixed at stage } s+1,\ \tau << \sigma\}$.

If a number is σ-extracted at some stage it remains so until τ-fixed by some τ, or cancelled as follower for σ^-. Likewise, a number remains σ-fixed until extracted or cancelled.

We define:

$$A_{s+1}(x) = \begin{cases} 1 & \text{if } x \text{ is } \sigma\text{-fixed, some } \sigma, \text{ at the end of stage } s+1 \\ 0 & \text{otherwise.} \end{cases}$$

This ends stage $s+1$.

We define

$$x \in A \Leftrightarrow \exists t \forall s > t (x \in A_s).$$

This completes the construction.

We first note that follower elements for different strings are distinct (since when a follower element is chosen for σ^- it avoids all follower elements of strings $\tau \underline{<<} \sigma^-$ and cancels all followers of strings $\tau >> \sigma^-$). Using this fact it is clear from the construction that no element is both extracted and fixed actually during the same stage of the construction. That is, A does contain the elements we intended in the definition.

Definition 2 We say σ is *persistent* iff σ is on the true path infinitely often and each string to the left of σ is on the true path at most finitely often.

Definition 3 We define the *leftmost path* L, by: $L = \{\sigma \mid \sigma \text{ is persistent }\}$.

Lemma 1 All requirements are satisfied.

Proof Consider σ, $lh(\sigma) = n+1, \sigma \in L$. As $\sigma \in L$ we can assume that after some stage no τ to the left of σ is ever again on the true path. We also inductively assume that for each $\tau \subsetneq \sigma$: (i) the set of numbers τ-fixed in A is the same at every stage after some stage $w+1$, (ii) the set of numbers τ-extracted from A is the same at all τ-true stages after some stage $w+1$, (iii) τ has at most finitely many followers ever assigned to it, (iv) there are at most finitely many changes in the following string assigned to τ.

Choose a stage $w+1$ large enough to both satisfy assumptions (i) and (ii) and such that no $\tau \subsetneq \sigma$ has any subsequent changes to its following string or follower.

Sublemma 1 If $lh(\sigma) = 2i+1$ then:

(a) $\exists t \forall s > t$ such that $s+1$ is a σ-true stage, σ's following string at stage $s+1 = \sigma$'s following string at stage $t+1$.

(b) The follower of σ does not change infinitely often.

(c) If $A = \Psi_i(B_i)$ and $B_i = \Theta_i(A)$ there is no stage after which we can never find a follower for σ.

Proof Assume stage $w+1$ as described above has been reached.

(a) Since every σ^--true stage is a τ-true stage for every $\tau \subset \sigma^-$, after some stage the set of numbers τ-extracted from A with $\tau << \sigma$ is the same at every σ^--true stage and the set of numbers τ-fixed in A with $\tau << \sigma$ is the same at every, (and in particular every σ-true) stage, by the induction assumption. Since $\tau << \sigma^- \Rightarrow \tau << \sigma$ this implies that $\exists t \forall s > t$, $s+1$ a σ-true stage, $E_{\sigma^-}(s) = E_{\sigma^-}(t)$ and $F_{\sigma^-}(s) = F_{\sigma^-}(t)$. Thus the set of σ^--compatible strings is the same at every σ-true stage after $t+1$. Eventually the least such string in W_i is enumerated in $W_{i,s}$, and this string will thereafter be chosen as σ's following string at every subsequent σ-true stage.

(b) There are four possible ways in which (x, D, E) could be cancelled as σ's follower, corresponding to the clauses I(i), II(i) and II(ii)(a) and (b). However, since no τ to the left of σ is ever on the true path again, only $\tau \subset \sigma$ can cause such a cancelling. If $\tau \subsetneq \sigma$, by the induction assumption σ's follower cannot be cancelled by it. If $\tau = \sigma$ then the numbers τ-extracted (via clause II(ii)(a)) are the negative elements of σ's following string. However, by (a) above, this will be unchanged after some stage and so can only cause at most finitely many cancellings of σ's follower. Thus after some stage σ's follower is never changed.

(c) Suppose at some stage we never subsequently have a follower for σ. Then, as σ is on the true path infinitely often, we must infinitely often get II(ii)(c). By the induction assumption, after some stage we always pick the same number x, say, to σ-fix in A via clause II(ii)(c). So x is fixed in A at infinitely many stages. x can only be extracted infinitely often by $\tau \subset \sigma$, since τ to the left of σ are only on the true path finitely often. But this contradicts our choice of x, and so $x \in A$.

Assume now that $A = \Psi_i(B_i)$ and $B_i = \Theta_i(A)$. Then $x \in \Psi_i(B_i)$. That is, there exist D and E such that $x \in \Psi_i(D), D \subset \Theta_i(E), D \subset B_i$ and $E \subset A$. However, we know that this cannot give us a follower for σ. So since x was chosen by the same criteria as a follower element it must be that some $y \in E - \{x\}$ is τ-extracted from A at infinitely many stages, contradicting the fact that $E \subset A$.

The sublemma is thus proved.

We now assume that $lh(\sigma) = 2i+1$, and consider requirement S_i. There are two cases:

Case (1) $\sigma(n) = 0$, where $n = 2i$. Then at every σ-true stage after which σ's following string α no longer changes we σ-fix all $x \in \alpha^+$ in A and σ-extract all $x \in \bar{\alpha}^+$ from A. As there are infinitely many σ-true stages we have $\bar{\alpha}^+ \cap A = \emptyset$. On the other hand, it may be that some $x \in \alpha^+$ is τ-extracted from A at infinitely many stages. This could happen via clauses I(i) and II(i). However, τ to the left of σ are only on the true path finitely often, so the only problem is with τ-extractions for $\tau \subset \sigma$. Since, by the induction assumption, such τ-extractions are the same at all τ-true, and, in particular, all σ-true stages, after some stage α would no longer be chosen as σ's following string, contradicting the choice of α.

Thus every x in α^+ is in A, so α is a beginning of A. That is, there is a beginning of A in W_i. So S_i is satisfied through its first disjunct.

Case (2) $\sigma(n) = 1$. This means there must be infinitely many σ^--true stages and at each of them, after some stage, there is no σ^--compatible $\alpha \in W_{i,s}$. Let $t+1$ be the stage after which the set of numbers τ-extracted from A with $\tau \underline{<<} \sigma^-$ is the same at every σ^--true stage and the set of numbers τ-fixed in A with $\tau \underline{<<} \sigma$ is the same at every stage. Let $E_{\sigma^-} = E_{\sigma^-}(t)$ and $F_{\sigma^-} = F_{\sigma^-}(t)$. Define $M = E_{\sigma^-} \cup F_{\sigma^-}$. The set of numbers τ-fixed in A for $\tau << \sigma$ at some stage $w+1$ is τ-fixed in A at all subsequent stages, by the induction assumption, so $F_{\sigma^-} \subset A$. Similarly, $E_{\sigma^-} \subset \bar{A}$.

Consider $\chi_A \upharpoonright maxM$. Suppose there was a string $\alpha \supset \chi_A \upharpoonright maxM$, $\alpha \in W_{i,s}$. Then

$$\alpha(x) = \begin{cases} 0 & \text{if} x \in E_{\sigma-} \\ 1 & \text{if } x \in F_{\sigma-} \end{cases}$$

so α is σ^--compatible at all σ^--true stages after some stage. This is a contradiction, so S_i is satisfied through its second disjunct.

Thus the construction satisfies the S_i requirements. We now consider the R_i requirements. We assume that $A = \Psi_i(B)$ & $B_i = \Theta_i(A)$ and show that in this case B_i cannot be Δ_2. By the sublemma, we can assume we have settled on a final follower for σ^-, where $lh(\sigma^-) = 2i + 2$. We again consider two cases:

Case (1) $\sigma(n) = 0$, where $n = 2i$. Consider σ^-'s final follower, (x, D, E). At each σ-true stage we σ-extract x from A unless x is τ-fixed in A by some $\tau << \sigma$. However, since x is picked to avoid such possibilities, this cannot happen unless x has been subsequently τ-fixed; but this would cause the cancellation of x as follower. Thus x is extracted from A at infinitely many stages, so $x \notin A$.

We have infinitely many σ-true stages and $\sigma(n) = 0$, so when a stage is reached after which σ^-'s follower no longer changes we have $D \subset B_i^{s+1}$ at each subsequent stage $s + 1$. So if B_i is Δ_2 we have $D \subset B_i$. But then since $x \in \Psi_i(D)$ we have $x \in A$. This is a contradiction, so B_i is not Δ_2.

Case (2) $\sigma(n) = 1$. When (x, D, E) was appointed as follower for σ^- all members of $E - \{x\}$ were fixed in A. So $E - \{x\} \subset A$, unless some element of it is later extracted. The only strings τ that could do this are $\tau \subset \sigma^-$, since any extraction by a string to the left of σ^- would cause cancellation of its follower. But in this case E was chosen to avoid the follower element and negative members of the following string of such a τ and, since any change to such followers or following strings would cause the cancellation of σ's follower, no number in $E - \{x\}$ can be extracted. Thus $E - \{x\} \subset A$.

Since $\sigma(n) = 1$ and σ is on the true path infinitely often we σ-fix x in A at infinitely many stages. So $x \in A$ if, after some stage, we no longer τ-extract x from A for any τ. The strings to the left of σ can only extract finitely often, so the only strings that could do this are those $\tau \subset \sigma$. The two clauses I(i) and II(i) could both, potentially, cause such extractions. Consider first clause I(i). x was chosen to avoid any elements τ-extracted by $\tau \subset \sigma$ and any change to τ's following string would cause cancellation of σ's follower, so x cannot be extracted in this way. Consider secondly clause II(ii). As follower elements for different strings are distinct the only string that can extract x infinitely often via this clause is $\sigma^{-\frown}0$. But this would contradict the fact that σ is on the leftmost path. Hence $x \in A$.

Taking the above two facts together gives us $E \subset A$. But then, since we are assuming $A = \Psi_i(B_i)$ and $B_i = \Theta_i(A)$ we have $D \subset B_i$. Since B_i is Σ_2 this means that for all stages t after some stage s, $D \subset B_i^t$. So at each such stage t we have $\tau(2i + 2) = 0$, if τ is on the true path. This contradicts the fact that $\sigma(2i + 2) = 1$, and σ lies on the true path infinitely often.

The construction therefore satisfies all the requirements.

Checking of the Induction Assumption. By the sublemma, σ's follower and following string do not change after some stage. The numbers possibly fixed in by σ are σ's follower and the positive elements of its following string. These are the same after some stage. They cannot be injured by τ-extractions for any τ to the left of σ as such τ are only acted upon finitely often. If $\tau \subsetneq \sigma$ every τ-true stage is also a σ-true stage, and by the induction assumption the τ-extractions are in force at every τ-true stage, and so are in force when σ's final follower and following string are picked. The numbers possibly extracted by σ are σ's follower and the negative elements of its following string; by the sublemma, these are the same after some stage.

Thus the induction holds and the theorem follows. □

The following corollary is immediate:

Corollary 3.2 There is a 1-generic set A with enumeration degree $\mathbf{a}$ such that $\mathbf{a} \leq_e 0'_e$ and $\mathbf{a}$ is not low.

4. FURTHER RESULTS AND QUESTIONS

In conclusion, we mention the following result:

T1. There is an enumeration degree $\mathbf{a} \leq_e 0'_e$ which contains a 1-generic set and bounds no minimal pair of enumeration degrees. (COOPER and COPESTAKE, in preparation).

We have not yet considered whether the main theorem of this paper holds for the e-degrees of 1-generic functions, that is:

Q1. Is there a 1-generic function with enumeration degree below $0'_e$ that is not low?

References

BEIGEL, R., GASARCH, W., OWINGS, J., 1987, Terse sets and verbose sets,**Recursive Function Theory Newsletter**, **36**:367.

CASE, J., 1971, Enumeration reducibility and partial degrees,**Ann. Math.**, **2**:419–439.

CHONG, C.T., Minimal degrees recursive in 1-generic degrees, to appear.

CHONG, C.T., DOWNEY, R.G., On degrees bounding minimal degrees, to appear.

CHONG, C.T., JOCKUSCH, C.G., JR., 1984, Minimal degrees and 1-generic sets below $0'$, **in** "Computation and Proof Theory", R. RICHTER, E. BORGER, W. OBERSCHELP, B. SCHINZEL, W. THOMAS, eds, Springer Lecture Notes in Mathematics **1104**, Berlin, Heidelberg, New York, Tokyo.

COHEN, P.J., 1963, The independence of the continuum hypothesis, I & II,**Proc. Nat. Acad. Sciences, U.S.A.**, **50**:105–150.

COOPER, S.B., 1982, Partial degrees and the density problem,**Jour. Symb. Logic**, **47**:854–859.

COOPER, S.B., 1984, Partial degrees and the density problem, II: the enumeration degrees of Σ_2 sets are dense,**Jour. Symb. Logic**, **49**:503–511.

COOPER, S.B., 1987, Enumeration reducibility using bounded information: counting minimal covers,**Z. Math. Logik und Grund. der Math.**, **33**:537–560.

COOPER, S.B., COPESTAKE, C.S., 1988, Properly-Σ_2 enumeration degrees, **Z. Math. Logik und Grund. der Math.**, to appear.

COPESTAKE, C.S., 1988, 1-genericity in the enumeration degrees,**Jour. Symb. Logic**, **53**.

FRIEDBERG, R., ROGERS, H., JR., 1959, Reducibility and completeness for sets of integers,**Z. Math. Logik und Grund. der Math.**, **5**:117–125.

GUTTERIDGE, L., 1971, Some results on enumeration reducibility, Ph.D. thesis, Simon Frazer University, U.S.A..

HAUGHT, C.A., 1985, Turing and truth table degrees of 1-generic and recursively enumerable sets. Ph.D. thesis, Cornell University, U.S.A..

JOCKUSCH, C.G., JR., 1980, Degrees of generic sets, **in** "Recursion Theory: its Generalisations and Applications", F. R. DRAKE, S. S. WAINER, eds, Cambridge University Press, Cambridge, London, New York, New Rochelle, Melbourne, Sydney.

LERMAN, M., 1983, "Degrees of unsolvability, local and global theory", Perspectives in mathematical logic, Springer-Verlag, Berlin, Heidelberg, New York, Tokyo.

MCEVOY, K., 1985, Jumps of quasi-minimal enumeration degrees, **Jour. Symb. Logic**, **50**:839–848.

MCEVOY, K., COOPER, S.B., 1985, On minimal pairs of enumeration degrees,**Jour. Symb. Logic**, **50**:983–1001.

MEDVEDEV, YU.T., 1955, Degrees of difficulty of the mass problem, **Doklady Academii Nauk SSSR**, **104**:501–504.

MOORE, B.B., 1974, Structure of the degrees of enumeration reducibility, Ph.D. thesis, Syracuse University.

POSNER, D., 1977, High degrees, Ph.D. thesis, University of California, Berkeley, U.S.A.

REMARKS ON DENJOY SETS

O. Demuth
Department of Computer Science, Charles University
Malostranske nam. 25, Prague 1, Czechoslovakia

Denjoy sets are sets of natural numbers corresponding to reals which are interesting from the point of view of the theory of differentiation of constructive real functions. In this paper, some results concerning the structure of T- and tt-degrees containing Denjoy sets of a special type are presented. Methods of recursion theory and those of constructive mathematical analysis are combined in the proofs.

Binary expansions of reals give us a many-to-many correspondence between reals and sets of natural numbers (NNs). For any set A of NNs, we denote by r_A the sum of the series $\sum_{x \in A} 2^{-x-1}$ and, for any real X, we denote by $\mathrm{Set}(X)$ the infinite set B of NNs for which $X - r_B$ is equal to an integer. Using reals to study sets of NNs we can restrict ourselves to reals from the closed unit interval $[0,1]$. A real X is said to be A-recursive if $\mathrm{Set}(X) \leq_T A$ holds. In constructive mathematics in Markov's sense we study, among others, constructive reals (i.e. codes of $\emptyset$-recursive reals) and everywhere (i.e. for any constructive real) defined constructive functions of a real variable (briefly: constructive functions). Let us remember that any constructive function is an algorithm transforming equal constructive reals into equal ones and it is constructively continuous (i.e. continuous with an $\emptyset$-recursive function being a corresponding modulus of continuity) at any constructive real[1]. Constructive functions constant on both $(-\infty, 0]$ and $[1, +\infty)$ we briefly call c-functions. It has turned out that, in constructive mathematical analysis, it is necessary to study also the behaviour of constructive functions in neighbourhoods of reals being non-$\emptyset$-recursive. In this connection, for any constructive function F

Mathematical Logic
Edited by P. P. Petkov
Plenum Press, New York, 1990

and for any real X, we have defined the lower (classical) derivate and the upper (classical) derivate of F at X (and denoted them by $\underline{D}F(X)$ and $\overline{D}F(X)$, respectively) using, on account of the continuity of constructive functions at constructive reals, values of F at rational numbers only.

A constructive function is said to be constructively uniformly (or Ø-uniformly) continuous if there is an Ø-recursive function being a modulus of its uniform continuity. For any constructive function F we denote by $R[F]$ a classical function of a real variable being a maximal (as to domain) continuous (with respect to its domain) extension of F. Let us recall:For any c-function F, (F is classically uniformly continuous) <=> ($R[F]$ is defined at any $\emptyset''$-recursive real) <=> ($R[F]$ is defined at any real) holds.

A set A of NNs is said to be a Denjoy set if there is no constructive function G for which $\underline{D}G(r_A) = +\infty$ holds. As proved by Demuth[2], for any Ø-uniformly continuous c-function F and any Denjoy set A, the Denjoy relations for Dini derivatives - see Thomson[3], p. 166 - hold for $R[F]$ and r_A, in particular,

$$\neg\neg(\underline{D}F(r_A) = -\infty \,\&\, \overline{D}F(r_A) = +\infty \ \vee\ -\infty < \underline{D}F(r_A) = \overline{D}F(r_A) < +\infty)$$

is valid. Obviously, only Denjoy sets can have this property.

Objects of constructive mathematical analysis used in this paper are either words in the alphabet Ξ or objects finitely codable by means of words in Ξ. Thus, for example, natural numbers, integers, rational numbers and constructive reals are words in Ξ and partial recursive functions (equivalently, elements of other mathematical models of intuitive algorithms) and their relativizations are codable by means of natural numbers. Let us fix two Markov algorithms: wd establishing a one-to-one correspondence between NNs and words in Ξ, and en being the algorithm inverse to wd.

The set of all NNs is denoted by N, the symbols s, t, u, v, w, x, y and z are variables for NNs, i and j for integers, a and b for rational numbers (RtNs), σ and τ for strings (i.e. finite sequences of NNs), A, B and C for sets of NNs, X and Y for reals and V for words in Ξ. $A \setminus B$ signifies the difference of the sets A and B and card(A) denotes the cardinality of A. We are supposed to have a fixed numbering of all binary strings (i.e. strings of 0's and 1's) - δ_x denotes the string with number x. We put $d(A) = \{\delta_x : x \in A\}$. For any set S of binary strings S^E denotes the class of all sets of NNs extending strings from S. Let $d^E(A)$ denote $(d(A))^E$.

We use a standard notation for indexing of all partial recursive functions (PRFs), recursively enumerable (r.e.) sets of NNs and those of their relativizations (A-PRFs, A-r.e. sets) - ϕ_y, W_y, W_y^s, ϕ_y^A, W_y^A, $W_y^{A,s}$. We denote by Ev the recursive predicate of variables x, y and z: evaluation of ϕ_x at y finishes during first z steps. Thus, ($\phi_x(y)$ is defined) $\Leftrightarrow$ $\exists z\, Ev(x,y,z)$ and $Ev(x,y,z) \Rightarrow Ev(x,y,z+w)$ hold for any NNs x, y, z and w. $\lambda xy\, \langle x,y\rangle$ denotes a recursive pairing function. $\simeq$ means: both sides are defined and equal, or both are undefined. For any mapping Ψ and for any object P of the corresponding type, $!\Psi(P)$ denotes (Ψ is defined at P).

Sacks introduced[4] Lebesgue measure for classes of sets of NNs. Obviously, for any set S of binary strings, the class S^E is measurable (let $\mu(S^E)$ denote its measure). The concept of $\emptyset$-measurability and its relativization have been introduced in constructive mathematics and recursion theory (cf. Demuth's paper[5]). Here, we remind the way of introducing these concepts for special classes of sets of NNs. A (total) function f is called a <u>modulus of measurability</u> of $d^E(W_y^A)$ if

$$\forall vw (f(v) \le w \Rightarrow |\mu(d^E(W_y^{A,f(v)})) - \mu(d^E(W_y^{A,w}))| \le 2^{-v})$$

holds. The class $d^E(W_y^A)$ is said to be <u>B-measurable</u> if $A \le_T B$ holds and there is a B-recursive function being a modulus of measurability of it. Let us notice that $d^E(W_y^A)$ is necessarily A'-measurable but it can be non-(A-measurable) (cf. Specker's Example). In fact[5], this class is A-measurable if and only if its measure is an A-recursive real. A class M of sets of NNs is said to be <u>of B-measure zero</u> if there are two B-recursive (or, equivalently, recursive) functions g and h of one variable such that, for any NN x, the class $d^E(W^B_{g(x)})$ contains M, its measure is less than 2^{-x} and the function $\phi^B_{h(x)}$ is a modulus of its measurability. Let us remind that, for any binary string σ and any B-measurable class $d^E(W_y^B)$ with measure less than $\mu(\{\sigma\}^E)$, there is a B-recursive set A in the difference $\{\sigma\}^E \setminus d^E(W_y^B)$ (cf. Demuth[5], Theorem 7).

A set S of binary strings is called a <u>covering</u> if S is r.e. and S^E contains all recursive sets. A set A is called

a) <u>semigeneric</u> if it is both non-recursive and contained in S^E for any covering S (see Demuth[6]);

b) an <u>AP-set</u> if there is a recursive function f such that, for any NN x, $d^E(W_{f(x)})$ contains A and its measure is less than 2^{-x};

c) an <u>NAP-set</u>[7,5] if it is not an AP-set.

Semigeneric sets and NAP-sets were studied in several papers[7,6,8,5]. According to Demuth's paper[9], any NAP-set is a Denjoy set. On the other hand, no set contained in a class of sets of $\emptyset$-measure zero (in particular, no recursive set) is a Denjoy set. We remind the following[9,6]: there is a recursive function e such that, for any NN x, $d(W_{e(x)})$ is a covering, $d^E(W_{e(x)})$ contains all AP-sets and its measure is less than 2^{-x}. Thus, (A is an AP-set) <=> $(A \in \bigcap_{x \in N} d^E(W_{e(x)}))$ holds. Consequently, any semigeneric set is an AP-set. In Demuth[10] we have constructed a few Denjoy sets being AP-sets, too. Some of them are semigeneric, others are non-semigeneric.

In the following, we want to study the structure of tt- and T-degrees containing sets being both Denjoy sets and AP-sets.

Remark 1. Let A be an AP-set. Then there is an A-recursive function f such that $A \in d^E(W^{f(x)}_{e(x)})$ holds for any NN x.
For any recursive function g there are recursive functions h_1 and h_2 such that, for any NNs z and v,

$$W_{h_1(z)} = \bigcup_{x=z+1}^{+\infty} W^{g(x)}_{e(x)} \quad \text{and} \quad \bigcup_{x=z+1}^{z+v+1} W^{g(x)}_{e(x)} \subseteq W^{h_2(z,v)}_{h_1(z)} \qquad \text{hold}$$

and, consequently, as $\mu(d^E(W^{g(x)}_{e(x)})) < 2^{-x}$ is valid for any NN x, the recursive function $\lambda y\, h_2(z,y)$ is a modulus of measurability of $d^E(W_{h_1(z)})$ and measure of this class is less than 2^{-z}. Hence, the class $\bigcap_{z \in N} d^E(W_{h_1(z)})$ is of $\emptyset$-measure zero and, as we already know, such class contains no Denjoy set. Let us notice: A is contained in the class unless f majorizes g almost everywhere (i.e. for any sufficiently large NN). Consequently, for any Denjoy set A being an AP-set there is an A-recursive function f majorizing any recursive function almost everywhere (equivalently, $\emptyset'' \le_T A'$ holds, cf. Epstein[11], Chapter XI, Theorem 11).

We have proved the following statement in Remark 1.

Theorem 2. For any Denjoy set A, being an AP-set, $\emptyset'' \le_T A'$ holds.

According to Demuth[8], we can use $\emptyset$-uniformly continuous c-functions for the study of tt-reducibility. In fact, a class of sets of $\emptyset$-measure zero, say Ω, having the following property, was constructed there:
For any sets A and B of NNs,
a) the proposition

$$(A \le_{tt} B \text{ holds}) \tag{1}$$

implies

(there is an $\emptyset$-uniformly continuous c-function F satisfying

$$\mathcal{R}[F](r_B) = r_A\ ), \tag{2}$$

whenever Ω does not contain B ;
b) (2) implies (1), whenever A is not in Ω.

Let us notice that Ω as a class of $\emptyset$-measure zero cannot contain any Denjoy set.

We want to introduce some constructive notions. For any set B of NNs and any NN y, the correspondence λV ($\underline{wd}$($\phi_y^B(\underline{en}(V))$)) will be called the B-algorithm with B-index y . Thus, any constructive function is an $\emptyset$-algorithm with the properties described above.

Let M be a set of words in the alphabet Ξ . A B-algorithm is said to be a B-sequence of

a) elements of M if it is applicable to any NN and transforms it into an element of M ;

b) C-algorithms if it is a B-sequence of C-indices of the corresponding algorithms.

In the following we shall present B-sequences of words (or algorithms) by their"members" using notation $\{\ldots\}_x^B$.

An RtN a is said to be a <u>ternary rational</u> if there are an integer i and an NN y fulfilling $a = i.3^{-y}$. For any NN x and any integer j we denote the closed interval $[(j-1).3^{-x},j.3^{-x}]$ by L[x,j] .

<u>Remark 3</u>. It will be useful to have a strongly oscillating $\emptyset$-uniformly continuous c-function for our study of tt-reducibility. We construct a partial recursive function $\bar{n}$ of two variables and an $\emptyset$-sequence $\{F_x\}_x^\emptyset$ of polygonal c-functions such that, for any NN x ,

(i) the set M_{2x} , where $M_y = \{z : 1 \le z \le 3^y\}$ for any NN y, is the domain and M_x the range of the function $\lambda y\, \bar{n}(x,y)$, in particular, $\bar{n}(0,1) \simeq 1$, and $\bar{n}(x+1,3^{2x}.(3.(v-1) + t-1) + y) \simeq (3^x.(v - 2^{-1}.(1-(-1)^t)) - (-1)^t.\bar{n}(x,y) + 2^{-1}.(1+(-1)^t))$ holds for any v and t from M_1 and any y from M_{2x} and, consequently, $card(\{z : \bar{n}(x,z) \simeq w\}) = 3^x$ is valid for any NN w from M_x ;

(ii) $F_x(0) = 0$, $F_x(1) = 1$, F_x is linear on L[2x,y] for any NN y from M_{2x} and maps the closed interval L[2w,v] onto $L[w,\bar{n}(w,v)]$ for any NNs w and v satisfying $w \le x \& v \in M_{2w}$.

Hence, $|F_x - F_{x+t}| \le 3^{-x}$ is valid for any NNs x and t and, consequently, we can construct an $\emptyset$-uniformly continuous c-function F being a limit of the $\emptyset$-sequence $\{F_x\}_x^\emptyset$. We have $F(0) = 0$, $F(1) = 1$; F maps the closed interval L[2x,y] onto $L[x,\bar{n}(x,y)]$ for any NNs x and $y \in M_{2x}$.

<u>Notation</u>. We assume that we have a fixed numbering of all strings. Let T_x denote the r.e. set of all strings which number is in the set W_x .

Let k be an NN, $2 \le k$.

1) For any string σ , let $Seg_k(\sigma)$ denote the closed rational interval $[a,a + k^{-m}]$, where m is the length of σ and

$$a = \sum_{x<m} \min(\sigma(x),k-1).k^{-x-1} .$$

2) T_x is said to be a <u>k-ary covering</u> if, for any recursive function f such that $f(0) = 1$ and $(f(y) - 1).k^{-y} < f(y+1).k^{-y-1} \le f(y).k^{-y}$ holds for every NN y, there are a string τ in T_x and an NN z satisfying $[(f(z) - 1).k^{-z}, f(z).k^{-z}] \subseteq Seg_k(\tau)$.

<u>Remark 4</u>. Let k be an NN, $2 \le k$. It is easy to show that a non-recursive set A of NNs is semigeneric if and only if, in any k-ary covering T_x, there is a string σ satisfying $r_A \in Seg_k(\sigma)$.

As we shall see, it is possible to replace c-functions by objects, being not so complicated as c-functions are, in our construction of Denjoy sets. We assume that we have a fixed indexing of all pairs of the type: an NN k and an increasing finite sequence of (reduced) ternary rationals $\{a_i\}_{i=0}^{3^k}$. Let k_n and $\{a_i^n\}_{i=0}^{3^{k_n}}$ denote the pair with index n for any NN n and let $k_0 = a_0^0 = 0$ and $a_1^0 = 1$. Let, for any NN x, $\Gamma(x)$ denote: $!\phi_x(0)$, $k_{\phi_x(0)} = 0$ and, for any NN y, $0 < y \& !\phi_x(y)$ implies $!\phi_x(y-1)$, $k_{\phi_x(y)} = y$ and $a_z^{\phi_x(y-1)} = a_{3z}^{\phi_x(y)}$ whenever $0 \le z \le 3^{y-1}$; let $\Gamma_0(x)$ denote: $\Gamma(x)$ and $\phi_x(0) \simeq 0$ are valid.

We construct $\emptyset$-algorithms Θ, Inc (the increment of Θ) and Quot (the differential quotient of Θ) transforming any word from their domain into a reduced ternary RtN and fulfilling the following conditions: for any NNs x, y, v and w and any integer i, $!\Theta(x*y*i) \Leftrightarrow !\phi_x(w)$ and $\Theta(x*y*i) \simeq a_v^{\phi_x(w)}$ hold, where $*$ is a letter from Ξ and $v.3^{-w}$ is a reduced ternary RtN equal to $\max(0,\min(i,3^y)).3^{-y}$, $Inc(x*y*i) \simeq (\Theta(x*y*i) - \Theta(x*y*i-1))$ and $Quot(x*y*i) \simeq Inc(x*y*i).3^y$ are valid.

<u>Remark 5</u>. 1) Let A be a non-recursive set and G a c-function satisfying $\underline{D}G(r_A) = +\infty$. Then, according to Demuth[10], there is a c-function H increasing on $[0,1]$ and such that $H(0) = 0$, $H(1) = 1$, $\underline{D}H(r_A) = +\infty$ and $\forall ab\ (0 \le a < b \le 1 \Rightarrow 2^{-1}.(b - a) \le H(b) - H(a))$. Starting from H, we can easily construct an NN t such that $\Gamma_0(t)$ holds, ϕ_t is total and, for the A-recursive function f fulfilling $r_A \in L[y,f(y)]$ for any NN y, $\lim_{y \to +\infty} Quot(t*y*f(y)) = +\infty$ is valid.

2) Let the function ϕ_x be total and satisfying $\Gamma(x)$. Then we can construct an $\emptyset$-sequence $\{S_y\}_y^{\emptyset}$ of constructive real numbers and an everywhere defined classical function J of a real variable increasing on $[0,1]$ and being an extension of the mapping transforming any ternary RtN $z.3^{-y}$ into $\Theta(x*y*z)$. This function is continuous at any real different from all S_w, $0 \le w$. In addition, for any real X from $(0,1)$ and the Set(X)-recursive function f,

where $f = \lambda y\ (\mu z(X \in L[y,z]))$, such that the Set(X)-sequence $\{Quot(x*y*f(y))\}_y^{Set(X)}$ is unbounded (hence, $\overline{D}J(X) = +\infty$ holds) there is, according to Demuth[2], a c-function G increasing on $[0,1]$ and satisfying $\underline{D}G(X) = +\infty$.

Using the s-m-n-theorem, we construct recursive functions β, γ, γ_0 and ψ of one variable and γ_1, ξ and η of two variables such that, for any NNs t, v, x and z,

a) $\gamma(x)$ satisfies the condition Γ and $\gamma_0(x)$, $\gamma_1(x,v)$, $\psi(x)$ and $\beta(x)$ satisfy the condition Γ_0; if $\Gamma_0(x)$ (or, as the case may be, $\Gamma(x)$) holds then we have $\phi_{\gamma_0(x)} = \phi_x$ (or $\phi_{\gamma(x)} = \phi_x$, respectively);

b) $!\phi_{\gamma(x)}(z) \Rightarrow \phi_x(z) \simeq \phi_{\gamma(x)}(z)$,

$!\phi_{\gamma_0(x)}(z) \Leftrightarrow !\phi_{\gamma_1(x,z+1)}(z)$, $\gamma_0(x) = \gamma_1(x,0)$ and

$\neg(!\phi_{\gamma_0(x)}(z)) \Rightarrow \phi_{\gamma_0(x)} = \phi_{\gamma_1(x,z+v+1)}$ hold;

c) if $!\phi_{\gamma_0(x)}(z)$ holds then we have the following: $\phi_{\gamma_0(x)}(z) \simeq \phi_{\gamma_1(x,z+1)}(z) \simeq \phi_x(z)$ is valid, $\phi_{\gamma_1(x,z+1)}$ is a total function and the mapping, transforming any ternary RtN $u.3^{-w}$ into the ternary RtN $\Theta(\gamma_1(x,z+1)*w*u)$, is linear on ternary RtNs from the closed interval $L[z,y]$ for any NN y;

d) $!\phi_{\gamma_0(x)}(2z) \Leftrightarrow !\phi_{\psi(x)}(2z) \Leftrightarrow !\phi_{\beta(x)}(z)$,

$!\phi_{\psi(x)}(2z+1) \Leftrightarrow !\phi_{\psi(x)}(2z+2)$ and $!\phi_{\eta(x,t)}(\langle z,v\rangle) \Leftrightarrow !\phi_{\gamma_0(x)}(2z)$ hold;

$\langle z,v\rangle \in W_{\xi(x,t)} \Leftrightarrow (!\phi_{\gamma_0(x)}(2z) \& Quot(\gamma_0(x) * 2z*v) > 3^{t+2} \&$

$\neg(\exists w)_{w<z}(\exists y)_{y \le 3^{2w}}(L[2z,v] \subseteq L[2w,y] \& Quot(\gamma_0(x)*2w*y) > 3^{t+2}))$,

on its domain, the function $\phi_{\eta(x,t)}$ is equal to the characteristic function of the set $W_{\xi(x,t)}$;

if $!\phi_{\gamma_0(x)}(2z)$ is valid then

$$a_y^{\phi_{\psi(x)}(2z)} = 2.3^{-1}.\sum_{j=0}^{z-1} 3^{-j}.\Theta(\gamma_1(x,2p_j+1)*2z*y) +$$

$3^{-z}.\Theta(\gamma_1(x,2z+1)*2z*y)$ and

$$a_y^{\phi_{\beta(x)}(z)} = \sum_{\substack{1 \le w \le 3^{2z} \\ \overline{n}(z,w) \le y}} Incr(\psi(x)*2z*w)$$, where $p_j \simeq \mu w(w=z \vee \exists u(L[2z,y] \subseteq L[2w,u] \& \langle w,u\rangle \in W_{\xi(x,j)}))$ for any NN j, hold for any NN y .

Remark 6. Let x be an NN. 1) The function $\phi_{\gamma_0(x)}$ (or, as the case may be, $\phi_{\gamma(x)}$) is total if and only if the function ϕ_x is total and, in addition, $\Gamma_0(x)$ (or $\Gamma(x)$, respectively) holds.

2) If, for an NN z, $!\phi_{\gamma_0(x)}(2z)$ is valid then $\phi_{\gamma_1(x,2z+1)}$ is total and, according to d),

$<y,v> \in W_{\xi(x,t)} \iff \phi_{\eta(\gamma_1(x,2z+1),t)}(<y,v>) \simeq 1$

holds for any NNs y, v and t, where $y \le z$.

3) Let $\phi_{\gamma_0(x)}$ be a total function. Then $\phi_{\psi(x)}$ and $\phi_{\beta(x)}$ are total, too, and, for any NNs w, y and t, $Quot(\psi(x)*w*y) > 2.(t+1)$ is valid whenever there are NNs z and v fulfilling $L[w,y] \subseteq L[2z,v]$ and $<z,v> \in W_{\xi(x,t)}$. Consequently, we have the following: Let g be a total function such that $g(0) = 1$ and $L[w+1,g(w+1)] \subseteq L[w,g(w)]$ holds for any NN w. Then there is just one set C of NNs fulfilling $r_C \in L[w,g(w)]$ for any NN w and g is a C-recursive function satisfying the following conditions a)-c).

a) If the C-sequence $\{Quot(\gamma_0(x)*2w*g(2w))\}_w^C$ is unbounded then

$$\lim_{w \to +\infty} Quot(\psi(x)*w*g(w)) = +\infty .$$

b) $Quot(\beta(x)*w*g(w)) = 3^{-w}. \sum_{\bar{n}(w,z)=g(w)} Quot(\psi(x)*2w*z)$ holds for any NN w.

c) If m is an NN being an upper bound of the C-sequence $\{Quot(\beta(x)*w*g(w))\}_w^C$ then, by b), the cardinality of the set $\{ y : \exists zv(<z,v> \in W_{\xi(x,t)} \& L[2w,y] \subseteq L[2z,v] \& \bar{n}(w,y) = g(w))\}$ is not greater than $m.(2t+2)^{-1}.3^w$ for any NNs w and t.

We want to have codes of linear combinations of mappings corresponding to the functions $\phi_{\gamma_0(x)}$, $0 \le x$. We assume that we have a fixed indexing of all triples of the type: an NN n, finite sequences $\{x_v\}_{v=0}^n$ of NNs and $\{b_v\}_{v=0}^n$ of positive ternary RtNs. We shall use the term "ε-index of the triple ...". Using the s-m-n-theorem, we construct recursive functions ε of one variable and α of two variables such that

a) for any NN n and any finite sequences $\{x_v\}_{v=0}^n$ of NNs and $\{b_v\}_{v=0}^n$ of positive ternary RtNs and for the ε-index t of this triple we have: $\Gamma(\varepsilon(t))$ is valid, $!\phi_{\varepsilon(t)}(y) \iff (\forall v)_{v \le n} (!\phi_{\gamma_1(v,x_v)}(y))$ and $!\phi_{\varepsilon(t)}(y) \Rightarrow \Theta(\varepsilon(t)*y*w) = \sum_{0 \le v \le n} b_v.\Theta(\gamma_1(v,x_v)*y*w)$ hold for any NNs y and w; thus, the function $\phi_{\varepsilon(t)}$ is total if and only if all the functions $\phi_{\gamma_1(v,x_v)}$, where $0 \le v \le n$, are total;

b) for any NNs s, t, u and x,

(i) $!\phi_{\alpha(s,t)}(<u,x>)$ holds if and only if there are a string σ and NNs v and y fulfilling

$$\sigma \in T_s \& u < v \& !\phi_{\varepsilon(t)}(v) \& L[v,y] \subseteq L[u,x_0] \cap Seg_3(\sigma) \& \forall wz(u < w \le v \& L[v,y] \subseteq L[w,z] \subseteq L[u,x_0] \Rightarrow Quot(\varepsilon(t)*w*z) \le Quot(\varepsilon(t)*u*x_0)) , \quad (3)$$

where

$$x_0 = \max(1,\min(x,3^u)); \quad (4)$$

(ii) if v and y are NNs satisfying $\phi_{\alpha(s,t)}(<u,x>) \simeq <v,y>$ then there is a string σ such that (3) and (4) are valid.

Remark 7. Let t be an NN such that the function $\phi_{\varepsilon(t)}$ is total. Then we can, for any NNs u and x, construct a recursive function f such that $f(u) = x_0$, where (4) holds, and $L[w+1,f(w+1)] \subseteq L[w,f(w)]$ and $u \le w \Rightarrow Quot(\varepsilon(t)*w*f(w)) \le Quot(\varepsilon(t)*u*f(u))$ are valid for any NN w. Hence, for any 3-shaped covering T_s, we can find a string σ in T_s and an NN w_0 such that the segments $Seg_3(\sigma)$ and $L[w_0,f(w_0)]$ are equal and, consequently, $\phi_{\alpha(s,t)}$ is defined at $<u,x>$. Thus, the function $\phi_{\alpha(s,t)}$ is total whenever $\phi_{\varepsilon(t)}$ is total and T_s is a 3-shaped covering.

Theorem 8. Let A be a set of NNs fulfilling $\emptyset'' \le_T A'$. Then there is a Denjoy set being both semigeneric and T-reducible to A.

Proof. According to Epstein[11] (Chapter XI, Theorem 11), there is an increasing A-recursive function f majorizing any recursive function almost everywhere (i.e. for any sufficiently large NN). Thus, for any NN x, ϕ_x is total if and only if there is an NN v satisfying $Ev(x,z,f(max(z,v)))$ for any NN z.
By Remarks 4 and 5, to prove the theorem, it is sufficient to construct an A-recursive function k of one variable such that $k(0) = 1$, $L[w+1,k(w+1)] \subseteq L[w,k(w)]$ holds for any NN w, the A-sequence $\{Quot(\gamma_0(x)*w*k(w))\}_w^A$ is bounded whenever $\phi_{\gamma_0(x)}$ is a total function and, in any 3-shaped covering T_s, there is a string σ fulfilling $L[w,k(w)] \subseteq Seg_3(\sigma)$ for any sufficiently large NN w.

Simultaneously with k, we shall construct A-recursive functions p and h of two variables, a partial A-recursive function g of three variables and A-sequences $\{n_s\}_s^A$ of NNs (an increasing one) and $\{\{b_{x,i}\}_{i=0}^{p(x,s)}\}_{<x,s>}^A$ of finite sequences of positive ternary RtNs such that, for any NNs s, x and w,

a) $\lambda t\, p(x,t)$ is a bounded non-decreasing function, $p(x,0) = 0$, $!g(x,s,w) \Leftrightarrow (w \le p(x,s) \& x \le n_s)$, the range of $\lambda t\, g(x,t, \lim_{v\to+\infty} p(x,v))$ is $\{0\}$ if and only if $\phi_{\gamma_0(x)}$ is a total function, the range of h is $\{0,1\}$, $\lambda t\, h(w,t)$ is non-decreasing, if $h(w,s) = 1$ holds then there is a string σ in T_w satisfying $L[n_s,k(n_s)] \subseteq Seg_3(\sigma)$;

b) if $x \le n_s \& w \le p(x,s)$ holds then $g(x,s,w) \le n_s+1$ and

(i) $w = p(x,s)$ holds and the function $\phi_{\gamma_1(x,n_s+1)}$ is total whenever $g(x,s,w) = 0$;

(ii) $g(x,s+1,w) = g(x,s,w)$ holds and the function $\phi_{\gamma_1(x,g(x,s,w))}$ is total whenever $1 \le g(x,s,w)$;

c) $0 < b_{x,i} \le 3^{-x-i-2}$ is valid whenever $x \le n_s \& 0 \le i \le p(x,s)$ holds;

d) for $x \le n_s$, $!\phi_{\varepsilon(t_{s,x})}(x)$ and $Quot(\varepsilon(t_{s,x})*x*k(x)) \le$

$$\sum_{j=0}^{n_s}\sum_{i=0}^{p(j,s)} b_{j,i}\cdot Quot(\gamma_1(j,g(j,s,i))*x*k(x)) \le \sum_{j=0}^{n_s}\sum_{i=0}^{p(j,s)} 3^{-j-i-2}$$

are valid, where $t_{s,x}$ denotes the ε-index of the triple x , $\{g(j,s,p(j,s))\}_{j=0}^{x}$ and $\{b_{j,p(j,s)}\}_{j=0}^{x}$.

We put $n_0 = 0$, $b_{0,0} = 3^{-2}$, $k(0) = 1$ and $p(x,0) = g(0,0,0) = h(x,0) = 0$ for any NN x . Clearly, the points b), c) and d) are valid for $s = 0$.

Our construction proceeds in stages.

Let s be an NN. Let the NN n_s and the values of the functions p and h be already constructed for any pair of NNs (x,t) , where $t \le s$, and let $s \le n_s$ and $p(x,s) \le s$ hold whenever $x \le n_s$, and $p(x,s) = h(x,s) = 0$ holds whenever $x > n_s$. In addition, let $k(x)$, $b_{x,i}$ and $g(x,s,i)$ be already defined for $x \le n_s \& 0 \le i \le p(x,s)$ and let the conditions described in the points b), c) and d) be fulfilled. Let $t_{s,x}$, where $x \le n_s$, be NNs described in d).

Stage s: Firstly, we try to contribute to guaranteeing semigenericity of the constructed set. We use Remark 7 here. Let x_0 be the least NN x satisfying the condition $(x \le n_s \& h(x,s)=0 \& Ev(\alpha(x,t_{s,x}),\langle n_s,k(n_s)\rangle, f(\langle n_s,k(n_s)\rangle)) \vee x=n_s+1)$. We put $h(x,s+1) = h(x,s)$ whenever either $x \ne x_0$ or $x > n_s$ holds. We go to (i), if $x_0 \le n_s$, and to (ii) otherwise.

(i) Let u and y be NNs fulfilling

$$\phi_{\alpha(x_0,t_{s,x_0})}(\langle n_s,k(n_s)\rangle) \simeq \langle u,y\rangle \ . \tag{5}$$

In particular, $n_s < u$, $L[u,y] \subseteq L[n_s,k(n_s)]$ and there is a string σ in T_{x_0} satisfying $L[u,y] \subseteq Seg_3(\sigma)$. We put $n_{s+1} = u$, $h(x_0,s+1) = 1$, $k(x) = \mu z(L[u,y] \subseteq L[x,z])$ whenever $n_s < x \le n_{s+1}$, $g(x,s+1,w) = g(x,s,w)$ whenever $w \le p(x,s) \& (x \le x_0 \vee x_0 < x \le n_s \& g(x,s,w) > 0)$, and $g(x,s+1,w) = n_s+1$ whenever $w \le p(x,s) \& x_0 < x \le n_s \& g(x,s,w)=0$. On account of (5) and b) we obtain

(ω) $!\phi_{\gamma_1(x,g(x,s+1,w))}(n_{s+1})$ holds for any $x \le n_s \& w \le p(x,s)$ and

$$\sum_{j=0}^{n_s}\sum_{i=0}^{p(j,s)} b_{j,i}\cdot Quot(\gamma_1(j,g(j,s+1,i))*x*k(x)) \le$$

$$\sum_{j=0}^{n_s}\sum_{i=0}^{p(j,s)} b_{j,i}\cdot Quot(\gamma_1(j,g(j,s+1,i))*n_s*k(n_s)) \quad \text{for} \quad n_s \le x \le n_{s+1}.$$

We go to substage (iii).

(ii) We put $n_{s+1} = n_s+1$ and, for any NNs x and w fulfilling $x \le n_s \& w \le p(x,s)$, we define $g(x,s+1,w) = g(x,s,w)$ whenever $(g(x,s,w) > 0 \vee$

$Ev(\gamma_0(x), n_{s+1}, f(n_{s+1})))$, and $g(x,s+1,w) = n_s + 1$ otherwise. It is easy to construct an NN $k(n_{s+1})$ such that $L[n_{s+1}, k(n_{s+1})] \subseteq L[n_s, k(n_s)]$ and (ω) are valid. We go to substage (iii).

(iii) We want to ensure that any NN x satisfying $x \le n_{s+1} \& Ev(\gamma_0(x), n_{s+1}, f(n_{s+1}))$ be considered. We put $p(x,s+1) = p(x,s) + 1$, $g(x,s+1,p(x,s+1)) = 0$ and

$$b_{x,p(x,s+1)} = 3^{-x-p(x,s+1)-2-n_{s+1}} , \tag{6}$$

if

$$x \le n_s \& g(x,s+1,p(x,s)) > 0 \& Ev(\gamma_0(x), n_{s+1}, f(n_{s+1})) \tag{7}$$

holds, and $p(x,s+1) = p(x,s)$ for any NN x not satisfying (7). For any NN x , $n_s < x \le n_{s+1}$, we put (6) and define: $g(x,s+1,p(x,s+1))$ is 0 whenever $Ev(\gamma_0(x), n_{s+1}, f(n_{s+1}))$ holds and 1 otherwise. We go on to stage $s+1$.

Note that our construction is recursive in A and that the function k fulfils all the conditions described at the beginning of our proof.

Corollary 9. There is a minimal T-degree containing a semigeneric Denjoy set.

Proof. Cooper proved (cf. Epstein[11], p. 222): For any set C of NNs satisfying $\emptyset' \le_T C$, there is a set A contained in a minimal T-degree and fulfilling $A' \equiv_T (A \text{ join } \emptyset') \equiv_T C$. To finish the proof it is sufficient to put $C = \emptyset''$ and use Cooper's result and Theorem 8.

Now, we present a few results concerning tt-reducibility of Denjoy sets.

Theorem 10. No AP-set being tt-reducible to an NAP-set can be a Denjoy set.

Proof. Let A be an AP-set being tt-reducible to an NAP-set B . Then, as we already know, the set B cannot be in any class of sets of $\emptyset$-measure zero, in particular, B is not in Ω . Thus, the statement (2) is valid. According to Demuth[5], Theorem 13, there are a non-decreasing (and, thus, $\emptyset$-uniformly continuous) c-function G and an NAP-set C such that $\mathcal{R}[G](r_C) = r_A$ and, according to the results[2] quoted above, $-\infty < \underline{D}G(r_C) = \overline{D}G(r_C) < +\infty$ holds. The set A is an AP-set and, hence, $\underline{D}G(r_C) \ne 0$ is excluded[12]. According to Lemma 3 from Demuth's paper[2], $\underline{D}G(r_C) = \overline{D}G(r_C) = 0$ implies: $Set(\mathcal{R}[G](r_C))$, i.e. A, is not a Denjoy set. The proof is finished.

Remark 11. Our concept of relativized measurability can be transferred to a situation described below. We shall use the c-function F and the partial recursive function $\bar{n}$ and their properties described in Remark 3.

Let B be a fixed non-recursive set of NNs and k the unique (B-recursive) function fulfilling $r_B \in L[x,k(x)]$ for any NN x . Let us note that, for any NNs y and z satisfying $1 \le z \le 3^{2y} \& \bar{n}(y,z) = k(y)$, there are just three NNs w such that $L[2y+2,w] \subseteq L[2y,z] \& \bar{n}(y+1,w) = k(y+1)$ holds. Thus, there is a one-to-one correspondence between $R[F]$-pre-images of the real r_B (or, equivalently, sets C of NNs fulfilling $R[F](r_C) = r_B$) and infinite paths of the full ternary tree (which, in turn, corresponds to the closed unit interval [0,1]). For any set A of NNs, we denote by $P^B(A)$ the class $\{ C : R[F](r_C) = r_B \& \exists uv(<u,v> \in A \& r_C \in L[2u,v])\}$ and by $weight^B(A)$ the supremum (or, equivalently, the limit) of the A-sequence

$$\{ 3^{-y}.card(\{ z : \exists uv(<u,v> \in A \& L[2y,z] \subseteq L[2u,v] \& \bar{n}(y,z) \simeq k(y)) \}) \}_y^A$$

being, by definition, the weight of $P^B(A)$ (relative to $P^B(\{ <0,1> \})$). We turn to sets A of a special type. Let C and D be sets of NNs. The class $P^B(W_t^C)$ is said to be D-weighable if (B join C) $\le_T$ D holds and there is a D-recursive function f being a modulus of weighability of $P^B(W_t^C)$, i.e. fulfilling $|weight^B(W_t^{C,f(v)}) - weight^B(W_t^{C,w})| \le 2^{-v}$ whenever $f(v) \le w$ holds. Thus, if $B \le_T C$ is valid, $P^B(W_t^C)$ is necessarily C'-weighable, yet can be non-C-weighable. Using standard methods of measure theory, we can prove the following:

1) If $P^B(W_t^C)$ is D-weighable and $weight^B(W_t^C) < weight^B(\{ <u,v> \})$ (which being non-zero equals 3^{-u}) then there is a D-recursive set E in the difference $P^B(\{<u,v>\}) \setminus P^B(W_t^C)$ (in particular, $R[F](r_E) = r_B$ and $r_E \in L[2u,v]$ are valid).

2) Let $B \le_T C$ hold and let g and h be C'-recursive functions such that $\sum_{x=h(v)+1}^{+\infty} weight^B(W^C_{g(x)}) < 2^{-v}$ is valid for any NN v . Then the class $P^B(A)$, where $A = \bigcup_{x \in N} W^C_{g(x)}$, is C'-weighable and $weight^B(A) \le \sum_{x \in N} weight^B(W^C_{g(x)})$.

Theorem 12. There is a covering $d(W_m)$ such that, for any Denjoy AP-set B , we can construct a Denjoy AP-set C being not covered by $d(W_m)$ and fulfilling $B \le_{tt} C \le_T B'$.

Proof. We shall use the recursive functions γ_0 , ψ , β and ξ . Their properties are discussed in Remark 6 and in the descriptions preceding it. In addition, we shall use Remark 3 and the $\emptyset$-uniformly continuous c-function F and the partial recursive function $\bar{n}$ defined in it as well as the weight introduced in Remark 11.

We construct NNs m_0 and m such that $W_{m_0} = \{ <y,z> : \exists t(y=4t+4 \& \phi_t(2y) \simeq z \& 1 \le z \le 3^{2y}) \}$ and $W_m = \{ w : \exists s(Seg_2(\delta_w) \subseteq \bigcup_{<y,z> \in W^s_{m_0}} L[2y,z])\}$.

It is easy to see that $d(W_m)$ is a covering.

Let B be a Denjoy AP-set. Then B is non-recursive and we construct a B-recursive function k and, on account of Theorem 2, a B'-recursive increasing function q such that $r_B \in L[x,k(x)]$ holds for any NN x and the set $\{x : \phi_{\gamma_0(x)}$ is total$\}$ is the range of q . For any NN s , we have $card(\{w : \exists yz(<y,z>\in W_{m_0} \& L[8s+8,w] \subseteq L[2y,z] \& \bar{n}(4s+4,w)=k(4s+4))\}) < 2^{-1}.3^{4s+1}$. Hence, according to Remark 11, the class $P^B(W_{m_0})$ is B'-weighable and $weight^B(W_{m_0}) < 3^{-3}$ holds. As follows from Remarks 5 and 6, there is a B'-recursive function f such that f(x) is an upper bound of the B-sequence $\{Quot(\beta(q(x))*y*k(y))\}_y^B$ for any NN x . Let g be a B'-recursive function fulfilling $g(x) = (f(x) + 1).3^{x+4}$ for any NN x . According to Remarks 6 and 11, the class $P^B(W_{\xi(q(x),g(x))})$ is B'-weighable and $weight^B(W_{\xi(q(x),g(x))}) < 3^{-x-4}$. As follows from Remark 11, the class $P^B(U)$, where $U = W_{m_0} \cup \bigcup_{x\in N} W_{\xi(q(x),g(x))}$, is B'-weighable and $weight^B(U) < 3^{-2}$. Consequently, there is a B'-recursive set C in $P^B(\{<0,1>\}) \setminus P^B(U)$. We obtain: $R[F](r_C) = r_B$ (thus, $B \leq_{tt} C$ is valid because B , as a Denjoy set, cannot be in the class Ω), C is not covered by $d(W_m)$ and, for any NN x , the number $3^{g(x)+2}$ is an upper bound of the C-sequence $\{Quot(\gamma_0(q(x))*2w*h(2w))\}_w^C$, where h is the C-recursive function satisfying $r_C \in L[y,h(y)]$ for any NN y . According to the part 1 of Remark 5, C must be a Denjoy set. Hence, by Theorem 10, C is necessarily an AP-set.

<u>Corollary 13</u>.Any semigeneric Denjoy set is tt-reducible to a Denjoy AP-set being not semigeneric.

<u>References</u>

1. B. A. Kushner, "Lectures on Constructive Mathematical Analysis," Amer. Math.Soc., Providence, R.I. (1984).
2. O. Demuth, A constructive analogue of Garg's theorem on Dini derivatives (Russian), <u>Comment.Math.Univ.Carolinae</u> 21:457 (1980).
3. B. S. Thomson, "Real Functions," Springer-Verlag, Berlin (1985).
4. G. E. Sacks, "Degrees of Unsolvability," Princeton University Press, Princeton, N.J. (1963).
5. O. Demuth, Remarks on the structure of tt-degrees based on constructive measure theory, <u>Comment.Math.Univ.Carolinae</u> 29:233 (1988).
6. O. Demuth, A notion of semigenericity, <u>Comment.Math.Univer.Carolinae</u> 28:71 (1987).
7. A. Kucera, Measure, Π_1^0-classes and complete extensions of PA, <u>in</u>:"Recursion Theory Week, Proceedings," Ebbinghaus et al., ed., Springer-Verlag, Berlin (1985).
8. O. Demuth, Reducibilities of sets based on constructive functions of a real variable, <u>Comment.Math.Univ.Carolinae</u> 29:143 (1988).
9. O. Demuth, On some classes of arithmetical real numbers (Russian), <u>Comment.Math.Univ.Carolinae</u> 23:453 (1982).
10. O. Demuth, An example of a construction of pseudonumbers by means of recursion theory (Russian), <u>in</u>:Prikladnaja Matematika 5, Yerevan State University, Yerevan (1987).

11. R. L. Epstein, "Degrees of Unsolvability: Structure and Theory," Springer-Verlag, Berlin (1979).
12. O. Demuth, On pseudodifferentiability of pseudouniformly continuous constructive functions with respect to functions of the same type (Russian), Comment.Math.Univ.Carolinae 24:391 (1983).

NORMAL MODAL LOGICS IN WHICH THE HEYTING PROPOSITIONAL CALCULUS CAN BE EMBEDDED

Kosta Došen

Matematički Institut
Knez Mihailova 35
Belgrade, Yugoslavia

INTRODUCTION

Let t(A) be the result of prefixing the necessity operator □ to every proper subformula, save conjunctions and disjunctions, of the formula A of the language of the Heyting propositional calculus H. It is well-known that H can be embedded by t in S4, i.e. A is provable in H iff t(A) is provable in S4. Esakia (1979), and also Blok (1976), have shown that S4Grz (defined below) is the maximal normal extension of S4 in which H can be embedded by t (as a matter of fact, we find in Esakia (1979) not t, but the translation which prefixes □ to every subformula; this translation is equivalent to t as far as S4 and its normal extensions are concerned).

It is not difficult to find the minimal normal modal propositional logic K4N, weaker than S4 (this logic, considered by Lemmon and Scott (1977, pp. 68-71), will be defined below), in which H can be embedded by t (cf. Došen 1981 and 1986). We may then ask whether S4Grz is also the only maximal normal extension of K4N in which we can embed H by t, i.e. whether it is true that H can be embedded by t in a normal modal propositional logic S iff S is between K4N and S4Grz. We shall show in this paper that methods of Esakia (1979) can be adapted to answer this question affirmatively.

This result depends essentially upon considering only <u>normal</u> modal propositional logics S. For nonnormal modal logics we may have minimal and maximal logics with respect to the embedding by t whose sets of theorems differ from those of K4N and S4Grz respectively (the nonmaximality of S4Grz for nonnormal modal logics was considered by Chagrov (1985)). Our result also depends upon using the translation t and not some analogous translation, which as far as S4 and its normal extensions are concerned is equivalent to t. We shall consider in this paper the difficulties which we encounter with these other translations.

The embeddings of H in K4N and related logics, which we shall consider in the next two sections, suggest that we may show H complete with respect to Kripke-style models in which the "accessibility" relation is not a quasi-ordering, but satisfies weaker conditions. After a section on these Kripke-style models, in the final section we shall make some brief comments on modal embeddings of Heyting first-order predicate logic and Heyting arithmetic, and on modal embeddings of classical logic.

Mathematical Logic
Edited by P. P. Petkov
Plenum Press, New York, 1990

Our basic nonmodal propositional language L will have countably many propositional variables, the binary connectives $\rightarrow$, $\wedge$ and $\vee$, and the unary connective $\neg$. The modal propositional language $L\Box$ will have in addition to what we have in L the unary connective $\Box$. For formulae of L or $L\Box$ we use the schematic letters $A, B, C, \ldots, A_1, \ldots$ As usual, $A \leftrightarrow B$ is defined as $(A \rightarrow B) \wedge (B \rightarrow A)$.

The Heyting propositional calculus in L will be denoted by H. The system K in $L\Box$ is the classical propositional calculus extended with $\Box(A \rightarrow B) \rightarrow (\Box A \rightarrow \Box B)$ and closed under the rules: modus ponens, substitution for propositional variables and necessitation (i.e. from A infer $\Box A$). We write $S' \subseteq S''$ when the theorems of the system S' are included among the theorems of the system S". A system S in $L\Box$ is normal iff $K \subseteq S$ and S is closed under the rules of K.

The normal system K4N will be obtained by extending K with $\Box A \rightarrow \Box\Box A$, $\neg\Box\neg(A \rightarrow A)$ and $\Box(\Box A \vee \Box B) \rightarrow (\Box A \vee \Box B)$. It is easy to show that S4, i.e. K plus $\Box A \rightarrow \Box\Box A$ and $\Box A \rightarrow A$, properly extends K4N. The normal system S4Grz is K extended with $\Box(\Box(A \rightarrow \Box A) \rightarrow A) \rightarrow A$. It is known that S4Grz properly extends S4 (see van Benthem and Blok 1978 and Boolos 1979, Chapter 13).

The translation t is a one-one mapping from L into $L\Box$ such that t(A) is the result of prefixing $\Box$ to every proper subformula of A save conjunctions and disjunctions. More precisely, t is defined as follows, via the translation s which prefixes $\Box$ to every subformula save conjunctions and disjunctions:

$s(A) = \Box A$, where A is a propositional variable,
$s(A \rightarrow B) = \Box(s(A) \rightarrow s(B))$,
$s(A \alpha B) = s(A) \alpha s(B)$, where α is $\wedge$ or $\vee$,
$s(\neg A) = \Box\neg s(A)$;

$t(A) = A$, where A is a propositional variable,
$t(A \beta B) = s(A) \beta s(B)$, where β is $\rightarrow$, $\wedge$ or $\vee$,
$t(\neg A) = \neg s(A)$.

We write $H \xrightarrow{t} S$ iff for every A in L we have that A is a theorem of H iff t(A) is a theorem of S, i.e. H can be embedded by t in S. The following lemma asserts that K4N is the minimal normal system in which H can be embedded by t:

Lemma 1. (1) $H \xrightarrow{t} K4N$.

(2) If S is normal and $H \xrightarrow{t} S$, then $K4N \subseteq S$.

Proof. (1) Let A be a theorem of H, and let s'(A) be obtained from A by prefixing $\Box$ to every subformula of A (including conjunctions and disjunctions). Then by induction on the length of proof of A in H we can show that s'(A) is a theorem of K4N. To obtain that t(A) is a theorem of K4N we remove superfluous necessity operators from s'(A) by using the fact that $\Box(\Box B \wedge \Box C) \leftrightarrow (\Box B \wedge \Box C)$ and $\Box(\Box B \vee \Box C) \leftrightarrow (\Box B \vee \Box C)$ are theorems of K4N, and that K4N is closed under replacement of equivalents and under the rules:

$$\frac{\Box(\Box B \rightarrow \Box C)}{\Box B \rightarrow \Box C}, \qquad \frac{\Box\neg\Box B}{\neg\Box B}.$$

To prove $H \xrightarrow{t} K4N$ it remains to observe that $K4N \subseteq S4$, and appeal to the well-known fact that $H \xrightarrow{t} S4$.

(2) The minimality of K4N follows from the fact that $\Box A \to \Box(\Box(\Box C \to \Box C) \to \Box A)$, $\neg\Box\neg\Box(\Box A \to \Box A)$ and $\Box(\Box(\Box C \to \Box C) \to (\Box A \vee \Box B)) \to (\Box A \vee \Box B)$, where A,B and C are propositional variables, are t-translations of theorems of H. q.e.d.

This lemma is tied up to the particular translation t and the particular primitive vocabulary we have assumed for L and L$\Box$. It is well-known that for embedding H in S4 we may also use the translation t' which prefixes $\Box$ to every proper subformula (including conjunctions and disjunctions). Indeed, in K4N and its normal extensions for every A in L we have that:

t(A) is provable iff t'(A) is provable
iff s(A) is provable
iff s'(A) is provable,

where s and s' are the two translations defined before Lemma 1 and in the proof of Lemma 1. To sum up, we have the following translations:

	prefixes $\Box$ to every
t	proper subformula save conjunctions and disjunctions
t'	proper subformula
s	subformula save conjunctions and disjunctions
s'	subformula

These various translations, which are not essentially different as far as K4N and its normal extensions are concerned, induce different minimal normal modal systems to replace K4N. Namely, the minimal normal modal system S such that:

$H \xrightarrow{t'} S$ is $Kt' = K + \Box A \leftrightarrow \Box\Box A,\ \neg\Box\neg(A \to A)$;

$H \xrightarrow{s} S$ is $Ks = K + \Box(\Box A \to \Box\Box A),\ \Box\neg\Box\neg(A \to A),\ \Box(\Box(\Box A \vee \Box B) \to (\Box A \vee \Box B))$;

$H \xrightarrow{s'} S$ is $Ks' = K + \Box(\Box A \leftrightarrow \Box\Box A),\ \Box\neg\Box\neg(A \to A)$.

To demonstrate this (cf. Došen 1981 and 1986) we proceed analogously to what we had for Lemma 1, save for the following. To show that if A is a theorem of H, then s'(A) is a theorem of Ks', we use the fact that Ks' is closed under the rule:

$$\frac{\Box(A_1 \leftrightarrow A_2) \qquad \Box B}{\Box B[A_1/A_2]}$$

where $B[A_1/A_2]$ is obtained from B by replacing zero or more occurrences of A_1 by A_2; we also use the fact that if A is a theorem of H, then in Ks' we can prove $\Box(s'(A) \to t'(A))$, i.e. $\Box s'(A) \to s'(A)$ (remember that H has the disjunction property, i.e. if B$\vee$C is provable in H, then either B or C is provable in H). To show that if A is a theorem of H, then s(A) is a theorem of Ks, we use the fact that the provability of B in K4N implies the provability of $\Box$B in Ks, and also the facts that B$\wedge$C is provable in H iff B and C are provable in H, and that B$\vee$C is provable in H iff B or C is provable in H. Finally, to show that if A is a theorem of H, then t'(A) is a theorem of Kt', we use the fact that the provability of A in H implies the provability of s'(A) in Kt'; this implies the provability of t'(A) in Kt' (remember again that H has the disjunction property).

With a different primitive vocabulary in L and L$\Box$ we may also end up with a minimal normal system different from K4N. For example, if we have the constant proposition $\bot$ as primitive, instead of $\neg$ (where $\neg$A is defined as $A \to \bot$), and if $\bot$ behaves in the translations t,t',s and s' as a propositional variable, then the minimal normal system replacing K4N in Lemma 1 will be

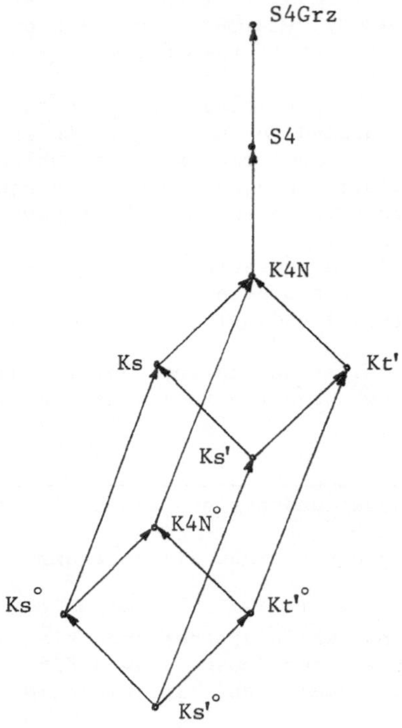

Fig. 1. Modal systems. (Arrows indicate proper inclusion.)

$K4N^{\circ} = K + \Box A \to \Box\Box A,\ \Box(\Box A \vee \Box B) \to (\Box A \vee \Box B)$. This difference arises because with $\bot$ primitive we have:

$$t(\neg A) = t(A \to \bot)$$
$$= s(A) \to \Box\bot,$$

whereas with $\neg$ primitive and $\bot$ defined as $\neg(A \to A)$ we have that $t(\neg A)$ is equivalent to $s(A) \to \bot$. With $\bot$ primitive, and the translations t', s and s', the minimal normal systems will be the respective systems in the list above with $\neg\Box\neg(A \to A)$ and $\Box\neg\Box\neg(A \to A)$ omitted. Let us denote these systems by Kt'°, Ks° and Ks'°. Then the modal systems which we have considered make the chart of Fig. 1.

MODAL ALGEBRAS

Let $HA = \langle H, \cap, \cup, \to, 1, 0 \rangle$ be a Heyting algebra (called pseudo-Boolean algebra by Rasiowa and Sikorski (1963)), and let $TB = \langle B, \cap, \cup, -, 1, 0, I \rangle$ be a topological Boolean algebra (where - is Boolean complement and I an interior operator). If $HA(B) = \{a \in B: Ia = a\}$, and for a and b in HA(B) we define $a \to b$ as $I(-a \cup b)$, then $HA(TB) = \langle HA(B), \cap, \cup, \to, 1, 0 \rangle$ is a Heyting algebra (which Esakia (1979) calls the stencil of TB).

By using a well-known construction of McKinsey and Tarski (inspired by Stone; see Rasiowa and Sikorski 1963, pp.128-130), we can embed a given Heyting algebra $HA = \langle H, \cap, \cup, \to, 1, 0 \rangle$ in a topological Boolean algebra $TB(HA) = \langle TB(H), \cap, \cup, -, 1, 0, I \rangle$ generated by H, where for every $a \in TB(H)$ there are $b_1, \ldots, b_n, c_1, \ldots, c_n \in H$ such that $a = (-b_1 \cup c_1) \cap \ldots \cap (-b_n \cup c_n)$ and

$Ia=(b_1 \to c_1)\cap\ldots\cap(b_n \to c_n)$. It can then be shown that HA(TB(HA)) is isomorphic with HA. For every TB we can prove the following lemma:

<u>Lemma 2</u>. TB(HA(TB)) is isomorphic to a subalgebra of TB.

<u>Proof</u>. For TB=<B,∩,∪,−,1,0,I> let $B^*=\{a\in B: \exists b_1,\ldots,b_n,c_1,\ldots,c_n\in B$ $(b_1=Ib_1$ & ... & $c_n=Ic_n$ & $a=(-b_1\cup c_1)\cap\ldots\cap(-b_n\cup c_n))\}$. It is easy to show that TB*=<B*,∩,∪,−,1,0,I> is a subalgebra of TB. It remains to check that TB* is isomorphic with TB(HA(TB)) (a detailed proof may be found in Maksimova and Rybakov 1974, Lemmata 3.3 and 3.4). q.e.d.

So, we may always consider that TB(HA(TB)) is a subalgebra of TB, but not necessarily isomorphic with TB. If TB(HA(TB)) is isomorphic with TB, following Esakia (1979), we call TB a <u>stenciled</u> topological Boolean algebra.

We write TB⊨A iff for every valuation v from L□ into B we have v(A)=1 in TB, and we write TB⊨S iff for every theorem A of the system S we have TB⊨A. The essential result of Esakia which we need is the following:

<u>Lemma 3</u>. (Esakia 1979, Corollary 4.10) If A is not a theorem of S4Grz, then there is a finite stenciled TB such that it is not the case that TB⊨A.

We shall call QTB=<B,∩,∪,−,1,0,I> a <u>quasi-topological</u> Boolean algebra iff <B,∩,∪,−,1,0> is a Boolean algebra and for every a,b∈B we have:

$I(a\cap b)=Ia\cap Ib$, $I1=1$, $Ia=IIa$, $I0=0$, $I(Ia\cup Ib)=Ia\cup Ib$.

Every topological Boolean algebra is quasi-topological (it satisfies moreover $Ia\leqslant a$), but not the other way round. It is easy to verify that A is a theorem of K4N iff for every QTB we have QTB⊨A; namely, the Lindenbaum algebra of K4N is a freely generated QTB.

If HA(QTB) is defined analogously to HA(TB), we can check that for every QTB the algebra HA(QTB) is a Heyting algebra (this is contained in the fact that H can be embedded in K4N by s'). For every QTB we can also prove the following analogue of Lemma 2:

<u>Lemma 4</u>. TB(HA(QTB)) is isomorphic to a subalgebra of QTB.

The proof of this lemma proceeds quite analogously to the proof of Lemma 2. Note that without I0=0 this proof might be blocked, since we might be unable to show that 0∈B*. And without $I(Ia\cup Ib)=Ia\cup Ib$, we might be unable to show that B* is closed under ∪ (and under −).

Note also that in QTB* (obtained as TB* in the proof of Lemma 2) we have for every a∈B* that $Ia\leqslant a$, though in QTB this is not the case for every a. Indeed, QTB*, which is isomorphic with TB(HA(QTB)), is a topological Boolean algebra. Lemma 4 yields as a corollary that for every quasi-topological Boolean algebra QTB there is a topological Boolean algebra TB which is a subalgebra of QTB and such that HA(QTB) is isomorphic with HA(TB). (Compare this with the fact, mentioned by Lemmon and Scott (1977, pp.70-71), that K4N can be axiomatized by extending K with □A→□□A and □B→B, where every propositional variable of B is within the scope of a □.)

Let S be a normal system such that K4N ⊆ S, and let VS={QTB: QTB⊨S}. The algebras in VS make a variety (because for every theorem A of S we can ask from our QTB's that they satisfy a=1, where a is obtained from A by translating logical with algebraic symbols). Let now HA(VS)={HA(QTB): QTB∈VS}. We can prove the following (cf. Blok and Dwinger 1975, Theorem 4.1):

<u>Lemma 5</u>. HA(VS) is closed under homomorphic images and subalgebras.

Proof. For closure under homomorphic images, suppose QTB∈VS and f: HA(QTB)⟶HA is an onto homomorphism. By Lemma 4, we have that TB(HA(QTB)) is a subalgebra of QTB, and since VS is a variety, TB(HA(QTB))∈VS. The homomorphism f can naturally be extended to a homomorphism g from TB(HA(QTB)) onto TB(HA) (for $a=(-b_1\cup c_1)\cap\ldots\cap(-b_n\cup c_n)$, take $g(a)=(-f(b_1)\cup f(c_1))\cap\ldots\cap(-f(b_n)\cup f(c_n))$). Since VS is a variety, TB(HA)∈VS. But HA is isomorphic with HA(TB(HA)). So, HA∈HA(VS).

For closure under subalgebras, suppose QTB∈VS and HA is a subalgebra of HA(QTB). Then TB(HA) is a subalgebra of TB(HA(QTB)). Since TB(HA(QTB)) is a subalgebra of QTB, we have that TB(HA) is a subalgebra of QTB, and since VS is a variety, TB(HA)∈VS. But HA is isomorphic with HA(TB(HA)), and, hence, HA∈HA(VS). q.e.d.

It is easy to show that HA(VS) is also closed under direct products; so, HA(VS) is in fact a variety.

By "countable" in the following two lemmata we understand "finitely or infinitely countable" (countability is assumed in these lemmata because the language L is assumed to be countable; without this assumption about L, we could prove analogous lemmata without the assumption of countability).

Lemma 6. If $H\xrightarrow{t}S$, then for every countable HA there is a QTB∈VS such that HA(QTB) is isomorphic with HA.

Proof. Suppose $H\xrightarrow{t}S$, and let s[Lind(S)]={[s(A)]: [s(A)]∈Lind(S)}, where Lind(S) is the Lindenbaum algebra of S. Of course, Lind(S)∈VS. The Heyting algebra s[S]=<s[Lind(S)],∧,∨,⟶,⊤,⊥>, where [s(A)]⟶[s(B)] is defined as □(¬[s(A)]∨[s(B)]), is a subalgebra of HA(Lind(S))∈HA(VS). So, by Lemma 5, s[S]∈HA(VS). On the other hand, s[S] can be shown isomorphic with Lind(H), the Lindenbaum algebra of H. We can define f: Lind(H)⟶s[S] by f([A])=[s(A)]. That f is a one-one mapping is shown as follows:

f([A])=f([B]) iff [s(A)]=[s(B)]
iff s(A)↔s(B) is provable in S
iff t(A→B) and t(B→A) are provable in S
iff A↔B is provable in H, since we have $H\xrightarrow{t}S$
iff [A]=[B].

It follows easily that f is a homomorphism and onto. So, Lind(H)∈HA(VS).

Since Lind(H) is a free Heyting algebra, there is a homomorphism from Lind(H) onto an arbitrary countable HA. According to Lemma 5, HA∈HA(VS). q.e.d.

Lemma 7. If $H\xrightarrow{t}S$, then every countable stenciled topological Boolean algebra belongs to VS.

Proof. Suppose $H\xrightarrow{t}S$, and let TB be a countable stenciled topological Boolean algebra. Then HA(TB) is a countable Heyting algebra, and by Lemma 6, there is a QTB∈VS such that HA(TB) is isomorphic with HA(QTB). Since TB is isomorphic with TB(HA(TB)), which is isomorphic with TB(HA(QTB)), we obtain by Lemma 4 that TB is a subalgebra of QTB. Since VS is a variety, TB∈VS. q.e.d.

We are now ready to prove our generalization of the theorem of Esakia and Blok:

Theorem. Let S be normal. Then $H\xrightarrow{t}S$ iff K4N ⊆ S ⊆ S4Grz.

Proof. Suppose $H\xrightarrow{t}S$. Then by Lemma 1, we have K4N ⊆ S. If A is not a theorem of S4Grz, then by Lemma 3 there is a finite stenciled TB such that not TB⊨A; hence, by Lemma 7, TB∈VS, and it follows that A is not a theorem of S. So, S ⊆ S4Grz. For the other direction it is enough to appeal to the fact that H can be embedded by t in K4N and S4Grz. q.e.d.

This method of proving our theorem depends essentially upon using the particular translation t and the particular primitive vocabulary of L and L□. To see that this is indeed the case, consider the normal modal systems from the previous section which are properly contained in K4N. These systems lack either $\neg\Box\neg(A\rightarrow A)$ or $\Box(\Box A\vee\Box B)\rightarrow(\Box A\vee\Box B)$, and, hence, in the analogues of our quasi-topological Boolean algebras we would not have either I0=0 or I(IaUIb)= IaUIb. As we have remarked after Lemma 4, the lack of these principles might block our proof. So, we leave open the question what form an analogue of our theorem should take with one of the translations t', s or s', or with ⊥ primitive, instead of ¬.

KRIPKE-STYLE MODELS

The embedding of H in K4N and in weaker normal modal systems suggests that we may obtain a completeness proof for H with respect to Kripke-style models in which the "accessibility" relation is not a quasi-ordering, but satisfies weaker conditions.

Let us first consider modal Kripke models with respect to which K4N may be shown complete (cf. Lemmon and Scott 1977, pp.68-71). These models are of the form $\langle X,R,v_0\rangle$, where X is a nonempty set of "worlds" (we shall use $x,y,z,\ldots,x_1,\ldots$ as variables ranging over X), R is a binary relation over X which satisfies:

(1) $\forall x,y,z((xRy \,\&\, yRz) \Rightarrow xRz)$, i.e. R is transitive,
(2) $\forall x\exists y(xRy)$, i.e. R is serial,
(3) $\forall x,y_1,y_2((xRy_1 \,\&\, xRy_2) \Rightarrow \exists z(xRz \,\&\, zRy_1 \,\&\, zRy_2))$,

and the basic valuation v_0 maps the propositional variables of L□ into PX, i.e. the power set of X. As usual, v_0 is extended to a valuation $v\colon L\Box\rightarrow PX$ by the following recursive clauses:

$v(A)= v_0(A)$, where A is a propositional variable,
$v(A\rightarrow B)= (X-v(A))\cup v(B)$,
$v(A\wedge B)= v(A)\cap v(B)$,
$v(A\vee B)= v(A)\cup v(B)$,
$v(\neg A)= X-v(A)$, $v(\Box A)= \{x\colon \forall y(xRy \Rightarrow y\in v(A))\}$.

A formula A holds in a model $\langle X,R,v_0\rangle$ iff $v(A)= X$.

The corresponding models for H, which we shall call Ht models, are of the form $\langle X,R,v_0\rangle$, where X and R are as above, and v_0, which maps the propositional variables of L into PX, satisfies the following condition for every propositional variable A and every $x\in X$:

$$x\in v_0(A) \Leftrightarrow \forall y(xRy \Rightarrow y\in v_0(A)).$$

In ordinary Kripke models for H this condition is usually assumed only from left to right, because the converse holds trivially when R is reflexive. But in Ht models the implication from right to left is not automatically satisfied. A basic valuation v_0 is extended to a valuation $v\colon L\rightarrow PX$ by the following usual recursive clauses:

$v(A)= v_0(A)$, where A is a propositional variable,
$v(A\rightarrow B)= \{x\colon \forall y(xRy \Rightarrow (y\in v(A) \Rightarrow y\in v(B)))\}$,

$v(A \wedge B) = v(A) \cap v(B)$,
$v(A \vee B) = v(A) \cup v(B)$,
$v(\neg A) = \{x: \forall y(xRy \Rightarrow y \notin v(A))\}$.

As before, A holds in an Ht model iff $v(A) = X$.

Then we can prove by induction on the complexity of A that for every formula A of L and every $x \in X$ the following holds:

(Heredity) $x \in v(A) \Longleftrightarrow \forall y(xRy \Rightarrow y \in v(A))$.

In proving Heredity, the transitivity of R is used in the cases when A is of the form $B \rightarrow C$ and $\neg B$, the seriality of R is used when A is of the form $\neg B$, whereas condition (3) for R is used when A is of the form $B \vee C$.

With the help of Heredity, we can easily verify by induction on the length of proof of A that if A is provable in H, then A holds in every Ht model. The converse, i.e. completeness, follows immediately from completeness with respect to ordinary Kripke models for H, and the fact that every ordinary Kripke model for H is an Ht model.

We may also expect to obtain models for H from models for our normal modal systems weaker than K4N, in which H can be embedded by various translations. These models for H would roughly correspond to the modal models as Ht models correspond to models for K4N. However, in these Kripke-style models for H we would have to modify in some cases the clauses for v, and in some cases the Heredity condition and the definition of holding in a model. In models which correspond to translations where $\Box$ is not omitted before disjunctions (namely, in models which correspond to Kt', Ks', Kt'° and Ks'°), the clause for $v(A \vee B)$ would be:

$v(A \vee B) = \{x: \forall y(xRy \Rightarrow (y \in v(A)$ or $y \in v(B)))\}$

rather than $v(A \vee B) = v(A) \cup v(B)$. In models which correspond to translations where $\Box$ is not prefixed only to proper subformulae (namely, in models which correspond to Ks, Ks', Ks° and Ks'°), Heredity would be replaced by the following conditional Heredity:

$\exists z(zRx) \Rightarrow (x \in v(A) \Longleftrightarrow \forall y(xRy \Rightarrow y \in v(A)))$

and holding in a model would be redefined as follows: A holds in a model iff $\forall x(\exists z(zRx) \Rightarrow x \in v(A))$. In models which correspond to systems based on $\bot$ primitive (namely, in models which correspond to the systems with the superscript °), the clause for $v(\bot)$ would be:

$v(\bot) = \{x:$ not $\exists y(xRy)\}$

rather than $v(\bot) = \emptyset$. (Since these last models need not have a serial R, we may have in them "blind worlds" in which $\bot$ holds. Heredity will guarantee that in these worlds every formula of L holds too; conditional Heredity will guarantee the same thing for blind worlds x such that $\exists z(zRx)$. This resembles the modified Kripke models of Veldman (1976) with their "exploding worlds".)

So, these various weak models for H would bring in some complications. On the other hand, for Ht models the clauses for v, as well as the Heredity condition and the definition of holding in a model, are exactly as for ordinary Kripke models for H. The only difference is in the conditions for R.

To conclude we note as a curiosity that the $\rightarrow$ and $\rightarrow, \wedge$ fragments of H can be shown sound and complete with respect to models $\langle X, R, v_0 \rangle$ for which everything is as for Ht models save that R is only transitive. However, it would be wrong to conclude from this that these two fragments of H can be embedded by t in K4, i.e. K + $\Box A \rightarrow \Box\Box A$.

CONCLUDING COMMENTS

We shall close this paper with some brief comments on the embeddings of the Heyting first-order predicate calculus in modal first-order predicate logics, on the embeddings of Heyting arithmetic in modal extensions of Peano arithmetic, and, finally, on the embeddings of classical logic in modal extensions of Heyting's logic.

Let L_1 be the first-order language which has individual constants and variables, predicate constants, the propositional connectives of L and the quantifiers $\forall x$ and $\exists x$. The language $L_1\Box$ has $\Box$ in addition to that. First-order K in $L_1\Box$ is the classical first-order predicate calculus extended with $\Box(A \rightarrow B) \rightarrow (\Box A \rightarrow \Box B)$ and closed under: modus ponens, necessitation and universal generalization. A system in $L_1\Box$ is normal iff it includes the theorems of first-order K and is closed under its rules.

The translation t_1': $L_1 \rightarrow L_1\Box$ prefixes like t' a $\Box$ to every proper subformula of a formula A of L_1, whereas t_1: $L_1 \rightarrow L_1\Box$ prefixes $\Box$ to every proper subformula save conjunctions, disjunctions and subformulae with an initial existential quantifier. Then it is not difficult to prove that the minimal normal first-order system in which the Heyting first-order predicate calculus can be embedded by t_1' is first-order K extended with $\Box A \leftrightarrow \Box\Box A$ and $\neg\Box\neg(A \rightarrow A)$ (to prove that we use, besides the disjunction property, the analogous existence property of the Heyting predicate calculus). With t_1 instead of t_1' this minimal normal first-order system will be $K4N_1$, which is first-order K extended with $\Box A \rightarrow \Box\Box A$, $\neg\Box\neg(A \rightarrow A)$, $\Box(\Box A \vee \Box B) \rightarrow (\Box A \vee \Box B)$ and $\Box\exists x\Box A \rightarrow \exists x\Box A$. With $\bot$ primitive, instead of $\neg$, we would omit $\neg\Box\neg(A \rightarrow A)$ from these minimal systems.

With translations from L_1 into $L_1\Box$ analogous to s and s' matters are not so straightforward (as explained in Došen 1986). It is the lack of a principle like the Barcan formula which produces difficulties in finding the minimal normal first-order systems in which the Heyting first-order predicate calculus can be embedded by these translations.

Next, let us mention that first-order Heyting arithmetic can be embedded by a translation analogous to t_1 in modal extensions of first-order Peano arithmetic, with the additional operator $\Box$, which lie in between the $K4N_1$ extension of Peano arithmetic and the S4 extension of Peano arithmetic. To demonstrate that the provability of A in Heyting arithmetic implies the provability of the translation of A in $K4N_1$ Peano arithmetic, we proceed analogously to what we had for Lemma 1(1). That the provability of the translation of A in S4 Peano arithmetic implies the provability of A in Heyting arithmetic was shown recently (see Flagg and Friedman 1986 for an elegant proof). Similar embeddings of Heyting arithmetic in appropriate modal extensions of Peano arithmetic contained in S4 Peano arithmetic can be proved with translations analogous to other modal translations we have considered. Can we prove such an embedding for the S4Grz extension of Peano arithmetic?

Besides the modal embeddings of Heyting's logic considered in this paper there is another famous type of embedding connected with Heyting's logic. Namely, classical logic can be embedded in Heyting's logic by various forms of the double-negation translation. Underlying this type of embedding there is also a modal translation.

Classical logic can be embedded by the translation s' into S5-like extensions of Heyting's logic, and in the case of propositional logic we can easily determine the minimal normal modal extension of H (where "normal" is understood relative to H) in which we can embed the classical propositional calculus C by s'. This is the system $H5p^-$, obtained by extending H with the modal postulates of Ks' and $\Box(\Box A \vee \Box\neg\Box A)$ (see Došen 1986). To determine the maximal normal extension of H in which we can embed C by s' is a straightforward matter (we have nothing like the complications connected with S4Grz). This is the system $C_{triv} = C + \Box A \leftrightarrow A$, obtained by extending H with $\Box(\Box A \vee \Box\neg\Box A)$ and $\Box A \leftrightarrow A$. This maximality of C_{triv} is proved like the fact that all consistent normal extensions of $H + \neg\Box\neg(A \to A)$ are included in C_{triv} (see Došen 1985, Lemma 1). The system C_{triv} is a conservative extension of C in L, but not of H in L. Can we find a maximal system (not necessarily unique) among the normal extensions of H in which C can be embedded by s', which are conservative extensions of H in L?

One such maximal conservative normal extension of H is the system $H_{dn} = H + \Box A \leftrightarrow \neg\neg A$. The embedding of C in H_{dn} by s' amounts to the simplest double-negation translation, where double negation is prefixed to every subformula. The translation s' is uneconomical for embedding C in H_{dn}: if $\Box$ in s'(A) is omitted in front of $\to, \wedge$ and $\neg$, we obtain a formula equivalent in H_{dn}. But the economy brought up by the translations t, t' and s is not now available. Of course, for embedding C in C_{triv} the economy can be total: all necessity operators are superfluous.

However, not all normal extensions of H, conservative with respect to H in L, in which we can embed C by s', are included in H_{dn}. One such extension which is not included in H_{dn} is obtained by adding $\Box A \to A$ to $H5p^-$. (That this system is conservative with respect to H in L may be proved with the help of models investigated in Ono 1977, Sotirov 1984, Došen 1985 and 1986a.) The economy brought up by the translations t, t' and s is now available, as well as a more thorough economy which omits every $\Box$ except those prefixed to propositional variables.

The general form of the embeddings considered here is the following. We have two nonmodal systems S' and S" such that S' is a proper subsystem of S", and we are able to show that:

(i) S' can be embedded by a modal translation in S" plus some modal postulates,

and vice versa:

(ii) S" can be embedded by a modal translation in S' plus some modal postulates.

Embeddings of H in modal systems with the nonmodal base C are of type (i), whereas embeddings of C in modal systems with the nonmodal base H are of type (ii). (The embeddings of classical and Heyting's logic into "linear logic" envisaged by Girard (1987) are like embeddings of type (ii).) For both types, one direction of our embeddings, that one which from the provability of A in the nonmodal system infers the provability of the translation of A in the modal system, is usually proved by a straightforward induction on the length of proof. The other direction is in principle more difficult to prove for type (i), because for type (ii) we usually have the following simple procedure. Our modal extension of S' must contain among other modal postulates the modal translations of theorems of S" missing from S'. To show that the provability of the modal translation of A in this extension of S' implies the provability of A in S", we use the fact that our modal extension of S' is included in S" plus $\Box A \leftrightarrow A$, and that this last system is a conservative extension of S". This simple procedure is not available for embeddings of type (i).

REFERENCES

van Benthem, J.F.A.K., and Blok, W.J., 1978, Transitivity follows from Dummett's axiom, Theoria, 44:117-118.

Blok, W.J., 1976, "Varieties of Interior Algebras", dissertation, University of Amsterdam.

Blok, W.J., and Dwinger, Ph., 1975, Equational classes of closure algebras I, Indag. Math., 37:189-198.

Boolos, G., 1979, "The Unprovability of Consistency: An Essay in Modal Logic", Cambridge University Press, Cambridge.

Chagrov, A.V., 1985, Varieties of logical matrices (in Russian), Algebra i Logika, 24:426-489 (English translation in: Algebra and Logic, 24:278-325).

Došen, K., 1981, Minimal modal systems in which Heyting and classical logic can be embedded, Publ. Inst. Math. (Beograd) (N.S.), 30 (44):41-52.

Došen, K., 1985, Models for stronger normal intuitionistic modal logics, Studia Logica, 44:39-70.

Došen, K., 1986, Modal translations and intuitionistic double negation, Logique et Anal. (N.S.), 29:81-94.

Došen, K., 1986a, Higher-level sequent-systems for intuitionistic modal logic, Publ. Inst. Math. (Beograd) (N.S.), 39(53):3-12.

Esakia, L.L., 1979, On the variety of Grzegorczyk algebras (in Russian), in: "Issledovaniya po neklassicheskim logikam i teorii mnozhestv", Nauka, Moscow, 257-287 (Math. Rev. 81j:03097; according to references in this paper, the results presented were announced in 1974).

Flagg, R.C., and Friedman, H., 1986, Epistemic and intuitionistic formal systems, Ann. Pure Appl. Logic, 32:53-60.

Girard, J.-Y., 1987, Linear logic, Theoret. Comput. Sci., 50:1-102.

Lemmon, E.J., and Scott, D.S., 1977, "An Introduction to Modal Logic: The 'Lemmon Notes'", Blackwell, Oxford.

Maksimova, L.L., and Rybakov, V.V., 1974, On the lattice of normal modal logics (in Russian), Algebra i Logika, 13:188-216 (English translation in: Algebra and Logic, 13:105-122).

Ono, H., 1977, On some intuitionistic modal logics, Publ. Res. Inst. Math. Sci. (Kyoto), 13:687-722.

Rasiowa, H., and Sikorski, R., 1963, "The Mathematics of Metamathematics", Państwowe Wydawnictwo Naukowe, Warsaw.

Sotirov, V.H., 1984, Modal theories with intuitionistic logic, in: "Mathematical Logic", Proceedings of the Conference Dedicated to Markov, Bulgarian Academy of Sciences, Sofia, 139-171.

Veldman, W., 1976, An intuitionistic completeness theorem for intuitionistic predicate logic, J. Symbolic Logic, 41:159-166.

LATTICES ADEQUATE FOR INTUITIONISTIC PREDICATE LOGIC

Wojciech Dzik

Institute of Mathematics
The Silesian University
Katowice, Poland

In this paper we show that there are many examples of adequate models for intuitionistic predicate logic (IPL) which appear in algebra and logic. We present a class of lattices, which are complete Heyting algebras, such that every lattice from this class is adequate for IPL in the sense that for every formula which cannot be derived in IPL and for every lattice L from the class there is a model of IPL over L in which the formula is not valid.

This class of lattices includes, among others, examples from algebra such as the lattice of all congruences of any countable bounded distributive lattice which is not weakly atomic, the lattice of all filters (ideals) of any countable Boolean algebra which is not atomic and examples from logic such as the lattice of all first order theories extending a given essentially undecidable theory, the lattice of all first-order (classical) theories and the lattice of all purely implicative theories.

The most commonly known lattice which is adequate for IPL is the Lindenbaum-Tarski algebra of equivalence classes of formulas of IPL. In [7] A. Tarski proved that the algebra of all open subsets of a separable metric space without isolated points is adequate for intuitionistic propositional logic. More recently, I.Moerdijk in [5] showed that the algebra of open subset of any metrizable space without isolated points is adequate for IPL. His result will be used here.

We will apply the notion of realization of a first-order language in a Heyting algebra and the notion of validity of a formula in a Heyting algebra analogous to that described by H.Rasiowa and R.Sikorski in [6], p.414.

A Heyting algebra A is called <u>adequate</u> for IPL if, for any formula φ, φ has a proof in IPL iff φ is valid in A.

1. For the standard notions of lattice theory we refer to Gratzer [4]. A lattice L is <u>infinitely join-distributive</u> if, for every $X \subseteq L$,

$$a \wedge \bigvee X = \bigvee (a \wedge x : x \in X).$$

An element a is called <u>cocompact</u> in L, $a \in Cc(L)$, if unit 1 is compact in the sublattice of all $x \geqslant a$. A lattice L is said to have a countable <u>basis</u> if there is a countable subset M of L such that for every element a from L there exists an $X \subseteq M$ such that $a = \bigvee X$.

Mathematical Logic
Edited by P. P. Petkov
Plenum Press, New York, 1990

Let us recall that if L is a complete infinitely join-distributive lattice then L determines a complete Heyting algebra, where the relative pseudocomplement of x to y is defined as

$$x \Longrightarrow y = \bigvee (z \in L : z \wedge x \leq y) ,$$

and the pseudocomplement of x as $-x = x \Longrightarrow 0$.

If we say that a complete infinitely join-distributive lattice is adequate for IPL we mean that the corresponding Heyting algebra is adequate for IPL.

THEOREM 1. Let L be a complete atomless and infinitely join-distributive lattice which has a countable basis and such that

$$(*) \qquad x = \bigvee (z \in L : -z \vee x = 1, \ -z \in Cc(L) \ )$$

for every $x \in L$. Then L is adequate for IPL.

In order to prove Theorem 1 we will prove some lemmata.

A filter F in a lattice L is completely prime if, for every $A \subseteq L$, $\bigvee A \in F$ implies that $a \in F$, for some $a \in A$.

LEMMA 1. For every $a < 1$ and $b \in Cc(L)$ such that $b \leq a$, there exists a maximal element $p \in L$ such that $a \leq p < 1$. Moreover,

$P_p = \{ x \in L : x \nleq p \}$ is a completely prime filter such that $a \notin P_p$.

Proof. Apply Kuratowski-Zorn lemma to the set $\{ x \in L : a \leq x < 1 \}$.

LEMMA 2. (cf. 1). Let L be a complete lattice satisfying (∗). If $b \nleq a$ then there exists a completely prime filter P such that $b \in P$ and $a \notin P$.

Proof. Let $b \nleq a$. Then there is an element z such that $-z \in Cc(L)$, $-z \vee b = 1$ and $-z \vee a \neq 1$. Otherwise, by (∗) , we would get $b \leq a$, which is impossible. Since $-z \leq -z \vee a < 1$ and $-z \in Cc(L)$, by Lemma 1 , there exists a completely prime filter P such that $-z \vee a \notin P$. But $-z \vee b = 1 \in P$ hence $b \in P$ and $a \notin P$.

It is known that if $\mathcal{T}$ is a family of all open subsets of a topological space X, then $\mathcal{T}$ is a complete lattice and in the corresponding complete Heyting algebra the following holds:

$\bigvee$ is $\bigcup$, union of sets, $\bigwedge$ is $\mathrm{Int} \bigcap$, i.e. $\bigwedge \mathcal{C} = \mathrm{Int}(\bigcap \mathcal{C})$, $\wedge$ is $\cap$,

$U \Rightarrow V = \mathrm{Int} ((X - U) \cup V)$, $-U = \mathrm{Int} (X - U)$, $1 = X$, $0 = \emptyset$.

Let us define a prime spectrum ΣL of a lattice L, which is the space of completely prime filters of L with the topology $\mathcal{T}_L$ of open sets of the form

$$s (a) = \{ P \in \Sigma L : a \in P \} , \ a \in L, \text{ (cf. [1]).}$$

LEMMA 3.(Representation lemma). Let L be a complete infinitely join-distributive lattice satisfying (∗). Then the prime spectrum of L, ΣL, is a locally compact Hausdorff space and the lattices L and $\mathcal{T}_L$ are isomorphic. The function s defined above is an isomorphism and :

1°. $a \leqslant b$ iff $s(a) \subseteq s(b)$, $a,b \in L$

2°. $s(\bigvee A) = \bigcup \{s(a) : a \in A\}$

3°. $s(\bigwedge A) = \operatorname{Int} \bigcap \{s(a) : a \in A\}$

In particular, the complete Heyting algebras L and $\mathcal{T}_L$ are isomorphic.

Proof. 1°. If $a \not\leqslant b$ then, by Lemma 2, there exists a filter $P \in \Sigma L$ such that $a \in P$ and $b \notin P$. Hence $P \in s(a) - s(b) \neq \emptyset$, thus $s(a) \not\subseteq s(b)$. The converse is obvious. We have also shown that s is one-to-one. By the definition of $\mathcal{T}_L$, s is "onto". 2° is obvious.

3°. Let $P \in \operatorname{Int} \bigcap \{s(a) : a \in A\}$. There is an element $b \in P$ such that for all $a \in A$, a neighbourhood $s(b) \subseteq s(a)$. By 1°, $b \leqslant a$, for all $a \in A$, i.e. $b \leqslant \bigwedge A$. Hence $\bigwedge A \in P$. For the converse put $b = \bigwedge A$. To show that ΣL is locally compact observe first that $\overline{s(c)} \subseteq s(b)$ iff $\Sigma L = s(b) \cup (\Sigma L - \overline{s(c)}) = s(-c \vee b)$, for $b,c \in L$. Hence $\overline{s(c)} \subseteq s(\bigvee A)$ implies that $\Sigma L = s(\bigvee(-c \vee a : a \in A))$, i.e. $1 = \bigvee(-c \vee a : a \in A)$. Since $-c \in Cc(L)$ and $-z \leqslant -z \vee a$ it follows that $1 = -c \vee a_1 \vee \ldots \vee a_n$, for some $a_1, \ldots, a_n \in A$, therefore $\Sigma L = s(-c \vee a_1 \vee \ldots \vee a_n)$, that is, $\overline{s(c)} \subseteq s(a_1) \cup \ldots \cup s(a_n)$. We have shown that, if $-c \in Cc(L)$, then $\overline{s(c)}$ is compact. By (*) it follows that each point of the space ΣL has an open neighbourhood U such that $\overline{U}$ is compact. It is easy to check that ΣL is a Hausdorff space since $s(c)$ and $s(-c)$, for some c such that $-c \in Cc(L)$, are two disjoint and open neighbourhoods of two different points of the space ΣL.

LEMMA 4. If, in addition to the assumptions of Lemma 3

1°. the lattice L is atomless, then ΣL has no isolated pionts,

2°. the lattice L has a countable basis, then ΣL is a metrizable space.

Proof. 1°. Observe that P is an isolated point iff there exists exactly one a such that $\{P\} = s(a)$. Moreover, by Lemma 2, if P is isolated then the above a is an atom of L.

2°. We use two theorems from Engelking [3] : Every locally compact space is a Tychonoff space (hence, it is regular). A space with a countable basis is metrizable iff it is a regular space.

Proof of Theorem 1. In [5] I.Moerdijk showed that the complete Heyting algebra of open subsets of the binary Kripke tree can be completely embedded into the algebra of open subsets of a metrizable space without isolated points and hence the latter is adequate for IPL. Since isomorphism preserves adequacy, by Lemma 3, L is adequate for IPL.

COROLLARY 1. If L is a complete atomless and infinitely join-distributive lattice in which the unit element is compact and which has a countable basis of complemented elements, then L is adequate for IPL.

Proof. Since 1 is compact, every element of L is cocompact. Observe that for any set A of complemented elements of L the formula (✱) holds for a $= \bigvee A$ and the rest of the proof follows.

COROLLARY 2. If the assumption in Theorem 1 and Corollary 1 that L is atomless is replaced by "L is not atomic" , then L is still adequate for IPL.

Proof. Recall from the proof of Lemma 4, that if P is an isolated point then there is an atom $a \in L$ such that $\{P\} = s(a)$. Now, if $P \in s(x)$ for any $x \in L$, and P is isolated then, by Lemma 3, $a \leqslant x$, for some atom a. Since L is not atomic it follows that there is an element $x \in L$ such that s(x) is an open set without isolated points. Hence s(x) is a metrizable open subspace of ΣL, without isolated points, and, by the result [5] quoted above, the algebra of open subsets of s(x) is adequate for IPL. It is known (for example, cf. [5] , Proposition 2.1 i) that, if Y is an open subspace of X, then the adequacy for IPL is preserved if one goes from Y to X.

2.EXAMPLES. Now we apply the results of the previous chapter to lattices of theories, lattices of congruences and lattices of filters.

THEOREM 2. The following lattices are adequate for IPL :

1°. The lattice of all theories extending a given theory which has no finitely axiomatizable complete extensions.

2°. The lattice of all theories extending an essentially undecidable theory (e.g. Peano arithmetic, ZF set theory).

3°. The lattice of all first-order theories (in a countable language with at least one binary relation symbol).

4°. The lattice of all extensions (closed with respect to Modus Ponens only) of classical propositional logic.

5°. The lattice similar to that in 4° but in purely implicative language.

6°. The lattice of all congruences of a countable bounded distributive lattice which is not weakly atomic.

7°. The lattice of all filters (ideals) of a countable Boolean algebra which is not atomic.

Proof. 1°. Consider a countable first-order language and let L_T be the set of all classical theories (i.e.sets of formulas containing classical predicate tautologies and closed with respect to Modus Ponens and Generalisation) extending a given theory. Then $\langle L_T, \subseteq \rangle$ is a complete infinitely join-distributive lattice with compact unit and a countable basis of complemented elements which have the form of finitely axiomatizable extensions of T. Let Ax denote all finitely axiomatizable consistent extensions of T. Since there is no complete theories in Ax, every $T' \in Ax$ can be essentially extended to $T'' \in Ax$. Now the set $Ax \cup \{\text{all formulas}, T\}$ forms a Boolean lattice. Taking complements one may prove that for every $T' \in Ax$ there is an essential subtheory $T'' \in Ax$. Hence L_T is atomless and, by Corollary 1, adequate for IPL.

6°. It is known (cf. eg. [4]) that the lattice Con L of all congruences of a lattice L is complete, algebraic and distributive. Compact elements of Con L of a bounded, i.e. with 0 and 1, distributive lattice L form a Boolean lattice, cf.[4],p.86. It follows that Con L has countable basis of complemented elements. The greatest element is compact since it equals to $\Theta(0,1)$. By the following theorem of Crawley and Dilworth, [2], p.78, Con L is not atomic : if a lattice L is distributive and Con L is atomic, then L is weakly atomic. Hence Con L satisfies all assumptions of Corollary 2. Remark: observe that L here cannot be algebraic.

REMARK. The assumption that a Boolean algebra (basis, lattice) in 7° (in Theorem 1, in 6° , respectively) is countable, is essential. The Boolean algebra RO([0,1]) of regular open subsets of the closed unit interval is atomless, complete and uncountable. The lattice of ideals (filters, congruences) of any complete Boolean algebra satisfies the Stone law: $-a \vee --a = 1$, (Tarski [7]), hence it cannot be adequate for intuitionistic logic, although the rest of assumptions are satisfied.

REFERENCES

1. B.Banaschewski and C.J.Mulvey , Stone-Cech compactification of locales I, Houston Journal of Math.vol.6, No.3,(1980),p.301.

2. P.Crawley and R.P.Dilworth, Algebraic theory of lattices, Prentice Hall, Englewood Cliffs, N.J. (1973).

3. R.Engelking, General Topology, PWN, Warsaw, (1977).

4. G.Gratzer, General Lattice Theory, Akademie-Verlag, Berlin (1978).

5. I.Moerdijk, Some topological spaces which are universal for intuitionistic predicate logic, Nederl. Akad. Wetensch. Ind.Math.,44,(1982),No.2, 227 - 235.

6. H.Rasiowa and R.Sikorski, Mathematics of the Metamathematics,PWN, Warsaw (1970).

7. A.Tarski, Sentential calculus and Topology, in : Logic, Semantics, Metamathematics : Papers from 1923 to 1938 (tran.J.H.Woodger), Oxford University Press, Oxford (1956).

A NOTE ON BOOLEAN MODAL LOGIC*

George GARGOV

Ling. Modelling Lab.
CICT - BAS
25a Acad. G. Bonchev Str.
Sofia 1113, Bulgaria

Solomon PASSY

Sector of Logic
Faculty of Mathematics
boul. Anton Ivanov 5
Sofia 1126, Bulgaria

ABSTRACT

We present a proof of a theorem mentioned in an earlier paper "Modal environment for Boolean speculations", devoted to the study of extended modal languages containing the so-called "window" or "sufficiency" modal operator ⊡. The theorem states that a particular axiom system for the poly-modal logic encompassing union, intersection and complement of relations (a Boolean analog of the propositional dynamic logic of Pratt, Fischer, Ladner and Segerberg) is complete for the standard Kripke semantics. Moreover this system modally defines the standard semantics - so in the terminology of the present paper the axiomatics is adequate. On the other hand our logic has the finite model property. Thus a fragment of second order logic, rather powerful with respect to expressiveness, turns out to be decidable.

0. INTRODUCTION

0.1. In a recent paper [GPT] we mentioned in passing the poly-modal logic of families of binary relations closed under the Boolean operations of union, intersection and complement. For this logic, called Boolean modal logic, an axiom system was given in [GPT] (with only a hint of the proof of completeness). Lately, similar investigations have been carried out in the area of dynamic logics where one is concerned with the input-output behaviour of programs (represented by binary relations). A typical case is the propositional dynamic logic PDL of Pratt, Fischer, Ladner and Segerberg (cf. e.g. [Har]). PDL is the logic of families of binary relations closed under the regular operations: union, composition and iteration (reflexive and transitive closure), so the analogy between our logic and PDL is obvious.

The problem with the Boolean modal logic, as compared with PDL, is that while union, composition and iteration are modally expressible (or modally definable - for a precise formulation see below) in the usual poly-modal language and the defining formulas are natural candidates for axioms, the operations of intersection and complement are not - hence there are no natural axioms for these operations, at least in the basic language.

To overcome the problems arising, we extend the modal language and its semantics with an additional modality ⊡, cf. [GPT], to obtain a suitable

*) Research partially supported by the Bulgarian Committee for Science, contracts # 56, # 247.

Mathematical Logic
Edited by P. P. Petkov
Plenum Press, New York, 1990

framework for the axiomatization of BML. Actually, there exists a complete axiomatization (see below) even in the classical modal language, but still we find the introduction of the extension justified, as explained in the concluding section.

0.2. Let us briefly review the main points of the Kripke approach to the semantics of modal languages. Assume for simplicity that the language has only one modal connective $\Box$. The formulas of the language are interpreted in models $\mathcal{M} = (W,R,V)$, over frames $\Phi = (W,R)$, where $W \neq \emptyset$, $R \subseteq W\times W$; V assigns to each formula A a subset of W according to a set of well known rules. We say that A is satisfied at $x \in W$, if $x \in V(A)$ and write $x \models A$. In this familiar notation it is clear that, given a model $\mathcal{M}$, the fact that $x \models A$ holds is equivalent to the truth of a first order formula in (W,R) - a formula in the first order language with predicates $u=v$, $R(u,v)$, and $P_k(v)$ for each propositional variable p_k. By K we shall denote the minimal modal logic.

The truth of a formula A in a model $\mathcal{M} = (W,R,V)$, $\mathcal{M} \models A$, is $\forall x \in W\ (x \models A)$; validity in a frame $\Phi = (W,R)$, $\Phi \models A$, is truth in every model over Φ. Thus validity of a modal formula A is equivalent to the truth of a second-order formula in the relational structure (W,R).

Let now Σ be a class of frames, Γ - a set of modal formulas.

<u>Definition</u>

a. $LG(\Sigma) =_{DF} \{A : \Phi \models A$, for all frames $\Phi \in \Sigma\}$;

b. $FR(\Gamma) =_{DF} \{\Phi : \Phi \models A$, for all formulas $A \in \Gamma\}$.

c. $K+\Gamma =_{DF} \{A : A$ is provable in K from instances of members of $\Gamma\}$.

The set $LG(\Sigma)$ is called <u>the logic of the class Σ</u> - it contains all the theorems of K and is also <u>normal</u>, i.e. closed under the rule of necessitation. If $\Sigma = FR(\Gamma)$ we say, cf. [Ben], that Σ is <u>modally definable</u> by the set Γ. If $K+\Gamma = LG(\Sigma)$, we say that Γ <u>axiomatizes</u> Σ. (Speaking of an axiomatization Γ of Σ, we shall tacitly assume that Γ is recursively enumerable.)

Given a set of frames Σ, two particular modal problems are:
1. Find, an axiomatization of Σ, if any, i.e. find Γ with $K+\Gamma = LG(\Sigma)$;
2. Find, a "modal definition" of Γ, if any, i.e. find Γ with $FR(\Gamma) = \Sigma$.

<u>Definition</u> We say that Γ is <u>an adequate axiomatization</u> for Σ, if both, Γ axiomatizes Σ and Γ modally defines Σ: $FR(\Gamma) = \Sigma$ and $LG(\Sigma) = K+\Gamma$.

EXAMPLES. Some well known facts illustrate the above notions:

1. The logic K is an adequate axiomatization of the class of all frames.

2. Most of the famous modal logics such as K, K4, T, B, S4, S5, etc. in most cases do not adequately axiomatize the corresponding classes of frames. For example: a) K (K4) adequately axiomatizes all (transitive) frames, but also axiomatizes all finite (transitive) frames, which axiomatization is therefore not adequate; b) the S5 axioms modally define the class of frames Φ where R is an equivalence relation (so S5 is an adequate axiomatization of this class), but at the same time S5 is the logic (hence inadequate) of the class of frames where R is the universal relation: $\Phi = (W, W\times W)$. Sahlqvist [Sal] has given many other examples of this sort.

Our aim in the present paper is to give an adequate axiomatization of the logic of Boolean operations over relations. Of course from results about modal definability in ordinary poly-modal languages (cf. e.g. [Gor2]) it follows that such an axiomatization is impossible without an extension of the language.

In the next section, we formulate an axiomatization of Boolean modal logic over K, which is clearly non adequate. The completeness proof employs the so-called copying construction the idea of which is due essentially to D. Vakarelov, and partially to T. Tinchev. Different versions of this ("important", cf. [GPT]) construction have been used in recent works on modal axiomatizations, cf. [Gor2, 3], [GPT], [Pen], [Vak1, 3]).

In the second section, an adequate axiomatization is given in the extension with the window operator $\boxdot$. It should be noted that while there can be given an axiomatization of the sort $K^{\sim}+\Gamma$ (where $K^{\sim}$ is the basic logic in the new language) we prefer to introduce a special new inference rule, for reasons discussed in the concluding remarks.

1. BOOLEAN MODAL LOGIC OVER K

1.1. The language of Boolean modal logic is based on infinitely many modal operators corresponding to relational terms. The notion of relational term is defined inductively as follows:

1. atomic relational terms π_1, π_2, ... are relational terms;
2. ∇ is a relational term;
3. if α and β are relational terms, so are and $\alpha\cap\beta$, $\alpha\cup\beta$ and $-\alpha$.

The set of relational terms is denoted by TM.

Formulas A, B, C are defined inductively (as usual):

- propositional variables p, q, r, ... and 0 (falsum) are formulas;
- formulas are closed under $\wedge$, $\vee$, $\neg$, and $\rightarrow$;
- if A is a formula and α is a relational term, then $\Box_\alpha A$ is a formula.

The set of all formulas is denoted by FOR; we assume the following:

$\Diamond_\alpha A =_{DF} \neg\Box_\alpha\neg A$, $1 =_{DF} \neg 0$, $A\leftrightarrow B =_{DF} (A\rightarrow B)\wedge(B\rightarrow A)$, $\lambda =_{DF} -\nabla$.

SEMANTICS. Formulas are interpreted in models $\mathcal{M} = (W, R, V)$ where:

- W is a non-empty set, R: TM $\rightarrow 2^{W\times W}$, and V: FOR $\rightarrow 2^W$ have the following properties:
- $R(\nabla) = W \times W$;
- $R(\alpha\cap\beta) = R(\alpha)\cap R(\beta)$;
- $R(\alpha\cup\beta) = R(\alpha)\cup R(\beta)$;
- $R(-\alpha) = W\backslash R(\alpha) = -R(\alpha)$;
- $V(0) = \emptyset$;
- $V(A\wedge B) = V(A) \cap V(B)$; etc.
- $V(\Box_\alpha A) = \{x\text{: } \forall y(xR(\alpha)y \text{ implies } y\in V(A))\}$.

Adhering to the common usage described in the Introduction we call a formula A satisfied at a state (or a world) x of W iff $x\in V(A)$. A is true in a model if it is satisfied everywhere in the set W. Finally, A is valid if it is true in every model.

NOTE. The above defined notion of a model is rather special as it includes requirements for the relations $R(\alpha)$, which are not modally definable over K. To underline the importance of this fact we call these models standard models. Thus the above validity is in fact validity with respect to standard models. In the process of proving completeness for the system axiomatizing the valid formulas we will need a more general notion of a model.

AXIOMATICS. The set of all valid formulas can be generated by the following axiomatic system BML which is based on classical propositional logic and has as additional axiom schemes and rules:

1. The K axiomatics of $\Box_\alpha$, for each α:
   $\Box_\alpha(A \rightarrow B) \rightarrow (\Box_\alpha A \rightarrow \Box_\alpha B)$
   If $\vdash A$, then $\vdash \Box_\alpha A$ (the necessitation rule).

2. Basic BML axioms:

$\Box_\alpha A \land \Box_\beta A \to \Box_{\alpha\cup\beta} A$

$\Box_\nu A \to A$

$\Box_\nu A \to \Box_\nu \Box_\nu A$

$A \to \Box_\nu \Diamond_\nu A$

$\Box_\lambda 0$.

For the relational terms α and β, we say that α is <u>included in</u> β, denoted by $\alpha \subseteq \beta$, if such an inclusion holds between the respective expressions, over the Boolean algebra based on the atomic relational terms, ν and the operations $\cap$, $\cup$, $-$; in other words $\alpha \subseteq \beta$, iff the terms α and $\alpha\cap\beta$ (or β and $\alpha\cup\beta$) are equivalent in this algebra.

3. The Boolean axiom, BA:

$\Box_\alpha A \to \Box_\beta A$, if $\beta\subseteq\alpha$.

Since $\subseteq$ is a decidable relation, BML has a recursive set of axioms.

<u>Lemma</u> BML $\vdash \Box_{\alpha\cup\beta} A \to \Box_\alpha A \land \Box_\beta A$

BML $\vdash \Box_\alpha A \lor \Box_\beta A \to \Box_{\alpha\cap\beta} A$

<u>Lemma</u> BML is sound with respect to standard models.

<u>Proof</u> Straightforward verification.

1.2. Our next task is to show that the system BML is complete w.r.t. standard models. In order to do this we introduce - in the spirit of [GPT] - the notion of a generalized model and prove first a completeness theorem for BML with respect to the new class of models and then as a second step we show that if there is a generalized model refuting a formula A, non-provable in BML, then this model can be converted into a standard one (modally equivalent to the original) so A will be refutable in a standard model, too. Under this conversion, generally speaking, the size of the model increases, but not "too much": a finite model is converted into a finite standard model. This fact allows us to prove that BML has the finite (standard) model property and is hence decidable.

<u>Definition</u> A Kripke model $\mathcal{M} = (W,R,V)$ for our language is called a <u>g-model</u>, if the following conditions are met:

g1) $R(\alpha) \subseteq R(\beta)$, if $\alpha \subseteq \beta$;

g2) $R(\nu) = W \times W$;

g3) $R(\alpha\cup\beta) \subseteq R(\alpha) \cup R(\beta)$.

<u>Lemma</u> In each g-model $\mathcal{M} = (W,R,V)$ it is the case that

$R(\alpha\cup\beta) = R(\alpha) \cup R(\beta)$,

$R(\alpha\cap\beta) \subseteq R(\alpha) \cap R(\beta)$, and

$R(\alpha)\cup R(-\alpha) = W^2$.

Thus g-models differ from the standard ones since they are based on frames where only $\cup$ is a "real" union, but the operations denoted by $\cap$ and $-$ correspond only partially to "real" intersection and complement: g-models are models based on <u>non-standard frames</u>.

<u>Theorem 1</u> If a formula A is not a theorem of BML, then it has a g-countermodel.

<u>Proof</u> Apply the well known canonical model construction of Scott-Makinson-Segerberg to the logic BML: consider maximal consistent theories x, y, z, ... and for each $\alpha\in TM$ set $xR(\alpha)y$ iff $\forall B(\Box_\alpha B\in x$ implies $B\in y)$. Set also (in order to define truth in the canonical model): $x\in V(B)$ iff $B\in x$. Since A is not provable, there is an x_0 such that not $A\in x_0$. Take now the submodel of the canonical model, generated by x_0. Call it $\mathcal{M}_0 = (W_0,R_0,V_0)$. It is fairly easy to show that:

i) M_0 is a g-model:
For example $\Box_\alpha A \wedge \Box_\beta A \rightarrow \Box_{\alpha\cup\beta}A$ and BA guarantee that in the canonical model (and hence in M_0) $R_0(\alpha\cup\beta) = R_0(\alpha) \cup R_0(\beta)$; the axioms about ∇ and BA make $R(\nabla)$ an equivalence relation greater than all other relations in the canonical model, so in the generated model ∇ is interpreted as the square of the set of worlds.

ii) all formulas in BML are true in M_0, but A is refuted at x_0 - This is the familiar truth lemma, cf. e.g., [Ben]. QED

Having a g-model where BML holds but A is refuted, the next step is to find a finite such g-model. To this end we may apply the filtration technique of Lemmon-Scott-Segerberg. Let Γ be a finite set of formulas, closed under subformulas and satisfying additionally the conditions:

- $A\in\Gamma$;
- if $\Box_{\alpha\cup\beta}B\in\Gamma$ or $\Box_{\alpha\cap\beta}B\in\Gamma$, then $\Box_\alpha B$ and $\Box_\beta B$ belong to Γ, too;
- if $\Box_{-\alpha}B\in\Gamma$, then $\Box_\alpha\neg B\in\Gamma$.

The minimal filtration M of M_0 with respect to Γ is defined as follows:

- $|x| = \{y$: for all $B\in\Gamma$, $x\in V_0(B)$ iff $y\in V_0(B)\}$, $W = \{|x| : x\in W_0\}$;
- for any relational term α appearing in Γ we set
  $R(\alpha) = \{(|x|,|y|)$: for some $x'\in|x|$ and $y'\in|y|$, $x'R_0(\alpha)y'\}$

It is a standard exercise to prove that M is a g-model (as far as the relational terms occuring in Γ are concerned) and that the non-provable formula A is refuted at $|x_0|$.

Thus we have the finite g-model property for BML, in other words:

<u>Theorem 2</u> If a formula A is not a theorem of BML, then it is refuted in a finite g-model.

1.3. Next we show how a g-model refuting a given formula A can be transformed into a standard model with the same property. Our procedure will be such that a finite g-model will generate a finite standard model. This will ensure the finite completeness of BML with respect to standard models.

We shall need a special operation defined for models of the Boolean modal language. Consider a model $M = (W,R,V)$ and a family $\mathfrak{F}_k = \{M : k\in I\}$ where I is some index set and all models M_k are isomorphic copies of M with pairwise disjoint sets of worlds W_k, so there is, for any $k\in I$, an isomorphism φ_k: $M_k \rightarrow M$ (in particular φ_k maps W_k onto W). Let $\varphi =_{DF} \cup\varphi_k$.

<u>Definition</u> A Kripke model $M_{\mathfrak{F}} = (W_{\mathfrak{F}},R_{\mathfrak{F}},V_{\mathfrak{F}})$, where $W_{\mathfrak{F}} = \cup\{W_k: k\in I\}$, is called <u>an $\mathfrak{F}$-power</u> of M if the following conditions are met:

1. $xR_{\mathfrak{F}}(\alpha)y$ implies $\varphi(x)R(\alpha)\varphi(y)$, for $x,y \in W_{\mathfrak{F}}$;
2. $\varphi(x)R(\alpha)y$ implies $\exists z\in W_{\mathfrak{F}}(xR_{\mathfrak{F}}(\alpha)z$ and $\varphi(z)=y)$, for $x\in W_{\mathfrak{F}}$, $y\in W$;
3. $x\in V_{\mathfrak{F}}(p)$ iff $\varphi(x)\in V(p)$, for a propositional variable p and $x\in W_{\mathfrak{F}}$.

<u>Basic lemma (truth in an $\mathfrak{F}$-power)</u> If $M_{\mathfrak{F}}$ is an $\mathfrak{F}$-power of M, then for any formula A and all $x\in W_{\mathfrak{F}}$:

$x\in V_{\mathfrak{F}}(A)$ iff $\varphi(x)\in V(A)$.

<u>Proof</u> This follows from the observation that the $\mathfrak{F}$-power is p-morphically mapped, by φ, onto the original model, and the theorem about p-morphisms, cf. e.g. [Ben]. QED

The lemma shows that $\mathfrak{F}$-powers are modally equivalent to the base model. The (non-deterministic) construction of an $\mathfrak{F}$-power is very useful when we want to obtain a modally equivalent model with some special properties of the relational part $R_{\mathfrak{F}}$.

Examples a) An example of this construction is the maximal $\mathcal{F}$-power of $\mathcal{M}$, for a given family $\mathcal{F}$ of pairwise disjoint isomorphic copies of $\mathcal{M}$ which is defined by postulating

$xR_{\mathcal{F}}(\alpha)y$ iff $\varphi(x)R(\alpha)\varphi(y)$ for any $\alpha \in TM$ and $x,y \in W_{\mathcal{F}}$.

b) The minimal $\mathcal{F}$-power of $\mathcal{M}$ is defined with the help of a function $\mu: W_{\mathcal{F}} \to I$ for which $\mu(x)=k$ iff $x \in W_k$. Here we set: $xR_{\mathcal{F}}(\alpha)y$ iff $\mu(x)=\mu(y)$ and $\varphi(x)R(\alpha)\varphi(y)$.

c) If, in a model $\mathcal{M} = (W,R,V)$, one has for the relational terms α, β

$R(\alpha \cap \beta) \subseteq R(\alpha) \cap R(\beta)$,

then we can define an $\mathcal{F}$-power of $\mathcal{M}$, where the intersection of α and β is real, i.e. where $R_{\mathcal{F}}(\alpha\cap\beta)=R_{\mathcal{F}}(\alpha)\cap R_{\mathcal{F}}(\beta)$. Take $\mathcal{F} = \{\mathcal{M}_1, \mathcal{M}_2\}$ and let $R_{\mathcal{F}}$ be defined for α and β as follows:

$xR_{\mathcal{F}}(\alpha)y$ iff $\varphi(x)R(\alpha)\varphi(y)$ and
if $\varphi(x)R(\beta)\varphi(y)$ and not $\varphi(x)R(\alpha\cap\beta)\varphi(y)$, then $\mu(y) = 1$;
$xR_{\mathcal{F}}(\beta)y$ iff $\varphi(x)R(\beta)\varphi(y)$ and
if $\varphi(x)R(\alpha)\varphi(y)$ and not $\varphi(x)R(\alpha\cap\beta)\varphi(y)$, then $\mu(y) = 2$;

for all other relational terms take the maximal definition from Example (a). This new model is with real intersection at α and β, but the relationships between them and the rest of the relational terms in the model are possibly violated, i.e. there is no guarantee that for instance if $\mathcal{M}$ was a g-model the new one will remain a g-model.

To prove the claim about $\alpha\cap\beta$ let us assume $xR_{\mathcal{F}}(\alpha\cap\beta)y$. Then $\varphi(x)R(\alpha\cap\beta)\varphi(y)$ cf. the definition in Example (a), and by the assumption on $\mathcal{M}$ we have $\varphi(x)R(\alpha)\varphi(y)$ and $\varphi(x)R(\beta)\varphi(y)$. Now by the definitions of $R_{\mathcal{F}}(\alpha)$, $R_{\mathcal{F}}(\beta)$, one has $xR_{\mathcal{F}}(\alpha)y$ and $xR_{\mathcal{F}}(\beta)y$. In this way we obtain $R_{\mathcal{F}}(\alpha\cap\beta) \subseteq R_{\mathcal{F}}(\alpha)\cap R_{\mathcal{F}}(\beta)$.

Conversly, if $xR_{\mathcal{F}}(\alpha)y$ and $xR_{\mathcal{F}}(\beta)y$, then $\varphi(x)R(\alpha)\varphi(y)$ and $\varphi(x)R(\beta)\varphi(y)$. Now if not $xR_{\mathcal{F}}(\alpha\cap\beta)y$ is assumed, then, by the definition in Example (a), we have not $\varphi(x)R(\alpha\cap\beta)\varphi(y)$. Since the left-hand sides in the definition of $R_{\mathcal{F}}$ are true, the right-hand sides are also true, therefore the implications must hold, but then the antecedents of these implications being true, the consequents must be true, too. Thus we get $1 = 2$ – a contradiction. So $R_{\mathcal{F}}(\alpha)\cap R_{\mathcal{F}}(\beta) \subseteq R_{\mathcal{F}}(\alpha\cap\beta)$. The two inclusions give $R_{\mathcal{F}}(\alpha\cap\beta) = R_{\mathcal{F}}(\alpha)\cap R_{\mathcal{F}}(\beta)$.

d) If, in a model $\mathcal{M} = (W,R,V)$, one has for the relational terms α and β

$R(\alpha)\cup R(\beta)=W\times W$,

then we can define an $\mathcal{F}$-power of $\mathcal{M}$ where $R_{\mathcal{F}}(\alpha) = -R_{\mathcal{F}}(\beta)$. To this end consider $\mathcal{F} = \{\mathcal{M}_1,\mathcal{M}_2\}$ and set

$xR_{\mathcal{F}}(\alpha)y$ iff $\varphi(x)R(\alpha)\varphi(y)$ and
if $\varphi(x)R(\beta)\varphi(y)$, then $\mu(x)\neq\mu(y)$.

For $R_{\mathcal{F}}(\beta)$ the definition is symmetric (interchanging the places of α and β, and of $\neq$ and $=$). In these circumstances we claim that

$R_{\mathcal{F}}(\alpha) = -R_{\mathcal{F}}(\beta)$.

Indeed for arbitrary $x,y \in W_{\mathcal{F}}$ one cannot have $xR_{\mathcal{F}}(\alpha)y$ and $xR_{\mathcal{F}}(\beta)y$: from the definition it is seen that in such a case we would have both $\varphi(x)R(\alpha)\varphi(y)$ and $\varphi(x)R(\beta)\varphi(y)$, thus $\mu(x)\neq\mu(y)$ and $\mu(x)=\mu(y)$ – a contradiction. On the other hand $xR_{\mathcal{F}}(\alpha)y$ or $xR_{\mathcal{F}}(\beta)y$ – this also easily follows from the definition of $\mathcal{M}_{\mathcal{F}}$ (where the interpretations of the remaining relational terms are given maximally).

So we have a model where the interpretation of α is the complement of the interpretation of β. Applying this construction to a pair α, $-\alpha$ one can get (starting from a g-model $\mathcal{M}$) models $\mathcal{M}_{\mathcal{F}}$ where $R_{\mathcal{F}}(-\alpha) = -R_{\mathcal{F}}(\alpha)$. Again it should be mentioned that this process does not in general preserve the property of g-modelhood.

When looking for a finite standard model of a formula A, non-provable in BML, it should be taken into account that the only relational terms whose interpretations affect the truth of A are the ones built from the atomic members of TM occuring in A. The interpretations of other relational terms are irrelevant. So we need to worry only about a finite number of TM members.

Let us assume then, for simplicity of presentation, that the only atomic relational terms occuring in the refutable formula A are π and δ. Models will be presented partially - we shall describe only the part relevant to A. Thus in a model we are going to define $R(\alpha)$ only for α built from π and δ. We define the basic relations: $R_1 = R(\pi\cap\delta)$, $R_2 = R(-\pi\cap\delta)$, $R_3 = R(\pi\cap-\delta)$, and $R_4 = R(-\pi\cap-\delta)$.

By Theorem 2 we have a finite g-model $\mathcal{M} = (W,R,V)$ refuting A, if A is not provable in the logic BML. Under the above assumptions this means in particular that:

0. for any α built from π and δ, $R(\alpha)$ is a union of some of the basic relations;

1. the union of all four basic relations yields $W\times W$;

On the other hand the model need not be standard, i.e. pairs of worlds (s,t) are possible such that

2. s and t are connected by at least two basic relations (in such a case we say that there is a conflict at the pair (s,t)).

We are next going to define an $\mathcal{F}$-power of $\mathcal{M}$, where $\mathcal{F}$ will be a four element family $\{\mathcal{M}_1, \mathcal{M}_2, \mathcal{M}_3, \mathcal{M}_4\}$. The definition will aim at "killing" all conflicts .

Definition The model $\mathcal{M}_{\mathcal{F}} = (W_{\mathcal{F}}, R_{\mathcal{F}}, V_{\mathcal{F}})$ is defined as follows:

$xR_{\mathcal{F}}(\pi\cap\delta)y$ iff $\varphi(x)R_1\varphi(y)$ and

if $\varphi(x)R\ \varphi_k(y)$ then $\mu(y)\neq k$ (k= 2, 3, 4).

The relations $R_{\mathcal{F}}(\pi\cap-\delta)$, etc. are defined symmetrically. (Note that this definition is a generalization of Example (d) above.) The rest of the relational terms built from π and δ get their interpretations in the following way: for a term α we define $R_{\mathcal{F}}(\alpha)$ to be that union of basic relations $R_{\mathcal{F}}(\pi\cap\delta)$, $R_{\mathcal{F}}(\pi\cap-\delta)$, $R_{\mathcal{F}}(-\pi\cap\delta)$, $R_{\mathcal{F}}(-\pi\cap-\delta)$, which comes from the disjunctive normal form representation of α in the Boolean algebra based on TM. As for $V_{\mathcal{F}}$ - it is defined for variables p by $x\in V_{\mathcal{F}}(p)$ iff $\varphi(x)\in V(p)$.

Lemma The model $\mathcal{M}_{\mathcal{F}}$ is indeed an $\mathcal{F}$-power of $\mathcal{M}$.

Proof We have to check conditions 1 - 3 of the definition of an $\mathcal{F}$-power. The proof proceeds by induction on $\alpha\in$TM (built from π and δ). Assume $xR_{\mathcal{F}}(\alpha)y$. If α is one of the basic terms ($\pi\cap\delta$, $-\pi\cap\delta$, $\pi\cap-\delta$, $-\pi\cap-\delta$), then from the definition it follows immediately that $\varphi(x)R(\alpha)\varphi(y)$. Again, an inspection of the definition will show that condition 2 is also fulfilled.

For a term α which is compound, we have that $R_{\mathcal{F}}(\alpha)$ = some union of basic relations. This clearly implies that $\varphi(x)R(\alpha)\varphi(y)$ if $xR_{\mathcal{F}}(\alpha)y$. Moreover, if $\varphi(x)R(\alpha)y$, then for some of the basic relations ($R(\pi\cap\delta)$ or $R(-\pi\cap\delta)$, etc.) $\varphi(x)$ and y will be in that relation. Using the property of basic relational terms established above we get that there is a z such that $\varphi(z)=y$ and x and z are in relation, corresponding to the basic one in $\mathcal{M}_{\mathcal{F}}$. That implies $xR_{\mathcal{F}}(\alpha)z$ and we are done with the second condition for an $\mathcal{F}$-power.

The third condition is also (obviously) fulfilled. QED

In this model $\mathcal{M}_{\mathcal{F}}$ (which is modally equivalent to the original one, at least in respect to formulas, relevant to the refuted formula A) the interpretations of relational terms built from π and δ have some nice properties:

Lemma In $\mathcal{M}_{\mathcal{F}}$, all pairs of worlds x, y have empty sets of conflicts.

Proof Look at the definition!

Corollary In $\mathcal{M}_{\mathcal{F}}$, the basic relations are pairwise disjoint.

<u>Lemma</u> The $\mathfrak{F}$-power is a standard model (w.r.t. the fragment concerning the formula A, but such partially standard models can obviously be extended to standard models for the full language).
<u>Proof</u> This is obvious in the light of the lack of conflicts.

Thus we have established that A has a standard countermodel since $\mathcal{M}_{\mathfrak{F}}$ is modally equivalent to $\mathcal{M}$. Moreover, A is refutable in a finite model. It should be noted that when k atomic relational terms occur in A, then the construction needs 2^k isomorphic copies of the original g-model.

<u>Theorem 3</u> BML is complete with respect to standard models, has the finite model property and is decidable with exponential complexity.

NOTE. Decidability is not more than exponentially complex: we need only check satisfiability in g-models since any such model can be transformed into a standard one.

2. BOOLEAN MODAL LOGIC OVER K~

2.1. In [GPT] an extension K~ of K is studied with the "sufficiency" modal operator $\boxplus$, and K~ is shown to be more appropriate than K for certain metamathematical purposes. The poly-K~ version, of $\Box_\alpha$ and $\boxplus_\alpha$, for several α (call it from now on BML~), was formulated, and a theorem [GPT, Section 2, last theorem] was stated, claiming BML~ complete for poly-modal frames of Boolean closed families of relations. We prove in this section that BML~ indeed presents an adequate axiomatization of Boolean closed modalities. We do not repeat the "copying"-completeness proof, nor do we refer to it: we do, instead, translate BML~ into BML, and refer to the completeness of the latter.

SYNTAX. Add to the above language a modality $\boxplus_\alpha$, for $\alpha \in TM$.

SEMANTICS. In models (W, R, V) interpret $\boxplus$ as follows
- $V(\boxplus_\alpha A) = \{x : \forall y(y \in V(A) \text{ implies } xR(\alpha)y)\}$.

Define now (following [GPT]) BML~ as the system based on classical propositional logic and having as additional axioms and rules:

1. The K~ (cf. [GPT]) axioms concerning $\Box_\alpha$ and $\boxplus_\alpha$, for each α:
   $\Box_\alpha(A \to B) \to (\Box_\alpha A \to \Box_\alpha B)$
   $\boxplus_\alpha A \wedge \boxplus_\alpha(\neg A \wedge B) \to \boxplus_\alpha B$
   $[U_\alpha]A \to A$
   $[U_\alpha]A \to [U_\alpha][U_\alpha]A$
   $A \to [U_\alpha]\langle U_\alpha\rangle A$
   (here $[U_\alpha]A =_{DF} (\Box_\alpha A \wedge \boxplus_\alpha\neg A)$, and $\langle U_\alpha\rangle =_{DF} \neg[U_\alpha]\neg$).

   If $\vdash A$, then $\vdash \Box_\alpha A$ (necessitation rule)
   If $\vdash A$, then $\vdash \boxplus_\alpha\neg A$ ("sufficiency" rule)

2. Basic BML~ axioms :
   $\Box_{\alpha\cup\beta}A \leftrightarrow \Box_\alpha A \wedge \Box_\beta A$
   $\boxplus_{\alpha\cap\beta}A \leftrightarrow \boxplus_\alpha A \wedge \boxplus_\beta A$
   $\Box_{-\alpha}A \leftrightarrow \boxplus_\alpha\neg A$
   $\boxplus_{-\alpha}A \leftrightarrow \Box_\alpha\neg A$
   $[U_\alpha]A \leftrightarrow \Box_{\mathbf{1}}A$
   $\boxplus_{\mathbf{1}}\bot$

3. The special BML~ rules:
   If $\vdash A(p)$, then $\vdash A(B)$: the substitution rule;
   If $\vdash \Box_\alpha p \to (\Box_\beta p \to \Box_\gamma p)$, then $\vdash \Box_\alpha p \to (\boxplus_\gamma\neg p \to \boxplus_\beta\neg p)$: Boolean Rule, BR.

The logic BML~ has the following important property:

<u>Proposition</u> BML~ modally defines the class of standard frames among all poly-K~ frames, i.e. a poly-modal frame for K~ is with real intersection, union and complement iff BML~ is valid in it.
<u>Proof</u> Straightforward verification. For example, the validity of $\boxplus_{\alpha\cap\beta}p \leftrightarrow \boxplus_{\alpha}p \wedge \boxplus_{\beta}p$ guarantees that the frame under consideration is with real intersection of relations, i.e. one has $R(\alpha\cap\beta)=R(\alpha)\cap R(\beta)$. QED

Thus if BML~ should be complete, then it will give an <u>adequate</u> (in the new language) axiomatization of the class of Boolean frames.

<u>Corollary</u> BML~ is sound.

2.2. Completeness will be established by reducing BML~ to BML.

<u>Definition</u> Let us define a translation σ from the language of BML~ into the modal language of BML (which is obviously a sublanguage of the former). The translation σ acts as follows:

$\sigma(p)=p$, for propositional variables p,

$\sigma(0)=0$, σ commutes with the Boolean connectives and $\Box_{\alpha}$, and

$\sigma(\boxplus_{\alpha}A) = \Box_{-\alpha}\neg\sigma(A)$.

NOTE. BML~ $\vdash A \leftrightarrow \sigma(A)$.

<u>Key fact</u> BML~ $\vdash A$ iff BML $\vdash \sigma(A)$.
<u>Proof</u> The proof of only-if-part goes by induction on the proof of A in BML~:

First we must show that an axiom A of BML~ translates into a theorem $\sigma(A)$ of BML, which is a routine matter. Next we should check that rules of BML~ translate into admissible rules of BML: this is obvious for Modus Ponens and the rule of necessitation, while the sufficiency rule translates into a necessitation rule. As for the special BML~ rules - the substitution is not problematic, and BR translates into a rule which is obviously admissible:

if $\vdash \Box_{\alpha}p \rightarrow (\Box_{\beta}p \rightarrow \Box_{\tau}p)$, then $\vdash \Box_{\alpha}p \rightarrow (\Box_{-\tau}p \rightarrow \Box_{-\beta}p)$.

Thus all theorems of BML~ translate into theorems of BML.

For the if-direction, it is enough to establish the claim for a $\boxplus$-free formula A: the general case is then simply reducible to this special instance, in virtue of BML~ $\vdash A \leftrightarrow \sigma(A)$.

To this purpose we argue by induction on the proof of A in BML: if A is an (instance of an) axiom then A is provable in BML~, too. The only non-trivial problem here is the axiom BA. An instance of this axiom can be established in BML~ by means of some particular cases (expressing modally the familiar laws of commutativity, associativity, distributivity, etc., i.e. all the axioms of Boolean algebra - in a relativised form):

$\Box_{\alpha}p \rightarrow (\Box_{\beta\cap\tau}p \leftrightarrow \Box_{\tau\cap\beta}p)$;

$\Box_{\alpha}p \rightarrow (\Box_{\beta\cap(\tau\cap\delta)}p \leftrightarrow \Box_{(\beta\cap\tau)\cap\delta}p)$;

....

$\Box_{\alpha}p \rightarrow (\Box_{(\delta\cup\beta)\cap\tau}p \leftrightarrow \Box_{(\delta\cap\tau)\cup(\beta\cap\tau)}p)$.

All these particular cases are provable in BML~ (mainly utilizing BR). We give an example here (and leave the rest to the interested reader): The last formula of the above list is provable in BML~ -

$\Box_{\alpha}p \rightarrow (\Box_{(\delta\cup\beta)\cap\tau}p \rightarrow \Box_{(\delta\cap\tau)\cup(\beta\cap\tau)}p)$ is established by an application of BR to the axiom concerning $\boxplus$ for a term with $\cap$ as main functor and the axiom for the union.

$\Box_{\alpha}p \rightarrow (\Box_{(\delta\cap\tau)\cup(\beta\cap\tau)}p \rightarrow \Box_{(\delta\cup\beta)\cap\tau}p)$ is proved as follows: first we use the axiom for union and get that the antecendent is conjunction of $\Box_{\delta\cap\tau}p$ and $\Box_{\beta\cap\tau}p$, from each of the conjuncts we turn (by means of axioms relating $\Box$ and $\boxplus$) to formulas starting with $\boxplus$; applying the $\cap$-axioms and BR we obtain the desired result. QED

From the above fact it follows that in the setting of many relations and the full spectrum of Boolean operations the window and box modalities become interdefinable. Note that this holds only for standard frames, i.e. frames with real Boolean operations on relations.

Now the completeness of BML~ can be established from the completeness of BML as follows: if A is not provable in BML~, then $\sigma(A)$ is not a theorem of BML, so there is a countermodel (standard!) for $\sigma(A)$. But it is very easy to verify that this model (viewed as a K~ model) is a countermodel for A. Thus we have:

Theorem 4 BML~ gives a complete and hence an adequate axiomatization of the logic of the Boolean operations on binary relations.

NOTE. BML~ clearly enjoys the finite model property, hence is decidable.

3. CONCLUDING REMARKS.

The theorems for which we have given proofs in the above two sections may be considered interesting in at least two respects:

1. BML is a close analog of PDL and can serve the same purposes in some areas of theoretical computer science (cf. e.g. the work of Vakarelov [Vak2] where it is pointed out that the modal logics of families of relation closed under different systems of operations play an important role in the theory of information systems and knowledge representation).

2. BML~ axiomatizes the same class of frames, but the axiomatization is now adequate: Boolean closed modal frames are fully described (defined and axiomatized simultaneously) in the class of poly-modal frames with infinitely many independent relations. We find this the appropriate moment to attract the attention to some advantages of the adequate axiomatizations which are achieved in the language of K~: for example the class of the frames $(W, W \times W)$ is adequately axiomatized over K~ by the axiom ⊡1. Especially in applications, when starting from a given semantics, it is always preferable to have this semantics not only axiomaized, but also determined by the axiomatization.

Another contribution is the method of transforming generalized models into standard ones by means of $\mathfrak{F}$-powers, which allows some new completeness proofs which were not accessible before. BML and BML~ serve just as picturesque examples of this "new" techniques, versions of which seem to have been independently reinvented by Sahlqvist [Sal], Shehtman [She], and Vakarelov and Tinchev [private communication].

As for the promised in the introduction discussion of the role of rules of proof, which are redundant in the basic system (be it K or K~) we mention the following incompleteness phenomenon: it is not always the case that if Γ defines a class Σ, then K+Γ axiomatizes LG(Σ). With the extensions of K~ such situation is typical, cf. [Gor2, 3], so the axiomatization of K~ is lacking in this respect. If we add to poly-K~ the Boolean rule BR of BML~, which is admissible, then we have a system which is "more complete" in the sense that in more cases if Γ modally defines Σ over the language of K~, then K~+Γ+BR axiomatizes LG(Σ). (This problem will be treated in detail in a forthcoming paper.) On the other hand, in contrast with BA, BR does not appeal to the Boolean structure of the relational terms, and reduces the Boolean peculiarities to the □-⊡ nature of our language, which therefore might be distinguished as adequate for the Boolean modal logic.

As for the history of intersection and negation, we should mention [Har] and [PT], and for more details, the references from [GPT]. The history of ⊡

is, more or less, outlined in [GPT], and the topic is developed in [Hum] and [Gor1, 2, 3]. A related story concerns the "complement-free" study of intersection. Here we should mention a special case in which the atomic terms are equivalence relations, and the operations are intersection and transitive-closure-of-union: this logic of equivalence relations, called DAL, was introduced by Fariñas del Cerro and Orlowska [FO], and later studied in [Gar1, 2] and [AT]. In the present paper, we use a $\subseteq$-version of BA, just as in [AT], instead of the equivalence-version used in the draft:

$\Box_\alpha A \leftrightarrow \Box_\beta A$, whenever $\alpha\subseteq\beta$ & $\beta\subseteq\alpha$.

Finally, we acknowledge with thanks the helpful remarks of the anonimous referees, and the fruitful influence of the non-anonimous part of Sofia non-classical logic group: Dimiter Vakarelov, Slavjan Radev, Tinko Tinchev, and particularly of Valentin Goranko who made stimulating comments.

REFERENCES

[AT] D. Arhangel'skij and M. Taiclin, On a Logic for Data Analysis - Abstract (in Russian), Abstracts of the 9-th All-Union Conference on Mathematical Logic, Leningrad, September, 1988.

[Ben] J. van Benthem, Modal Logic and Classical Logic, Bibliopolis, Napoli, 1986.

[FO] L. Fariñas del Cerro and E. Orlowska, DAL - A Logic for Data Analysis, Theor. Comp. Sci., 35, 1985, 251-264.

[Gar1] G. Gargov, Two Completeness Theorems in the Logic for Data Analysis, ICS-PAS Reports # 581, Warsaw, 1986.

[Gar2] G. Gargov, Modal Logics of Families of Equivalence Relations (to appear).

[GPT] G. Gargov, S. Passy and T. Tinchev, Modal Environment for Boolean Speculations, in: Mathematical Logic and Its Applications (ed. D. Skordev), Proc. of the 1986 Goedel Conference, Plenum Press, New York, 1987, 253-263.

[Gor1] V. Goranko, Modal Definability in Enriched Languages, Notre Dame Journal of Formal Logic, to appear.

[Gor2] V. Goranko, Definability and Completeness in Poly-Modal Logics, Ph.D. Thesis, Sofia University, January 1988 (in Bulgarian).

[Gor3] V. Goranko, Completeness and Incompleteness in the Bi-modal Base $\mathfrak{L}(R,-R)$, this volume.

[Har] D. Harel, Dynamic Logic, in: Handbook of Philos. Logic, vol. II (eds. D. Gabbay & F. Guenthner), D. Reidel. Publ. Comp., 1984, 605-714.

[Hum] I.L. Humberstone, The Modal Logic of 'All and Only', Notre Dame Journal of Formal Logic, 28, 1987, 2, 177-188.

[PT] S. Passy and T. Tinchev, PDL with Data Constants, Inf. Proc. Lett. 20, 1985, 35-41.

[Pen] W. Penczek, A Temporal Logic for Event Structures, this volume.

[Sal] H. Sahlqvist. Completeness and Correspondence in First and Second Order Semantics for Modal Logic, in: S. Kanger (ed.), Proceedings of the third Scandinavian logic symposium, Uppsala 1973, North-Holland, Amsterdam, 1975.

[Seg] K. Segerberg, A Completeness Theorem in the Modal Logic of Programs, Notices of the AMS, 24, 6(1977), A-552.

[She] V. Shehtman, Two-Dimensional Modal Logics (in Russian), Matematicheskie Zametki, 23, 1978, 759-772.

[Vak1] D. Vakarelov, S4 and S5 Together - S4+5 (Abstract), in: Proc. of the VIII Int. Congress LMPS'87, Moscow 1987, vol. 5, part 3, 222.

[Vak2] D. Vakarelov, Modal Logics for Knowledge Representation Systems, to appear in: Proc. of SCT'87, Sixth Symposium on Computation Theory, Wendisch - Rietz, DDR.

[Vak3] D. Vakarelov, Modal Characterization of the Classes of Finite and Infinite Quasi-ordered Sets, this volume.

COMPLETENESS AND INCOMPLETENESS IN THE BIMODAL BASE $\mathcal{L}(R,-R)$

Valentin Goranko

Sector of mathematical logic
Faculty of Mathematics and Computer Science, Sofia University
boul. Anton Ivanov 5, Sofia 1126, BULGARIA

INTRODUCTION[1,2]

The paper deals with a modal language $\mathcal{L}(R,-R)$, having an ordinary modality $\boxplus$ (dual - $\Diamond$) with an usual Kripke-semantics $x \vDash \boxplus\varphi$ iff $\forall y(Rxy \rightarrow y \vDash \varphi)$ and an additional modality $\boxminus$ (dual - $\Diamond$), with the same semantics however over the complement $-R$ of R: $x \vDash \boxminus\varphi$ iff $\forall y(-Rxy \rightarrow y \vDash \varphi)$. Such a modality has been considered by some authors in different contexts - see e.g. [Hum] and [GPT], where the completeness theorems for the minimal normal $\mathcal{L}(R,-R)$- logic are independently proved. This language appears as a special case of the notion <u>polymodal base</u>, introduced by the author in [Gor]. This notion combines a polymodal language $\mathcal{L}(\Box_1,\dots,\Box_n)$ with a set of formulae Φ, having a usual relational semantics over structures $\langle W,R_1,\dots,R_n\rangle$ (frames) and a theory T in some language (for definiteness - first-order) for such structures. We shall denote such a base $\mathcal{L}_T(R_1,\dots,R_n)$. The models of the theory T will be called <u>standard frames</u> of this base. In particular, when the theory T determines some of the relations $R_1,\dots,R_n$ by means of the rest of them, the polymodal base becomes an enriched [poly]modal language. A typical example of it provides the modal language for tense logics - it is a bimodal base with a theory T_{-1} having a single axiom (-1) $\forall xy(R_1xy \leftrightarrow R_2yx)$ and standard frames $\langle W,R,R^{-1}\rangle$ - it is an enriched modal language for $\langle W,R\rangle$. Another example is the language in question $\mathcal{L}(R,-R)$ being a bimodal base with theory T_- with an axiom (-) $\forall xy(R_1xy \leftrightarrow -R_2xy)$ and standard frames $\langle W,R,-R\rangle$. However, there exists an important distinction between the two

1 Research partially supported by the Committee for Science at the Council of Ministries of Bulgaria, Contract No. 933.

2 The author is grateful to the referees for the useful remarks.

Mathematical Logic
Edited by P. P. Petkov
Plenum Press, New York, 1990

bases: The axiom (-1) is modally definable while (-) is not. This creates considerable differences between the expressive possibilities ($\mathcal{L}$(R,-R) is a strong, with respect to definability, modal language; e.g. each universal formula for R and = is definable in it (see [Gor])) and the axiomatizing procedures in them. The aim of this paper is to suggest a general technique for proving completeness and to apply it in concrete situations.

PRELIMINARIES

We shall briefly recall or introduce some notions concerning polymodal logic and especially polymodal bases. The basic notions of modal logic (as given in the initial chapters of [Seg], [HC] or [Ben2], in particular: valuation and model over a given frame; general (first-order) frame; forcing ($\models$) and validity in a model (frame, general frame), canonical frame, model and general frame etc. will be expected to be familiar. Their generalizations in a polymodal language are trivial. We shall also deal with the natural generalizations of the basic frame constructions: generated subframe, p-morphism, ultrafilter extensions (ue) and Stone representation (SR) with a little specification: the notion 'generated subframe' of a given frame will be reserved for the ones generated from one world and the others will be called simply subframes. If F and G are frames and $F \cong ue(G)$ then G will be called an <u>ultrafilter contraction of</u> F.

Let C and D be classes of frames for a given polymodal language with a set of formulae Φ. Define:

- <u>modal theory of</u> C: $Th_{mod}(C) \rightleftharpoons \{\varphi \in \Phi \ / \ C \models \varphi\}$;
- <u>modally definable closure of</u> C <u>in</u> D : $[C]_D$ consists of all frames from D, in which $Th_{mod}(C)$ is valid. When D is the class of all frames of the given language, the closure will be denoted by [ ].

Let $\mathcal{L}=\mathcal{L}_T(R_1,\ldots,R_n)$ be a fixed polymodal base with a class of standard frames FR(T). The models over standard frames will be called <u>standard models</u>. The frames from [FR(T)] will be called <u>basic frames</u> and their generated subframes - <u>total frames</u>. <u>Basic $\mathcal{L}$-logic</u> is the logic $K_{\mathcal{L}} = Th_{mod}(FR(T))$; each simple extension of $K_{\mathcal{L}}$ is an <u>$\mathcal{L}$-logic</u>. $\mathcal{L}$-logic L is <u>complete with respect to a class of frames (models)</u> C iff $\forall \varphi \in \Phi (L \vdash \varphi \Leftrightarrow C \models \varphi)$. A frame F is a <u>frame for</u> L (<u>L-frame</u>) iff $\forall \varphi \in \Phi (L \vdash \varphi \Rightarrow F \models \varphi)$. An <u>L-model</u> is defined in the same manner.

Some notions concerning completeness. An $\mathcal{L}$-logic L is: <u>trivially complete</u> if L is complete with respect to the class of all L-models; <u>standardly complete</u> if L is complete with respect to the class of the standard L-models; <u>basically complete</u> if L is complete with respect to the class of all L-frames; <u>normally complete</u> if L is complete with respect to the class of the standard L-frames. Trivial completeness is provided by the canonical model and always holds, so it is not interesting as distinct from the proving method. The interesting notion of completeness in the classical modal logic is basic completeness. However in polymodal bases the goal is normal completeness, which guarantees the adequacy of the axiomatics with respect to the special semantics specified at the base. The purpose of the paper is to provide namely the normal completeness. So the adjective "normal" will be omitted and completeness will mean normal completeness from now on.

Some more notions: the logic L' is an <u>weak extension</u> of L if L' is an extension of L and has the same standard frames as L. (Obviously, L has proper weak extensions iff L is incomplete). The complete weak extension of L will be called a <u>completion of</u> L and denoted by c(L). Let $\mathcal{L}_1$ and $\mathcal{L}_2$ be polymodal languages such that $\mathcal{L}_1 \subseteq \mathcal{L}_2$, $K_{\mathcal{L}_1}$ and $K_{\mathcal{L}_2}$ - their minimal logics and $\Gamma \subseteq \Phi_{\mathcal{L}_1}$. Then the $\mathcal{L}_2$-logic $K_{\mathcal{L}_2}+\Gamma$, (axiomatized over $K_{\mathcal{L}_2}$ with Γ) will be called a <u>minimal extension of</u> $K_{\mathcal{L}_1}+\Gamma$ <u>in the language</u> $\mathcal{L}_2$. It is easy to prove that the minimal extensions are conservative.

Now, let us recall some notions from modal logic. A general frame $\mathfrak{F}=\langle W,R_1,\dots,R_n,\mathbb{W}\rangle$ is <u>descriptive</u> iff for each ultrafilter u in $\mathbb{W}$:

1) $\exists w\in W$: $\cap u=\{w\}$ and

2) $\cap\{\langle R_i\rangle X \;/\; X\in u\} \subseteq \langle R_i\rangle(\cap u)$ for $i=1,\dots,n$.

A <u>set of formulae</u> Γ is <u>canonical</u> iff for each descriptive frame $\mathfrak{F}=\langle F,\mathbb{W}\rangle$: $\mathfrak{F}\vDash\Gamma \Rightarrow F\vDash\Gamma$. A <u>logic</u> L is <u>canonical</u> if is axiomatized with a canonical set of formulae. The canonical logics are complete, since the canonical model generates a descriptive frame, but the canonicity is a condition stronger than truth in the canonical frame ; the logics with the latter property will be called (following Segerberg) <u>natural</u>.

Now our strategy will be the following. The proof of completeness will be split into two stages: first - basic completeness and second - [normal] completeness. The first stage is attacked with traditional methods of the modal logic - the canonical model method when the logic is natural, or with an appropriate supplementary construction otherwise (see [Seg], [HC]). The following lemma guarantees transferring of canonicity (hence the basic completeness) to minimal extensions.

LEMMA 1. A minimal extension of a canonical logic is canonical.

PROOF Let $L^\sim$ be a minimal extension of the canonical logic L and, for convenience, the base of L be $\mathcal{L}=\mathcal{L}_T(R_1,\dots,R_k)$ and of $L^\sim$ - $\mathcal{L}^\sim=\mathcal{L}_{T^\sim}(R_1,\dots,R_n)$, n>k. Then an $\mathcal{L}^\sim$-frame $\langle W,R_1,\dots,R_n\rangle$ is an $L^\sim$-frame iff $\langle W,R_1,\dots,R_k\rangle$ is

L-frame. Let $\mathfrak{F}^{\sim}=\langle W,R_1,\ldots,R_n,\mathbb{W}\rangle$ be a descriptive $L^{\sim}$-frame. Then $\mathfrak{F}=\langle W,R_1,\ldots,R_k,\mathbb{W}\rangle$ is a descriptive L-frame $\Rightarrow$ $\langle W,R_1,\ldots,R_k\rangle$ is an L-frame $\Rightarrow$ $\langle W,R_1,\ldots,R_n\rangle$ is an $L^{\sim}$-frame $\Rightarrow$ $L^{\sim}$ is canonical. #

It is not clear when the naturalness is transferred into minimal extension, but, as Fine has proved (see [Fin]) basic completeness (in particular naturalness) + first order definability implies canonicity.

The second stage, having a basic completeness result, is to prove (normal) completeness. So, let L be a basically complete logic of some base and φ be a non-theorem of L, refuted in a basic L-frame F. Let $FR_b(L)$ /$FR_t(L)$, $FR_s(L)$/ be the class of basic /total, standard/ L-frames. If L is complete, then $F\in[FR_s(L)]=FR_b(L)$. So if we know the description of the modally definable closure as a sequence of closure operations then the way back, starting from F, will bring us to a standard frame G, refuting φ. The way can be shortened, e.g. if we start from a total frame, refuting φ. Conversely if the procedure always goes through this proves the completeness of L; if it does not, this may show us the reasons for the incompleteness and show the missing axioms.

<u>Completeness, incompleteness and completions in $\mathcal{L}(R,-R)$</u>

The basic $\mathcal{L}(R.-R)$-logic, denoted in [GPT] by $K^{\sim}$ is axiomatized there with the S5-axioms for the modality $\blacksquare$ (dual - $\blacklozenge$), defined by $\blacksquare p \leftrightarrows \boxplus p \wedge \boxminus p$: (r) $\blacksquare p \rightarrow p$, (t) $\blacksquare p \rightarrow \blacksquare\blacksquare p$ and (s) $p \rightarrow \blacksquare\blacklozenge p$. So the basic $\mathcal{L}(R,-R)$-frames are those $\langle W,R_1,R_2\rangle$ in which $R_1\cup R_2$ is an equivalence relation and the total $\mathcal{L}(R,-R)$-frames - those in which $R_1\cup R_2$ is an universal relation. Denote the class of all total $\mathcal{L}(R,-R)$-frames by $\mathbb{C}t$. The proof of the standard completeness of $K^{\sim}$, exposed in [GPT] is directly transferred to any simple extension. However the problem for completeness is rather more complicated. We shall apply the idea sketched above for both proving completeness and completing basically complete extensions of $K^{\sim}$ in the case when the axioms are first-order definable formulae. Let L be a basically complete simple extension of $K^{\sim}$ with first-order definable axioms and φ be a non-theorem of L. Then φ is refuted in a total L-frame G. If L is complete then $G\in[FR_s(L)]_{\mathbb{C}t}$, which consists of all ultrafilter contractions of p-morphic images of members of $\mathcal{L}(R,-R)$ (see [Gor]). Hence the refuting standard L-frame is to be constructed as a p-morphic inverse-image of an ultrafilter extension of G. Conversely if for an arbitrary non-theorem φ and total frame G, refuting φ, we succeed to find in such a way standard L-frame F then $F\not\models\varphi$ since $G\in[F]$. So the completeness of L will be established. Really, as we shall see, the procedure establishing completeness is still simpler.

LEMMA 2.
If $\mathfrak{F}=\langle F,\mathbb{W}\rangle$ is a descriptive frame then F is a p-morphic image of ue(F).

PROOF: $\mathfrak{F}^+\leq F^+ \Rightarrow \mathfrak{F}\cong SR(\mathfrak{F}^+)$ is a p-morphic image of $SR(F^+)$ (see [Gol]), hence the assertion follows. #

LEMMA 3. Each generated subframe of a canonical general frame for $\mathcal{L}$(R.−R)-logic is descriptive.

PROOF Let L be an $\mathcal{L}$(R.−R)-logic, Γ' be a maximal L-consistent set (L-CS) and $\mathfrak{F}'=\langle W,R_1,R_2,\mathbb{W}\rangle$ be the subframe of the canonical general L-frame $\mathfrak{F}^L=\langle W^L,R_1^L,R_2^L,\mathbb{W}^L\rangle$ that is generated from Γ', i.e. $W=(R_1^L\cup R_2^L)(\Gamma')$; $R_i=R_i^L|_W$; $\mathbb{W}=\{|\varphi|' \;/\; \varphi\in\Phi\}$, where $|\varphi|'=\{\Gamma\in W \;/\; \varphi\in\Gamma\}$. Let us note that $\langle W,R_1,R_2\rangle$ is total. Let u be an ultrafilter in W. We shall verify the conditions i and ii from the definition of descriptive frame:

i) Put $\Delta\leftrightharpoons\{\varphi / |\varphi|'\in u\}$.

− Δ is a maximal L-CS: let $\varphi_1,\ldots,\varphi_n\in\Delta$. Then $|\varphi_1|'\cap\ldots\cap|\varphi_n|'\in u \Rightarrow$ there exists $\Gamma\in W$: $\varphi_1,\ldots,\varphi_n\in\Gamma \Rightarrow \varphi_1\wedge\ldots\wedge\varphi_n\in\Gamma \Rightarrow L\nvdash \varphi_1\wedge\ldots\wedge\varphi_n\rightarrow\bot \Rightarrow \Delta$ is L-CS; Δ is maximal: if $\varphi\notin\Delta$, then $\{\Gamma \;/\; \varphi\in\Gamma\}\notin u \Rightarrow \{\Gamma \;/\; \neg\varphi\in\Gamma\}\in u \Rightarrow \neg\varphi\in\Delta$;

− $\Delta\in W$: Let us assume that $\boxplus\Gamma'\nsubseteq\Delta$ and $\boxminus\Gamma'\nsubseteq\Delta$, i.e. there exist ψ_+ and ψ_- such that $\boxplus\psi_+\in\Gamma'$ & $\neg\psi_+\in\Delta$ & $\boxminus\psi_-\in\Gamma'$ & $\neg\psi_-\in\Delta \Rightarrow (\neg\psi_+\wedge\neg\psi_-)\in\Delta \Rightarrow |\neg\psi_+\wedge\neg\psi_-|'\in u \Rightarrow$ there exists $\Gamma\in W$, such that $(\neg\psi_+\wedge\neg\psi_-)\in\Gamma \Rightarrow \boxplus\Gamma'\nsubseteq\Gamma$ & $\boxminus\Gamma'\nsubseteq\Gamma$: $\lightning$. So $\Delta\in u$; if $\Delta'\in u$ then for every $\varphi\in\Delta(|\varphi|'\in u) \Rightarrow \Delta'\in|\varphi|' \Rightarrow \varphi\in\Delta' \Rightarrow \Delta\subseteq\Delta' \Rightarrow \Delta=\Delta'$, i.e. $\cap u=\{\Delta\}$.

ii) Let $\Gamma\in\{\Diamond_+|\varphi|' \;/\; |\varphi|'\in u\} = \cap\{|\Diamond_+\varphi|' \;/\; |\varphi|'\in u\}$ and $\cap u=\{\Delta\}$. If we assume that $\Gamma\notin\Diamond_+\{\Delta\}$, then $\boxplus\Gamma\nsubseteq\Delta \Rightarrow$ there exists φ such that $\boxplus\varphi\in\Gamma$ and $\neg\varphi\in\Delta$, i.e. $|\neg\varphi|'\in u \Rightarrow \Gamma\in|\Diamond_+\neg\varphi|' \Rightarrow \neg\boxplus\varphi\in\Gamma$: $\lightning$. So $\cap\{|\Diamond_+\varphi|' \;/\; |\varphi|'\in u\}\leq\Diamond_+\cap u$. Analogously for $\Diamond_-$. #

Now let L be a complete extension of K~ with first-order formulae, φ be a non-theorem of L and G be a φ-refuting generated subframe of the canonical L-frame. Then, as a consequence of the lemmata 2 and 3 and the reasoning, preceding them, G is a p-morphic image of a standard L-frame F, i.e. the φ-refuting standard L-frame is to be constructed as a p-morphic inverse-image of G.

The following definition has a technical importance for unifying the further exposition.

DEFINITION. Let $F^{\sim}=\langle W^{\sim},R_1,R_2\rangle$ be a total frame. A <u>standard extension of $F^{\sim}$</u> is any standard frame $F=\langle W,R,-R\rangle$ for which: $W=\bigcup_{x\in W^{\sim}}\{x\}\times s_x$, where s_x are (index) sets and the relation R is defined as follows: $R\langle x,i\rangle\langle y,j\rangle$ iff $R_1xy\&(R_2xy \Rightarrow \mathcal{P}\langle x,i\rangle\langle y,j\rangle)$, where $\mathcal{P}$ is a binary predicate in $W^{\sim}$, which satisfies the condition:

(@) $(R_1\cap R_2)xy \Rightarrow \forall i\in s_x(\exists j\in s_y\mathcal{P}\langle x,i\rangle\langle y,j\rangle \;\&\; \exists j\in s_y\neg\mathcal{P}\langle x,i\rangle\langle y,j\rangle)$. #

Note. It immediately follows from the definition of R that $-R\langle x,i\rangle\langle y,j\rangle$ iff $R_2xy\&(R_1xy \Rightarrow \neg\mathcal{P}\langle x,i\rangle\langle y,j\rangle)$. #

LEMMA 4. The standard frame F is isomorphic to a standard extension of F~ iff F~ is a p-morphic image of F. #

PROOF

1) If F is a standard extension of F~ then we shall prove that the mapping $f:F \longrightarrow F^\sim$, defined by $f(\langle x,i\rangle)=x$ is a p-morphism:

i) $R\langle x,i\rangle\langle y,j\rangle \Rightarrow R_1xy$; $-R\langle x,i\rangle\langle y,j\rangle \Rightarrow R_2xy$ – by the definition.

ii) If $R_1f(\langle x,i\rangle)y$, then according to (@) there exists $j\in s_y$ such that $\mathcal{P}\langle x,i\rangle\langle y,j\rangle$ hence $R\langle x,i\rangle\langle y,j\rangle$. Analogously it follows from $R_1f(\langle x,i\rangle)y$ that $-R\langle x,i\rangle\langle y,j\rangle$.

2) Conversely, let $F=\langle W,R,-R\rangle$ and $f:F \longrightarrow F^\sim$ be a p-morphism. We shall prove that F is isomorphic to a standard extension F' of F~. Let $x\in W^\sim$. Put $s_x \rightleftharpoons f^{-1}(x)$ and define predicate $\mathcal{P}$: $\mathcal{P}\langle x,i\rangle\langle y,j\rangle$ iff Rij. The condition (@): Since f is a p-morphism, then R_1xy implies that for every $i\in s_x$ there exists $j\in s_y$ such that Rij; R_2xy implies that for every $i\in s_x$ there exists $j\in s_y$ such that -Rij. The mapping f': $F' \longrightarrow F$, defined by $f'(\langle x,i\rangle)=i$ is an isomorphism. #

COROLLARY 5 . If $F^\sim \models p$ and F is a standard extension of F~, then $F \models p$.#

So if L is a logic with a natural first-order axiomatization in order to prove the completeness of L it is sufficient to find a standard extension, which is an L-frame, for each total L-frame. To this aim we have to construct the carrier of the extension and the predicate $\mathcal{P}$, satisfying the condition @. Actually the standard extensions are natural generalizations of the construction 'copying' proposed by Vakarelov and applied in [GPT] (for proving the completeness of K~), [GP], [Vak1] and [Vak2] for axiomatizing of some relational structures formalizing knowledge representation systems.

Some results

Now let us consider some simple examples applying the above reasoning.

THEOREM 6 The logics

i) $Ver_n^\sim = K^\sim + \boxplus^n\bot$,

ii) $Seq_n^\sim = K^\sim + \blacklozenge^n\top$,

iii) $Triv^\sim = K^\sim + p\leftrightarrow\boxplus p$ are complete.

PROOF Ver_n, Seq_n and Triv are canonical ([Seg1]), hence their minimal extensions are basically complete. It only remains to construct the corresponding standard extensions.

i) Let us note that the $Ver_n^\sim$-frames are those in which there are no

R_1-chains with a length n ([HC]). Let $F^{\sim}=\langle W^{\sim},R_1,R_2\rangle$ be a total $Ver_n{}^{\sim}$-frame. Put $W \leftrightharpoons W^{\sim}\times\{0,1\}$ and $\mathcal{P}\langle x,i\rangle\langle y,j\rangle$ iff $i=j$. The condition @ holds: if $i\in\{0,1\}$ then $\mathcal{P}\langle x,i\rangle\langle y,i\rangle \& \neg\mathcal{P}\langle x,i\rangle\langle y,1-i\rangle$. The obtained standard frame F is a $Ver_n{}^{\sim}$-frame: if $R^n\langle x,i\rangle\langle y,j\rangle$ then $R_1^n xy$.

ii) The $Seq_n{}^{\sim}$-frames are those in which from every point an R_1- chain with a length n starts. Let $F^{\sim}=\langle W^{\sim},R_1,R_2\rangle$ is a total $Seq_n{}^{\sim}$- frame. We shall use the construction from i). Then F will be a $Seq_n{}^{\sim}$-frame: if $\langle x,i\rangle\in W$, then there exists an R_1-chain with a length n and first point x in $F^{\sim}$: $x,x_1,\dots,x_{n-1} \Rightarrow \langle x,i\rangle,\langle x_1,i\rangle,\dots,\langle x_{n-1},i\rangle$ is an R-chain with a length n in F.

iii) The $Triv^{\sim}$-frames are $\langle W^{\sim},=,R_2\rangle$. Now the above construction does not apply. Put $W' \leftrightharpoons \{x\in W^{\sim} / R_2xx\}$ and $W \leftrightharpoons W'\times\{0,1\} \cup (W^{\sim}\setminus W')\times\{0\}$. The predicate $\mathcal{P}$ is defined as before and a $Triv^{\sim}$-frame F is obtained: $R\langle x,i\rangle\langle x,i\rangle$ and if $R\langle x,i\rangle\langle y,j\rangle$, then $x=y$; if $\neg R_2xx$ then $i=j=0$; if R_2xx then it follows from the definition of R that $i=j$. In both cases $\langle x,i\rangle=\langle y,j\rangle$. #

Note that $Triv^{\sim}$ axiomatizes in essence the modality $[\neq]$ (which is $\boxminus$).

The impression that completeness is almost directly transferred into minimal extensions is deceptive. On the contrary, as we shall see here the incompleteness events are quite usual and this is because the language $\mathcal{L}(R,-R)$ can express the same things in essentially different manners (with non-equivalent formulae).

EXAMPLE: The logic $KB^{\sim}=K^{\sim} + p\to\blacksquare\blacklozenge p$ is incomplete. Indeed, the $KB^{\sim}$-frames are those $\langle W,R_1,R_2\rangle$ in which R_1 is a symmetric relation. In the standard $KB^{\sim}$-frames it automatically follows that R_2 $(=-R_1)$ is also symmetric, which is expressed by the formula $p\to\boxminus\blacklozenge p$. However this formula is not true in all $KB^{\sim}$-frames, hence is not provable in $KB^{\sim}$. #

A great number of interesting modal logics are axiomatized by so called modal reduction principles (see [Ben1]). A typical example is the formula $A^{m,n}_{j,k} = \Diamond^m\Box^n p\to\Box^j\Diamond^k p$, which expresses the condition $C^{m,n}_{j,k}$: $\forall x\forall y\forall z\exists t(R^m xy\wedge R^j xz \to R^n yt\wedge R^k zt)$ ([LS]). The formula $A^{m,n}_{j,k}$ is canonical and the logic $L^{m,n}_{j,k} = K + A^{m,n}_{j,k}$ is complete ([HC,3.1]).

We shall introduce some denotations which will be used further on. Let $\langle W,R_1,R_2\rangle$ be a total frame and $x\in W$. We shall say that x has:

- <u>entry defect</u> (denoted $d_i(x)$) if $\exists y\in W(R_1yx\&R_2yx)$;
- <u>exit defect</u> (denoted $d_e(x)$) if $\exists y\in W(R_1xy\&R_2xy)$.

Put $D_i(W)\leftrightharpoons\{x\in W / d_i(x)\}$; $D_e(W)\leftrightharpoons\{x\in W / d_e(x)\}$.

THEOREM 7

i) The logic $\tilde{L}^{m,n}_{j,k} = K^{\sim} + \blacklozenge^m\blacksquare^n p\to\blacksquare^j\blacklozenge^k p$ is complete in the following cases:

i1) $\tilde{L}^{m,0}_{0,m} = K^{\sim} + \blacklozenge^{m}p \to \blacklozenge^{m}p$ and $\tilde{L}^{0,m}_{m,0} = K^{\sim} + \boxplus^{m}p \to \boxplus^{m}p$;

i2) $m+n+j+k = 1$;

i3) $n+k \geq 2$.

ii) In all remaining cases $\tilde{L}^{m,n}_{j,k}$ is incomplete and is completed by an additional axiom $B^{m,n}_{j,k}$ as follows:

ii.1) $\tilde{L}^{m,1}_{0,0} = K^{\sim} + \blacklozenge^{m}\boxplus p \to p$, $m \geq 1$. $B^{m,1}_{0,0} = \Diamond\blacklozenge^{m-1}\boxminus p \to p$;

ii.2) $\tilde{L}^{m,0}_{0,0} = K^{\sim} + \blacklozenge^{m}p \to p$, $m \geq 2$. $B^{m,0}_{0,0} = \blacklozenge^{m-2}(\blacklozenge\boxminus p \wedge \Diamond\boxminus p) \to p$;

ii.3) $\tilde{L}^{m,1}_{j,0} = K^{\sim} + \blacklozenge^{m}\boxplus p \to \boxplus^{j}p$, $j \geq 1$ & $m+j \geq 2$. $B^{m,n}_{j,k} = \blacklozenge^{m}\Diamond p \to \boxplus^{j-1}\Diamond p$;

ii.4) $\tilde{L}^{m,0}_{j,0} = K^{\sim} + \blacklozenge^{m}p \to \boxplus^{j}p$, $m \geq 1$, $j \geq 1$.

$B^{m,0}_{j,0} = \blacklozenge(\Diamond p \wedge \Diamond p) \to ((\blacklozenge^{j}p \to \boxplus^{m-1}\Diamond p) \wedge (\blacklozenge^{m}p \to \boxplus^{j-1}\Diamond p))$.

ii.5) The remaining cases $\tilde{L}^{0,0}_{j,1}$, $\tilde{L}^{0,0}_{j,0}$, $\tilde{L}^{m,0}_{j,1}$ are correspondingly equivalent to ii.1-ii.3.

PROOF Let us note that $\tilde{L}^{m,n}_{j,k}$ and $\tilde{L}^{j,k}_{m,n}$ are equivalent. The $\tilde{L}^{m,n}_{j,k}$-frames are those, for which R_1 satisfies the condition $C^{m,n}_{j,k}$. $\tilde{L}^{m,n}_{j,k}$ is basically complete since $L^{m,n}_{j,k}$ is canonical. Let us construct the corresponding standard extensions.

i.1) the formulae $A^{m,0}_{0,m}$ and $A^{0,m}_{m,0}$ are tautologies and the assertion follows from the completeness of K.

i.2) a) $\tilde{L}^{1,0}_{0,0} = K^{\sim} + \blacklozenge p \to p$; $C^{1,0}_{0,0}$: $Rxy \to x=y$. Let $F^{\sim} = \langle W^{\sim}, R_1, R_2 \rangle$ be a total $\tilde{L}^{1,0}_{0,0}$-frame. Put $W \rightleftharpoons W^{\sim} \times \{0\} \cup D_i(W) \times \{1\}$ and $\mathcal{P}uv$ iff $u=v$ for every $u,v \in W$. The condition @: $(R_1 \cap R_2)xy \to x=y \;\&\; d_i(x)$. Then $\mathcal{P}\langle x,i\rangle\langle x,i\rangle \;\&\; \neg\mathcal{P}\langle x,i\rangle\langle x,1-i\rangle$. $C^{1,0}_{0,0}$ holds in F. Let $R\langle x,i\rangle\langle y,1\rangle$. Then $R_1xy \to x=y$. If $i \neq 1$ then $d_i(x) \to \exists z(R_1 \cap R_2)zx \to z=x \to R_2xx$, i.e. $R_2xy \to \mathcal{P}\langle x,i\rangle\langle y,1\rangle \to \langle x,i\rangle = \langle y,1\rangle$: $\lightning$.

b) $\tilde{L}^{0,1}_{0,0} = K^{\sim} + \boxplus p \to p$, $C^{0,1}_{0,0}$: $\forall x Rxx$. Put W as in a) and $\mathcal{P}\langle x,i\rangle\langle y,1\rangle$ iff $i=1$. The condition @ holds and for every $u \in W$: $\mathcal{P}uu \to Ruu$.

c) $\tilde{L}^{0,0}_{1,0}$ and $\tilde{L}^{0,0}_{0,1}$ are equivalent correspondingly to the a) and b).

i.3) a) $n=k=1$. $C^{m,1}_{j,1}$: $\forall x \forall y \forall z \exists t(R^{m}xy \,\&\, R^{j}xz \to Ryt \,\&\, Rzt)$. Put W as in i.2 and $\mathcal{P}\langle x,i\rangle\langle y,1\rangle$ iff $j=0$. The condition @: if $(R_1 \cap R_2)xy$ then $d_i(x)$ and $\mathcal{P}\langle x,i\rangle\langle y,0\rangle \;\&\; \neg\mathcal{P}\langle x,i\rangle\langle y,1\rangle$. The condition $C^{m,1}_{j,1}$:

Let $R^{m}\langle x,i\rangle\langle y,1\rangle$ and $R^{j}\langle x,i\rangle\langle z,s\rangle$. Then $R_1xy \;\&\; R^{j}xz \to$ there exists t such that $R_1yt \;\&\; R_1zt \to R\langle y,1\rangle\langle t,0\rangle \;\&\; R\langle z,s\rangle\langle t,0\rangle$.

b) $n \geq 2$. Put $W \rightleftharpoons W^{\sim} \times \{\langle 0,0\rangle\} \cup D_i(W^{\sim}) \times \{\langle 1,0\rangle\} \cup D_e(W^{\sim}) \times \{\langle 0,1\rangle\} \cup (D_i(W^{\sim}) \cap D_e(W^{\sim})) \times \{\langle 1,1\rangle\}$. The elements of W will be denoted by triples $\langle x,i,e\rangle$; put $c(\langle x,i,e\rangle) \rightleftharpoons x$, $i(\langle x,i,e\rangle) \rightleftharpoons i$, $e(\langle x,i,e\rangle) \rightleftharpoons e$. Now define $\mathcal{P}uv$ iff $i(v)=e(u)$. The condition @: If $(R_1 \cap R_2)xy$, then $d_e(x) \,\&\, d_i(y) \to$ for every $\langle i,e\rangle$: $\mathcal{P}\langle x,i,e\rangle\langle y,e,0\rangle \;\&\; \neg\mathcal{P}\langle x,i,e\rangle\langle y,1-e,0\rangle$. $C^{m,n}_{j,k}$: Let $R^{m}uv \;\&\; R^{j}uw$. Then $R_1^{m}c(u)c(v) \;\&\; R_1^{j}c(u)c(w) \to$ there exists $r \in W^{\sim}$ such that $R_1^{n}c(v)r \;\&\; R_1^{k}c(w)r$. Let us make the following observation. If $n \geq 2$ then $R_1^{n}c(u_1)c(u_2)$ implies $R^{n}u_1u_2$: Let $r_1,\dots,r_{n-1} \in W^{\sim}$ are such that $R_1c(u_1)r_1,\dots,R_1r_{n-1}c(u_2)$. Then define $t_1,\dots,t_{n-1} \in W$, such that $c(t_s)=r_s$ for $s=1,\dots,n-1$ and $i(t_s)=0$ for $s=2,.\ n-1$

(when $n\geq 3$), $e(t_s)=0$ for $s=1,\ldots,n-2$ (when $n\geq 3$) and it remains to define $i(t_1)$ and $e(t_{n-1})$. If $R_2c(u_1)r_1$, then $d_i(t_1)$ and put $i(t_1)=e(u_1)$ else $i(t_1)=0$. Analogously if $R_2r_{n-1}c(u_2)$, then $d_e(r_{n-1})$ and put $e(t_{n-1})=i(u_2)$ else $e(t_{n-1})=0$. So it is insured that $Ru_1t_1,\ldots,Rt_{n-1}u_2$, i.e. $R^nu_1u_2$. Now: if $k=0$ then $c(w)=r$ and put $t=w$. Then R^kwt and according to the above observation R^nvt. If $k=1$ then put $t=\langle r,e(w),0\rangle$ and analogously R^nvt and Rwt.

Finally if $k\geq 2$ put $t=\langle r,0,0\rangle$. According to the observation R^nvt and R^kwt.

c) $k\geq 2$ – analogously to b). So in all cases the standard frame F is an $\tilde{L}^{m,n}_{j,k}$-frame, hence the completeness of $\tilde{L}^{m,n}_{j,k}$ follows.

ii) First we shall prove that the additional axiom $B^{m,n}_{j,k}$ is true in the class of the standard $\tilde{L}^{m,n}_{j,k}$-frames.

ii.1) It is immediately proved that in the formula $B^{m,1}_{0,0}=\lozenge\blacklozenge^{m-1}\boxminus p\rightarrow p$ expresses in $\mathbb{C}_t$ the condition $\beta^{m,1}_{0,0}$: $R_2yx\wedge R_1^{m-1}xz\rightarrow R_2zy$, which in a standard frame is equivalent to $C^{m,1}_{0,0}$: $R^mxy\rightarrow Ryx$, $m\geq 1$.

ii.2) In $\mathbb{C}_t$ the formula $B^{m,0}_{0,0}=\blacklozenge(\blacklozenge p\wedge\lozenge\boxminus p)\rightarrow p$, $m\geq 2$ expresses $\beta^{m,0}_{0,0}=(R_1^{m-2}xy\wedge R_1yz_1\wedge R_2yz_2)\rightarrow(R_2z_1x\vee R_2z_2x)$, which in a standard frame follows from $C^{m,0}_{0,0}$: $R^mxy\rightarrow x=y$. If $R_2=-R_1$ and if assume $R_1^{m-2}xy\wedge R_1yz_1\wedge -R_1yz_2\wedge R_1z_1x\wedge R_1z_2x$, then it follows $R_1^mz_1z_2\Rightarrow z_1=z_2\Rightarrow R_1yz_1\wedge -R_1yz_1$: $\lightning$.

ii.3) In $\mathbb{C}_t$ the formula $B^{m,1}_{j,0}=\blacklozenge^m\lozenge p\rightarrow\boxplus^{j-1}\lozenge p$, $m+j\geq 2$, $j\geq 1$ expresses $\beta^{m,1}_{j,0}$: $R_1^mxy\wedge R_1^{j-1}xt\wedge R_2yz\rightarrow R_2tz$, which in $\mathbb{C}_s$ is equivalent to $C^{m,1}_{j,0}$: $R^mxy\wedge R^jxz\rightarrow Ryz$.

ii.4) It is convenient to split $B^{m,0}_{j,0}$ in two formulae:

$\blacklozenge(\blacklozenge p\wedge\lozenge p)\rightarrow(\blacklozenge^jp\rightarrow\boxplus^{m-1}\lozenge p)$ and $\blacklozenge(\blacklozenge p\wedge\lozenge p)\rightarrow(\blacklozenge^mp\rightarrow\boxplus^{j-1}\lozenge p)$, $m,j\geq 1$.

The first means in $\mathbb{C}_t$ $R_1tu_1\wedge R_2tu_2\wedge R_1^jxy\wedge R_1^{m-1}xv\rightarrow R_2vu_1\vee R_2vu_2\vee R_2vy$, which in $\mathbb{C}_s$ follows from $C^{m,0}_{j,0}$: $R^mxy\wedge R^jxy\rightarrow y=z$. The second formula is verified analogously. Now, in order to prove the basic completeness of $\tilde{L}^{m,n}_{j,k}+B^{m,n}_{j,k}$ it is sufficient to ascertain that $B^{m,n}_{j,k}$ are canonical. Let Γ' be a maximal $(\tilde{L}^{m,n}_{j,k}+B^{m,n}_{j,k})$-CS and $F'=\langle W',R_1,R_2\rangle$ be the subframe, generated from Γ', of the canonical frame for the above logic.

We shall show that in cases ii.1–ii.4 the condition $\beta^{m,n}_{j,k}$ holds in F':

ii.1) Let for some $x,y,z\in W'$: R_2yx & $R_1^{m-1}xz$ and $\boxminus\varphi\in z$.

Then $\blacklozenge^{m-1}\boxminus\varphi\in x\Rightarrow\lozenge\blacklozenge^{m-1}\boxminus\varphi\in y\Rightarrow\varphi\in y\Rightarrow R_2zy$.

ii.2) Let $R_1^{m-2}xy$ & R_1yz_1 & R_2yz_2. If we assume that $\neg R_2z_1x$ & $\neg R_2z_2x$ then there exist φ,ψ: $\boxminus\varphi\in z_1$ & $\varphi\notin x$ & $\boxminus\psi\in z_2$ & $\psi\notin x$. Let $\chi=\varphi\vee\psi$. Then $\boxminus\chi\in z_1$ & $\boxminus\chi\in z_2\Rightarrow\blacklozenge\boxminus\chi\wedge\lozenge\boxminus\chi\in y\Rightarrow\blacklozenge^{m-2}(\blacklozenge\lozenge\chi\wedge\lozenge\boxminus\chi)\in x\Rightarrow\chi\in x$: $\lightning$.

ii.3) Let R_1xy & $R_1^{j-1}xt$ & R_2yz and $\boxminus\varphi\in t$.

Then $\blacklozenge^{j-1}\boxminus\varphi\in x\Rightarrow\boxplus^m\boxminus\varphi\in x\Rightarrow\boxminus\varphi\in y\Rightarrow\varphi\in z\Rightarrow R_2tz$.

ii.4) Let R_1tu_1 & R_2tu_2 & R_1^jxy & $R_1^{m-1}xv$ and let us assume $\neg R_2vu_2$ & $\neg R_2vu_2$ & $\neg R_2vy$. Then there exist φ,ψ,χ such that $\boxminus\varphi\in v$ & $\varphi\notin u_1$, $\boxminus\psi\in v$ & $\psi\notin u_2$, $\boxminus\chi\in v$ & $\chi\notin y$. Let $\theta\leftrightharpoons\varphi\wedge\psi\wedge\chi$. Then $\boxminus\theta\in v$ and $\theta\notin u_1,u_2,y\Rightarrow\neg\theta\in u_1,u_2\Rightarrow\blacklozenge\neg\theta\wedge\lozenge\neg\theta\in t$

$\Rightarrow \blacklozenge(\Diamond\neg\theta\wedge\Diamond\neg\theta)\in x \Rightarrow \blacklozenge^{j}\neg\theta\rightarrow\boxplus^{m-1}\Diamond\neg\theta\in x \Rightarrow \blacklozenge^{m-1}\boxminus\theta\rightarrow\boxplus^{j}\theta\in x$; $\boxminus\theta\in v \Rightarrow \blacklozenge^{m-1}\boxminus\theta\in v \Rightarrow$ $\boxplus^{j}\theta\in x \Rightarrow \theta\in y$: $\lightning$.

Now we shall proof the completeness of $\tilde{L}^{m,n}_{j,k} + B^{m,n}_{j,k}$.
In all cases put $W \leftrightharpoons W^{\sim}\times\{0\} \cup D_i(W^{\sim})\times\{1\}$.

ii.1) Define $\mathcal{P}\langle x,i\rangle\langle y,l\rangle$ iff $i=1$. The condition @ holds. Now we shall prove $C^{m,1}_{0,0}$. Let $R^{m}\langle x,i\rangle\langle y,l\rangle$. Then $R_1^{m}xy \Rightarrow R_1yx$. Let in addition R_2yx. Then there exist $\langle y_1,l_1\rangle,\dots,\langle y_{m-1},l_{m-1}\rangle$ such that
$\langle x,i\rangle R\langle y_1,l_1\rangle\dots R\langle y_{m-1},l_{m-1}\rangle R\langle y,l\rangle$ and therefore $xR_1y_1\dots y_{m-1}R_1y$.

ii.2) The predicate $\mathcal{P}$ is defined as in ii.1.
$C^{m,0}_{0,0}$: Let $R^{m}\langle x,i\rangle\langle y,l\rangle$. It follows from $R_1^{m}xy$ that $x=y$. If $\neg d_i(x)$ then $i=l=0$ $\Rightarrow \langle x,i\rangle=\langle y,l\rangle$. Let $d_i(x)$, i.e. there exists z such that $(R_1\cap R_2)zx \Rightarrow R_1^{m}zy_{m-1}$ $\Rightarrow z=y_{m-1} \Rightarrow (R_1\cap R_2)y_{m-1}y \Rightarrow \mathcal{P}\langle y_{m-1},l_{m-1}\rangle\langle y,l\rangle \Rightarrow l_{m-1}=1$. Put $y_0\leftrightharpoons x$, $y_m=y$.
Inductively on k we shall prove $R_2y_ky_{k+1}$ for $k=0,\dots,m-1$.

- $k=0$: Let V be a valuation, such that $V(p)=R_2(x)$. Then $x\models\boxminus p \Rightarrow$ $y_{m-1}\models\blacklozenge\boxminus p\wedge\Diamond\boxminus p \Rightarrow y_1\models\blacklozenge^{m-2}(\blacklozenge\boxminus p\wedge\Diamond\boxminus p) \Rightarrow y_1\models p \Rightarrow R_2y_0y_1$.

- let $R_2y_ky_{k+1}$ and V be a valuation, such that $V(p)=R_2(y_{k+1})$. Then $y_{k+1}\models\boxminus p \Rightarrow y_k\models\blacklozenge\boxminus p\wedge\Diamond\boxminus p \Rightarrow y\models\blacklozenge^{k}(\blacklozenge\boxminus p\wedge\Diamond\boxminus p) \Rightarrow y_{k+2}\models\blacklozenge^{m-2}(\blacklozenge\boxminus p\wedge\Diamond\boxminus p) \Rightarrow y_{k+2}\models p \Rightarrow$ $R_2y_{k+1}y_{k+2}$. The induction is completed. So $\mathcal{P}\langle x,i\rangle\langle y_1,l_1\rangle$ &...& $\mathcal{P}\langle y_{m-1},l_{m-1}\rangle\langle y,l\rangle \Rightarrow i=l_1,\dots,l_{m-1}=l \Rightarrow \langle x,i\rangle=\langle y,l\rangle$.

ii.3) Put $\mathcal{P}\langle x,i\rangle\langle y,l\rangle$ iff $l=0$. If $(R_1\cap R_2)xy$ then $d_i(y)$ and the condition @ holds. $C^{m,1}_{j,0}$: let $R^{m}\langle x,i\rangle\langle y,l\rangle$ and $R^{j}\langle x,i\rangle\langle z,s\rangle$. Then there exist $\langle y_1,l_1\rangle,\dots,\langle y_{m-1},l_{m-1}\rangle,\langle z_1,k_1\rangle,\dots,\langle z_{j-1},k_{j-1}\rangle$ such that
$\langle x,i\rangle R\langle y_1,l_1\rangle\dots R\langle y,l\rangle$ and $\langle x,i\rangle R\langle z_1,k_1\rangle\dots R\langle z,k\rangle$. It follows from $R_1^{m}xy$ and $R_1^{j}xz$ that R_1yz. Let in addition R_2yz. We shall prove that $R_2z_{j-1}z$: let V be a valuation, such that $V(p)=R_2(z_{j-1})$. Then $z_{j-1}\models\boxminus p \Rightarrow x\models\blacklozenge^{j-1}\boxminus p \Rightarrow x\models\boxplus^{j-1}\Diamond\neg p \Rightarrow$ $x\models\blacklozenge^{m}\Diamond\neg p \Rightarrow x\models\boxplus^{m}\boxminus p \Rightarrow y\models\boxminus p \Rightarrow z\models p$. So $R_2z_{j-1}z$ and it follows from $R_1z_{j-1}z$ that $\mathcal{P}\langle z_{j-1},k_{j-1}\rangle\langle z,k\rangle \Rightarrow k=0 \Rightarrow \mathcal{P}\langle y,l\rangle\langle z,k\rangle \Rightarrow R\langle y,l\rangle\langle z,k\rangle$.

ii.4) The predicate $\mathcal{P}$ is defined as in ii.3. $C^{m,0}_{j,0}$: let $R^{m}\langle x,i\rangle\langle y,l\rangle$ and $R^{j}\langle x,i\rangle\langle z,k\rangle$. Then R_1xy & $R_1^{j}xz \Rightarrow y=z$. If $\neg d_i(y)$ then $k=l=0$. Let $d_i(y)$, i.e. there exists t such that $(R_1\cap R_2)ty$. Let $\langle x,i\rangle R\langle y_1,l_1\rangle\dots$ $R\langle y_{m-1},l_{m-1}\rangle R\langle y,l\rangle$ and $\langle x,i\rangle R\langle z_1,k_1\rangle\dots\langle z_{j-1},k_{j-1}\rangle R\langle z,k\rangle$. We shall prove that $R_2y_{m-1}y$ and $R_2z_{j-1}z$. Let V be a valuation, such that $V(p)=\{y\}$. Then $t\models\blacklozenge p\wedge\Diamond p \Rightarrow x\models\blacklozenge(\blacklozenge p\wedge\Diamond p) \Rightarrow x\models\blacklozenge^{j}p\rightarrow\boxplus^{m-1}\Diamond p$; it follows from $y=z$ that $z\models p \Rightarrow x\models\boxplus^{m-1}\Diamond p$ $\Rightarrow y_{m-1}\models\Diamond p \Rightarrow$ there exists y': $R_2y_{m-1}y'$ and $y'\models p \Rightarrow y'=y \Rightarrow R_2y_{m-1}y$. Analogously $R_2z_{j-1}z$. Then $R\langle y_{m-1},l_{m-1}\rangle\langle y,l\rangle$ & $R\langle z_{j-1},k_{j-1}\rangle\langle z,k\rangle \Rightarrow \mathcal{P}\langle y_{m-1},l_{m-1}\rangle\langle y,l\rangle$ and $\mathcal{P}\langle z_{j-1},k_{j-1}\rangle\langle z,k\rangle$, $\Rightarrow l=0=k \Rightarrow \langle y,l\rangle=\langle z,k\rangle$. Finally we shall proof the incompleteness $\tilde{L}^{m,n}_{j,k}$ in the case ii.1-ii.4 proving that $\tilde{L}^{m,n}_{j,k} \not\vdash B^{m,n}_{j,k}$. To this end we shall construct a total $L^{m,n}_{j,k}$-frame $\langle W,R_1,R_2\rangle$, in which $B^{m,n}_{j,k}$ is refuted:

ii.1) $W=\{x,y\}$, $R_1=W^2$, $R_2=\{\langle x,y\rangle\}$. $C^{m,1}_{0,0}$ trivially holds but a valuation V, such that $V(p)=\{y\}$ refutes $B^{m,n}_{j,k}$: $x\nVdash p$ but R_2xy, $R_1^{m-1}yy$ and $R_2(y)\leq V(p) \Rightarrow x\Vdash\Diamond\blacklozenge^{m-1}\boxminus p$.

ii.2) $W=\{x_1,\ldots,x_m\}$, $R_1=\{\langle x_1,x_2\rangle,\langle x_2,x_3\rangle,\ldots,\langle x_m,x_1\rangle\}$, $R_2=\{\langle x_{m-1},x_m\rangle\}$ $\cup\ W^2\backslash R_1$. The condition $C^{m,0}_{0,0}$ holds but a valuation V, such that $V(p)=\{y\}$ refutes $B^{m,n}_{j,k}$: If $m=0$ then $j\geq 2$. $y\Vdash p \Rightarrow y\Vdash\Diamond p \Rightarrow y\Vdash\blacklozenge^m\Diamond p$ but $x\Vdash\boxminus\neg p \Rightarrow y\Vdash\blacklozenge^{j-1}\boxminus\neg p$.

If $m>0$ then $y\Vdash\Diamond p \Rightarrow x\Vdash\blacklozenge^m\Diamond p$ but $x\Vdash\blacklozenge^{j-1}\boxminus\neg p$.

ii.4) $W=\{x,y\}$, $R_1=\{\langle x,y\rangle\langle y,y\rangle\}$, $R_2=W^2\backslash\{\langle y,y\rangle\}$. $C^{m,0}_{j,0}$ holds but a valuation V, such that $V(p)=\{y\}$ refutes $B^{m,n}_{j,k}$: $y\Vdash p$ and $(R_1\cap R_2)xy \Rightarrow x\Vdash\blacklozenge p\wedge\Diamond p \Rightarrow y\Vdash\blacklozenge(\blacklozenge p\wedge\Diamond p)$ but $y\Vdash\blacklozenge^j p$ and $y\Vdash\boxminus\neg p \Rightarrow y\Vdash\blacklozenge^{m-1}\boxminus\neg p \Rightarrow y\nVdash\blacklozenge^j p\rightarrow\boxplus^{m-1}\Diamond p$.

The proof of the theorem is completed. #

Let us examine, as consequences of the theorem, the minimal extensions over K~ of some well-known modal logics:

1. The logic of the reflexive frames $T^\sim= K^\sim + \boxplus p\rightarrow p = L^{\sim 0,1}_{0,0}$ is complete.
2. The logic of the serial frames $D^\sim= K^\sim + \boxplus p\rightarrow\blacklozenge p = L^{\sim 0,1}_{0,1}$ is complete.
3. The logic $G1^\sim= K^\sim + \blacklozenge\boxplus p\rightarrow\boxplus\blacklozenge p = L^{\sim 1,1}_{1,1}$. is complete.
4. The logic of the symmetric frames $KB^\sim= K^\sim + p\rightarrow\boxplus\blacklozenge p = L^{\sim 0,0}_{1,1}$ is incomplete (as we have already seen) and is completed by $\Diamond\boxminus p\rightarrow p$.
5. The logic of the transitive frames $K4^\sim= K^\sim + \boxplus p\rightarrow\boxplus\boxplus p = L^{\sim 0,1}_{2,0}$ is incomplete and is completed by $\Diamond p\rightarrow\boxplus\Diamond p$.
6. The logic $E^\sim= K^\sim + \blacklozenge\boxplus p\rightarrow\boxplus p = L^{\sim 1,1}_{1,0}$ is incomplete and is completed by $\blacklozenge\Diamond p\rightarrow\Diamond p$.

Let us note that as consequences of th. 7 we can obtain results about completeness and completing logics, axiomatized by a set of formulae of the examined type, but only in the case, when the completeness of each of them is proved by one and the same construction, e.g. formulae from ii.1, ii.2, and possibly $\boxplus p\rightarrow p$, or from ii.3 and ii.4.

EXAMPLE: The logic of the reflexive and symmetric frames is axiomatized by the axioms $\boxplus p\rightarrow p$, $p\rightarrow\boxplus\blacklozenge p$ and $p\rightarrow\boxminus\Diamond p$. #

However concerning other combinations, e.g. like S4~=K4~ + T~ or S5~ = S4~ + B~ /= T~+E~, see [HC]/ we can not draw a direct conclusion from th. 7.

THEOREM 8 i) S4~ is incomplete;

ii) The completion of S4~ is $c(S4^\sim)=S4^\sim + \blacklozenge\boxminus p\rightarrow\boxminus p$.

PROOF

i) $S4^\sim\vdash_{\mathbb{C}_s}\blacklozenge\boxminus p\rightarrow\boxminus p$, but $S4^\sim\nvdash_{\mathbb{C}_t}\blacklozenge\boxminus p\rightarrow\boxminus p$: this formula expresses in total frames $\forall x\forall y\forall z(R_1xy\wedge R_2xz\rightarrow R_2yz)$ and in standard ones - simply transitivity. The total frame $F=\langle\{x,y\},\{x,y\}^2,\{\langle x,y\rangle\}\rangle$ is an S4~-frame refuting the above formula: in it R_1xy & R_2xy & $\neg R_2yy$.

ii) c(S4~) is basically complete, according to lemma 1, since S4 is canonical. Let F~=<W~,R_1,R_2> be a total c(S4~)-frame, i.e. R_1 is reflexive and transitive and R_1xy & R_2xz ⇒ R_2yz. The standard extension: W ⇌ W~×{0} ∪ D_i(W~)×{1,2,3}; Put 𝒫<x,i><y,j> iff (<x,i>=<y,j> or j is an even number). The condition (@): Let (R_1∩R_2)xy. Then d_i(y) ⇒ 𝒫<x,i><y,0> & ¬𝒫<x,i><y,j> where j∈{1,3}, j≠i.

Obviously R is reflexive; transitivity: let R<x,i><y,j> and R<y,j><z,k>. Then R_1xy & R_1yz ⇒ R_1xz. Let in addition R_2xz. Then R_1xy & R_2xz ⇒ R_2yz ⇒ 𝒫<y,j><z,k>. If <y,j>=<z,k> then R<x,i><z,k>, else k is even ⇒ 𝒫<x,i><z,k> and again R<x,i><z,k>. #

So we are convinced that at least in the minimal extensions incompleteness is as usual as the completeness. Still the hope remains that a join of complete /completed/ minimal extensions will also be complete. Alas, this hope is also refuted. A counter-example: as we have seen the logics T~ and c(E~)=E~+◈◈p→◈p are complete. Their join, let us denote it by S5~', is a weak extension of S5~ since S5=E + T, but the next assertion shows that S5~' is incomplete (of course S5~ also is).

THEOREM 9 i) The completion of S5~ is c(S5~) = S5~ + p→⊟◈p + ◈⊟p→⊟p;
ii) S5~' ⊬ p→⊟◈p.

PROOF i) c(S5~) is basically complete. Let <W~,R_1,R_2> be a c(S5~)-frame. Standard extension: W ⇌ W~×{0} ∪ D_i(W~)×{1}; 𝒫<x,i><y,j> iff i=j. The condition @ has already been checked. R is reflexive; symmetry: Let R<x,i><y,j>. Then R_1xy, hence R_1yx. If in addition R_2yx, then R_2xy (from p→⊟◈p) ⇒ 𝒫<x,i><y,j> ⇒ i=j ⇒ 𝒫<y,j><x,i> ⇒ R<y,j><x,i>. Transitivity: Let R<x,i><y,j> and R<y,j><z,k>. Then R_1xy and R_1yz ⇒ R_1xz. If in addition R_2xz then R_2zy; R_1xy & R_2xz ⇒ R_2yz ⇒ 𝒫<y,j><z,k> ⇒ j=k; R_1xz ⇒ R_1zx; R_1zx & R_2zy ⇒ R_2xy ⇒ 𝒫<x,i><y,j> ⇒ i=j ⇒ i=k ⇒ 𝒫<x,i><z,k> ⇒ R<x,i><z,k>.

ii) The frame F=<{x,y,z},{x,y,z}^{2},{<x,z>,<y,z>,<z,z>}> is an S5~'-frame: the formula ◈◈p→◈p expresses in total frames R_1xy∧R_2yz→R_2xz. However the relation R_2 is not symmetric ⇒ F⊭p→⊟◈p ⇒ S5~'⊬p→⊟◈p. #

Indeed the technique of the adduced proofs is quite monotonous, in some places even boring. That is why it would be better to find some generalizations reducing the tedious repetitions.

Further on we shall briefly sketch completeness results axiomatizing some orderings, proved in a similar manner.

THEOREM 10 i) The logic of weak linearity
S4.3~ = S4~ + ⊞(⊞p→q)∨⊞(⊞q→p) is incomplete.

ii) Its completion $c(S4.3^{\sim})$ is axiomatized as follows:

Axioms:	Corresponding conditions in total frames:
(ref) $\boxplus p \to p$	R_1xx
(tran) $\boxplus p \to \boxplus\boxplus p$	$R_1xy \wedge R_1yz \to R_1xz$
(tran') $\blacklozenge\boxminus p \to \boxminus p$	$R_1xy \wedge R_2xz \to R_2yz$
(wlin) $\boxplus(\boxplus p \to q) \vee \boxplus(\boxplus q \to p)$	$R_1xy \wedge R_1xz \to R_1yz \vee R_1zy$
(wlin') $\boxminus p \to \boxplus(q \vee \boxminus(\boxplus q \to p))$	$R_1xy \wedge R_2yz \to R_2xz \vee R_1zy$
(wlin") $\boxminus p \wedge \Diamond(q \wedge \Diamond(\neg p \wedge q)) \to \blacksquare\Diamond q$	$R_2xy \wedge R_2yz \to R_2xz \vee R_2ty \vee R_2tz$.

PROOF The correctness in total frames and naturalness of the proposed axioms are directly verified. So $c(S4.3^{\sim})$ is basically complete. Now let $F^{\sim} = \langle W^{\sim}, R_1, R_2\rangle$ be a total $c(S4.3^{\sim})$-frame. In order to construct the corresponding standard extension we shall define relations ρ and η in $W^{\sim}$:

ρxy iff R_1xy & R_1yx.

ρ is an equivalence relation in $W^{\sim}$ since R_1 is a quasi-ordering.

ηxy iff ρxy and if $x \neq y$ then:

i) R_2xy & R_2yx

ii) there exists $t \in W$ such that $\neg R_2tx$ & $\neg R_2ty$.

η is also an equivalence in $W^{\sim}$. Now the standard extension:

$W \leftrightharpoons W^{\sim}\times\{0\} \cup D_i(W^{\sim})\times\mathbb{N}$; $\mathcal{P}\langle x,i\rangle\langle y,j\rangle$ iff $j=0$ or ηxy & $i \geq j$.

(@): Let $(R_1 \cap R_2)xy$. Then $\mathcal{P}\langle x,i\rangle\langle y,0\rangle$ & $\neg\mathcal{P}\langle x,i\rangle\langle y,i+1\rangle$. The proof of weak linearity of R goes as usual. #

THEOREM 11

i) The class of the asymmetric frames, defined by the condition $\forall xy(Rxy \to \neg Ryx)$ is axiomatized over $K^{\sim}$ with the formula $(asym_s)$ $p \to \boxplus\Diamond p$;

ii) The class of the antisymmetric frames, defined by the condition $\forall xy(Rxy \wedge Ryx \to x=y)$ is axiomatized over $K^{\sim}$ with the formulae $(asym_w)$ $\blacklozenge(\boxminus p \wedge p) \to p$ and $(asym_-)$ $\blacklozenge(\boxminus p \wedge q) \to \boxminus(p \vee \Diamond q)$;

iii) The class of the completely antisymmetric frames, defined by the condition $\forall xy(x \neq y \to (Rxy \leftrightarrow \neg Ryx))$ is axiomatized over $K^{\sim}$ with the formulae (asym) $(\blacklozenge(\boxminus p \wedge p) \vee \Diamond(\boxplus p \wedge p)) \to p$, $(asym_+)$ $\Diamond(\boxplus p \wedge q) \to \boxplus(p \vee \blacklozenge q)$ and $(asym_-)$ $\blacklozenge(\boxminus p \wedge q) \to \boxminus(p \vee \Diamond q)$.

Adding the axiom (irref) $\boxminus p \to p$ axiomatizes the class of the "tournaments" - irreflexive and completely antisymmetric frames.

PROOF

The correctness and naturalness of the axiomatics are proved as usual.

The standard extension:

i) $W \leftrightharpoons W^{\sim}\times\mathbb{N}$; $\mathcal{P}\langle x,i\rangle\langle y,j\rangle$ iff $i<j$.

(@): If $(R_1 \cap R_2)xy$ then $\mathcal{P}\langle x,i\rangle\langle y,i+1\rangle$ & $\neg\mathcal{P}\langle x,i\rangle\langle y,0\rangle$.

ii) $W \leftrightharpoons W^{\sim}\times\{0\} \cup D_i(W^{\sim})\times\mathbb{N}$; $\mathcal{P}\langle x,i\rangle\langle y,j\rangle$ iff $i<j$.

iii) $W \leftrightharpoons W^{\sim}\times\{0\} \cup D_i(W^{\sim})\times\mathbb{Z}$. Let $\leq'$ be some linear ordering in $W^{\sim}$. Put $\mathcal{P}\langle x,i\rangle\langle y,j\rangle$ iff $i<j$ or $(i=j \Rightarrow x\leq' y)$.

(@): If $(R_1 \cap R_2)xy$ then $\mathcal{P}\langle x,i\rangle\langle y,i+1\rangle$ & $\neg\mathcal{P}\langle x,i\rangle\langle y,i-1\rangle$. #

COROLLARY 12

i) The class of the strongly connected frames, defined by the condition $\forall xy(\neg Ryx \rightarrow Rxy)$ is axiomatized over $K^{\sim}$ with the formula $(conn_s)$ $p \rightarrow \boxplus\Diamond p$.

ii) The class of frames, satisfying the condition "trichotomy": $\forall xy(x=y \vee Rxy \vee Ryx)$ is axiomatized with the formulae (trih) $\Diamond(\boxplus p \rightarrow p) \rightarrow p$ and $(asym_+)$ $\Diamond(\boxplus p \wedge q) \rightarrow \boxplus(p \vee \Diamond q)$. #

THEOREM 13

i) The logic SPO, axiomatized over $K^{\sim}$ with $(asym_s)$, (tran) and (tran') is complete with respect to the class of the strict partial orderings.

ii) The logic PO, axiomatized over $K^{\sim}$ with $(asym_w)$, (ref), (tran) and (tran') is complete with respect to the class of the partial orderings.

PROOF

The logics SPO and PO are natural. The standard extensions:

i) An equivalence relation ρ is defined in $W^{\sim}$:

ρxy iff $x=y \vee R_1xy \& R_1yx$.

Now put $W \leftrightharpoons W^{\sim}\times\mathbb{N}$ and $\mathcal{P}\langle x,i\rangle\langle y,j\rangle$ iff $(\rho xy \;\&\; i<j) \vee (\neg\rho xy \;\&\; j>0)$.

(@): If $(R_1 \cap R_2)xy$ then $\mathcal{P}\langle x,i\rangle\langle y,i+1\rangle$ & $\neg\mathcal{P}\langle x,i\rangle\langle y,0\rangle$.

The obtained frame is a strict partial ordering.

ii) Let ρ be an equivalence relation in $W^{\sim}$, defined by:

ρxy iff R_1xy & R_1yx and let $\leq'$ be some linear ordering in $W^{\sim}$.

Put $W \leftrightharpoons W^{\sim}\times\{0\} \cup D_i(W^{\sim})\times\mathbb{Z}$ and

$\mathcal{P}\langle x,i\rangle\langle y,j\rangle$ iff $(\rho xy \;\&\; (i<j \vee (i=j \;\&\; x\leq' y))) \vee (\neg\rho xy \;\&\; j\geq 0)$.

(@): If $(R_1 \cap R_2)$ then $\mathcal{P}\langle x,i\rangle\langle y,i+1\rangle$ & $\neg\mathcal{P}\langle x,i\rangle\langle y,-1-|i|\rangle$.

The obtained frame is a partial ordering. #

THEOREM 14

i) The logics SLO, axiomatized over $K^{\sim}$ with $(asym_s)$, (tran), (tran') and (trih) is complete with respect to the class of the strict linear orderings.

ii) The logic LO, axiomatized over $K^{\sim}$ by (asym), (ref), (tran) and (tran') is complete with respect to the class of the linear orderings.

PROOF

The logics SLO and LO are natural. The standard extensions:

i) Let $<$ be a strict linear ordering in $W^{\sim}$.

Put $W \leftrightharpoons W^{\sim}\times\{0\} \cup D_i(W^{\sim})\times\mathbb{Z}$ and $\mathcal{P}\langle x,i\rangle\langle y,j\rangle$ iff $i<j \vee (i=j \;\&\; x<y)$.

(@): If $(R_1 \cap R_2)$ then $\mathcal{P}\langle x,i\rangle\langle y,i+1\rangle$ & $\neg\mathcal{P}\langle x,i\rangle\langle y,i-1\rangle$.

The obtained frame is a strict linear ordering.

ii) Let $\leq'$ be a linear ordering in $W^{\sim}$.

Put W and $\mathcal{P}$ as in i, but applying $\leq'$ instead of $<$.

The obtained frame is a linear ordering. #

A structure $\langle I,R\rangle$, where I is a system of open intervals on the real line and $R(x_1,x_2)(y_1,y_2)$ iff $x_2<y_1$ is called an <u>interval ordering.</u> The following result (Mirkin, [3ык]) characterizes the interval ordering: $\langle I,R\rangle$ is an interval ordering iff R is a strict partial ordering satisfying the condition $\forall x \forall y (R(x)\subseteq R(y) \vee R(y)\subseteq R(x))$.

THEOREM 15 The class of the interval orderings is axiomatized over SPO with the following formulae:

Axioms	Corresponding conditions in $\mathbb{C}_t$
(int) $(\blacklozenge p\wedge\Diamond q)\to\blacksquare(\blacklozenge p\vee\Diamond p)$	$\forall x\forall y\forall z\forall t(R_1xz\wedge R_2xt \to R_1yz\vee R_2yt)$
(int_+) $(\boxplus p\wedge\Diamond q)\to\blacksquare(\boxplus p\vee\Diamond p)$	$\forall x\forall y\forall z\forall t(R_1xz\wedge R_1yt \to R_1xt\vee R_1yz)$
(int_-) $(\boxminus p\wedge\Diamond q)\to\blacksquare(\boxminus p\vee\Diamond p)$	$\forall x\forall y\forall z\forall t(R_2xz\wedge R_2yt \to R_2xt\vee R_2yz)$

PROOF INT $\leftrightharpoons$ SPO + int + int_+ + int_-. It is verified in the usual way thatthe axioms express the corresponding conditions in total frames and INT is a natural logic. Now let us provide a standard extension. Define an equivalence relation ρ in $W^{\sim}$: ρxy iff $x=y$ or $R_1xy \& R_1yx$. Put $W \leftrightharpoons W^{\sim}\times\mathbb{N}$ and $\mathcal{P}\langle x,i\rangle\langle y,j\rangle$ iff $(\rho xy \ \&\ i<j) \vee (\neg\rho xy \ \&\ j>0)$. (@): Let $(R_1\cap R_2)xy$. Then $\mathcal{P}\langle x,i\rangle\langle y,i+1\rangle \ \&\ \neg\mathcal{P}\langle x,i\rangle\langle y,0\rangle$. The obtained frame is an interval ordering. #

Finally let us note that the exposed technique can be adapted in appropriate manners in other bases, e.g. $\mathcal{L}(R,-R,R^{-1},-R^{-1})$ etc.

I am grateful to Dimiter Vakarelov, George Gargov and Solomon Passy for fruitful discussions on the problems of the paper. Also I thank Mark Brown from the Syracuse University for some stylistic remarks.#

REFERENCES

[Ben1] van Benthem J.F.A.K., Modal reduction principles, <u>JSL</u>, <u>41</u>(1976), 301-312.

[Ben2] van Benthem J.F.A.K., "Modal Logic and Classical Logic", Bibliopolis, Napoli, 1986.

[Fin] Fine K., Some connections between elementary and modal logic, <u>in</u>: Proceedings of the third Scandinavian logic symposium, Uppsala 1973, S. Kanger (ed.), North-Holland, Amsterdam, 1975, 15-31.

[Gol] Goldblatt R.I., Metamathematics of modal logic, Reports on Math. logic, 6(1976),41-77 and 7(1976),21-52.

[Gor] Goranko V., Modal definability in enriched languages, Notre Dame Journal of Formal Logic, to appear.

[GPT] Gargov G., S. Passy, T. Tinchev, Modal environment for Boolean speculations, in: "Mathematical Logic", D. Skordev (ed.), Plenum Press, New York, 1988.

[GP] Gargov G, S. Passy, A Note on Boolean Modal Logic, this volume

[HC] Hughes G.E. & M.J.Cresswell, "A Companion to Modal Logic", Methuen, London, 1984.

[Hum] Humberstone I., Inaccessible worlds, Notre Dame Journal of Formal Logic, 24(1983), 346-352.

[Seg] Segerberg K., "An Essay in Classical Modal Logic", Filosofiska Studier 13, Uppsala, 1971.

[Vak1] Vakarelov D., S4 + S5 together - S4+5, LMPS'87, Vol.5, Moskow,1987

[Vak2] Vakarelov D., Modal logics for knowledge representation systems, to appear.

[Зык] ЗЫКОВ А.А. "Основы теории графов", Наука, Москва, 1987.

A TEMPORAL LOGIC FOR EVENT STRUCTURES *

Wojciech Penczek

Institute of Computer Science
Polish Academy of Sciences
Warsaw, PKiN, P.O. Box 22

INTRODUCTION

The formalism of temporal logic has been suggested as an appropriate tool for specifying and proving properties of distributed programs. It has become clear that the modalities of temporal logic are well suited for capturing the dynamic properties of distributed programs and systems. Originally, temporal logic was designed in order to analyse and reason about time sequences in general (for example, by Emerson and Halpern, 1985, 1986; Lamport, 1980; Gabbay et al., 1980; Pnueli, 1981). In most of the papers in the area of temporal logic concurrency is represented in terms of an arbitrary nondeterministic interleaving. Because of that the difference between concurrency and non-determinism is lost. This is quite acceptable for many purposes, but not always, as shown by Mazurkiewicz et al. (1988). A major consequence, however, is that one is forced to attach formulas to the global states of a distributed system (program). In general, it is very difficult, if not impossible, to observe such global states; parts of the global state may be changing simultaneously due to independent actions carried out on two separate locations. So, we need a formalism which deals only with local states. In this formalism we incorporate operators representing the relations of causality and conflict. As the most appriopriate model we choose an event structure. One such a formalism was put forward by Lodaya and Thiagarajan (1987). In that formalism some subclass of event structures was considered as models. Our model is an event structure as general as possible. Strictly speaking, a model is a run in an event structure. We show that inevitable and invariant properties can be expressed in our formalized language.

An event structure is a partially ordered set of event occurrences together with a symmetric conflict relation. The ordering relation models causality, whereas the conflict relation expresses conflicts between event occurrences. Two event occurrences that are neither comparable nor in conflict, may occur concurrently. In this sense event structures provide explicit

---

* The full version of this paper will appear in Fundamenta Informaticae.

Mathematical Logic
Edited by P. P. Petkov
Plenum Press, New York, 1990

and separate representation of sequence, choice and concurrency. Event structures arise naturally from the net theory as shown by Nielsen et al. (1981). Winskel (1982) has obtained a non-interleaved semantics of CCS - like languages using event structures. Hence, there is a good reason to hope that these objects can serve as an adequate model of distributed systems.

In the next section we introduce the notion of event structures. Inevitability and invariancy for event structures are defined in section 2. The syntax and semantics of the Event Structure Logic (ESL for short) are presented in section 3. The axiom system is presented in section 4, where we also argue for the soundness of this system w.r.t. the chosen semantics. Many of our axioms are standard (compare S4 and KB logics), quoted from the book of Gabbay (1976). In section 5 we show the completeness of our axiom system using a Rasiowa-Sikorski (1970) style proof and a Vakarelov (1987) method for combined modalities. Next, we show that ESL does not possess the finite model property and that ESL is decidable - section 6. In section 7 we give a simple example of a specification of system properties using ESL. In the concluding section we discuss possible extensions of our logic.

This paper is based on the report of Penczek (1987), where the reader can find the detailed proof of the completeness theorem.

1. EVENT STRUCTURES

The notion of an event structure has been introduced and studied by Winskel (1980). An event structure represents the behaviour of a distributed system by means of a set of event occurrences, a causality relation that partially orders the event occurrences and a conflict relation which reflects the choices available to the system.

Definition 1.1
An <u>event structure</u> is a triple $ES = (E, \leq, \#)$, where
i) E is a set of <u>events</u>,
ii) $\leq \subseteq E \times E$ is a partial order, called the <u>causality relation</u>,
iii) $\# \subseteq E \times E$ is an irreflexive, symmetric relation, called the <u>conflict relation</u>,
iv) For any e, e', e'' in E: $e \# e' \leq e''$ implies $e \# e''$./ / /

The last clause in the definition captures the intuition that in the past of any event no two events can be in a conflict. We refer to that property as to the <u>conflict preservation</u>. Notice that the relations $\leq$ and $\#$ are disjoint by iii) and iv).

Example 1.2
Consider an event structure in Fig. 1. (taken from the paper of Lodaya and Thiagarajan (1987)). The events have been labelled to reflect the behaviour of a <u>p</u>roducer communicating via an unbounded buffer to a <u>c</u>onsumer. The producer can <u>s</u>top after producing zero or more items.

For convenience, in Fig.1. we have just indicated the generating elements of the causality relation (as directed arcs) and the conflict relation (as squiggly lines). Thus $e_1 \leq e_3$ because $e_1 \leq e_2$ and $e_2 \leq e_3$. Moreover, $e'_1 \# e''_2$ because $e'_1 \# e_1$ and $e_1 \leq e''_2$. Finally, e''_1 and e_2 can occur concurrently because

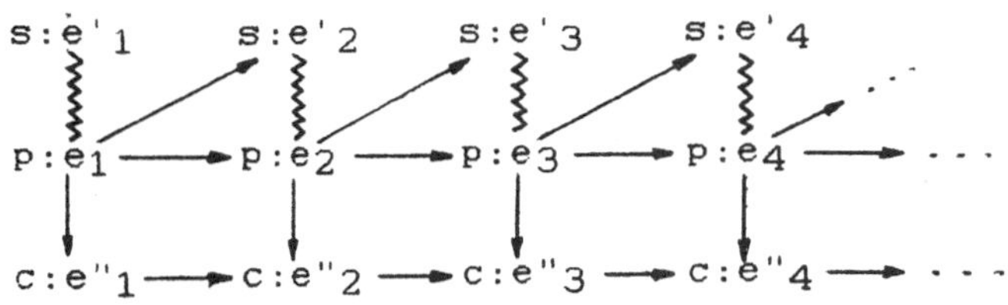

Fig. 1. An event structure PC - (producer - consumer)

they are causally incomparable and are not in a conflict one with another.

We now give some notions related to event structures which are necessary to understand a behaviour of a system described by an event structure.

Definition 1.3
Let $ES = (E, \leq, \#)$ be an event structure and let $R \subseteq E$.

i) R is <u>conflict-free</u> iff $(R \times R) \cap \# = \emptyset$.
ii) R is <u>backward closed</u> iff for any $e \in R$ and $e' \in E$, $e' \leq e$ implies $e' \in R$.
The <u>backward closure</u> of e, denoted $\downarrow e$, is defined by $\downarrow e = \{e' \in E \mid e' \leq e\}$.
The <u>forward closure</u> of e, denoted $\uparrow e$, is defined by $\uparrow e = \{e' \in E \mid e \leq e'\}$.
iii) R is a <u>local state</u> iff there is e in E s.t. $R = \downarrow e$.
iv) R is a <u>global state</u> iff R is conflict-free and backward closed.
v) R is a <u>run</u> iff R is a maximal global state (w.r.t. the set-theoretical inclusion).
vi) $R_{ES} = \{R \subseteq E \mid R \text{ is a run}\}$.///

As explained by Lodaya and Thiagarajan (1987), an event structure $(E, \leq, \#)$ can be considered as a structure of local states whereas $(S, \subseteq)$, where S is the set of global states, is a structure of global states.

For the event structure shown in Fig.1. $\{e_1, e''_1\}$, $\{e_1, e''_1, e'_2\}$ and $\{e_1, e_2, e_3, ..\}$ are global states, but $\{e_1, e'_1\}$ and $\{e''_1, e_2\}$ are not. $R = \{e_1, e''_1, e_2, e''_2, e_3, e''_3, ..\}$ and $R_3 = \{e_1, e''_1, e_2, e''_2, e'_3\}$ are runs, but $\{e_1, e''_1, e''_2, e''_3, ..\}$ is not. $R_{ES} = \{R_j \mid j \geq 1\} \cup \{R\}$, where $R_1 = \{e'_1\}$, $R_j = \downarrow e''_{j-1} \cup \{e'_j\}$ for $j \geq 2$.

Fact 1.4
Let ES be an event structure. The set R_{ES} is not empty.
Proof. By Kuratowski-Zorn lemma.///

We formulate two conditions which are equivalent to the definition of a run. We will use them in proofs of theorems instead of Definition 1.3v).

Lemma 1.5
Let $ES = (E, \leq, \#)$ be an event structure s.t. $E \neq \emptyset$, and $R \subseteq E$. Then, the following three conditions are equivalent:

i) R is a run,
ii) $e \in R$ iff $(\forall e' \in E)$ $(e \# e'$ implies $e' \notin R)$,
iii) $(\forall e \in E)$ $card(X_e \cap R) = 1$, where $X_e = \{e' \in E \mid e \# e'\} \cup \{e\}$.

Proof. Omitted.///

2. INVARIANCY AND INEVITABILITY

In this section we will define two most important properties for event structures: invariancy and inevitability. It is known that the most interesting properties of concurrent systems are combinations of invariancy and inevitability. It should be noticed that one can easily define and express invariancy in each model of a concurrent system, but, as Mazurkiewicz et al. (1988) have shown, in order to express inevitability we need a partial order framework.

Definition 2.1
Let $ES=(E,\leq,\#)$ be an event structure, $B,Q \subseteq E$ and $e \in E$.

i) Q is <u>inevitable from e</u> iff $(\forall R \in R_{ES})$ ($e \in R$ implies $\uparrow e \cap R \cap Q \neq \emptyset$).

ii) Q is an <u>invariant from e</u> iff $(\forall R \in R_{ES})$ ($e \in R$ implies $\uparrow e \cap R \subseteq Q$).

iii) Q is <u>inevitable from B</u> iff Q is inevitable from each e in B.

iv) Q is an <u>invariant from B</u> iff Q is an invariant from each e in B.///

For the event structure shown in Fig.1.: $\{e''_1\}$ and $\{e_2,e'_2\}$ are inevitable from e_1. Sets $\{e'_2,e'_3,e'_4,...\}$ and $\{e_2,e_3,e_4,...\}$ are not inevitable from e_1. $\{e''_1,e''_2,e''_3,...\}$ is an invariant from e''_1 and is inevitable from $\{e_1,e_2,e_3,..\}$.

3. THE FORMALIZED LANGUAGE OF EVENT STRUCTURE LOGIC (ESL)

Syntax

We fix a countable set of atomic propositions $P=\{p_1,p_2,...\}$ and we also fix one atomic <u>run</u> proposition δ. We assume that $\delta \notin P$ and set $P'= P \cup \{\delta\}$. Let p ranges over P'. The run proposition will be used for marking a run of an event structure. We shall use also the logical connectives $\neg$ and $\vee$, as well as modalities G, H and Z. The set Form of formulas of ESL can now be built up inductively.

Definition 3.1

i) Every member of P' is a formula.

ii) If α and β are formulas, then so are $\neg\alpha$, $\alpha \vee \beta$, $G\alpha$, $H\alpha$ and $Z\alpha$.///

We let α,β and φ range over Form, the set of formulas. It will be convenient to have available the following derived logical connectives and modalities.

Definition 3.2

i) $\alpha \wedge \beta$ =(def) $\neg(\neg\alpha \vee \neg\beta)$,

ii) $\alpha \rightarrow \beta$ =(def) $\neg\alpha \vee \beta$,

iii) $\alpha \equiv \beta$ =(def) $(\alpha \rightarrow \beta) \wedge (\beta \rightarrow \alpha)$,

iv) $F\alpha$ =(def) $\neg G \neg\alpha$,

v) $P\alpha$ =(def) $\neg H \neg\alpha$,

vi) $D\alpha$ =(def) $\neg Z \neg\alpha$.///

Semantics

Definition 3.3
A <u>general structure</u> is a quadruple $GS=(E,R,\leq,\#)$, where

i) E is a set of events,
ii) $\leq \subseteq E \times E$ is reflexive and transitive,
iii) $\# \subseteq E \times E$ is symmetric,
iv) For any e,e',e'' in E: $e \# e' \leq e''$ implies $e \# e''$,
v) $R \subseteq E$ satisfies the condition:
$e \in R$ iff $(\forall e' \in E)(e \# e'$ implies $e' \notin R)$.///

From the above definition, it follows that the relation # is irreflexive on R and E is not empty.

Definition 3.4
A <u>general frame</u> is a general structure GS.
A <u>standard frame</u> is a quadruple $(E,R,\leq,\#)$, where $ES=(E,\leq,\#)$ is an event structure s.t. $E \neq \emptyset$ and $R \in R_{ES}$.///

Note that a standard frame is a general frame s.t. $\leq$ is antisymmetric and # is irreflexive on E\R.

We define the semantics for ESL by means of the notion of a model determined by a standard or general frame. The general frame is used only to make the proof of completeness of ESL more clear.

Definition 3.5
A <u>standard (general) model</u> is an ordered pair $M = (S,V)$, where
i) $S=(E,R,\leq,\#)$ is a standard (general, resp.) frame,
ii) $V: P' \longrightarrow 2^E$ is a valuation function from the set P' into the set of all subsets of E such that $R = V(\delta)$.///

Let $M=(S,V)$ be a standard or general model with $S=(E,R,\leq,\#)$. Let $e \in E$ and $\alpha \in$ Form. Now members of E will be called states. Then the notion of α being true at a state e in the standard or general model M, written $e,M \models \alpha$, is defined inductively:

Definition 3.6
i) $e,M \models p$ iff $e \in V(p)$,
ii) $e,M \models \neg\alpha$ iff $e,M \not\models \alpha$,
iii) $e,M \models \alpha \vee \beta$ iff $e,M \models \alpha$ or $e,M \models \beta$,
iv) $e,M \models G\alpha$ iff $(\forall e' \in E)(e \leq e'$ implies $e',M \models \alpha)$,
v) $e,M \models H\alpha$ iff $(\forall e' \in E)(e' \leq e$ implies $e',M \models \alpha)$,
vi) $e,M \models Z\alpha$ iff $(\forall e' \in E)(e \# e'$ implies $e',M \models \alpha)$.///

We note down the standard notions of satisfiability and validity.

Definition 3.7
i) A set L of formulas is <u>satisfiable</u> in a standard (general) model iff there exists a standard (general, resp.) model $M=(S,V)$ with $S=(E,R,\leq,\#)$ and $e \in E$ such that $e,M \models \alpha$ for all $\alpha \in L$. We say that α is satisfiable, if $\{\alpha\}$ is satisfiable.
ii) For any model $M=(S,V)$, where $S=(E,R,\leq,\#)$
$M \models \alpha$ iff $e,M \models \alpha$ for all $e \in E$.
iii) α is <u>valid</u> in all standard (general) models, in symbols $\models_s \alpha$ ($\models_g \alpha$, resp.) iff $M \models \alpha$ for all standard (general, resp.) models M.
iv) For any model $M=(S,V)$, where $S=(E,R,\leq,\#)$
$M \models L$ iff $e,M \models \alpha$ for all $e \in E$ and for all $\alpha \in L$.
v) A formula α is a <u>semantical consequence</u> of a set L of formulas for standard (general) models, in symbols $L \models_s \alpha$ ($L \models_g \alpha$, resp.) iff $M \models L$ implies $M \models \alpha$ for each standard (general, resp.) model M.///

To show what can be expressed in our language we focus our attention on a fixed event structure. Then, let $ES=(E,\leq,\#)$ be an event structure. We define a class of models corresponding to ES: $M_{ES} = \{(S,V) \mid (S,V)$ is a standard model s.t. $S=(E,R,\leq,\#)$, where $R \in R_{ES}\}$.

Definition 3.8
We say that a formula α is valid in ES, written $ES \models \alpha$, iff $(\forall M \in M_{ES})\ M \models \alpha$.///

Let $\alpha \in$ Form. Then $Q_\alpha = \{e \in E \mid e,M \models \alpha$ for each $M \in M_{ES}\}$ is the subset of E satisfying α.

Theorem 3.9
The following three equivalences hold:
i) $ES \models (\alpha \wedge \delta) \rightarrow F(\beta \wedge \delta)$ iff Q_β is inevitable from Q_α.
ii) $ES \models \alpha \rightarrow G\beta$ iff Q_β is an invariant from Q_α.
iii) $ES \models \delta$ iff ES is conflict-free (i.e., there is only one run in ES).
Proof. Obvious.///

Our approach is related to linear and branching time logic, because we defined the validity of a formula in an event structure ES with respect to its validity in all runs in ES or w.r.t. the whole event structure ES. Thus properties of possibility are also expressible in ESL, namely $ES \models \alpha \rightarrow F\beta$ means that Q_β is possible from Q_α. This kind of properties is expressible in branching logic but not in linear logic as shown by Lamport (1980). In the next section we will give the axiom system for ESL.

4. THE AXIOM SYSTEM FOR ESL

Most of our axioms are versions of standard temporal logic axioms taken from the book of Gabbay (1976). There are also some new axioms, which characterize the relationship between the causality and conflict relations, and describe the properties of a run. We first present the axiom scheme in full and then provide some explanatory remarks.

Axioms

A0.	All formulas having the form of tautologies of the classical propositional calculus.	
A1.	$G(\alpha \rightarrow \beta) \rightarrow (G\alpha \rightarrow G\beta)$	
A2.	$H(\alpha \rightarrow \beta) \rightarrow (H\alpha \rightarrow H\beta)$	(deductive closures)
A3.	$Z(\alpha \rightarrow \beta) \rightarrow (Z\alpha \rightarrow Z\beta)$	
A4.	$G\alpha \rightarrow \alpha$	(reflexivity of $\leq$)
A5.	$H\alpha \rightarrow \alpha$	(reflexivity of $\leq^{-1}$)
A6.	$G\alpha \rightarrow GG\alpha$	(transitivity of $\leq$)
A7.	$H\alpha \rightarrow HH\alpha$	(transitivity of $\leq^{-1}$)
A8.	$\alpha \rightarrow GP\alpha$	(relating past and future)
A9.	$\alpha \rightarrow HF\alpha$	(relating past and future)
A10.	$\alpha \rightarrow ZD\alpha$	(symmetry of #)
A11.	$Z\alpha \rightarrow ZG\alpha$	(conflict preservation)
A12.	$\delta \rightarrow Z\neg\delta$	(conflict-freeness of R)
A13.	$Z\neg\delta \rightarrow \delta$	(maximality of R)

Conflict-freeness of R implies irreflexivity of # on R. Note that there are no temporal formulas which could represent

the following two conditions:
*) $\leq$ is antisymmetric.
**) # is irreflexive on $E \setminus R$.
That lack of such formulas causes the main difficulty in axiomatizing ESL in the class of standard models.

Inference rules

(MP) If $\vdash \alpha$ and $\vdash (\alpha \rightarrow \beta)$, then $\vdash \beta$ (modus ponens).
(TG.a) If $\vdash \alpha$, then $\vdash G\alpha$.
(TG.b) If $\vdash \alpha$, then $\vdash H\alpha$ (generalization rules).
(TG.c) If $\vdash \alpha$, then $\vdash Z\alpha$.

The notion of a <u>proof</u> of a formula is defined as usual. We will say that a formula α is <u>derivable</u> from a set L of formulas, in symbols $L \vdash \alpha$, if there is a proof of α from L. A formula α is a <u>theorem</u> of a logic ESL, denoted $\vdash \alpha$ if α can be derived from the empty set of formulas using the axioms and the inference rules. The set L of formulas is <u>consistent</u> if a formula of the form $\alpha \wedge \neg\alpha$ is not derivable from L. We say that α is <u>consistent</u>, if $\{\alpha\}$ is consistent.

Theorem 4.1 (Soundness)
For any formula α in Form, the following conditions hold:
i) if $\vdash \alpha$, then $\models_g \alpha$ and $\models_s \alpha$,
ii) if $L \vdash \alpha$, then $L \models_g \alpha$ and $L \models_s \alpha$.///
Proof. Omitted.///

Clearly, by definitions of a general and standard model, $\models_g \alpha$ implies $\models_s \alpha$ for any α in Form.
In the next section we will prove completeness of our axiom system w.r.t. the general and standard models. It will be shown that standard and general models are not distinguishable in our formal language which reflects that the conditions *) and **), mentioned after the set of axioms, are not expressible in ESL.

5. COMPLETENESS OF THE AXIOM SYSTEM

At the beginning we show that our axiom system is complete w.r.t. general models. We use a Rasiowa-Sikorski (1970) style proof.

Theorem 5.1 (Completeness w.r.t. the general models)
For any formula α in Form, the following conditions hold:
i) if $\models_g \alpha$, then $\vdash \alpha$,
ii) if $L \models_g \alpha$, then $L \vdash \alpha$.
Proof. (sketch) ii) Let L be a consistent set of formulas. Let $\alpha \approx \beta$ iff $L \vdash \alpha \rightarrow \beta$ and $L \vdash \beta \rightarrow \alpha$, for $\alpha, \beta \in$ Form. If $\alpha \in$ Form, then $[\alpha] \in \mathrm{Form}/\!\approx$. We take the Lindenbaum algebra $(\mathrm{Form}/\!\approx, \cap, \cup, -, \mathbf{1}, \mathbf{0})$ of ESL s.t. $L \vdash \alpha$ iff $[\alpha] = \mathbf{1}$ for each formula α. Let $\mathbf{F}$ be a family of all ultrafilters in this algebra. Next, we define a canonical general frame $GS_0 = (E_0, R_0, \leq_0, \#_0)$ and a canonical general model $GM_0 = (GS_0, V_0)$ as follows:
$E_0 = \mathbf{F}$,
$R_0 = \{A \in \mathbf{F} \mid [\delta] \in A\}$,
$\leq_0 = \{(A,B) \in \mathbf{F} \times \mathbf{F} \mid \{[\alpha] \mid [G\alpha] \in A\} \subseteq B\}$,
$\#_0 = \{(A,B) \in \mathbf{F} \times \mathbf{F} \mid \{[\alpha] \mid [Z\alpha] \in A\} \subseteq B\}$,
$A \in V_0(p)$ iff $[p] \in A$, where $p \in P'$ and $A \in \mathbf{F}$.
As usually, we show that $\#_0$ is symmetric (A10), $\leq_0$ is reflexive (A4) and transitive (A6), $A \#_0 B \leq_0 C$ implies $A \#_0 C$ (A11) and

that $A \in R_0$ iff $(\forall B \in \mathbf{F})$ $(A \#_0 B$ implies $B \notin R_0)$ (A12, A13 and the definition of R_0). We prove by induction on the length of a formula that for each $\alpha \in$ Form and each $A \in \mathbf{F}$: $A, GM_0 \models \alpha$ iff $[\alpha] \in A$. Next, we show that, if $L \nvdash \alpha$, then there is A in $\mathbf{F}$ s.t. $A, GM_0 \models \neg\alpha$ and $GM_0 \models L$, whence $L \nvDash_g \alpha$. i) follows from ii).///

We now show that our axiom system is complete w.r.t. the class of all standard models.

Theorem 5.2
For any formula α in Form, the following conditions hold:
i) if $\models_s \alpha$, then $\models_g \alpha$,
ii) if $L \models_s \alpha$, then $L \models_g \alpha$.
Proof. (sketch) ii). Suppose that for some general model $GM=(GS,V)$, where $GS=(E,R,\leq,\#)$ and $GM \models L$, there is a state e in E s.t. $e, GM \nvDash \alpha$. We shall construct another (standard) model $SM = (SS,V')$, where $SS = (E',R',\leq',\#')$ s.t. $SM \models L$ and for some e' in E' we have $e', SM \nvDash \alpha$.
Let $E' = E \times \mathbf{Z}$, where $\mathbf{Z}$ is the set of integers. We abbreviate: $e_j = (e,j)$ for $e \in E$ and $j \in \mathbf{Z}$. Let:
$R' = R \times \mathbf{Z}$,
$\leq' = \{(e_i, e'_j) \mid (e = e'$ and $i = j)$ or $(e \leq e'$ and $i < j$ and $j \equiv i (\mathrm{mod}\ 2)\}$,
$\#' = \{(e_i, e'_j) \mid e \# e'$ and $j \equiv i+1 (\mathrm{mod}\ 2)\}$, where
$k \equiv 1 (\mathrm{mod}\ 2)$ iff $|k-1|$ is an even number.
$V'(p) = \{e_j \mid j \in \mathbf{Z}$ and $e \in V(p)\}$, where $p \in P'$.

Lemma 5.3
The following conditions hold:
i) the relation $\leq'$ is a partial order in E',
ii) the relation $\#'$ is irreflexive in E',
iii) the relation $\#'$ is symmetric in E',
iv) for any e_i, e'_j, e''_k in E': $e_i \#' e'_j \leq' e''_k$ implies $e_i \#' e''_k$,
v) for any e_j in E': $e_j \in R'$ iff $(\forall e'_i \in E')(e_j \#' e'_i$ implies $e'_i \notin R')$.
Proof. Omitted.///

Lemma 5.4
The model $SM = (SS,V')$ is standard.
Proof. The frame $SS = (E',R',\leq',\#')$ is standard by Lemma 5.3 and $V'(\delta)=R'$ by definitions of R' and V'.///

Let us observe some relationships between $\leq$, $\#$ and $\leq'$, $\#'$.

Lemma 5.5
For any e, $e' \in E$, the following implications hold:
i) $e \leq e'$ implies $\forall i\ \exists j\ e_i \leq' e'_j$,
ii) $e_i \leq' e'_j$ implies $e \leq e'$,
iii) $e \leq e'$ implies $\forall i\ \exists j\ e_j \leq' e'_i$,
iv) $e \# e'$ implies $\forall i\ \exists j\ e_i \#' e'_j$,
v) $e_i \#' e'_j$ implies $e \# e'$.
Proof. Directly by the definitions of $\leq'$ and $\#'$.///

Now, we are going to show the correspondence between the standard model SM and the general model GM.

Lemma 5.6
For any formula α in Form, the following equivalences hold:
i) $(\forall j \in \mathbf{Z})$ $(e, GM \models \alpha$ iff $e_j, SM \models \alpha)$,
ii) $GM \models \alpha$ iff $SM \models \alpha$.

Proof. i) The proof is by induction on the length of a formula using Lemma 5.5. ii) follows immediately from i).///

Now, the proof of Theorem 5.2 can be completed as follows: ii) Since we have $e,GM \not\models \alpha$ and $GM \models L$, then by Lemma 5.6, $SM \models L$ and $e_j,SM \not\models \alpha$ for some $j \in \mathbf{Z}$. Thus α is not true in the model $SM = (SS,V')$ which is standard by Lemma 5.4 and in which L is true. i) follows from ii).///

Theorem 5.7 (Completeness w.r.t. the standard models)
For any formula α in Form, the following conditions hold:
i) if $\models_S \alpha$, then $\vdash \alpha$,
ii) if $L \models_S \alpha$, then $L \vdash \alpha$.
Proof. By Theorems 5.1 and 5.2.///

6. FINITE MODEL PROPERTY AND DECIDABILITY

In this section we will show that ESL has the finite model property w.r.t. general models and does not have this property w.r.t. standard models. Next we will prove that ESL is decidable w.r.t. the standard models.

Theorem 6.1
ESL does not have the finite model property w.r.t. the standard models.
Proof. To prove this theorem, we will show that there is a formula α in Form s.t. α is true in each finite standard model of ESL and α is not a theorem. Then $\neg\alpha$ is a consistent formula which is not satisfiable in any finite model.///

Lemma 6.2
Let $\alpha = G(G(p\rightarrow Gp)\rightarrow p)\rightarrow p$, where $p \in P$. Then, for each finite standard model M, $M \models \alpha$.
Proof. Note that α is equivalent to the formula $G(F(p\wedge F\neg p)\vee p)\rightarrow p$. Now, let $M = (SS,V)$ be a standard model with $SS = (E,R,\leq,\#)$. Consider an arbitrary $e \in E$. Assume that $e,M \models G(F(p\wedge F\neg p)\vee p)$ and suppose that $e,M \models \neg p$. Then $e,M \models F(p\wedge F\neg p)$. Thus there is e_1 in E s.t. $e \leq e_1$, $e \neq e_1$ and $e_1,M \models \neg p\wedge G(F(p\wedge F\neg p)\vee p)$. In such a way we can construct an infinite sequence of states $e_1,e_2,\ldots$ of E, s.t. $\forall\, 1 \leq i < j$: $e_i \leq e_j$ ($e_i \neq e_j$ because $\leq$ is acyclic) contradicting that E is finite.///

Lemma 6.3
A formula $\alpha = G(G(p\rightarrow Gp)\rightarrow p)\rightarrow p$, where $p \in P$, is not a theorem of ESL.
Proof. We shall construct an (infinite) standard model for ESL in which $\neg\alpha$ is satisfiable.
Let $E = \{e_j \mid j \in \mathbf{Z}\}$,
$R = E$,
$\leq\, = \{(e_i,e_j) \mid i \leq j\}$,
$\# = \emptyset$.
Let $M = (SS,V)$, where $SS = (E,R,\leq,\#)$ and V is such that $e_j \in V(p)$ iff $j\equiv 1 (\mathrm{mod}\ 2)$, where $p \in P$ and $V(\delta) = E$. Note that $e_0,M \models G(F(p\wedge F\neg p)\vee p)\wedge\neg p$, i.e., $e_0,M \models \neg\alpha$ and M is clearly a standard model.///

However, ESL does not have the finite model property w.r.t. standard models; nevertheless ESL is decidable w.r.t. that class of models. To show this we will prove that ESL has the finite model property w.r.t. the general models. Next, by Theorem 5.2

we will obtain decidability of ESL w.r.t. standard models.

Theorem 6.4
ESL has the finite model property w.r.t. the general models.
Proof. We have to show that for each consistent formula there is a finite general model in which this formula is satisfiable. Let $\alpha \in$ Form be a consistent formula. By Theorem 5.1, α is satisfiable in a general model. Thus, by the construction given in the proof of Theorem 5.1, there is a canonical general model $M=(S,V)$, where $S=(E,R,\leq,\#)$, $R=V(\delta)$, and $o,M \models \alpha$ for a certain $o \in E$. Using ideas taken from the papers of Fischer and Ladner (1979); Kozen and Parikh (1981), we will construct a finite model for α. The first step is to extend $\{\alpha\}$ to the smallest set $C(\alpha)$ of formulas having the following closure properties:
1) $\alpha \in C(\alpha)$,
2) if $\beta \vee \varphi \in C(\alpha)$, then $\beta \in C(\alpha)$ and $\varphi \in C(\alpha)$,
3) if $\neg\beta \in C(\alpha)$, then $\beta \in C(\alpha)$,
4) if $X\beta \in C(\alpha)$, then $\beta \in C(\alpha)$, where X denotes G, H or Z,
5) if $Z\beta \in C(\alpha)$, then $G\beta \in C(\alpha)$,
6) $\delta \in C(\alpha)$.
Note that $C(\alpha)$ is finite.
Let $Cl(\alpha) = \{[\beta] \mid \beta \in C(\alpha)\}$. Clearly $Cl(\alpha)$ is also a finite set.
Let $M'=((E',R',\leq',\#'),V')$, where:
$E' = \{e \cap Cl(\alpha) \mid e \in E\}$,
$R' = \{e \cap Cl(\alpha) \mid e \in R\}$,
$t \leq_1 s$ iff for some $e,f \in E$, $e \leq f$ and $t = e \cap Cl(\alpha)$,
$s = f \cap Cl(\alpha)$,
$t \#_1 s$ iff for some $e,f \in E$, $e \# f$ and $t = e \cap Cl(\alpha)$,
$s = f \cap Cl(\alpha)$,
$\leq'$ is the transitive closure of $\leq_1$,
$\#' = ((\leq')^{-1})\circ(\#_1)\circ(\leq')$, where $\circ$ denotes the composition of relations.
$s \in V'(p)$ iff $[p] \in s$, where $p \in P'$ and $s \in E'$.

Lemma 6.5
M' is a finite general model such that for each formula β in $C(\alpha)$ and for each s in E': $s,M' \models \beta$ iff $[\beta] \in s$.
Proof. Omitted.///

We now complete the proof of Theorem 6.4. We know that $o,M \models \alpha$. Hence $[\alpha] \in o$. Let $o' = o \cap Cl(\alpha)$. Thus $[\alpha] \in o'$. By Lemma 6.5, $o',M' \models \alpha$, where M' is a finite general model.///

Theorem 6.6
ESL is decidable w.r.t. the general models.
Proof. ESL is finitly axiomatizable and has the finite model property.///

Theorem 6.7
ESL is decidable w.r.t. the standard models.
Proof. By Theorems 6.6 and 5.2.///

7. AN EXAMPLE OF USING ESL TO SPECIFY PROPERTIES OF CONCURRENT SYSTEMS

It is not the aim of this paper to develop methods of using our language for specifying and proving properties of concurrent systems. Nevertheless, we would like to show how one can specify some properties of a system from Example 1.2. The most important thing we want to express is that whenever the system produces an item, then this item will be (inevitably) consumed (see INEV1) and (also inevitably) either the items will be produced and consumed for ever or the system will stop (see INEV2). In our formulas we will use propositions p, c and s corresponding to the labels in the event structure PC. Such a proposition is true at the given state if it is the label of this state.

$$INEV1 = (p \wedge \delta) \rightarrow F(c \wedge \delta)$$
$$INEV2 = (p \wedge \delta) \rightarrow (F(s \wedge \delta) \vee G(\delta \rightarrow (p \vee c)))$$

It is easy to see that each run in PC meets the formulas INEV1 and INEV2. We can also specify an invariant property. INV1 says that whenever the system consumes an item, then only consumption is possible in the future. Another explanation of INV1 could be that consumption causes only subsequent consumption.

$$INV1 = c \rightarrow Gc$$

Using the conflict operator we can specify that production is in conflict only with stopping, but not with consuming, (see CONF1), and that consuming is also in conflict only with stopping, but not with production, (see CONF2).

$$CONF1 = p \rightarrow Zs$$
$$CONF2 = c \rightarrow Zs$$

Notice that a specification of a system PC could be more "accurate" if we have other modalities available, for example Next-State operator. We could then express that after producing, the item will be inevitably consumed in the next state and that no action is possible after stopping. We continue the discussion of possible extensions of our language in the next section.

8. CONCLUSIONS

In this paper we have obtained a temporal characterization of the event structures with runs. Our major goal has been to use temporal logic to express non-global properties of distributed systems. This work should be considered as a first step to obtain a logic for specifying concurrent systems in a partial order framework of local states. Our language must be expanded, if any other assumptions about a model are to be made. For example, if we want to specify discrete systems, then we should introduce a Next-State operator. We can also introduce the operators Until and Prior, similarly to Reisig (1987). Then we will have to look for the right axioms corresponding to the semantics of these new operators and a new completeness proof will have to be constructed. We should also give a method for constructing a set of specific axioms for a specificated system, i.e., axioms which restrict the class of possible models to the class of models corresponding to the given event structure.

Another task is to use our logic for specifying non-global properties of C/E-systems. There exist only languages for expressing the global properties of such systems; for example Tuominen (1986). A short review of existing formal languages for specifying concurrent systems is given by Reisig (1987). Furthermore, we want to underline that practical applications of temporal logic involve, in general, first order concepts, which implies the need for extending ESL to a first-order language.

Acknowledgements: The author wishes to thank Prof. Antoni Mazurkiewicz for his help in the preparation of this paper.

REFERENCES

Emerson, E. A., and Halpern, J. Y., 1986, "Sometimes" and "Not Never" Revisited: On branching versus Linear Time Temporal Logic, Journal of ACM, Vol. 33 No. 1, 151:178.

Emerson, E. A., and Halpern, J. Y., 1985, Decision Procedures and Expressiveness in the Temporal Logic of Branching Time, Journal of CSS, Vol. 30, 1:24.

Fischer, M. J., and Ladner, L. E., Propositional Dynamic Logic of Regular Programs, 1979, Journal of CSS, Vol. 18, No. 2, 194:211.

Gabbay, D. M., 1976, "Investigations in modal and tense logics with applications to problems in philosophy and linguistics", D. Reidel Publishing Company, Holland.

Gabbay, D. M., Pnueli, A., and Shelah, S., Stavi, J., 1980, On the Temporal Analysis of Fairness, 7 ACM Symp. on Principles of Progr. Logic, 163:173.

Kozen, D., and Parikh, R., 1981, An elementary proof of the completeness of PDL, TCS, Vol. 14(1), 113:118.

Lamport, L., 1980, "Sometime" is sometimes "not ever", On the Temporal Logic of Programs, 7 ACM Symp. on Principles of Progr. Logic, 174:185.

Lodaya, K., and Thiagarajan, P. S., 1987, "A Modal Logic for a Subclass of Event Structures", Report 220 of Computer Science Department Aarhus University.

Mazurkiewicz, A., 1986, "Complete processes and inevitability", Report 86-06, University of Leiden.

Mazurkiewicz, A., Ochmański, E., and Penczek, W., 1988, Concurrent Systems and Inevitability, to appear in TCS.

Nielsen, M., Plotkin, G., and Winskel., 1981, Petri Nets, Events Structures and Domains, Part I, TCS, Vol. 13(1), 85:109.

Penczek, W., 1987, "A temporal logic for event structures", Report 616, IPI PAN, Warsaw.

Pnueli, A., 1981, The temporal semantics of concurrent programs, TCS, Vol. 13(1), 45:61.

Reisig, W., 1987, "Towards Temporal Logic for True Concurrecy", GMD Report 277.

Rasiowa, H., Sikorski, R., 1970, "Mathematics of Metamathematics" PWN Warsaw.

Tuominen, H., 1986, "A Logical Basis for C/E-systems", Helsinki University of Technology, Report 35 (B).

Winskel, G., 1980, "Events in Computation", Ph.D. Thesis, Dept. of Comp. Science, University of Edinburgh.

Winskel, G., 1982, Event Structure Semantics for CCS and Related Languages, in: "Lecture Notes in Computer Science 140", Springer-Verlag, New York, London, 561:577.

Vakarelov, D., 1987, S4 and S5 together - S4+5, VIII Congres Logic, Methodology and Philosophy of Science, Moscow.

COMPLETENESS OF PROPOSITIONAL DYNAMIC LOGIC WITH INFINITE REPEATING

Jurate Sakalauskaite and Mars K. Valiev[1]

Institute of Mathematics and Cybernetics
Lithuanian Academy of Sciences
SU - 232600, Vilnius
[1]Institute for System Studies
Academy of Sciences
SU - D117312, Moscow

Completeness of a finite deductive system for PDL with infinite repeating is proved using the method of tableaux trees.

1. INTRODUCTION

Propositional dynamic logic with infinite repeating (ΔPDL) was introduced by Streett (1982) (Streett calls it PDL of looping) as a natural extension of PDL (Fisher and Ladner, 1979) by adding formulas of the form Δa, where a is a program. Δa expresses the possibility of infinite repating of execution of a from an initial state. In ΔPDL a useful property of infinite looping (Havel and Pratt, 1978) is also expressed. Moreover, formulas of propositional algorithmic logic (Mirkowska, 1978) can be easily transformed into those of ΔPDL.

By Streett (1982), a natural finite deductive system for ΔPDL (without tests) is proposed and the problem of its completeness is stated. Here we present the proof of completeness of this system. It is an improved version of the one presented by the first author (Sakalauskaite, 1986). This proof can also be applied (almost without changes) to prove the completeness of ΔPDL (with tests) and the variant of ΔPDL with deterministic interpretation of program variables. The idea of the proof has been inspired by D. Kozen's paper (Kozen, 1983). In this paper a propositional μ-calculus, extending ΔPDL, is introduced, and the completeness of a fragment of this calculus, obtained by some syntactic restriction, is proved. This fragment is strictly more expressive than PDL but it is not known whether it includes ΔPDL.

The proof presented here as well as that by Kozen is based on the notion of tableaux trees and can be outlined as follows. For formula r a tableaux tree T(r) is constructed inductively and it is proved that r is consistent (i.e., ¬r is not provable) iff T(r) contains a subtree A(r) with properties which enables A(r) to transform to a model of r. It can be pointed out that for ΔPDL and μ-calculus (in contrast to PDL) in order to extract A(r) it is not sufficient to delete inconsistent nodes from T(r) and we have to delete some consistent nodes too (a node of T(r) is consistent iff its label is a consistent set of formulas). For the

Mathematical Logic
Edited by P. P. Petkov
Plenum Press, New York, 1990

mentioned fragment of μ-calculus this problem was solved by Kozen (1983). The solution consists in transforming the consistent nodes, which must be deleted, into inconsistent ones by joining certain formulas to subformulas of formulas in nodes of T(r) in an appropriate manner. The procedure of joining is defined inductively beginning from the root parallelly along all the branches of T(r) and results in a tree $\tilde{T}(r)$ such that the nodes to be deleted are inconsistent. The definition of this procedure is essentially connected with the above mentioned syntactic restriction. In general, formulas of ΔPDL do not satisfy this restriction.

Here another way to obtain $\tilde{T}(r)$ is proposed. A sequence of transformations of T(r), which consist in joining formulas to subformulas in some parts of T(r), is defined. In general, intersection of these parts can be nonempty. $\tilde{T}(r)$ is the limit of this process. The correctness of the procedure, when subformulas are joined to a node several times, is guaranteed by choosing the sequence of parts for transformation in accordance with a certain structure of T(r).

2. SYNTAX AND SEMANTICS OF ΔPDL

We consider ΔPDL without tests only. Let us define the language of ΔPDL. It contains two sorts of variables: propositional ones P,Q,... and program ones A,B,... which are called atomic programs as well.

The formulas and programs are defined inductively as follows:

1) every atomic program is a program, every propositional variable is a formula;
2) if a, b are programs, p, q are formulas, then a;b, a ∪ b, a* are programs, ¬p, p ∨ q, [a]p, Δa are formulas.

Here we give only informal semantics of programs and formulas. Let M be an arbitrary nonempty set of states. Atomic programs are interpreted in ΔPDL as binary relations over M. a;b denotes the composition of programs a,b; a ∪ b denotes a nondeterministic choice of executing a or b, $a^* = Id \cup a \cup a^2 \ldots$, where Id denotes the identity program. The semantics of p ∨ q, ¬p is defined as in propositional calculus. [a]p is true in a state s iff p is true in all the states accessible from s by a. Δa is true in a state s iff there exists an infinite sequence of states $s_1,\ldots,s_i,\ldots$, such that $s_1 = s$ and s_{i+1} is accessible from s_i by a.

Let ∧, ⊃, ≡ denote boolean operators and ⟨a⟩ denote ¬[a]¬. Below we consider formulas in a positive normal form, i.e., formulas which contain ¬ only in front of propositional variables or in front of Δa, using as usual abbreviations p ∧ q, ⟨a⟩p for ¬(¬p ∨ ¬q), ¬[a]¬p, respectively.

3. DEDUCTIVE SYSTEM FOR ΔPDL

Let us consider the following deductive system denoted by S_Δ.

Axioms:

3.1 all the tautologies of propositional calculus;

3.2 [a](p ⊃ q) ⊃ ([a]p ⊃ [a]q);

3.3 [a;b]p ≡ [a][b]p;

3.4 [a ∪ b]p ≡ [a]p ∧ [b]p;

3.5 $[a^*]p \supset p \wedge [a][a^*]p$;

3.6 $\Delta a \supset \langle a \rangle \Delta a$.

Rules of inference:

3.7 $\dfrac{p; p \supset q}{q}$;

3.8 $\dfrac{p}{[a]p}$;

3.9 $\dfrac{p \supset [a]p}{p \supset [a^*]p}$;

3.10 $\dfrac{p \supset \langle a \rangle p}{p \supset \Delta a}$.

S_Δ is equivalent to the system presented by Streett (1982). In S_Δ the following theorems and derived rules can be proved.

3.11 $\langle a;b \rangle p \equiv \langle a \rangle \langle b \rangle p$;

3.12 $\langle a \cup b \rangle p \equiv \langle a \rangle p \vee \langle b \rangle p$;

3.13 $[a^*]p \equiv p \wedge [a][a^*]p$;

3.14 $\langle a^* \rangle p \equiv p \vee \langle a \rangle \langle a^* \rangle p$;

3.15 $\Delta a \equiv \langle a \rangle \Delta a$;

3.16 $\neg \Delta a \equiv [a] \neg \Delta a$;

3.17 $$\frac{\{q_i \supset (p \wedge [a](q \vee [a^*]p)) \mid 1 \leqslant i \leqslant n\}}{q_j \supset [a^*]p};$$

3.18 $$\frac{\{q_i \supset \langle a \rangle (q \vee \Delta a) \mid 1 \leqslant i \leqslant n\}}{q_j \supset \Delta a};$$

where $q = \bigvee_{i=1}^{n} q_i$, $1 \leqslant j \leqslant n$. Let $\vdash p$ denote that p is provable in S_Δ. Let X be a finite set of formulas. Let $\wedge X$ denote conjunction of formulas in X. We say that X is consistent iff not $\vdash \neg(\wedge X)$.

<u>Proposition 3.19</u>

Let $\{q_1,\ldots,q_n\}$ be a set of formulas and $q = \bigvee_{i=1}^{n} q_i$.

1) if $\{q_j, \langle a^* \rangle p\}$ for some $1 \leqslant j \leqslant n$ is consistent, then there exists k, $1 \leqslant k \leqslant n$ such that $\{q_k, p \vee \langle a \rangle (\neg q \wedge \langle a^* \rangle p)\}$ is consistent;

2) if $\{q_j, \neg \Delta a\}$ for some $1 \leqslant j \leqslant n$ is consistent, then there exists k, $1 \leqslant k \leqslant n$ such that $\{q_k, [a](\neg q \wedge \neg \Delta a)\}$ is consistent.

<u>Proof.</u> By derived rules 3.17, 3.18.

The proposition 3.19 is crucial in the proof of completeness of S_Δ presented below.

4. BASIC NOTIONS

Let r be a formula in the positive normal form. In the section we define a tableaux tree T(r) for r and other notions, used to prove the completeness of S_Δ.

T(r) is a labeled tree constructed inductively by applying the extension rules described below. Certain edges of T(r) will be labeled with atomic programs, others will be unlabeled. Each node s will be labeled with a finite multiset U_s of formulas.

Initially T(r) consists of a single node (root) labeled {r}. The tree T(r) is extended by applying the following five extension rules to the leaves in an order to be specified later.

4.1 ∧-rule. If $p_1 \wedge p_2 \in U_s$, then create node t with $U_t = (U_s \setminus \{p_1 \wedge p_2\}) \cup \cup \{p_1, p_2\}$ and unlabeled edge $s \to t$.

4.2 ∨-rule. If $p_1 \vee p_2 \in U_s$, then create two new nodes t_1, t_2 with $U_{t_i} = (U_s \setminus \{p_1 \vee p_2\}) \cup \{p_i\}$, $1 \leq i \leq 2$ and unlabeled $s \to t_1$, $s \to t_2$.

4.3 ≡-rule. If $p \in U_s$, $p \equiv q$ is one of the axioms 3.3, 3.4, or the theorems 3.11, 3.12, 3.13, 3.14, 3.15, 3.16, then create t with $U_t = (U_s \setminus \{p\}) \cup \{q\}$ and unlabeled edge $s \to t$.

4.4 $\langle\rangle$ -rule. For each $\langle A \rangle p \in U_s$, A is an atomic program, create node t with $U_t = \{p\} \cup \{q \mid [A]q \in U_s\}$ and edge $s \to t$ labeled by A.

4.5 Contracting rule. If U_s contains $n > 1$ occurrences of a formula p, then create t with U_t, which is obtained from U_s by dropping $n - 1$ occurrences of p, and the unlabeled edge $s \to t$.

We define a partial order relation denoted by α on the set of all occurrences of formulas from all notes of T(r). It is sufficient to define α only for the occurrences of formulas from the neighbor nodes.

<u>Definition 4.6</u>

Let $s \to t$ be an edge of T(r). Let x, y be occurrences of formulas p, q at nodes s, t, respectively. $x \,\alpha\, y$ iff one of the following conditions holds:

1) $p = q$, $p = q \vee p_1$, $p = p_1 \vee q$, $p = q \wedge p_1$, $p = p_1 \wedge q$;

2) $p = [a;b]p_1$ and $q = [a][b]p_1$;

3) $p = \langle a;b \rangle p_1$ and $q = \langle a \rangle \langle b \rangle p_1$;

4) $p = [a \cup b]p_1$ and $q = [a]p_1 \wedge [b]p_1$;

5) $p = \langle a \cup b \rangle p_1$ and $q = \langle a \rangle p_1 \vee \langle b \rangle p_1$;

6) $p = [a^*]p_1$ and $q = p_1 \wedge [a][a^*]p_1$;

7) $p = \langle a^* \rangle p_1$ and $q = p_1 \vee \langle a \rangle \langle a^* \rangle p_1$;

8) $p = \Delta a$ and $q = \langle a \rangle \Delta a$;

9) $p = \neg\Delta a$ and $q = [a]\neg\Delta a$;

10) $s \to t$ is labeled by A and ($p = \langle A \rangle p_1$ or $p = [A]p_1$) and $q = p_1$, A is an atomic program.

In the case of 1) the distinguished occurrence of subformula q of p in x is called a preimage of y. In other cases the distinguished occurrences of the subformula p_1 of p in x are called preimages of the distinguished occurrences of the subformula p_1 of q in y. Moreover, in cases of 6), 7), 8), 9), x is called a preimage of the occurrence of p in y. Since an occurrence and its preimage are occurrences of the same formula, say y, the notion of preimage can be naturally extended to the occurrences of subformulas of y.

Now we define a partial order relation $<$ on occurrences of subformulas in nodes of T(r) as a transitive closure of relation $<_0$ such that $x_1 <_0 y_1$ iff x_1, y_1 are occurrences of subformulas in occurrences x, y such that $x \alpha y$ and x_1 is a preimage of y_1.

The inductive definition of T(r) is as follows.

Definition 4.7

Initially T(r) consists of a root labeled {r}. In further stages at each leaf the contraction rule is applied whenever possible. Otherwise one of the Λ-, Λ-, ≡-rules is applied. The ≡-rule is applied to $q \in \{[a^*]p, \Delta a\}$ only if the following condition is satisfied:

If x is an occurrence of q in s, and $s_1,\ldots,s_n = s$ is a path in T(r), consisting of all nodes at which the contraction rule or Λ-, Λ-, ≡-rules are applied, then there exists no occurrence y of q in s_i, $1 \leq i < n$ such that $y < x$ and the ≡-rule is applied to y at s_i. If the contraction rule of any of the Λ-. Λ-. ≡-rules cannot be applied at s, then the $\langle\,\rangle$-rule is applied at s, whenever possible; otherwise s is a leaf of T(r).

Proposition 4.8

There exist at most $2^{|r|}$ different labels of nodes of T(r). The proof is similar to one in Fisher and Ladner (1979).

We define the notion of a trace of a formula of the form $\langle a^* \rangle p$ or $\neg\Delta a$.

Definition 4.9

Any sequence $x_0 < \ldots < x_j < \ldots$, where x_j is an occurrence of subformula q of the form $\langle a^* \rangle p$ or $\neg\Delta a$ in a formula in a node of T(r) is a trace of q.

The following notions of tableau and of alternative subtree with finite traces will be used to define models for r.

Definition 4.10

Any path $s_1,\ldots,s_n$ of T(r) is called a tableau of T(r) iff the following conditions are satisfied:

1) s_n is a node at which the $\langle\,\rangle$-rule is applied or s_n is a leaf of T(r);

2) s_1 is the root or a successor of a node at which the $\langle\,\rangle$-rule is applied;

3) for each $1 \leq i < n$ at s_i the $\langle\,\rangle$-rule is not applied.

Definition 4.11

A subtree V of T(r) is an alternative subtree of T(r) with finite traces iff the following conditions are satisfied:

1) the root of T(r) is the root of V.

2) if $s \in V$, $s \to t$ is the unique edge in T(r) beginning at s, then $t \in V$;

3) if $s \in V$, the V-rule is applied at s and $s \to t_1$, $s \to t_2$ are edges of $T(r)$ then $t_i \in V$ for some $1 \leqslant i \leqslant 2$;

4) if $s \in V$, the $\langle\rangle$-rule is applied at s and $s \to t_1, \dots, s \to t_k$ are all edges beginning at s in $T(r)$, then $t_1 \in V, \dots, t_k \in V$.

5) V does not contain any trace of formula q of the form $\langle a^* \rangle p$ or $\neg\Delta a$, which is infinite and consists of the occurrences of q such that the $\equiv$-rule is applied to each of them.

5. THE PROOF OF COMPLETENESS OF S_Δ

The following theorem asserts the completeness of the deductive system S_Δ for ΔPDL. Let r be in the positive normal form.

Theorem 5.1

The following are equivalent:

(i) $\neg r$ is not provable in S_Δ;

(ii) there exists an alternative subtree V of $T(r)$ with finite traces such that each node of V is consistent;

(iii) r is satisfiable.

5.2 Proof of (i) $\to$ (iii)

To give the main lines of the proof we need additional definitions. Let T_r^∞ denote the set of infinite traces of formulas of the form $\langle a^* \rangle p$ and $\neg\Delta a$ in $T(r)$ such that:

1) each trace $\alpha \in T_r^\infty$ includes only such occurrences to which the $\equiv$-rule is applied;

2) for each trace $\alpha \in T_r^\infty$, if $y_1 \in \alpha$, $y_2 \in \alpha$, and there is an occurrence x, $y_1 < x < y_2$ and the $\equiv$-rule is applied to x, then $x \in \alpha$.

Let $f(\alpha)$ denote the first occurrence of trace α. Let B^∞ denote set of branches such that each of them includes the root of $T(r)$ and contains a trace of T_r^∞.

We introduce binary (nonsymmetric in general) relation "α touches β" $\alpha \in T_r^\infty$, $\beta \in T_r^\infty$, as follows.

Definition 5.2.1

α touches β iff $f(\alpha) \neq f(\beta)$ and $f(\alpha)$ belongs to the path from the root to the node containing $f(\beta)$ and there exists an occurrence $x \in \alpha$ such that x occurs between some occurrences from β.

We define a subset T^∞ of T_r^∞ such that for each $\alpha, \beta \in T^\infty$ if $f(\alpha) \in \pi$, $f(\beta) \in \pi$ for some $\pi \in B^\infty$, then α does not touch β.

Let $\alpha \in T_r^\infty$ and α be a trace $x_0, x_1, \dots, x_j, \dots$ Let α^j denote a trace $x_j, x_{j+1}, \dots,$ from T_r^∞.

Definition 5.2.2

For each $\pi \in B^{\infty}$ take a trace, say α, from Tr^{∞} in π such that for each $\beta \in Tr^{\infty}$ which belongs to π $f(\beta)$ is not closer to the root than $f(\alpha)$.

Let A_{α} denote the subset of Tr^{∞}.

$\{\beta | \beta$ touches $\alpha\}$. If $A_{\alpha} = \phi$, then put α into T^{∞}.

If $A_{\alpha} \neq \phi$, then put α^{k} into T^{∞}, where k is the smallest number such that no trace from A_{α} touches α^{k}. (A_{α} is finite as follows from the assumption on α and König's lemma. This guarantees the existence of a such number k.)

Proposition 5.2.3

a) for any $\pi \in B^{\infty}$ there exists unique $\alpha \in T^{\infty}$ which is in π;

b) for $\alpha \in T^{\infty}$, $\beta \in T^{\infty}$ if $f(\alpha) \in \pi$, $f(\beta) \in \pi$ for some branch $\pi \in B^{\infty}$, then α does not touch β;

c) for $\alpha_1 \in T^{\infty}$, $\alpha_2 \in T^{\infty}$, if $\alpha_1 \cap \alpha_2 \neq \phi$, then $f(\alpha_1) = f(\alpha_2)$.

Proof is a straightforward examination of the definitions.

If $\alpha \in T^{\infty}$ and $f(\alpha)$ occurs at depth i in T(r), then we say that α has the priority $Q(\alpha) = i$ and $Q(x) = i$ for each $x \in \alpha$. If $\alpha \in T^{\infty}$, where α denotes a trace $x_0,\dots,x_j,\dots$, then we define $C(x_j) = j$.

Definition 5.2.4

A tree T' is called a copy of T(r), iff T' = T(r) or T' is obtained by joining some formulas to some occurrences of subformulas of the form $\langle a^* \rangle p$ or $\neg \Delta a$ in nodes of T(r).

Let T' be a copy of T(r). A node $\tilde{s} \in T'$ which is obtained from a node $s \in T(r)$ is called a copy of s (in T'). If $\tilde{s} \in T'$ is a copy of $s \in T(r)$ then each occurrence x of a subformula $\langle a^* \rangle p$ (resp. $\neg \Delta a$) at s has an image $\langle a^* \rangle p'$ or $\langle a^* \rangle p' \wedge y$ (resp. $\neg \Delta a$ or $\neg \Delta a \wedge y$) at $\tilde{s}$, where $p' = p$ or p' is obtained by joining formulas to subformulas of p; y is an adjoined formula. The distinguished occurrence of the subformula $\langle a^* \rangle p'$ (resp. $\neg \Delta a$) in this image is called a copy of x. If $\tilde{x}$ is a copy of x in T' and C(x) is defined, then we define $C(\tilde{x}) = C(x)$. The notion of the copies of traces in T' is defined similarly. If $\tilde{\alpha}$ is a copy of trace α, where $\tilde{\alpha}$ is in T', α is in T(r), then $\tilde{\alpha}$ is called a trace. If $\tilde{\alpha}$ is a trace in T', where $\tilde{\alpha}$ is a copy of α from T(r), and $Q(\alpha)$ is defined, then we define $Q(\tilde{\alpha}) = Q(\alpha)$ and $Q(\tilde{x}) = Q(\tilde{\alpha})$ for each occurrence $\tilde{x} \in \tilde{\alpha}$. Also if x_1, x_2 are copies in T' of occurrences x_1, x_2 from T(r) and $x_1 < x_2$, then we define $\tilde{x}_1 < \tilde{x}_2$.

From the definition 5.2.4 we obtain:

Proposition 5.2.5

If T' is a copy of T(r) and a node $s \in T'$ is a copy of $s \in T(r)$, then $\vdash (\wedge U_{\tilde{s}}) \supset (\wedge U_s)$.

Here U_s, $U_{\tilde{s}}$ denote as above labels of s, $\tilde{s}$ respectively.

The following two definitions introduce classes of copies of T(r) which are used to prove (i) $\rightarrow$ (ii).

Definition 5.2.6

1) A copy T' of T(r) is a copy with closed traces of priority i iff each branch of T' containing a trace α with $Q(\alpha) = i$ includes an inconsistent node;

2) a copy T' of T(r) is a copy with closed traces iff T' is a copy with closed traces of priority i for each $i \geq 1$.

Definition 5.2.7

A copy T' of T(r) preserves consistency iff for each consistent node $\tilde{s} \in T'$ the following holds:

a) if $\tilde{s} \to \tilde{t}$ is a unique edge beginning at $\tilde{s}$, then $\tilde{t}$ is consistent;

b) if $\tilde{s}$ is a copy of $s \in T(r)$, and the V-rule is applied at s and $\tilde{s} \to \tilde{t}_1$, $\tilde{s} \to \tilde{t}_2$ are edges beginning at $\tilde{s}$, then $\tilde{t}_i$ is consistent for some $1 \leq i \leq 2$.

c) if $\tilde{s}$ is a copy of $s \in T(r)$, the $\langle\,\rangle$-rule is applied at s and $\tilde{s} \to \tilde{t}_1, \ldots, \tilde{s} \to \tilde{t}_k$ are all edges beginning at $\tilde{s}$, then $\tilde{t}_1, \ldots, \tilde{t}_k$ are consistent.

The following proposition reduces the proof of (i) $\to$ (ii), where (i), (ii) are from Theorem 5.1, to a construction of a special copy $\tilde{T}$ of T(r).

Proposition 5.2.8

If there exists copy $\tilde{T}$ of T(r) which satisfies the conditions:

a) $\tilde{T}$ is a copy of T(r) with closed traces;

b) the root of $\tilde{T}$ is labeled $\{r\}$;

c) $\tilde{T}$ preserves consistency.

Then (i) $\to$ (ii). Proof consists in:

1) constructing (using b), c) and (i)) a subtree $\tilde{T}_1$ of $\tilde{T}$ such that $\tilde{T}_1$ contains only consistent nodes and satisfies the conditions 1), 2), 3), 4) from Definition 4.11;

2) verifying that a subtree V of T(r), which is a preimage of $\tilde{T}_1$, satisfies (ii).

Now we turn to the construction of copy $\tilde{T}$ of T(r) which satisfies the conditions a), b), c) from Proposition 5.2.8. $\tilde{T}$ is obtained as a limit of an inductively defined sequence $T_0 = T(r), \ldots, T_i, \ldots$ of copies of T(r). Any T_i satisfies the conditions b), c) above and for $i \geq 1$ the condition:

a') T_i is a copy with closed traces of priority i.

Moreover, the following two propositions hold true for T_i.

Proposition 5.2.9

There is a finite number (say n_{i+1}) of different labels of nodes in T_i.

Let α be a trace in T_i for which $Q(\alpha)$ is defined. We say that a trace β in T_i is an extension of α iff $f(\beta) = f(\alpha)$, α is a subsequence of β and the branch π containing β and beginning at a node including $f(\beta)$, does not include an occurrence z with $Q(z) < i$.

Proposition 5.2.10

1) Each trace α in T_i with $Q(\alpha) \geqslant i + 1$ consists of occurrences of the same formula;

2) an extension of each trace α in T_i with $Q(\alpha) \geqslant i + 1$ consists of occurrences of the same subformula.

To define T_i, $i \geqslant 1$ we extract some subtrees of T_{i-1} (denoted by W_x^u below) such that T_i is obtained from T_{i-1} by joining certain formulas (defined simultaneously with these subtrees) to subformulas in labels of nodes of these subtrees. Assume that T_{i-1} includes nonclosed traces with priority equal to i (otherwise we define $T_i = T_{i-1}$). We introduce some additional definitions.

Let x be an occurrence in T_{i-1} such that $C(x) = 0$, $x \in \alpha$ where α is nonclosed trace and $Q(\alpha) = 1$. x will be fixed up to Definition 5.2.13.

Let W_x denote a minimal subtree of T_{i-1} such that:

1) the root of W_x contains x;

2) each trace α with $f(\alpha) = x$ and $Q(\alpha) = i$ is in W_x;

3) each node s containing x_1, where $x < x_1$, and not containing y with $Q(y) < i$ belongs to W_x.

In fact 2) can be omitted from the definition. Condition 3) is used to prove Proposition 5.2.10.

Let E_x denote all nodes of W_x which contain occurrences from initial segments of the length $n_i + 1$ of nonclosed traces as in condition 2) of the definition of W_x. Let F_x denote the set of all labels of nodes in E_x. Let f be an enumeration of F_x and Φ_j^f denote the j-th element of F_x according to f. Let us define:

$$r_x^f(j) = \bigvee_{i=j}^{|F_x|} (\wedge \Phi_i^f);$$

$$q_x^f(j) = \begin{cases} p \vee \langle a\rangle(\neg r_x^f(j) \wedge \langle a^*\rangle p), & \text{if x is an occurrence of } \langle a^*\rangle p; \\ [a](\neg r_x^f(j) \wedge \neg\Delta a) & \text{if x is an occurrence of } \neg\Delta a. \end{cases}$$

By the definition of E_x and 1) of Proposition 5.2.10, each element of F_x contains the formula $\langle a^*\rangle p$, if x is an occurrence of $\langle a^*\rangle p$; otherwise it contains the formula $\neg\Delta a$. Since nodes in E_x contain occurrences from initial segments of nonclosed traces, each $s \in E_x$ is consistent. So by iterative repetition of the application of Proposition 3.19 we prove:

Proposition 5.2.11

There exists an enumeration $\tilde{f}$ of F_x such that for each $1 \leqslant j \leqslant |F_x|$ the set $\Phi_j^{\tilde{f}} \cup \{q_x^{\tilde{f}}(j)\}$ is consistent.

This proposition is crucial in defining the above mentioned subtrees W_x^u of W_x and to define formulas which are joined in nodes of W_x^u.

Let $\tilde{f}$ be an enumeration as in the Proposition 5.2.11. Let $E_x(j)$ denote

$$\{s \mid s \in E_x,\ U_s = \Phi_j^{\tilde{f}}\}$$

(U_s denotes a label of s in T_{i-1}).

Let x_1 be an occurrence in $u \in W_x$ such that x_1 belongs to any nonclosed trace α, where $Q(\alpha) = i$, $f(\alpha) = x$ and $C(x_1) < n_i + 1$. Let W_x^u denote the maximal subtree of W_x such that the following conditions are satisfied:

1) u is the root of W_x^u;

2) s is a leaf of W_x^u iff s includes an occurrence x_2, where $x_1 < x_2$, $C(x_2) = C(x_1) + 1$, $Q(x_2) = i$, or s is a leaf of W_x.

<u>Definition 5.2.12</u>

The procedure of closing traces in W_x of priority i is as follows: for each $1 \leqslant j \leqslant |F_x|$ and each $u \in E_x(j)$ join $\neg r_x^{\tilde{f}}(j)$ to any occurrence $\tilde{x}$ in $W_x^u \backslash \{u\}$, where $x < \tilde{x}$.

The copy T_i of T(r) is constructed from T_{i-1} as follows.

<u>Definition 5.2.13</u>

If there exists a nonclosed trace of priority i in T_{i-1}, then T_i is obtained by applying to W_x the procedure of Definition 5.2.12 for each occurrence x in T_{i-1} such that $C(x) = 0$, x belongs to a nonclosed trace α in T_{i-1} with $Q(\alpha) = i$; otherwise $T_i = T_{i-1}$.

So the sequence $T_0 = T(r), \ldots, T_i, \ldots$ of copies of T(r) is defined. While constructing T_i we change only labels of nodes at the depth $j < i$ in T_{i-1}. Thus there exists a limit, say $\tilde{T}$, of the sequence.

As seen from Proposition 5.2.8 to finish the proof of (i) → (ii) it suffices to verify:

<u>Theorem 5.2.14</u>

$\tilde{T}$ satisfies the conditions a), b), c) from Proposition 5.2.8.

Theorem 5.2.14 is a straightforward corollary of the following:

<u>Proposition 5.2.15</u>

For each $i \geqslant 1$ the copy T_i of T(r) satisfies the condition

a') T_i is a copy with closed traces of priority i;

and the conditions b), c) as in Theorem 5.2.14.

Moreover, recall that the construction of copy T_i for $i \geqslant 1$ depends on Propositions 5.2.9, 5.2.10. So we have to prove that T_i satisfies these propositions. The proofs of Proposition 5.2.15 and Propositions 5.2.9,

5.2.10 are carried on by induction on i making extensive use of the properties stated in 1) and 2) of Proposition 5.2.3; they are omitted here.

5.3 Proof of (ii) → (iii)

Assume (ii). Let $\mathcal{M}_V$ be the following model $<M_V, \rho_V, \pi_V>$, where:

1) M_V is the set of tableaux of V;

2) for each atomic program A: $\rho_V(A) = \{(\tau_1, \tau_2) | s \rightarrow t$ is an edge of V labeled A, where s (resp. t) is the last (resp. first) node of tableau τ_1 (resp. τ_2)$\}$;

3) for each propositional variable p: $\pi_V(p) = \{\tau | p$ occurs in the last node of $\tau\}$.

That r is satisfied in $\mathcal{M}_V$ (this proves (iii) is implied by the following tow propositions.

Proposition 5.3.1

There exists a tableau $\tau_0 \in M_V$ such that the root of V belongs to τ_0.

U_s denotes the label of node s.

Proposition 5.3.2

If $q \in U_s$, where $s \in V$ and τ is a tableau of V such that $s \in \tau$, then q is true in state τ of $\mathcal{M}_V$.

Proposition 5.3.1 is proved by using the restriction imposed on the $\equiv$-rule (Definition 4.7) and condition 5) from Definition 4.11 satisfied by V.

Proposition 5.3.2 is proved by the usual induction on the structure of q.

5.4 (iii) → (i)

This is the soundness of S_Δ and it is verified straightforwardly.

REFERENCES

Fisher, M. J. and Ladner, R. E., 1979, Propositional dynamic logic of regular programs, J. Comput. Syst. Sci., 18:194-211.

Harel, D. and Pratt, V.R., 1978, Nondeterminism in logics of programs, in: Proc. 5th ACM Symp. on Principles of Programming Languages, pp 203-213.

Kozen, D. 1983, Results on the propositional μ-calculus, Theor. Comp. Sci., 27:333-354.

Mirkowska, G., 1981, PAL-Proposition algorithmic logic, Fundamenta Informaticae, IV.3:675-760.

Sakalauskaite, J., 1986, "Completeness of propositional dynamic logic with infinite repeating" (in Russian), Ph.D Thesis, Institute of Mathematics and Cybernetics of Lithuanian Academy of Sciences, Vilnius.

Streett, R.S., 1982, PDL of looping and converse is elementary, decidable, Inform. and Contr., 54:121-141.

AN EQUIVALENCE BETWEEN POLYNOMIAL CONSTRUCTIVITY OF MARKOV's PRINCIPLE AND THE EQUALITY P=NP

V.Yu.Sazonov

Institute of Program Systems
of USSR Akademy of Sciences
Pereslavl'-Zalessky, 152140, USSR

0. FORMULATING RESULTS

As usually, P (resp., NP) stands for "deterministic (resp., nondeterministic) polynomial-time computable". We will consider P-functions and P-predicates over the set $\{0,1\}^*$ of finite binary strings. Let also PT (resp., HPT) denote classical (resp., intuitionistic) theory of polynomial-time computability, i.e. PT (HPT) = "all true" quantifier-free formulae, or axioms, built from &, $\vee$, $\neg$, $\Rightarrow$ and from P-predicates and P-functions, + classical (resp., intuitionistic) logic. Schemes

$$M: \neg\neg\exists xA \Rightarrow \exists xA \quad \text{and} \quad M_U: \neg\neg\exists x{\in}U.A \Rightarrow \exists x{\in}U.A,$$

with A quantifier-free and the predicate $U=_{df}\{1\}^*\subseteq\{0,1\}^*$, are Markov's rule and its "unary" restriction. Church's thesis ECT, or ECT(G,F) (or, respectively, its unary restriction ECT_U) is the scheme

$$\forall x(G(x) \Rightarrow \exists yF(x,y)) \Rightarrow \exists e\forall x(G(x) \Rightarrow !\{e\}(x) \,\&\, F(x,\{e\}(x)))$$

where $\{e\}(x)$ denotes the result (if any) of applying algorithm e to argument x, F is arbitrary and G is any almost negative (resp. U-negative) formula. Here, almost negative formulae are formulae in which $\vee$ and $\exists$ are applied only to quantifier-free subformulae, and U-negative formulae are such almost negative formulae whose all existential quantifiers are unary, i.e. are of the kind $\exists x{\in}U$. This almost negativity of G is necessary to escape formal inconsistency of ECT with HPT (as in the case of Heyting arithmetic). Call any theory C constructive if for any Harrop's formula H and arbitrary F

$$C \vdash H(\bar{x}) \Rightarrow \exists yF(\bar{x},y) \quad \text{implies} \quad C \vdash H(\bar{x}) \Rightarrow F(\bar{x},\{t\}(\bar{x}))$$

for some signature term t in C containing no variables. Note, that $\{t\}(\bar{x})$ is partial recursive, non-signature term. Also note, that in general these theory C and term t may involve some added to HPT, possibly "nonstandard" constants. By omitting antecedent H and $\Rightarrow$ or taking H=TRUE (and requiring t to be polynomial-time algorithm, when applied to $\bar{x}$), we come to the notion of (P-) $\exists$-constructivity of C.

Mathematical Logic
Edited by P. P. Petkov
Plenum Press, New York, 1990

Essentially, the paper contains two Main results: I. Principles ECT and M, in contrast to ECT_U and M_U, are constructive relative to theory HPT iff P=NP holds and II. Superpolynomial complexity of NP-complete problems accepting (do not mix with recognizing!) is not provable in HPT+ECT+M. More exactly, we have the following theorems (see also the beginning of Section 2) where the reader may concentrate his attention, say, on Theorem 2 (and on propositions 4a, 4b, 2a, 1a&1b⇒2c&3c and 4c&4d⇒2b stated and proved in Sect.2 from which considerable part of theorem 2 follows).

Theorems: 1. Theories $HPT \pm ECT_U \pm M_U$ are constructive and even P-∃-constructive. (Here ± means adding/nonadding a scheme.)

2. $HPT+ECT_U+M$ = $HPT+ECT+M_U$ = HPT+ECT+M is constructive iff it is (P-)∃-constructive iff P=NP. The same also holds for HPT+ECT.

3. Theory HPT+M is constructive iff P=NP. However, it is P-∃-constructive.

4. Theory HPT+ECT+M conservatively extends PT and HPT relative to ∀∃-sentences. So, its provably-recursive functions are just polynomial-time computable ones. However, HPT+ECT ⊢ "there exists a search algorithm e which, given any satisfiable propositional formula x, finds some truth evaluation y={e}(x) such that x under evaluation y is true".

Therefore, there exists no reasonable monotonic superpolynomial estimate f for which HPT+ECT+M proves that f is lower bound for time complexity of such search algorithms. The latter is also true for PT (see [Sazonov, 1980] and also corrections in [Sazonov, 1981, p.490] which mean that this unprovability statement remains open question for the stronger theory PT+ induction scheme Δ-Ind over formulas with lexicographically bounded quantifiers; cf. Sect.1).

1. PRELIMINARY CONSIDERATIONS AND DISCUSSION

This paper is an extended english version of [Sazonov, 1987] and the development of [Sazonov, 1980; 1981]. Here we only outline main ideas underlying results formulated above. Full proofs would require much more place. (Note, that in the op. cit. denotation T_0 is used in place of PT.)

Of course, this approach, which is quite independent and seems new (especially concerning the complexity-theoretic look at Markov´s principle and concerning our version of the notions of constructive vs. nonconstructive finite binary strings; see below), may be compared with numerous other investigations on weak theories such as Bounded Arithmetic due to S.R.Buss, C.Dimitracopoulos, H.Gaifman, L.A.S.Kirby, J.Krajicek, E.Nelson, R.J.Parikh, J.B.Paris, P.Pudlak, J.P.Ressayre, B.Solovay, G.Takeuti, A.J.Wilkie and others. See e.g. [Buss, 1986].

As stated in theorems 2 and 3 above, the constructivity property (in our setting) of Markov´s principle M, as well as of Church´s thesis ECT, is equivalent to the equality P=NP. Therefore, due to generally accepted hypothesis P≠NP we can conclude that these principles M and ECT, in contrast to their restrictions M_U and ECT_U, most likely become nonconstructive, if we consider them in the context of weak intuitionistic theory HPT of polynomial-time computability. Remember, that M and ECT are constructive relative to Heyting arithmetic HA which contains full scheme of induction in difference to our theories HPT and PT.

We insist on such a relativeness of constructivity notion because the stronger is corresponding theory, the more is the constructive power of algorithms which can be described in this theory, i.e. the wider is the class of corresponding provably (total) recursive functions. So, the sense of the word "constructive" may change. Also it can be argued that the absolute notion of total recursive functions is not at all correct because this notion is based on not completely clear (and, as it proves, even unnecessary [Sazonov, 1980]) abstraction of potential feasibility. We only can approximate this vague "class" of functions by the classes of P-functions, Kalmar elementary computable, primitively recursive functions, etc. (But does this "etc" mean something definite? Cf. [Yessenin-Vol′pin, 1959] for intriguing discussion on such an "etc" in the ordinary process of counting "1,2,3,etc".) This is one of the reasons why the notion "constructive" is non-absolute.

Analogously, it seems vague the notion of "all true" formulae mentioned in the definition of PT and HPT in Sect.0. However, this "all true" may be replaced by "suitable finite number of intuitively true" without any essential change of our results.

Only some weak form of induction axiom, namely quantifier-free (lexicographical or, provably equivalent, linear) induction, is deducible in HPT and PT (see below and [Sazonov, 1980; 1981]). So, it cannot be proved in PT that exponentiation 2^n is total function, and therefore 2^n should be considered as computable (by corresponding Turing machine) partial function,or function which may be equal to infinity. This weak constructive power of algorithms in HPT and PT seems very reasonable due to conventional negative attitude to the exponential complexity. Note, that $m=2^n$ (or, equuivalently, $\log m = n$) is P-predicate, so totalness of 2^n is expressible in the language of theories PT and HPT by the sentence

$$EXP =_{df} \forall n \exists m (m=2^n),$$

and we have PT $\nvdash$ EXP and therefore HPT $\nvdash$ EXP. Such theories in which EXP is not provable may be called exponentialless.

In fact, provably-recursive functions of theories HPT and PT (i.e. recursive functions whose totality can be proved in these theories) are just polynomial-time computable ones. This can be easily shown by using Harrop′s theorem for intuitionistic logic and, respectively, Herbrand′s theorem for classical logic, because nonlogical axioms of HPT and PT are quantifier-free.

The ordinary intuitive justification for constructivity of Markov′s principle $\neg\neg\exists x A(x) \Rightarrow \exists x A(x)$ essentially consists in the process of trying to verify A(x) for successively choosen values of x in some standard order. In our case x runs over finite binary strings. But which is the proper order of search? Can we take this order to be lexicographical

$$\emptyset < 0 < 1 < 00 < 01 < 10 < 11 < 000 < \ldots ?$$

Let, for example, A(x) be the equality $x=111\ldots1_{1000}$ (thousand times). Then we should undertake extraordinary large number ($>2^{1000}$) of trials in this lexicographical order to find x for which A(x)=TRUE. Instead, it seems better to use the evident nondeterministic Turing machine to guess (if we are lucky) the required x in feasible (about 1000) number of steps. These considerations witness that lexicographical order is not at all satisfactory here and that, indeed, some connection exists between constructivity of Markov′s principle and the equality P=NP.

However, we can justify Markov′s principle through some another, nonlexicographical enumeration of finite binary strings (if an additional axiom SA^e is postulated; see below). Among computable enumerations of finite strings there is even optimal one which is actually used in the proofs of our results formulated in Sect.0. This enumeration (or sequence) is represented by suitable P-function $\xi:\{0,1\}^* \to \{0,1\}^*$ which depends only on the length $|i|$ of the argument: $|i|=|j| \Rightarrow \xi_i=\xi_j$. Of course, we can, instead, consider ξ as the function from $\{1\}^*$ to $\{0,1\}^*$, unary strings $\emptyset,1,11,111,\ldots$ being identified with natural numbers. So, $|i|$ is considered as unary string of the same length as the string i. We give exact definition of ξ below (see (ξ) and also [Sazonov, 1980]). However, for the following it suffices to know only that ξ is polynomial (or P-) optimal, i.e. that for any other P-function $s:\{1\}^* \to \{0,1\}^*$ there exists P-function $f:\{1\}^* \to \{1\}^*$ sucn that $s_i = \xi_{f(i)}$ for all $i\in\{1\}^*$. Thus, ξ_i is universal or the quickest, up to "polynomials" f, enumeration of binary strings. Moreover, we can prove in HPT+EXP the sentence

$$\mathrm{DET} =_{df} \forall x\exists i(x = \xi_i).$$

which means that each binary string is constructive (or deterministic) i.e. is in the range of ξ. This easily follows from P-optimality of ξ applied to P-sequence $B:\{1\}^*\to\{0,1\}^*$, which is defined as the lexicographical enumeration of binary strings, and from provable in HPT equivalence

$$\mathrm{EXP} \Leftrightarrow \forall x\exists i(x=B_i).$$

However, both DET and ¬DET are consistent with PT+¬EXP [Sazonov, 1980]. So, we may consistently think both that all strings are constructive and that there exists some nonconstructive finite string. For example, PT + ¬EXP + DET may be just interpreted in PT + ¬EXP by taking $\{x|\exists i(x=\xi_i)\}$ as the domain of interpretation, which is closed under P-functions due to P-optimality of ξ.

More generally, we should consider P-optimal sequence ξ^e_i relativized to some finite binary string e so that any P-function s^e_i in binary e and unary i can be represented as $s^e_i = \xi^e_{f(e,i)}$ for some P-function f. Like DET we can define its weakening

$$\mathrm{RDET} =_{df} \exists e\forall x\exists i(x=\xi^e_i).$$

Then both RDET&¬DET and ¬RDET are consistent with PT (see [Sazonov, 1980]).

For any quantifier-free A(x,y) the sequence $\xi^{\langle e,y\rangle}_i$, $i=\emptyset,1,11,\ldots$, where ⟨e,y⟩ is some natural pairing P-function, gives us the evident search algorithm (depending on e,y and A) which finds x $(=\xi^{\langle e,y\rangle}_i)$ such that A(x,y). It is this algorithm (for suitable e) that we take as the "best" justification (or as Kleene's realization) of Markov's principle $\neg\neg\exists xA(x,y) \Rightarrow \exists xA(x,y)$ with parameter y.

However, it is unknown whether it is provable in exponentialless theories HPT and PT that the above described search algorithm does work for some particular e. We are unable to infer even the scheme

$$\exists e(\exists xA(x,y) \Rightarrow \exists iA(\xi^{\langle e,y\rangle}_i,y)),\ \text{A quantifier-free.}$$

Of course, we should formalize in our theory HPT what is an algorithm. This is done through P-function I(e,n) which denotes the result that initial Turing machine e gives on its output tape in step $|n|$ and through P-predicate H(e,n) which is defined true iff e halts in $|n|$ steps. (So, n may be considered as an unary string.) Here <u>initial Turing machine</u> e=<m,v> is (binary code of) the ordinary Turing machine m together with some initial binary information v to be written on the input tape. Also, let $e\#x =_{df} \langle m,\langle v,x\rangle\rangle$ be the result of adding new input information x to algorithm e. Note, that (codes of) initial Turing machines, which compute e.g. a given function, are some more economic than the ordinary, non-initial ones, what may be important in the context of weak, exponentialless theories. P-functions I, # and P-predicate H allow us to consider in HPT (much as the analogouos Kleene's predicate T and function U in Peano or in Heyting arithmetic) Turing computability, universal Turing machine, s-m-n theorem, fixed-point theorem etc. So, the beginnings of partial-recursive function theory can be easily developed without postulating EXP (and full induction scheme). In particular, we will use (informally) the ordinary Kleene's notation {e}(x) or even {e}:

$$\{e\} =_{df} I(e,\mu n\in U.H(e,n)) \quad \text{and} \quad \{e\}(x) =_{df} \{e\#x\},$$

which, of course, may be of undefined value and hence cannot be represented by any signature term.

Let SA(e) denote the assertion (from Theorem 4) "e is (binary code of) a search algorithm which transforms (binary code of) any satisfiable propositional formula x into some truth evaluation y={e}(x) under which x is true". More formally, let x[y] denote the truth value of propositional formula x under truth evaluation y and

$$SA(e) =_{df} \forall x(\exists y.x[y] \Rightarrow\ !\{e\}(x)\&x[\{e\}(x)]), \text{ or}$$
$$=_{df} \forall x(\exists y.x[y] \Rightarrow \exists n(H(e\#x,n)\&x[I(e\#x,n)])),$$
$$SA^e =_{df} \exists jSA(\xi_j^e) \quad \text{and} \quad SA =_{df} \exists eSA(e) \quad (\Leftrightarrow \exists eSA^e).$$

So, SA^e means that e <u>contains information</u> about some search algorithm, and SA means that there exists some search algorithm. Then for e considered as new constant for (information about) some unknown search algorithm and for any quantifier-free A we can prove

$$HPT+SA^e \vdash \exists xA(x,y) \Leftrightarrow \exists j\in U.A(\xi_j^{\langle e,y\rangle},y). \qquad (*)$$

This means that our search algorithm based on ξ can be used to justify Markov's principle M if (and only if) the axiom SA^e is postulated. (Here, of course, NP-completeness of SAT $=_{df} \{x|\exists y.x[y]=TRUE\}$ is used.)

Note, that SA is most likely unprovable in $HPT+ECT_U+M_U$ and also in PT: it follows from Theorem 1 (or from Harrop's theorem for intuitionistic logic in the case of HPT) that $HPT\pm ECT_U\pm M_U\vdash SA$ is equivalent to P=NP. For, let $\vdash$SA, then Theorem 1 implies $\vdash SA(e_o)$ for some particular algorithm e_o. Another application of Theorem 1 and conservativity assertion of Theorem 4 prove that e_o is (essentially) polynomial-time algorithm, which evidently gives P=NP. The converse is easy. However, it is unknown to the author whether PT$\vdash$SA implies P=NP.

Also, with the help of Harrop's (or Herbrand's) theorem it can be proved that P=NP is equivalent to the proposition: "every formula $\exists xA(x,y)$ with quantifier-free A is provably equivalent in HPT (or, respectively, in PT) to formula $\exists j \in U.A_1(j,y)$ with unary $\exists$ and some quantifier-free A_1 depending on A".

It is intuitively evident that HPT+RDET$\vdash$SA: the required search algorithm is based on the sequence ξ_i^e because the latter satisfies RDET. Further, let Δ-Ind denote lexicograhic induction scheme

$$F(\emptyset) \ \& \ \forall x<y(F(x) \Rightarrow F(x')) \Rightarrow F(y)$$

over formulae F with lexicographically bounded quantifiers $\forall x$<term, $\exists x$<term, where x′ mean "the lexicographical next to x". It is unknown, whether PT (or HPT) $\vdash$ Δ-Ind. However, quantifier-free lexicographic induction (i.e. one with quantifier-free F) is provable in HPT. Also, it seems very important the following open question: whether PT (or HPT $\pm$ ECT $\pm$ M) + Δ-Ind + RDET (or +DET, or +SA) $\vdash$ EXP? E.g, HPT+ECT+M + Δ-Ind $\vdash$ EXP? (Cf. corrections in [Sazonov, 1981, p.490] to [Sazonov, 1980].) If Δ-Ind is omitted then EXP is, indeed, unprovable (see above and [Sazonov, 1980]) and we can immediately conclude that theory PT does not prove the exponential or even superpolynomial complexity of search algorithms and of NP-complete problems accepting, i.e. that this theory does not prove e.g. the formula

$$SA(e) \Rightarrow \forall z \exists x>z \exists n>2^{|x|}(H(e\#x,n) \& \forall m<n.\neg H(e\#x,m)).$$

(Otherwise, we could successively infer e.g. in PT+DET that SA(e), for some particular e, the consequent of the above implication, and, finally, EXP hold.) The same is true for the theory HPT+ECT+M (cf. Theorem 4), as well. However, in contrast to what is said in [Berman, 1983], we are not able to conclude that in any exponentialless theory (e.g. in PT + Δ-Ind or in HPT $\pm$ M + Δ-Ind) exponential complexity of NP-complete problems is not provable. Further, it is unknown, whether HPT$\pm$ECT$_U\pm$M$_U$+ Δ-Ind is constructive?

As noted above, one should have in mind the possibility that constructive strings ξ_i do not exhaust all binary strings in the case of exponentialless theories such as PT and HPT. It is interesting that these strings can be characterized also as strings enumerated by partial recursive sequence $\{B_i\}$:

$$\text{HPT} \vdash \exists i(x=\xi_i) \Leftrightarrow \exists j(x=\{B_j\}) \quad (\Leftrightarrow \exists j \exists n(x = I(B_j,n) \ \& \ H(B_j,n))).$$

The binary strings enumerated by sequence B_i itself (- not by sequence $\{B_i\}$) we call short or feasible strings because HPT proves

$$\exists i(x=B_i) \ \& \ |y| \le |x| \Rightarrow \exists j(y=B_j),$$

i.e. "if x is short and y is shorter than x then y is short", but it cannot be proved in PT that $\forall x \exists i(x=B_i)$, i.e. that "all strings are short". So, we can characterize constructive strings as those which can be computed by short algorithms. Note, that PT+DET $\nvdash$ $\forall x \exists i(x=B_i)$. Also, it seems reasonable to call genuine computable those functions which are computable by short algorithms.

This should be compared with Kolmogorov′s complexity theory for finite objects [cf. Zvonkin and Levin, 1970] because, as required in this theory, codes of finite objects are desirable to be comparatively short, as well, and partial coding $\lambda x.\{x\}$ defined here may be shown to be additively optimal relative to all others partial recursive codings $\lambda x.\{e\}(x)$ of binary strings: $\{e\}(x) = \{e\#x\}$, with $|e\#x| \leq |x| + const(e)$. Note, that the latter inequality as well as given above characterization of constructive strings as generated by short algorithms holds due to our definition of algorithms as initial Turing machines and, additionally, due to choosing such (evident and well known) definition of pairing function $\langle u,v \rangle$ involved in #, etc. that $|\langle u,v\rangle|=2|u|+2+|v|$. Essentially, this is done as in Kolmogorov's theory, except our approach to the "short strings" via exponentialless theories.

Finally, let us present a "diagonal" definition of polynomial optimal sequence ξ (which inessentially differs from that in [Sazonov, 1980]) in terms of Turing machines:

$$\xi_i^e =_{df} I((B_i)\#e,i). \qquad (\xi)$$

However, it is more suggestive to begin with "three-dimentional" ξ, as in [Sazonov, 1980]: $\xi^e(i,j,t) =_{df} I((B_i)\#j\#e,t)$. In this case P-optimality is proved as $s_j^e = I((B_{i_o})\#j\#e,p_o(j,e)) = \xi^e(i_o,j,p_o(j,e))$ with i_o and p_o depending on any given P-function s. So, the triple $(i_o,j,p_o(j,e))$ of unary strings is polynomial-time computable as P-optimality of ξ requires. Further details are remained to the reader.

2.PROOF OUTLINES

To prove our theorems 1-4 let us divide them into the following exhausting parts (possibly with some additions).

1a. Theories $HPT \pm SA^e \pm M_U + ECT_U$ are constructive and theories $HPT \pm M_U + ECT_U$ are P-∃-constructive

1b. The same as 1a with ECT_U omitted.

2a. $HPT+ECT_U+M \vdash ECT$ and $HPT+ECT+M_U \vdash M$.

2b. If $HPT+ECT \pm M$ (or $HPT+SA \pm M$) is ∃-constructive then P=NP.

3a. If HPT+M is constructive then P=NP.

So, we see that constructivity of theories HPT+M, $HPT+ECT \pm M$ and $HPT+SA \pm M$ is rather doubtfull hypothesis. However, if we add to them axiom SA^e for some new constant e, then, surprisingly, 1a, 1b and 4b below imply that all they i.e. $HPT+SA^e \pm M \pm ECT$ become constructive (what seems to contradict 2b and 4a below). This is due to added to the language new signature constant e for (information about) unknown search algorithm which now can take part in the definition of constructivity notion given in Sect.0, i.e. e may occur in the term t from the definition. In other words, we can say that principles M and ECT (and SA) are constructive relative to some hypothetical search algorithm (or relative to a string containing information about such an algorithm) which may be involved in algorithms needed to guarantee constructivity of considered theories.

3b. HPT+M is P-∃-constructive.

2c&3c. If P=NP holds then $HPT \vdash SA$ and theories $HPT+SA \pm ECT \pm M$ = $HPT \pm ECT \pm M$ are both constructive and P-∃-constructive.

4a. $HPT+ECT \vdash SA$.

4b. $HPT+SA+M_U \vdash M$ and $HPT+SA+ECT_U \vdash ECT$.

4c. PT+SA conservatively extends PT relative to ∀∃-sentences.

4d. $HPT+SA^e+ECT_U+M_U$ or, equivalently, HPT+ECT+M conservatively extends PT and HPT relative to ∀∃-sentences (without constant e).

Note, that the following first five proofs, in fact, reduce theorem 2 to assertions on constructivity of HPT±ECT_U±M_U and on conservativity of HPT+ECT+M and PT+SA over PT relative to ∀∃-sentences.

Proof of 4a. It suffices to apply the following case of ECT:

$$\forall x(\exists y.x[y] \Rightarrow \exists y.x[y]) \Rightarrow \exists e\forall x(\exists y.x[y] \Rightarrow x[\{e\}(x)]),$$

because it has the form ∀x(F⇒F) ⇒ SA with the logical axiom in premise.□

Proof of 4b. The required reducing M to M_U and ECT to ECT_U is based on the equivalence (*) in Sect.1.□

Ptoof of 2a. The first statement of 2a is based on replacing all ∃ in G of ECT(G,F) by ¬¬∃, which is correct due to M, and then by ¬∀¬. This reduces ECT to ECT_U. The second part of 2a follows from 4b and 4a.□

Proof of 1a&1b⇒2c&3c. Evidently, P=NP implies HPT ⊢ SA. Therefore, P=NP with proved 4b give the equivalence in HPT of principles M and M_U and also of principles ECT and ECT_U (and conversely, as follows from Theorems 1 and 2). Then, constructivity assertions 2c&3c follow from 1a and 1b.□

Proof of 4c&4d⇒2b. Suppose, 4c&4d and constructivity hypothesis of 2b hold. By stated above 4a, the sentence SA $=_{df}$ ∃e.SA(e) is provable in the given theory. So, by supposed ∃-constructivity, SA(e_o) and hence ∃n(x[y] ⇒ H(e_o#x,n)) are provable , too, for some particular e_o. Then 4d or 4c, and Herbrand's theorem for classical logic can be used to show that there exists polynomial upper bound pol(|x|) for n and, therefore, that e_o is (essentially) polynomial-time search algorithm. This gives P=NP.□

Proof of 3a. First, M implies ¬∀y¬x[y] ⇒ ∃y.x[y], ¬∀y¬x[y] being Harrop's formula. Then, supposed constructivity of HPT+M gives some particular search algorithm e_o for which HPT+M⊢x[y]⇒!{e_o}(x)&x[{e_o}(x)] holds and hence PT ⊢ ∃n(x[y]⇒H(e_o#x,n)) holds, too. Then Herbrand's theorem guarantees that e_o is (essentially) polynomial-time search algorithm and P=NP.□

Proof of 4c. This statement on conservativity is evidently implied by the same conservativity result for theory PT+RDET over PT, where RDET $=_{df}$ $\exists e\forall x\exists i(x=\xi_i^e)$, because in this theory such sequence ξ_i^e gives the required search algorithm, and PT+RDET ⊢ SA. The conservativity of PT+RDET over PT is based, in its turn, on the natural interpretation of this theory in theory PT, where the domain of interpretation is $\{x|\exists i(x=\xi_i^e)\}$. This domain is closed under signature operations due to polynomial optimality of ξ.□

Proofs of 1a and 4d are based on the ordinary technique of Kleene's realizability as it presented e.g. in [Dragalin, 1979 or Troelstra, 1977]. However, some additional refinements are necessary. So, given any formula F, we define the corresponding formula xrF, x realizes F, whose free variables are those of F plus x. This can be done so that xrF will be not only almost negative, as usually (e.g. in op. cit.), but also U-negative formula (cf. Sect.0). Note, that the ordinary theory of realizability does not distinguish between almost negative and U-negative formulae. For example, define

$$xr(F \Rightarrow G) =_{df} \forall y(yrF \Rightarrow \exists z \in U.H(x\#y,z)) \; \& \\ \& \; \forall yz(yrF \; \& \; H(x\#y,z) \Rightarrow I(x\#y,z)rG).$$

Such U-negativity of xrF allows us to prove as usually, by induction on F, the important proposition (with ECT_U instead of ECT)

$$HPT+ECT_U \vdash F \Leftrightarrow \exists x(xrF).$$

Moreover, if F is U-negative then ECT_U is not used here, and if F is almost negative then ECT_U can be replaced by SA. Then we can prove

<u>Realizability Theorem</u>. If $HPT \pm SA^e \pm M_U + ECT_U \vdash F(e,\bar{x})$ then, with respective $\pm$, $HPT \pm SA^e \pm M_U$ (without ECT_U!) $\vdash \{t(e,\bar{x})\}rF(e,\bar{x})$ for some signature term t. Moreover, if SA^e is not used in the first proof then (it is not used in the second and) partial recursive term $\{t(e,\bar{x})\}$ can be replaced by some signature (polynomial-time computable) term $t'(\bar{x})$.

Suitable combining of this theorem with the immediately preceeding proposition gives the required constructivity assertions of 1a and conservativity of $HPT \pm SA^e \pm M_U + ECT_U$ (a) over $HPT+SA^e \pm M$ (here and below with respective $\pm$) relative to almost negative formulae, (b) over $HPT \pm SA^e \pm M$ relative to U-negative formulae and, finally, (c) over PT and HPT relative to $\forall\exists$-formulae without e (due to inclusion $HPT+SA^e+M \subseteq PT+SA^e$ and conservativity theorem 4c). This proves 1a and 4d.□

<u>Proof of 1b</u>. To prove constructivity properties of theories $HPT \pm SA^e \pm M_U$ ' and $HPT \pm M_U$ which do not contain ECT_U we must replace the above notion of realizability by some new one, namely, q-realizability [Troelstra, 1977], because we cannot use the general equivalence $F \Leftrightarrow \exists x(xrF)$ in the absence of ECT_U. Only for U-negative (resp., for almost negative) F we have

$$HPT \text{ (resp., } + SA^e) \vdash F \Leftrightarrow \exists x(xqF).$$

This equivalence is sufficient to prove the corresponding q-realizability theorem whose formulation may be obtained from the above realizability theorem by omitting ECT_U and substituting q in place of r. Again, we can infer (with somewhat different details) constructivity of theories $HPT \pm SA^e \pm M_U$ and P-$\exists$-constructivity of theories $HPT \pm M_U$.□

<u>Proof of 3b</u>. It is better to prove P-$\exists$-constructivity of HPT+M in the following, some more general setting: if (i) $HPT+M \vdash_{LJ} \Gamma \rightarrow \exists xF(x)$ then (ii) $HPT+M \vdash_{LJ} \Gamma' \rightarrow F(t)$ for some t, where LJ is Gentzen's intuitionistic sequent predicate calculus [Takeuti, 1975], F is an arbitrary formula, formulae of the list Γ have the form A, $\exists xA$ or $\neg\neg\exists xA$ with A quantifier-free, t is signature term and Γ' results from Γ by omitting in Γ all (existensial) quantifiers. Here M and nonlogical axioms of HPT are naturally considered as (new) initial sequents of LJ-proofs (i) and (ii).

Then, the natural structural induction over LJ-prpofs (i) succeeds if cut formulae [Takeuti, 1975] of (i) are only of the kind A, $\exists xA$ or $\neg\neg\exists xA$ with A quantifier-free. Fortunately, this restriction on LJ-proofs from HPT+M is inessential. This (and, therefore, 3b) follows immediately from the following natural and useful

Generalization of Gentzen's Cut Elimination Theorem [Sazonov, 1980]. Any proof $P:\mathfrak{I}\vdash S$ in Gentzen's sequent calculus LK or LJ [Takeuti, 1975] from some collection $\mathfrak{I}$ of new initial sequents (besides the ordinary initial sequents $A \to A$) such that $\mathfrak{I}$ is closed under term substitutions can be transformed into corresponding regular [Takeuti, 1975] proof $P':\mathfrak{I}\vdash S$ in which all cut formulae belong to $\mathfrak{I}$. In fact, one of cut formulae of each cut is introduced in P' by some new initial sequent.□

Note, that some analogous forms of Cut Elimination Theorem are presented in [Buss, 1986 and Girard, 1987].

The author is grateful to G.E.Mints for useful notes on the preliminary version of this paper.

REFERENCES

Berman P., 1983, Review on [Sazonov, 1980], M.R.,83j:68055.

Buss S.R., 1986, "Bounded Arithmetic", Bibliopolis, Napoli, 221 pp.

Dragalin A.G., 1979, "Mathematical Intuitionism. Introduction to proof theory", Nauka, Moscow, 256pp. (In Russian)

Girard J.-Y., 1987, "Proof Theory and Logical Complexity", Vol.1, Bibliopolis, Napoli, 503 pp.

Sazonov V.Yu., 1980, A logical approach to the problem "P=NP?", in: Lecture Notes in Computer Science, N88, Springer, New York, P.562-575. (An important correction to this paper is given in [Sazonov, 1981, P.490.])

Sazonov V.Yu., 1981, On existence of complete predicate calculus in metamathematics without exponentiation, in:Lecture Notes in Computer Science, N118, Springer, New York, P.483-490.

Sazonov V.Yu., 1987, An equivalence between polynomial constructivity of Markov's principle and the equality P=NP, in: "19th All-Union Algebraic Conference, Proceedings", part 2, L'vov, P.250-251. (In Russian) (A paper with the same title will be published in the Proceedings of Institute of Mathematics, Siberian Branch of USSR Akademy of Sciences, Novosibirsk, 1989, about 80 of typescript pages.)

Takeuti G., 1975, "Proof theory", North-Holland, Amsterdam.

Troelstra A., 1977, Aspects of constructive mathematics, in: "Handbook of Mathematical Logic", J.Barwise, ed., North-Holland, Amsterdam.

Yessenin-Vol'pin A.S., 1959, An analysis of potential feasibility, in: "Logic investigations", Moscow, P.218-262. (In Russian)

Zvonkin A.K. and Levin L.A., 1970, Complexity of finite objects and foundation of information and randomness notions through algorithms theory, Uspechi Mat. Nauk, Vol.25, N6, P.85-127. (In Russian)

EFFECTIVE ENUMERATIONS OF ABSTRACT STRUCTURES*

Alexandra A. Soskova
Mathematical Faculty
Sofia University
boul. A. Ivanov 5
Sofia 1126, Bulgaria

Ivan N. Soskov
Laboratory for Applied Logic
Sofia University
boul. A. Ivanov 5
Sofia 1126, Bulgaria

There is a close relationship between the enumerations which admits an abstract structure $\mathfrak{A}$ and the "effective" computable functions in $\mathfrak{A}$, see [1,2,3].

It is natural to suppose that some special properties of the computable functions will imply the existence of some "special" enumerations of $\mathfrak{A}$.

In the present paper a characterization of the class of all abstract structures which admit effective enumerations is obtained. Our approach is based on the notion of computability by means of recursively enumerable definitional schemes (REDS) [4].

1. Preliminaries.

Let $\mathfrak{A} = (B; \theta_1, \theta_2, \ldots, \theta_n; \Sigma_1, \Sigma_2, \ldots, \Sigma_k)$ be a partial structure, where B is an arbitrary denumerable set; $\theta_1, \theta_2, \ldots, \theta_n$ are partial functions of many arguments on B; $\Sigma_1, \Sigma_2, \ldots, \Sigma_k$ are partial predicates of many arguments on B. There n, k ≥ 0. The relational type of $\mathfrak{A}$ is the ordered pair $\langle\langle a_1, a_2, \ldots, a_n\rangle, \langle b_1, b_2, \ldots, b_k\rangle\rangle$ where each θ_i is a_i-ary and each Σ_j is b_j-ary.

An ordered pair $\langle\alpha, \mathfrak{B}\rangle$ is called <u>enumeration</u> of $\mathfrak{A}$ if α is a partial, surjective mapping of the set of all natural numbers N onto B and $\mathfrak{B} = (N; \varphi_1, \varphi_2, \ldots, \varphi_n; \sigma_1, \sigma_2, \ldots, \sigma_k)$ is a partial structure of the same

* Research partially supported by the Ministry of Culture, Science and Education, Contract # 933, 1988.

relational type as $\mathfrak{A}$ such that the following conditions hold :

(i) The domain of α ($Dom(\alpha)$) is closed with respect to the partial operations $\varphi_1, \dots, \varphi_n$;

(ii) For each i, $1 \le i \le n$, and all $x_1, \dots, x_{a_i}$ of $Dom(\alpha)$,

$\alpha(\varphi_i(x_1, \dots, x_{a_i})) \simeq \theta_i(\alpha(x_1), \dots, \alpha(x_{a_i}))$;

(iii) For each j, $1 \le j \le k$, and all $x_1, \dots, x_{b_j}$ of $Dom(\alpha)$,

$\sigma_j(x_1, \dots, x_{b_j}) \simeq \Sigma_j(\alpha(x_1), \dots, \alpha(x_{b_j}))$.

An enumeration $\langle \alpha, \mathfrak{B} = (N; \varphi_1, \varphi_2, \dots, \varphi_n; \sigma_1, \sigma_2, \dots, \sigma_k) \rangle$ of $\mathfrak{A}$ is said to be <u>effective</u> iff all basic functions $\varphi_1, \varphi_2, \dots, \varphi_n$ and all basic predicates $\sigma_1, \sigma_2, \dots, \sigma_k$ of $\mathfrak{B}$ are partial recursive.

We introduce bellow the so called definable in $\mathfrak{A}$ sets of natural numbers. In fact these sets coincide with the domains of the partial functions in N, which are computable by means of REDS, using the basic functions and predicates of $\mathfrak{A}$ and finitely many constants, derived from B.

In next section we shall prove that the structure $\mathfrak{A}$ admits an effective enumeration if and only if all definable in $\mathfrak{A}$ subsets of N are recursively enumerable.

Using almost the same methods as in [3], one can obtain and the following characterization of the definable in $\mathfrak{A}$ subsets of N :

A subset A of N is definable in $\mathfrak{A}$ iff for every enumeration $\langle \alpha, \mathfrak{B} \rangle$ of $\mathfrak{A}$, A is partial recursive in $\mathfrak{B}$.

We shall use the following notations. The letters s, t will denote arbitrary elements of B; x, y, v -- elements of N. We shall identify the partial predicates with the partial mappings which obtain values in $\{0,1\}$, taking 0 for true and 1 for false.

Let $\mathcal{L}$ be the first order language corresponding to the structure $\mathfrak{A}$ i. e. $\mathcal{L}$ consists of n functional symbols $f_1, \dots, f_n$ and k predicate symbols $T_1, T_2, \dots, T_k$ where each f_i is a_i-ary and each T_j is b_j-ary. Let T_0 be a new unary predicate symbol which is intended to represent the unary total predicate $\Sigma_0 = \lambda s.0$.

Let $\{\underline{X}_0, \underline{X}_1, \dots\}$ be a denumerable set of variables. We shall use the capital letters X, Y to denote variables.

If τ is a term in the language $\mathscr{L}$ then we shall write $\tau(X_1, X_2, \ldots, X_a)$ to denote that all variables of τ are among $X_1, X_2, \ldots, X_a$. If $\tau(X_1, X_2, \ldots, X_a)$ is a term, $s_1, \ldots, s_a$ are arbitrary elements of B, then by $\tau_{\mathfrak{A}}(X_1/s_1, \ldots, X_a/s_a)$ we shall denote the value, if it exists, of the term τ in the structure $\mathfrak{A}$ over the elements $s_1, \ldots, s_a$.

Termal predicates in the language $\mathscr{L}$ are defined by the inductive clauses:

If $T \in \{T_0, \ldots, T_k\}$, T is b-ary and $\tau^1, \ldots, \tau^b$ are terms, then $T(\tau^1, \ldots, \tau^b)$ is an (atomic) termal predicate;

If Π is an atomic predicate, then $\neg\Pi$ is a termal predicate;

If Π_1 and Π_2 are termal predicates, then $(\Pi_1 \& \Pi_2)$ is a termal predicate.

Let $\Pi(X_1, X_2, \ldots, X_a)$ be a termal predicate with variables among $X_1, X_2, \ldots, X_a$ and let $s_1, \ldots, s_a$ be arbitrary elements of B. The value $\Pi_{\mathfrak{A}}(X_1/s_1, \ldots, X_a/s_a)$ of Π over $s_1, \ldots, s_a$ in $\mathfrak{A}$ is defined by the inductive clauses:

If $\Pi = T_j(\tau^1, \ldots, \tau^{b_j})$, $0 \le j \le k$, then $\Pi_{\mathfrak{A}}(X_1/s_1, \ldots, X_a/s_a) \cong \Sigma_j(\tau^1_{\mathfrak{A}}(X_1/s_1, \ldots, X_a/s_a), \ldots, \tau^{b_j}_{\mathfrak{A}}(X_1/s_1, \ldots, X_a/s_a))$;

If $\Pi = \neg\Pi^1$, where Π^1 is an atomic predicate, then

$$\Pi_{\mathfrak{A}}(X_1/s_1, \ldots, X_a/s_a) \cong \begin{cases} 1 & \text{,if } \Pi^1_{\mathfrak{A}}(X_1/s_1, \ldots, X_a/s_a) \cong 0 ; \\ 0 & \text{,if } \Pi^1_{\mathfrak{A}}(X_1/s_1, \ldots, X_a/s_a) \cong 1 ; \\ \text{undefined,otherwise.} & \end{cases}$$

If $\Pi = (\Pi^1 \& \Pi^2)$, where Π^1 and Π^2 are termal predicates, then

$$\Pi_{\mathfrak{A}}(X_1/s_1, \ldots, X_a/s_a) \cong \begin{cases} \Pi^2_{\mathfrak{A}}(X_1/s_1, \ldots, X_a/s_a) & \text{,if } \Pi^1_{\mathfrak{A}}(X_1/s_1, \ldots, X_a/s_a) \cong 0; \\ 1 & \text{,if } \Pi^1_{\mathfrak{A}}(X_1/s_1, \ldots, X_a/s_a) \cong 1; \\ \text{undefined} & \text{,otherwise.} \end{cases}$$

If Π is a termal predicate and x is a natural number, then $(\Pi \supset x)$ is called <u>conditional term</u>.

Let $Q = (\Pi \supset x)$ be a conditional term with variables among $X_1, \ldots, X_a$ and let $s_1, \ldots, s_a$ be arbitrary elements of B. The value $Q_{\mathfrak{A}}(X_1/s_1, \ldots, X_a/s_a)$ of Q is defined by the equivalence

$$Q_{\mathfrak{A}}(X_1/s_1, \ldots, X_a/s_a) \cong y \leftrightarrow (\Pi_{\mathfrak{A}}(X_1/s_1, \ldots, X_a/s_a) \cong 0 \;\&\; y \cong x).$$

We shall suppose that an effective coding of the expressions of the language $\mathcal{L}$ is given. By $\ulcorner\tau\urcorner$ we shall denote the code of the term τ.

Let $A \subseteq N$. The set A is said to be <u>definable</u> in the structure $\mathfrak{A}$ iff for some r.e. set of conditional terms $\{Q^v\}_{v\in V}$ with variables among $X_1, \ldots, X_a$ and for some fixed elements $s_1, \ldots, s_a$ of B the equivalence

$x \in A \leftrightarrow \exists v(v \in V \ \& \ x \cong Q^v_{\mathfrak{A}}(X_1/s_1, \ldots, X_a/s_a))$ is true.

We point out two propositions, which follow easy from the definition above.

<u>Proposition 1.</u> Every r.e. subset of N is definable in $\mathfrak{A}$.

<u>Proposition 2.</u> If the structure $\mathfrak{A}$ admits an effective enumeration then every definable in $\mathfrak{A}$ subset of N is recursively enumerable.

2. The main result

In this section we shall prove the following theorem.

<u>Theorem.</u> The partial structure $\mathfrak{A}$ admits an effective enumeration iff every definable in $\mathfrak{A}$ set of natural numbers is recursively enumerable.

In the first direction the theorem follows from proposition 2.

To prove the theorem in the second direction let us suppose that every definable in $\mathfrak{A}$ set of natural numbers is recursively enumerable. We shall construct an effective enumeration $\langle\alpha, \mathfrak{B}\rangle$ of $\mathfrak{A}$. The enumeration $\langle\alpha, \mathfrak{B}\rangle$ will be a standard enumeration, in the sense of [3].

Suppose $\mathfrak{A} = (B; \theta_1, \theta_2, \ldots, \theta_n; \Sigma_1, \Sigma_2, \ldots, \Sigma_k)$.

Let $\langle , \rangle$ be an effective coding of the ordered pairs of natural numbers. Denote by φ_i^*, $1 \le i \le n$, the a_i-ary recursive function $\lambda x_1 \ldots x_{a_i}.\langle i, x_1, \ldots, x_{a_i}\rangle$.

Denote by N^0 the set $N \setminus (\mathrm{Range}(\varphi_1^*) \cup \ldots \cup \mathrm{Range}(\varphi_n^*))$ and suppose that α^0 is a partial surjective mapping of N^0 onto B.

Define the partial mapping α of N onto B by the inductive clauses :

If $x \in N^0$, then $\alpha(x) \cong \alpha^0(x)$;

If $x = \langle i, x_1, \ldots, x_{a_i}\rangle$, $\alpha(x_1) = s_1, \ldots, \alpha(x_{a_i}) \cong s_{a_i}$ and $\theta_i(s_1, \ldots, s_{a_i}) \cong t$, then $\alpha(x) \cong t$.

Let $D_1, \ldots, D_n$ be partial predicates in N satisfying the condition

(1) if $x_1, \ldots, x_{a_i} \in \mathrm{Dom}(\alpha)$ then

$D_i(x_1, \ldots, x_{a_i}) \cong 0 \leftrightarrow (\alpha(x_1), \ldots, \alpha(x_{a_i})) \in \mathrm{Dom}(\theta_i)$,

and let $\varphi_1, \ldots, \varphi_n$ be partial functions in N such that

$$\varphi_i(x_1, \ldots, x_{a_i}) \cong \begin{cases} \varphi_i^*(x_1, \ldots, x_{a_i}), & \text{if } D_i(x_1, \ldots, x_{a_i}) \cong 0; \\ \text{undefined}, & \text{otherwise.} \end{cases}$$

Let $\sigma_1, \ldots, \sigma_k$ be partial predicates in N, satisfying the condition

(2) if $x_1, \ldots, x_{b_j} \in \mathrm{Dom}(\alpha)$ then

$\sigma_j(x_1, \ldots, x_{b_j}) \cong \Sigma_j(\alpha(x_1), \ldots, \alpha(x_{b_j}))$, $j = 1, \ldots, k$.

Denote by $\mathfrak{B}$ the partial structure $(N; \varphi_1, \varphi_2, \ldots, \varphi_n; \sigma_1, \sigma_2, \ldots, \sigma_k)$.

Every enumeration $\langle \alpha, \mathfrak{B} \rangle$, obtained by the method, described above, is called <u>standard</u> <u>enumeration.</u>

It is clear that a standard enumeration $\langle \alpha, \mathfrak{B} \rangle$ is effective if $D_1, \ldots, D_n, \sigma_1, \ldots, \sigma_k$ are partial recursive. We shall construct the desired effective enumeration of $\mathfrak{A}$ as a standard enumeration.

We begin with some properties of the standard enumerations.

Suppose a standard enumeration $\langle \alpha, \mathfrak{B} \rangle$ is fixed.

From the definition of the mapping α it follows :

<u>Proposition</u> <u>3.</u> Let $1 \le i \le n$, and $x_1, \ldots, x_{a_i}$ are natural numbers. Then

$\alpha(\langle i, x_1, \ldots, x_{a_i} \rangle) \cong \theta_i(\alpha(x_1), \ldots, \alpha(x_{a_i}))$.

<u>Proposition</u> <u>4.</u> Every standard enumeration is an enumeration of $\mathfrak{A}$.

Proof. We have to show the validity of the conditions (i) - (iii) from the definition of enumeration of $\mathfrak{A}$.

The condition (ii) follows from (1) and proposition 3.

The condition (iii) follows from (2).

To prove (i), suppose that $1 \le i \le n$, $x_1, \ldots, x_{a_i} \in \mathrm{Dom}(\alpha)$ and $\varphi_i(x_1, \ldots, x_{a_i}) \cong y$. Then, $D_i(x_1, \ldots, x_{a_i}) \cong 0$ and $y = \langle i, x_1, \ldots, x_{a_i} \rangle$. Hence, $\theta_i(\alpha(x_1), \ldots, \alpha(x_{a_i}))$ is defined and, by proposition 3, $\alpha(y)$ is defined.

We shall assume that an effective coding of the finite sets of natural numbers is fixed. By E_v we shall denote the finite set with code v.

Let ϱ be a recursive function whose values are codes of finite sets of natural numbers and such that if $x \in N^0$ then $E_{\varrho(x)} = \{x\}$ and if $x = \langle i, x_1, \dots, x_{a_i} \rangle$ then $E_{\varrho(x)} = E_{\varrho(x_1)} \cup \dots \cup E_{\varrho(x_{a_i})}$.

Using induction on $|x|$, where $|x| = 0$ if $x \in N^0$, and $|\langle i, x_1, \dots, x_{a_i} \rangle| = |x_1| + \dots + |x_{a_i}| + 1$, $1 \le i \le n$, one can prove the following useful property of ϱ:

Proposition 5. If $x \in \mathrm{Dom}(\alpha)$, then $E_{\varrho(x)} \subseteq \mathrm{Dom}(\alpha^0)$.

Denote by $\mathfrak{B}^*$ the structure $(N; \varphi_1^*, \dots, \varphi_n^*)$.

Proposition 6. There exists an effective way to define for every natural number x, every sequence $y_1, \dots, y_r$ containing all elements of $E_{\varrho(x)}$ and every sequence of distinct variables $Y_1, \dots, Y_r$ a term $\tau(Y_1, \dots, Y_r)$, such that $\tau_{\mathfrak{B}^*}(Y_1/y_1, \dots, Y_r/y_r) = x$.

Proof. The proof follows from the definition of the functions $\varphi_1^*, \dots, \varphi_n^*$.

An immediate consequence of proposition 6 is the following.

Proposition 7. There exists an effective way to define for every finite sequence $x_1, \dots, x_a$ of natural numbers, every finite sequence $y_1, \dots, y_r$ containing all elements of $E_{\varrho(x_1)} \cup \dots \cup E_{\varrho(x_a)}$ and every sequence of distinct variables $Y_1, \dots, Y_r$, terms $\tau^1, \dots, \tau^a$ with variables among $Y_1, \dots, Y_r$, such that

$$x_1 = \tau^1_{\mathfrak{B}^*}(Y_1/y_1, \dots, Y_r/y_r),$$

$$\dots$$

$$x_a = \tau^a_{\mathfrak{B}^*}(Y_1/y_1, \dots, Y_r/y_r).$$

Note: It is easy to check that if the elements of the sequence $y_1, \dots, y_r$ are different and each y_l, $1 \le l \le r$, belongs to $E_{\varrho(x_1)} \cup \dots \cup E_{\varrho(x_a)}$ then the terms $\tau^1, \dots, \tau^a$ are unique.

Proposition 8. Let $\tau(Y_1, \dots, Y_r)$ be a term and $y_1, \dots, y_r$ be arbitrary elements of $\mathrm{Dom}(\alpha)$. Then

$$\alpha(\tau_{\mathfrak{B}^*}(Y_1/y_1, \dots, Y_r/y_r)) \cong \tau_{\mathfrak{A}}(Y_1/\alpha(y_1), \dots, Y_r/\alpha(y_r)).$$

Proof. Induction on τ. If $\tau = Y_l$, $1 \le l \le r$, then the equality is evident.

Let $\tau = f_i(\tau^1, \ldots, \tau^{a_i})$, $1 \le i \le n$. Then, $\alpha(\tau_{\mathfrak{B}^*}(Y_1/y_1, \ldots, Y_r/y_r)) \cong$
$\alpha(\langle i, \tau^1_{\mathfrak{B}^*}(Y_1/y_1, \ldots, Y_r/y_r), \ldots, \tau^{a_i}_{\mathfrak{B}^*}(Y_1/y_1, \ldots, Y_r/y_r)\rangle) \cong$
$\theta_i(\alpha(\tau^1_{\mathfrak{B}^*}(Y_1/y_1, \ldots, Y_r/y_r)), \ldots, \alpha(\tau^{a_i}_{\mathfrak{B}^*}(Y_1/y_1, \ldots, Y_r/y_r))) \cong$
$\theta_i(\tau^1_{\mathfrak{A}}(Y_1/\alpha(y_1), \ldots, Y_r/\alpha(y_r)), \ldots, \tau^{a_i}_{\mathfrak{A}}(Y_1/\alpha(y_1), \ldots, Y_r/\alpha(y_r))) \cong$
$\tau_{\mathfrak{A}}(Y_1/\alpha(y_1), \ldots, Y_r/\alpha(y_r))$.

Now, we are ready to define an effective enumeration $\langle \alpha, \mathfrak{B} \rangle$ of $\mathfrak{A}$. The enumeration $\langle \alpha, \mathfrak{B} \rangle$ will be a standard enumeration. First we shall define the partial mapping α^0 of N^0 onto B, and hence, the mapping α. After that we shall define partial recursive predicates $D_1, \ldots, D_n$, satisfying the condition (1) and partial recursive predicates $\sigma_1, \ldots, \sigma_k$, satisfying (2).

Let Φ be a binary partial recursive function, which is universal for the unary partial recursive functions. As usual, if χ is an unary partial recursive function and $\chi = \lambda x.\Phi(h,x)$ then the number h we shall call index of χ.

Let N^0_m, $m \ge 0$, denote the subset of N^0 which consists of all codes of $n + k + 2$ -- tuples $\langle 0, m, h_1, \ldots, h_n, c_1, \ldots, c_k \rangle$, where $h_1, \ldots, h_n, c_1, \ldots, c_k$ are arbitrary natural numbers. Let $\tilde{N} = \bigcup_{m=0}^{\infty} N^0_m$.

Let $t_0, t_1, \ldots, t_m, \ldots$ be an arbitrary enumeration of B.

For every $m \ge 0$ we shall choose an unique element $\underline{x}_m$ of N^0_m and take $\alpha^0(\underline{x}_m) = t_m$. Let $m \ge 0$, and let $\underline{X}_0, \ldots, \underline{X}_m$ be the first m+1 variables. Denote by $\mathcal{T}_m$ the set of all terms of the language $\mathcal{L}$ with variables among $\underline{X}_0, \ldots, \underline{X}_m$. Suppose $1 \le i \le n$. Let K^m_i be the recursive set which consists of all numbers $\langle \ulcorner \tau^1 \urcorner, \ldots, \ulcorner \tau^{a_i} \urcorner \rangle$, where $\tau^1, \ldots, \tau^{a_i}$ are elements of $\mathcal{T}_m$.

Define the partial function χ^m_i by the conditions :

(i) $\chi^m_i(x)$ is undefined if $x \notin K^m_i$.

(ii) Let $x \in K^m_i$ and $x = \langle \ulcorner \tau^1 \urcorner, \ldots, \ulcorner \tau^{a_i} \urcorner \rangle$. Then
$\chi^m_i(x) \cong \Sigma_0(\theta_i(\tau^1_{\mathfrak{A}}(\underline{X}_0/t_0, \ldots \underline{X}_m/t_m), \ldots, \tau^{a_i}_{\mathfrak{A}}(\underline{X}_0/t_0, \ldots \underline{X}_m/t_m)))$.

It is clear that the graph of χ^m_i is definable in $\mathfrak{A}$, and hence, χ^m_i is partial recursive.

Analogously, we define the partial recursive functions $\xi^m_1, \ldots, \xi^m_k$. Let

$1 \le j \le k$ and let K_j^m be the recursive set consisting of all numbers $\langle \ulcorner \tau^1 \urcorner, \ldots, \ulcorner \tau^{b_j} \urcorner \rangle$, where $\tau^1, \ldots, \tau^{b_j} \in \mathcal{T}_m$.

Let $\xi_j^m(x)$ be undefined for $x \notin K_j^m$ and if $x \in K_j^m$ and $x = \langle \ulcorner \tau^1 \urcorner, \ldots, \ulcorner \tau^{b_j} \urcorner \rangle$ then

$\xi_j^m(x) \cong \Sigma_j(\tau^1_{\mathfrak{U}}(\underline{X}_0/t_0, \ldots, \underline{X}_m/t_m), \ldots, \tau^{b_j}_{\mathfrak{U}}(\underline{X}_0/t_0, \ldots, \underline{X}_m/t_m))$.

Now, let $h_1, \ldots, h_n, c_1, \ldots, c_k$ be some fixed indexes of $\chi_1^m, \ldots, \chi_n^m, \xi_1^m, \ldots, \xi_k^m$, respectively. Take $\underline{x}_m = \langle 0, m, h_1, \ldots, h_n, c_1, \ldots, c_k \rangle$.

Thereby the definition of α^0, and hence, the definition of α is completed.

Our second task is the definition of the predicates $D_1, \ldots, D_n, \sigma_1, \ldots, \sigma_k$.

Denote by st the recursive function $\lambda x.(x)_1$. It is clear that for $x \in N_m^0$, $st(x) = m$.

Let $1 \le i \le n$. We begin the definition of D_i.

Let $x_1, \ldots, x_{a_i}$ be arbitrary elements of N. Let $y_1, \ldots, y_r$ be the elements of $E_{\mathcal{g}(x_1)} \cup \ldots \cup E_{\mathcal{g}(x_{a_i})}$. If $\{y_1, \ldots, y_r\}$ is not a subset of $\tilde{N}$ then $\{y_1, \ldots, y_r\}$ is not a subset of $Dom(\alpha^0)$ and, therefore, by proposition 5, at least one of $x_1, \ldots, x_{a_i}$ is not an element of $Dom(\alpha)$.

Let $D_i(x_1, \ldots, x_{a_i})$ be undefined in this case.

Suppose that $\{y_1, \ldots, y_r\} \subseteq \tilde{N}$. If there exist two different elements y_{l_1} and y_{l_2} of $\{y_1, \ldots, y_r\}$ such that $st(y_{l_1}) = st(y_{l_2})$ then at least one of y_{l_1} and y_{l_2} does not belong to $Dom(\alpha^0)$ and therefore, at least one of $x_1, \ldots, x_{a_i}$ does not belong to $Dom(\alpha)$. Let $D_i(x_1, \ldots, x_{a_i})$ be undefined.

Now, suppose that $\{y_1, \ldots, y_r\} \subseteq \tilde{N}$ and $st(y_1) < st(y_2) < \ldots < st(y_r)$. Let $st(y_l) = m_l$, $l = 1, \ldots, r$ and $y_r = \langle 0, m_r, h_1, \ldots, h_n, c_1, \ldots, c_k \rangle$. Let χ be the partial recursive function with index h_i. By proposition 7, there is an effective way to choose the terms $\tau^1, \ldots, \tau^{a_i}$ with variables among $\underline{X}_{m_1}, \ldots, \underline{X}_{m_r}$ so that

$x_1 = \tau^1_{\mathfrak{B}^*}(\underline{X}_{m_1}/y_1, \ldots, \underline{X}_{m_r}/y_r),$

(3) $\ldots$

$x_{a_i} = \tau^{a_i}_{\mathfrak{B}^*}(\underline{X}_{m_1}/y_1, \ldots, \underline{X}_{m_r}/y_r).$

Let $D_i(x_1, \ldots, x_{a_i}) \cong \begin{cases} 0, & \text{if } \chi(\langle \ulcorner\tau^1\urcorner, \ldots, \ulcorner\tau^{a_i}\urcorner\rangle) \cong 0; \\ \text{undefined,} & \text{otherwise.} \end{cases}$

It is clear that D_i is partial recursive. We shall prove that D_i satisfies the condition (1).

Let $x_1, \ldots, x_{a_i}$ be elements of $\mathrm{Dom}(\alpha)$. We have to prove that

$\theta_i(\alpha(x_1), \ldots, \alpha(x_{a_i}))$ is defined $\leftrightarrow D_i(x_1, \ldots, x_{a_i}) \cong 0.$

Let $\{y_1, \ldots, y_r\} = E_{\vartheta(x_1)} \cup \ldots \cup E_{\vartheta(x_{a_i})}$. By proposition 5, $\{y_1, \ldots, y_r\} \subseteq \mathrm{Dom}(\alpha^0)$, and hence, $\{y_1, \ldots, y_r\} \subseteq \tilde{N}$ and $\mathrm{st}(y_{l_1}) \neq \mathrm{st}(y_{l_2})$ for $1 \leq l_1 < l_2 \leq r$.

Let $\mathrm{st}(y_l) = m_l$, $l = 1, \ldots, r$. We can suppose that $m_1 < m_2 \ldots < m_r$. From the definition of α^0 it follows that $\alpha^0(y_l) = t_{m_l}$, $l = 1, \ldots, r$. Let $y_r = \langle 0, m_r, h_1, \ldots, h_n, c_1, \ldots, c_k\rangle$ and let χ be the partial recursive function with index h_i. By the definition of α^0 it follows that $\chi = \chi_i^{m_r}$.

Let $\tau^1, \ldots, \tau^{a_i}$ be the terms with variables among $\underline{X}_{m_1}, \ldots, \underline{X}_{m_r}$, satisfying (3). Then

$\theta_i(\alpha(x_1), \ldots, \alpha(x_{a_i}))$ is defined $\leftrightarrow$

$\theta_i(\alpha(\tau^1_{\mathfrak{B}^*}(\underline{X}_{m_1}/y_1, \ldots, \underline{X}_{m_r}/y_r)), \ldots, \alpha(\tau^{a_i}_{\mathfrak{B}^*}(\underline{X}_{m_1}/y_1, \ldots, \underline{X}_{m_r}/y_r)))$ is defined $\leftrightarrow$

(by proposition 8)

$\theta_i(\tau^1_{\mathfrak{A}}(\underline{X}_{m_1}/t_{m_1}, \ldots, \underline{X}_{m_r}/t_{m_r}), \ldots, \tau^{a_i}_{\mathfrak{A}}(\underline{X}_{m_1}/t_{m_1}, \ldots, \underline{X}_{m_r}/t_{m_r}))$ is defined $\leftrightarrow$

$\Sigma_0(\theta_i(\tau^1_{\mathfrak{A}}(\underline{X}_{m_1}/t_{m_1}, \ldots, \underline{X}_{m_r}/t_{m_r}), \ldots, \tau^{a_i}_{\mathfrak{A}}(\underline{X}_{m_1}/t_{m_1}, \ldots, \underline{X}_{m_r}/t_{m_r}))) \cong 0 \leftrightarrow$

$\chi_i^{m_r}(\langle \ulcorner\tau^1\urcorner, \ldots, \ulcorner\tau^{a_i}\urcorner\rangle) \cong 0 \leftrightarrow D_i(x_1, \ldots, x_{a_i}) \cong 0.$

The definition of the predicates $\sigma_1, \ldots, \sigma_k$ is similar to the definition of $D_1, \ldots, D_n$.

Let $1 \leq j \leq k$ and $x_1, \ldots, x_{b_j}$ be arbitrary natural numbers. Let $y_1, \ldots$

..., y_r be the elements of $E_{g(x_1)} \cup \ldots \cup E_{g(x_{b_j})}$. If $\{y_1, \ldots, y_r\}$ is not a subset of $\tilde{N}$ or for some different y_{l_1} and y_{l_2}, $st(y_{l_1}) = st(y_{l_2})$, then $\sigma_j(x_1, \ldots, x_{b_j})$ is undefined.

Suppose that $\{y_1, \ldots, y_r\} \subseteq \tilde{N}$, $st(y_1) < st(y_2) < \ldots < st(y_r)$, $st(y_l) = m_l$, $l = 1, \ldots, r$, and $y_r = \langle 0, m_r, h_1, \ldots, h_n, c_1, \ldots, c_k \rangle$. Let ξ be the partial recursive function with index c_j.

Take the terms $\tau^1, \ldots, \tau^{b_j}$ with variables among $X_{m_1}, \ldots, X_{m_r}$ such that

$$x_1 = \tau^1_{\mathfrak{B}^*}(X_{m_1}/y_1, \ldots, X_{m_r}/y_r),$$

...

$$x_{b_j} = \tau^{b_j}_{\mathfrak{B}^*}(X_{m_1}/y_1, \ldots, X_{m_r}/y_r).$$

$$\text{Take } \sigma_j(x_1, \ldots, x_{b_j}) \cong \begin{cases} 0, & \text{if } \xi(\langle \ulcorner \tau^1 \urcorner, \ldots, \ulcorner \tau^{b_j} \urcorner \rangle) \cong 0 ; \\ 1, & \text{if } \xi(\langle \ulcorner \tau^1 \urcorner, \ldots, \ulcorner \tau^{b_j} \urcorner \rangle) \cong 1 ; \\ \text{undefined, otherwise.} \end{cases}$$

It is clear that the predicates $\sigma_1, \ldots, \sigma_k$ are partial recursive. A proof similar to the previous one shows that for all $x_1, \ldots, x_{b_j}$ from $Dom(\alpha)$, $\sigma_j(x_1, \ldots, x_{b_j}) \cong \Sigma_j(\alpha(x_1), \ldots, \alpha(x_{b_j}))$.

Thereby the proof of the theorem is completed.

3. Two examples

The first example shows that there exist structures which do not admit effective enumerations.

Consider the structure $\mathfrak{A} = (N; S, \varphi; =)$, where $S = \lambda x.x+1$; φ is an arbitrary function which is unary and it is not partial recursive. In order to prove that $\mathfrak{A}$ does not admit an effective enumeration we shall construct a definable in $\mathfrak{A}$ subset of N which is not recursively enumerable. Let $V = \{ \langle n,m \rangle ; n \in N, m \in N \}$.

For every $v \in V$ define the conditional term $Q^v(X)$ as $(\varphi(S^n(X)) = S^m(X) \supset \langle n,m \rangle)$.

Let $A \subseteq N$ and $y \in A \leftrightarrow \exists v(v \in V \ \& \ Q^v_{\mathfrak{A}}(X/0) \cong y)$.

It is clear that A is definable in $\mathfrak{A}$. Since φ is not partial recursive, and hence, the graph of φ is not recursively enumerable, A is not recursively enumerable.

An effective enumeration $\langle \alpha, \mathfrak{B} \rangle$ of a structure $\mathfrak{A}$ is said to be <u>strong effective</u> if Dom(α) is recursively enumerable. It is an interesting task to find necessary and sufficient conditions for the existence of strong effective enumerations of a structure $\mathfrak{A}$.

Our second example shows that there exist structures which admit effective enumerations but do not admit strong effective ones. The idea of this example belongs to Angel Ditchev.

Let $\mathfrak{A} = (B; \Sigma_1, \Sigma_2)$ be a structure, where B is an infinite but denumerable set; Σ_1, Σ_2 are binary predicates, totally defined on B and $\Sigma_2(s,t) = 0 \Leftrightarrow s \neq t$.

Say that there exists a $\underline{k}$ - cycle for Σ_1 if for some distinct elements $t_1, t_2, \ldots, t_k$ of B, $\Sigma_1(t_1,t_2) \& \ldots \& \Sigma_1(t_k,t_1) = 0$.

Let W be a subset of N which is not recursively enumerable.

Suppose that the predicate Σ_1 is choosen so that there exists a $\underline{k}$ - cycle for Σ_1 iff $\underline{k} \in W$.

Suppose that $\langle \alpha, \mathfrak{B} = (N; \sigma_1, \sigma_2) \rangle$ is a strong effective enumeration of $\mathfrak{A}$. Then there exists a $\underline{k}$ - cycle for Σ_1 iff for some elements $y_1, \ldots, y_k$ of Dom(α), $\alpha(y_i) \neq \alpha(y_j)$, $1 \leq i < j \leq k$, and $\sigma_1(y_1,y_2) \cong 0 \&, \ldots, \& \sigma_1(y_k,y_1) \cong 0$. Hence, there exists a $\underline{k}$ - cycle for Σ_1 iff $\exists y_1 \ldots \exists y_k (y_1 \in \in \mathrm{Dom}(\alpha) \& \ldots \& y_k \in \mathrm{Dom}(\alpha) \& \forall i \forall j (1 \leq i < j \leq k \Rightarrow \sigma_2(y_i,y_j) \cong 0) \& \sigma_1(y_1,y_2) \cong 0 \& \ldots \& \sigma_1(y_k,y_1) \cong 0)$. From here it follows that the set $\{ k :$ there exists a $\underline{k}$ - cycle for $\Sigma_1\}$ is recursively enumerable. This fact contradicts the choice of W.

On the other hand, there exists an effective enumeration of $\mathfrak{A}$. Indeed, let A be a definable in $\mathfrak{A}$ subset of N. We shall prove that A is recursively enumerable.

Let $\{Q^v\}_{v \in V}$ be a r.e. set of conditional terms with variables among $X_1, \ldots, X_a$ and $s_1, \ldots, s_a$ be fixed elements of B, such that $x \in A \Leftrightarrow \exists v(v \in V \& x \cong Q^v_{\mathfrak{A}}(X_1/s_1, \ldots, X_a/s_a))$.

Let E_1 and E_2 be finite subsets of N, consisting of all ordered pairs $\langle i,j \rangle$ such that $1 \leq i \leq a$, $1 \leq j \leq a$ and $\langle i,j \rangle \in E_1 \Leftrightarrow \Sigma_1(s_i,s_j) = 0$, $\langle i,j \rangle \in E_2 \Leftrightarrow \Sigma_2(s_i,s_j) = 0$. Let T_1 and T_2 be the binary predicate symbols which represent Σ_1 and Σ_2 in the language $\mathcal{L}$ corresponding to $\mathfrak{A}$, respectively.

Now, we can replace all occurrences of $T_l(X_i, X_j)$ in the definition of A with $\langle i,j\rangle \in E_l$ and all occurrences of $\neg T_l(X_i, X_j)$ with $\langle i,j\rangle \notin E_l$ for l = 1,2. Since E_1 and E_2 are finite, the obtained definition of A shows that A is recursively enumerable.

Acknowledgments

The authors are indebted to the referees of the paper for the valuable remarks.

References

1. Y. N. Moschovakis, Abstract computability and invariant definability, J. Symb. Logic, 34:4 (1969).
2. I. N. Soskov, Definability via enumerations. J. Symb. Logic (to appear).
3. I. N. Soskov, Computability by means of effectively definable schemes and definability via enumerations, Submitted for publication.
4. J. C. Shepherdson, Computation over abstract structures, in: "Logic Colloquium'73", H. E. Rose, J. C. Shepherdson eds., North-Holland, Amsterdam, (1975), pp. 445 - 513.

MODAL CHARACTERIZATION OF THE CLASSES OF FINITE AND INFINITE QUASI-ORDERED SETS*

Dimiter Vakarelov

Laboratory for Applied Logic
Sofia University
Boul. Anton Ivanov 5
1126 Sofia, Bulgaria

0. Introduction

To formulate the main aim of this paper we will begin with some definitions and notations. A system $\underline{U}=(U,\leq)$, where $U\neq\emptyset$ and $\leq$ is a binary relation in U, is called quasi-ordered set - qoset for short, - if the relation $\leq$ is reflexive and transitive one. By QO, QO_{fin} and QO_{inf} we will denote respectively the class of all qosets, all finite qosets and all infinite qosets. If U is a qoset and in addition the relation $\leq$ is an antisymmetric one /$x\leq y$ & $y\leq x \rightarrow x=y$/ then U is called partially ordered set - poset for short. By PO, PO_{fin} and PO_{inf} we denote respectively the class of all posets, all finite posets, and all infinite posets. It is a well known fact that the modal logic S4 is complete in the classes QO and QO_{fin}. So QO and QO_{fin} cannot be separated by S4. However, the situation is different when we consider posets: S4 is complete in the class PO_{inf} but not in the class PO_{fin}, and S4Grz / the so called Grzegorczik system/ is complete in PO_{fin} but not in PO_{inf} /see [1]/. Roughly speaking, we will say in this case, that S4 is a modal characterization of PO_{inf} and S4Grz is a modal characterization of PO_{fin}.

The main aim of this paper is to find modal logics L_1 and L_2 such that L_1 is complete in the class QO_{inf} but not in QO_{fin} and L_2 - complete in QO_{fin} but not in QO_{inf}. L_1 and L_2 cannot be monomodal logics, because then its modalities should be S4 ones and by the above results S4 cannot separate QO_{fin} from QO_{inf}. We will show that the above puzzle can be solved positively if we use a bi-modal language with two modalities - $[\leq]$ and $[\equiv]$ -

*Research partially supported by the Committee for Science at the Council of Ministers, Contract # 56, 1987.

with the usual relational semantics. The corresponding relation for $[\leq]$ is the quasi-ordering $\leq$ and the corresponding relation for $[\equiv]$ is the equivalence relation $\equiv$ determined by $\leq$ in the following way:

(*) $x\equiv y$ iff $x\leq y$ and $y\leq x$.

Since $[\leq]$ is an S4-modality and $[\equiv]$ is an S5-modality, then the set of all formulas in this bi-modal language, true in the class QO, will contain the logics S4 and S5. This suggests to call the new logic S4+5. Our first aim is to axiomatize S4+5. Then we will show that S4+5 is complete not only in QO, but in QO_{inf} and that it is not complete in QO_{fin}. So S4+5 is the required logic L_1. Then we will find an exttension of S4+5, denoted S4+5gGrz, which will play the role of L_2.

1. Modal characterization of infinite quasi-ordered sets

We shall use a bi-modal language $\underline{L}([\leq],[\equiv])$ containing an infinite set VAR of propositional variables; classical Boolean connectives $\neg$, $\wedge$, $\vee$, $\Rightarrow$, $\Leftrightarrow$; two modal box operations $[\leq]$ and $[\equiv]$; and (,) - parentheses. The definition of the set FOR of all formulas is the usual one.

A relational structure $\underline{U}=(U,\leq,\equiv)$, where $U\neq\emptyset$ and $\leq$ and $\equiv$ are binary relations in U, will be called a frame. We will use the usual Kripke interpretation of $\underline{L}$ in $\underline{U}$. Namely, let $v:VAR \longrightarrow 2^U$ be a valuation, which assigns to each variable $p\in VAR$ a subset $v(p)\subseteq U$. The satisfiability relation $x \Vdash_v A$ /"the formula A is true in $x\in U$ at the valuation v"/ is defined inductively:

$x \Vdash_v A$ iff $x\in v(A)$ for $A\in VAR$,

$x \Vdash_v \neg A$ iff $x \not\Vdash_v A$,

$x \Vdash_v A\wedge B$ iff $x \Vdash_v A$ and $x \Vdash_v B$,

$x \Vdash_v A\vee B$ iff $x \Vdash_v A$ or $x \Vdash_v B$,

$x \Vdash_v [R]A$ iff $(\forall y)(xRy \longrightarrow y \Vdash_v A)$ for $R\in\{\leq, \equiv\}$.

A formula A is true in the frame $\underline{U}$ if for any $x\in U$ and valuation v we have $x \Vdash_v A$. A is true in the class Σ of frames if A is true in any frame of Σ. The set $L(\Sigma)=\{A\in FOR/ A$ is true in $\Sigma\}$ will be called the logic of Σ.

Our aim now is to axiomatize the logic L(QO) considering any qoset $(U,\leq)$ as a frame $(U,\leq,\equiv)$, in which the relation $\equiv$ is defined by the condition (*) $x\equiv y$ iff $x\leq y$ and $y\leq x$.

In the next table the conditions (1) - (7) characterize the relations $\leq$ and $\equiv$ and the modal formulas A1 - A6 are the corresponding modal translations of the conditions (1) - (6). Note that there is no formula which is a modal translation of the condition (7) and this is the main difficulty in the axiomatization of the logic L(QO).

(1) $x \leq x$ A1. $[\leq]A \Rightarrow A$

(2) $x \leq y$ & $y \leq z \longrightarrow x \leq z$ A2. $[\leq]A \Rightarrow [\leq][\leq]A$

(3) $x \equiv x$ A3. $[\equiv]A \Rightarrow A$

(4) $x \equiv y \longrightarrow y \equiv x$ A4. $A \vee [\equiv]\neg[\equiv]A$

(5) $x \equiv y$ & $y \equiv z \longrightarrow x \equiv z$ A5. $[\equiv]A \Rightarrow [\equiv][\equiv]A$

(6) $x \equiv y \longrightarrow x \leq y$ A6. $[\leq]A \Rightarrow [\equiv]A$

(7) $x \leq y$ & $y \leq x \longrightarrow x \equiv y$ ----------

Let gQO - the class of generalized qosets - be the class of all frames $(U, \leq, \equiv)$ satisfying the conditions (1) - (6). It is not so difficult to axiomatize the logic L(gQO) with the help of the axioms A1 - A6. Then we will show that $L(gQO)=L(QO)=L(QO_{inf})$. Namely this logic will be denoted by S4+5.

<u>Axioms and rules for S4+5.</u>

(BooL) All or enough Boolean tautologies,

(Mod) $[R](A \Rightarrow B) \Rightarrow ([R]A \Rightarrow [R]B)$, $R \in \{\leq, \equiv\}$

A1 - A6

(MP) $A, A \Rightarrow B/B$, $(N\leq)$ $A/[\leq]A$

<u>Theorem 1.1.</u>/Completeness theorem for S4+5/ For any formula A of S4+5 the following conditions are satisfied:

(i) A is a theorem of S4+5,

(ii) A is true in the class gQO,

(iii) A is true in the class QO,

(iv) A is true in the class QO_{inf}.

<u>Proof.</u>/ A sketch of the proof of this theorem is given in [2]/. The implications (i) $\longrightarrow$ (ii) $\longrightarrow$ (iii) $\longrightarrow$ (iv) are obvious. The proof of (ii) $\longrightarrow$ (i) can be given by standard canonical-model-techniques, dealing with maximal consistent sets of formulas /see for e.g. [1] and [3]/.

(iv) $\longrightarrow$ (ii). Suppose A_0 is not true in some generalized qoset $(U, \leq, \equiv)$. Then there exist a valuation v and $c \in U$ such that $c \not\Vdash_v A_0$. Now we shall construct an infinite qoset $\underline{U}'=(U', \leq', \equiv')$ and a valuation v' such that for some $c' \in U'$: $c' \not\Vdash_{v'} A_0$.

Define $U'=U \times N$ where $N=\{0, 1, 2, \ldots\}$. We abbreviate: $x_i =_{def} (x,i)$ for $x \in U$ and $i \in N$. This means that U' is a disjoint union of ω copies of U and x_i is the image of x in the i-th copy.

The relations $\leq'$ and $\equiv'$ we define in several steps:

$x_i \leq_1 y_j$ iff $j=i$ & $x \leq y$ & $(y \leq x \longrightarrow x \equiv y)$,

$x_i \leq_2 y_j$ iff $j=i+1$ & $x \leq y$ & $y \leq x$ & $x \not\equiv y$,

$\leq' = (\leq_1 \cup \leq_2)^*$ where * is the operation of reflexive and transitive closure of the relation.

$x_i \equiv' y_j$ iff $x_i \leq' y_j$ and $y_j \leq' x_i$,

$v'(A)=\{x_i / i\in N \text{ and } x\in v(A)\}$ for $A\in VAR$.

The intuitive idea of defining $\le'$ in this way is the following. It is possible for some pair $\{x,y\}$ in a generalized qoset $\underline{U}$ to satisfy the conditions $x\le y$, $y\le x$, $x\not\equiv y$. We will call such pairs defective /qosets do not have defective pairs/. The purpose is to repair $\underline{U}$, repairing defecive pairs in such a way as to preserve the validity of formulas. The definition of $\le_1$ says that $x_i\le_1 y_j$ holds iff x_i, y_j are in one and same copy, $x\le y$ and $\{x,y\}$ is not a defective pair. The definition of $\le_2$ says that $x_i\le_2 y_j$ iff $\{x,y\}$ is a defective pair and $j=i+1$. So in this two steps we "repair" the defects but destroy the transitivity, which is "repaired" in the third step. The following lemmas will show that the construction is good.

<u>Lemma</u> <u>1.2.</u> (i) $x_i\le_1 x_i$,

(ii) $x_i\le_1 y_i \;\&\; y_i\le_1 z_i \longrightarrow x_i\le_1 z_i$

(iii) $x_i\le' y_i \longleftrightarrow x_i\le_1 y_i$,

(iv) $i\ne j \longrightarrow x_i\le' y_j \longleftrightarrow \exists k\ne 0,\ j=i+k \;\&\; \exists z^{00}, z^{01}, \dots z^{k0}, z^{k1}\in U$

$z^{00}=x,\ z^{k1}=y,\ z_i^{00}\le_1 z_i^{01}\le_2 z_{i+1}^{10}\le_1 z_{i+1}^{11}\le_2 \dots \le_2 z^{k0}\le_1 z_{i+1}^{k1}$,

(v) $x_i\le y_j \longrightarrow x\le y \;\&\; i\le j$,

(vi) $x_i\equiv' y_j \longleftrightarrow x\equiv y \;\&\; i=j$.

<u>Proof.</u> Condition (i) is obvious because $\{x,x\}$ is not a defective pair and we have $x\le x$.

Proof of (ii). Suppose $x_i\le_1 y_i$, $y_i\le_1 z_i$. Then we have

a/ $x\le y \;\&\; (y\le x \longrightarrow x\equiv y)$,

b/ $y\le z \;\&\; (z\le y \longrightarrow y\equiv z)$.

We have to show

c/ $x\le z \;\&\; (z\le x \longrightarrow x\equiv z)$.

From $x\le y$ and $y\le z$ we obtain $x\le z$. Suppose $z\le x$. Then by $x\le y$ from a/ we obtain $z\le y$ and by b/ we get $y\equiv z$. From $y\le z$ /b/ and $z\le x$ we obtain $y\le x$ and by a/ we get $x\equiv y$. Then $x\equiv y$ and $y\equiv z$ give $x\equiv z$, which had to be proved.

Conditions (iii) and (iv) follow directly from the definition of $\le'$, (i) and (ii). Condition (v) is a consequence of the definition of $\le'$, (iii) and (iv).

Proof of (vi),($\longrightarrow$). Suppose $x_i\equiv' y_j$. Then $x_i\le' y_j$, $y_j\le' x_i$. From (v) we obtain $i\le j$ and $j\le i$, so $i=j$. Then by (iii) we have $x_i\le_1 y_j$, $y_j\le_1 x_i$ and by the definition of $\le_1$ – $x\le y \;\&\; (y\le x \longrightarrow x\equiv y)$ and $y\le x \;\&\; (x\le y \longrightarrow x\equiv y)$. This implies $x\equiv y$.

For (vi),($\longleftarrow$) suppose $x\equiv y$ and $i=j$. Then we have $x\le y$, $y\le x$ and by the definition of $\le_1$ – $x_i\le_1 y_i$ and $y_i\le_1 x_i$. By (iii) this infers $x_i\le' y_i$ and $y_i\le' x_i$, so $x_i\equiv' y_i$.

<u>Lemma</u> <u>1.3.</u> Let $R\in\{\le, \equiv\}$. Then

(i) $xRy \longrightarrow \forall i\exists j\ x_i R' y_j$,

(ii) $x_i R' y_j \longrightarrow xRy$.

Proof. $(R=\leq)$, (i). Suppose $x\leq y$. If $\{x,y\}$ is not a defective pair then $x_i\leq' y_i$. Otherwise $x_i\leq' y_{i+1}$. Condition (ii) for this case follows from lemma 1.2(v).

$(R=\equiv)$. Both conditions follow from lemma 1.2(vi).

Lemma 1.4. For any $A\in R$, $x\in U$ and $i\in N$ the following equivalence holds:

$$x \Vdash_v A \text{ iff } x_i \Vdash_{v'} A.$$

Proof. We will proceed by induction on the complexity of A. For $A\in VAR$ this is true by the definition of v'. Boolean combinations of formulas do not present difficulties. So let $A=[R]B$, $R\in\{\leq,\equiv\}$ and let the assertion for B holds by induction hypothesis (i.h.).

$(\longrightarrow)$ Suppose $x\Vdash_v [R]B$ and $x_i R' y_j$. We have to show that $y_j \Vdash_{v'} B$. By lemma 1.3(ii) we have xRy, and then $y\Vdash_v B$ and by i.h. $y_j\Vdash_{v'} B$.

$(\longleftarrow)$ Suppose $x_i \Vdash_{v'} [R]B$ and xRy. We have to show that $y\Vdash_v B$. By lemma 1.3(i) we have that there exists j such that $x_i R' y_j$. Then, since $x_i \Vdash_{v'} [R]B$, we obtain $y_j\Vdash_{v'} B$ and by the i.h. - $y\Vdash_v B$.

Now we can complete the proof of theorem 1.1 in the following way. We have that for some $c\in U$ $c\not\Vdash_v A_0$. Then by lemma 1.4 we have that $c_i\not\Vdash_v A_0$ and hence A_0 is not true in the qoset $(U',\leq',\equiv')$, which is infinite. ■

Now we shall show that S4+5 is not complete in the class QO_{fin}. For that purpose consider the following formula which I call generalized Grzegorczik formula

gGrz: $[\leq]([\leq]([\equiv]A \Rightarrow [\leq]A) \Rightarrow A) \Rightarrow A$

/The original Grzegorczik formula is Grz: $\Box(\Box(A \Rightarrow \Box A) \Rightarrow A) \Rightarrow A$, see [1]/.

Theorem 1.5. (i) gGrz is true in QO_{fin},

(ii) gGrz is not a theorem of S4+5,

(iii) gGrz is not true in QO_{inf},

(iv) S4+5 is not complete in QO_{fin}.

Proof. (i) Suppose that gGrz is not true in some qoset $\underline{U}=(U,\leq,\equiv)$. Then we shall show that U is infinite. This will prove that gGrz is true in all finite qosets.

Suppose $x_0\in U$ and $x_0\not\Vdash_v$ gGrz. Then $x_0\not\Vdash_v A$ and

(*) $\quad x_0\Vdash_v [\leq]([\leq]([\equiv]A \Rightarrow [\leq]A) \Rightarrow A)$.

Since $x_0\leq x_0$ we obtain $x_0\Vdash_v [\leq]([\equiv]A \Rightarrow [\leq]A \Rightarrow A$. Since $x_0\not\Vdash_v A$ we have $x_0\not\Vdash_v [\leq]([\equiv]A \Rightarrow [\leq]A)$. Then for some x_1 s.t. $x_0\leq x_1$, $x_1\Vdash_v [\equiv]A$ and $x_1\not\Vdash_v [\leq]A$. Hence there exists x_2 such that $x_1\leq x_2$ and $x_2\not\Vdash_v A$. From here we obtain that $x_0\not\equiv x_1$. Otherwise we will have $x_1\equiv x_0$ and by $x_1\Vdash_v [\equiv]A$ - $x_1\Vdash_v A$ - a contradiction. In the same way we obtain $x_1\not\equiv x_2$. Define $x<y$ iff $x\leq y$ & $x\not\equiv y$. It is easy to see that $<$ is a transitive and irreflexive relation. Hence we obtain $x_0<x_1<x_2$. From transitivity we have $x_0\leq x_2$ and by

(*) we obtain that $x_2 \Vdash_v [\leq]([\equiv]A \to [\leq]A) \to A$ and since $x_2 \nVdash_v A$ we can repeat the above procedure with x_2 to obtain x_3 and x_4 such that $x_2 < x_3 < x_4$. Repeating this procedure infinitely many times we obtain an infinite sequence $x_0 < x_1 < x_2 < \ldots$. Applying transitivity and irreflexivity of $<$ we infer that all members of this sequence are different and hence that U is infinite.

Proof of (ii) and (iii). Let $U=\{0, 1\}$, $\leq = U \times U$, $x \equiv y$ iff $x=y$ and $v(A)=\{0\}$, $A \in VAR$. Then $(U, \leq, \equiv)$ is a generalized qoset and $0 \nVdash_v$ gGrz. By theorem 1.1 gGrz is not a theorem of S4+5, and hence is not true in QO_{inf}. This shows (iii).

Condition (iv) follows from (i) and (ii).■

Corollary 1.6. There is no formula A of S4+5 with the following property:

For any generalized qoset $\underline{U}=(U, \leq, \equiv)$ A is true in $\underline{U}$ iff the condition (7) / $x \leq y$ & $y \leq x \longrightarrow x \equiv y$/ holds in $\underline{U}$.

Proof. Suppose that such a formula exists. Since there exists a qoset $\underline{U}_0$ in which the condition (7) does not hold then A is not true in $\underline{U}_0$ and hence $A \notin L(gQO)$. But (7) holds in all qosets, so $A \in L(QO)$. Thus $L(QO) \neq L(gQO)$, which contradicts theorem 1.1.■

Another nice fact for the logic S4+5 is that it is a conservative extension of S4 and S5 /the proof is given in [2]/. We will give without proof the following theorem.

Theorem 1.7. S4+5 is complete in the class of finite generalized qosets and hence is decidable.

The proof of theorem 1.7 can be obtained by a modification of Lemmon filtration of S4 and S5 / see for e.g. [1]/.

As a consequence of theorem 1.7 we obtain that in the proof of theorem 1.1 the infinite disjoint union $U \times N$ cannot be replaced by a finite one. Otherwise, applying first theorem 1.7 we will obtain that S4+5 is complete in QO_{fin}, contrary to theorem 1.5.

Hence the first part of our aim is fulfilled - we have found a logic L_1 complete in QO_{inf} but not in QO_{fin}, namely L_1=S4+5.

2. Modal characterization of finite quasi-ordered sets

In this section we will show that a natural candidate for L_2 is an extension of S4+5 with the axiome gGrz, denoted here by S4+5gGrz.

Theorem 2.1. /Completeness theorem for S4+5gGrz/ For any formula A of S4+5gGrz the following conditions are equivalent:

(i) A is a theorem of S4+5gGrz,

(ii) A is true in the class QO_{fin}.

Proof. The implication (i)$\longrightarrow$(ii) is easy - all axioms of S4+5 are true in QO_{fin} , and by theorem 1.5 gGrz is also true in QO_{fin}.

(ii)$\longrightarrow$(i). Suppose that this implication does not hold. Then for some formula A_0 we have: A_0 is not a theorem of S4+5gGrz but A_0 is true in the class QO_{fin}. We will obtain a contradiction by constructing a finite qoset in which A_0 is not true. For that purpose we shall apply techniques, called selective filtration, which I learned from Gargov and Kirov's paper [4]. In the next proof, however, we use a version, which is a modification of Hughes & Cresswell's subordination method, described in [3].

Suppose now that A_0 is not a theorem of S4+5gGrz. Then, applying standard canonical-model-techniques dealing with maximal consistent sets of formulas, we can construct - as in the proof of implication (ii)$\longrightarrow$(i) of theorem 1.1 - an infinite generalized qoset $\underline{U}=(U,\leq,\equiv)$ and a valuation v such that for some $x_0 \in U$ we have $x_0 \not\Vdash_v A_0$. Now we shall construct a finite qoset $\underline{U}'=(U',\leq',\equiv')$, a valuation v' and an element $\alpha_0 \in U'$ such that $\alpha_0 \not\Vdash_{v'} A_0$.

The elements of U' will be the vertices of a tree, which we will construct by stages. After each stage we will obtain a tree, which will extend the previous one by adding new edges and new vertices. Simultaneously we will define a function Γ, acting on vertices, such that for any vertex α, $\Gamma(\alpha) \in U$. If α and β are connected by an edge we will denote this by $\alpha\Sigma\beta$. We will also define in the process of the construction two relations between vertices - $\leq_0$ and $\equiv_0$ - satisfying the following condition:

(Σ) $\alpha\Sigma\beta$ iff only one of the following relations holds: $\alpha\equiv_0\beta$, $\alpha\leq_0\beta$, $\beta\equiv_0\alpha$, $\beta\leq_0\alpha$.

A sequence, finite or infinite, of different vertices $(\alpha_1,\alpha_2,\alpha_3,\ldots)$ will be called a path if $\alpha_1\Sigma\alpha_2\Sigma\alpha_3\Sigma\ldots$. If $(\alpha_1,\alpha_2,\alpha_3,\ldots)$ is a path and R is a relation between vertices then this path will be called an R-path if $\alpha_1 R\alpha_2 R\alpha_3 R\ldots$.

Stage 0. The root of the tree T is denoted by α_0. The resulting tree T_0 is $\{\alpha_0\}$ - without edges. Define $\Gamma(\alpha_0)=x_0$.

Suppose we have constructed the trees T_0, $T_1,\ldots,T_n$, and Γ, $\leq_0$, $\equiv_0$ are defined, satisfying the condition (Σ).

Stage n+1. Let Δ be the smallest set of formulas satisfying the following conditions:

(Δi) $A_0 \in \Delta$,

(Δii) If $A\in\Delta$ and B is a subformula of A then $B\in\Delta$,

(Δiii) If $[\leq]A\in\Delta$ then $[\equiv]A\in\Delta$.

Then for each endpoint $\alpha\in T_n$ define the following sets:

$S(\alpha,[\equiv])=\{[\equiv]A\in\Delta / \ \Gamma(\alpha) \not\Vdash_v [\equiv]A,\ \Gamma((\alpha) \Vdash_v A$ and there is no $\equiv_0$-path

$(\beta_0,\beta_1,\ldots,\beta_k)$ such that $\beta_k=\alpha$, $\beta_0,\ldots,\beta_k\in T_n$ and $\Gamma(\beta_0)\nVdash_V A\}$.

$S(\alpha,[\leq])=\{[\leq]A\in\Delta/\ \Gamma(\alpha)\nVdash_V [\leq]A,\ \Gamma(\alpha)\Vdash_V [\equiv]A\}$.

Now we will proceed as follows.

Case 1. Let $[\equiv]A\in S(\alpha,[\equiv])$. Then $\Gamma(\alpha)\nVdash_V [\equiv]A$ and hence there exists $y\in U$ such that $\Gamma(\alpha)\equiv y$ and $y\nVdash_V A$. Let β be an object, not used before in the process of the construction of the tree . We take β as a new vertex and define: $\alpha\Sigma\beta$, $\alpha\equiv_0\beta$, $\Gamma(\beta)=y$. The same we do with all members of the set $S(\alpha,[\equiv])$.

Case 2. Let $[\leq]A\in S(\alpha,[\leq])$. Then $\Gamma(\alpha)\nVdash_V [\leq]A$ and $\Gamma(\alpha)\Vdash_V [\equiv]A$. We will find a $y\in U$ such that: $\Gamma(\alpha)\leq y$, $y\nVdash_V A$ and $y\Vdash_V [\leq]([\equiv]\to[\leq]A)$. We do this as follows. Since $\Gamma(\alpha)\nVdash_V [\leq]A$ then there exists y_1 such that $\Gamma(\alpha)\leq y_1$ and $y_1\nVdash_V A$. Then since $y_1\Vdash_V$ gGrz we obtain $y_1\nVdash_V$ $[\leq]([\leq]([\equiv]A\to[\leq]A)\to A)$. From here we obtain that for some $y\in U$ such that $y_1\leq y$ we have $y\Vdash_V [\leq]([\equiv]A\to[\equiv]A)$ and $y\nVdash_V A$. Then from $\Gamma(\alpha)\leq y_1\leq y$ we obtain $\Gamma(\alpha)\leq y$. Take a new vertex β, not used before, and define: $\alpha\Sigma\beta$, $\alpha\leq_0\beta$, $\Gamma(\beta)=y$. The same we do for all members of $S(\alpha,[\leq])$.

These two procedures we apply to all endpoints of T_n and in this way we obtain T_{n+1}. If for all endpoints the sets $S(\alpha,[\leq])$ and $S(\alpha,[\equiv])$ are empty, then $T_{n+1}=T_n$ and the construction stops. But in any case we define the tree T as the union of all T_n.

Lemma 2.2. (i) T is finitely branched tree.

(ii) All paths of T from the root are finite.

(iii) T is a finite tree.

Proof. (i). Since Δ is finite set of formulas the same are the sets $S(\alpha,[\leq])$ and $S(\alpha,[\equiv])$, so in the proces of the construction to each vertex α we can put only a finite number of edges.

(ii). First we shall show that in T there are no infinite $\equiv_0$-paths. Suppose that $\alpha_1\equiv_0\alpha_2\equiv_0\alpha_3\equiv_0\ldots$ is an infinite $\equiv_0$-path. Then from case 1 of the construction of the tree we have $\Gamma(\alpha_1)\equiv\Gamma(\alpha_2)\equiv\Gamma(\alpha_3)\equiv\ldots$ and there exist formulas $[\equiv]A_1$, $[\equiv]A_2$, $[\equiv]A_3,\ldots$ from the set Δ such that for any i=1, 2,... we have $\Gamma(\alpha_i)\nVdash_V [\equiv]A_i$, $\Gamma(\alpha_{i+1})\nVdash_V A_i$ and for any $j\leq i$ – $\Gamma(\alpha_j)\Vdash_V A_i$. From the last two conditions we obtain that all members of the sequence A_1, $A_2,\ldots$ are different. But this is impossible, because all these formulas are from the finite set Δ.

Suppose now that $(\alpha_0,\alpha_1,\ldots)$ is a path from the root α_0. Then from the construction of the tree we have that $\alpha_0R_0\alpha_1R_1\alpha_2R_2\ldots$.and for each i we have $R_i\in\{\equiv_0,\leq_0\}$. Since $x\equiv y$ implies $x\leq y$ then we have:

(*) $\Gamma(\alpha_0)\leq\Gamma(\alpha_1)\leq\Gamma(\alpha_2)\leq\ldots$

We will show that in the sequence R_0, R_1, $R_2,\ldots$ there is no infinite subsequence R_{i_0}, R_{i_1}, $R_{i_2}\ldots$ such that all R_{i_k} are $\leq_0$. Suppose so. Then, remining the case 2 from the construction of the tree, for the vertices α_{i_0}, α_{i_1}, α_{i_2} ... we have formulas $[\leq]A_{i_0}$, $[\leq]A_{i_1}$, $[\leq]A_{i_2}$ ... from the set Δ such that for k=0,1... we have :

$$\Gamma(\alpha_{i_k}) \Vdash\!\!\!\not\,\frac{}{v} [\leq]A_{i_k},\ \Gamma(\alpha_{i_k}) \Vdash\frac{}{v} [\equiv]A_{i_k},\ \text{and}$$

$$(**)\quad \Gamma(\alpha_{i_k+1}) \Vdash\frac{}{v} [\leq]([\equiv]A_{i_k} \Rightarrow [\leq]A_{i_k}).$$

Since from (*) we have $\Gamma(\alpha_{i_k+1}) \leq \Gamma(\alpha_{i_l})$ for $k<l$, then from here and (**) we obtain $\Gamma(\alpha_{i_l}) \Vdash\frac{}{v} [\equiv]A_{i_k} \Rightarrow [\leq]A_{i_k}$. This condition is true in the following two cases:

case 1: $\Gamma(\alpha_{i_l}) \Vdash\!\!\!\not\,\frac{}{v} [\equiv]A_{i_k}$ or

case 2: $\Gamma(\alpha_{i_l}) \Vdash\frac{}{v} [\leq]A_{i_k}$.

But also we have that $\Gamma(\alpha_{i_l}) \Vdash\!\!\!\not\,\frac{}{v} [\leq]A_{i_l}$ and $\Gamma(\alpha_{i_l}) \Vdash\frac{}{v} [\equiv]A_{i_l}$. Then in both cases we have that $A_{i_k} \neq A_{i_l}$. This shows that all members of the infinite sequence A_{i_0}, A_{i_1},... are different, which is impossible, because all they are from the finite set Δ.

Now, from the above two results it is easy to conclude that each path from the root is finite. It follows from here that all paths are finite.

(iii) It follows from (i) and (ii) by König's lemma that the tree T is finite.■

Now we define in T the following new relations:

$\alpha\equiv_1\beta$ iff $\alpha\equiv_0\beta$ or $\beta\equiv_0\alpha$,

$\alpha\leq_1\beta$ iff $\alpha\leq_0\beta$ or $\alpha\equiv_0\beta$ or $\beta\equiv_0\alpha$.

<u>Lemma</u> <u>2.3.</u> For any two vertices α and β:

(i) $\alpha\Sigma\beta$ iff $\alpha\equiv_1\beta$ or $\alpha\leq_1\beta$ or $\beta\leq_1\alpha$,

(ii) $\alpha\equiv_1\beta$ iff $\alpha\leq_1\beta$ and $\beta\leq_1\alpha$.

<u>Proof.</u> (i) and (ii)($\longrightarrow$) are obvious.

(ii)($\longleftarrow$). Suppose $\alpha\leq_1\beta$ and $\beta\leq_1\alpha$. Then we have : ($\alpha\leq_0\beta$ or $\alpha\equiv_0\beta$ or $\beta\equiv_0\alpha$)&($\beta\leq_0\alpha$ or $\beta\equiv_0\alpha$ or $\alpha\equiv_0\beta$). By the distributivity law this condition is equivalent to the following one: ($\alpha\leq_0\beta$ & $\beta\leq_0\alpha$) or ($\alpha\equiv_0\beta$ or $\beta\equiv_0\alpha$). It follows from (Σ) that the first disjunct is impossible, so it remains only the second part of the disjunction, which implies $\alpha\equiv_1\beta$.■

Now we define the structure $\underline{U}'=(U',\leq',\equiv')$ in the following way:

U' is the set of all vertices of the tree T,

$\alpha R'\beta$ iff $\alpha=\beta$ or there is an R-path from α to β, $R\in\{\leq_1,\equiv_1\}$.

$v'(A)=\{\alpha\in U'/\Gamma(\alpha)\Vdash\frac{}{v} A\}$, $A\in VAR$.

<u>Lemma2.4.</u> The structure $\underline{U}'=(U',\leq',\equiv')$ is a finite qoset.

<u>Proof.</u> It follows from lemma 2.2 that U'is a finite set. Obviously $\leq'$is a quasi-ordering in U'. We shall show that $\alpha\equiv'\beta$ iff $\alpha\leq'\beta$ & $\beta\leq'\alpha$.

(⟶) Suppose $\alpha\neq\beta$ /the case $\alpha=\beta$ is obvious/ and $\alpha\equiv'\beta$. Then there exists a $\equiv_1$-path $\alpha\equiv_1\alpha_1\equiv_1\alpha_2\equiv_1\ldots\equiv_1\alpha_n\equiv_1\beta$ from α to β. From the definition of $\equiv_1$ we have that $\alpha\leq_1\alpha_1\leq_1\ldots\leq_1\alpha_n\leq_1\beta$ and $\beta\leq_1\alpha_n\leq_1\ldots\leq_1\alpha_1\leq_1\alpha$. This shows that we have $\alpha\leq'\beta$ and $\beta\leq'\alpha$.

(⟵) Suppose $\alpha\leq'\beta$ and $\beta\leq'\alpha$. and $\alpha\neq\beta$. Since T is a tree, then there exists a unique path from α to β: $\alpha\Sigma\alpha_1\Sigma\ldots\Sigma\alpha_n\Sigma\beta$. From $\alpha\leq'\beta$ and $\beta\leq'\alpha$ we have also $\alpha\leq_1\alpha_1\leq_1\ldots\leq_1\alpha_n\leq_1\beta$ and $\beta\leq_1\alpha_n\leq_1\ldots\leq_1\alpha_1\leq_1\alpha$. Then from lemma 2.3 we obtain that $\alpha\equiv_1\alpha_1\equiv_1\ldots\equiv_1\alpha_n\equiv_1\beta$, which shows that $\alpha\equiv'\beta$.■

Lemma 2.5. For any $\alpha,\beta\in U'$ and $R\in\{\leq,\equiv\}$ the following condition is true: $\alpha R'\beta \longrightarrow \Gamma(\alpha)R\Gamma(\beta)$.

The proof follows immediately from the definition of relations R' and the construction of the tree.■

Lemma 2.6. /The truth lemma/ For any formula A of Δ and $\alpha\in U'$ the following equivalence holds:

$$\alpha\Vdash_{v'} A \text{ iff } \Gamma(\alpha)\Vdash_{v} A.$$

Proof. We will proceed by induction on the complexity of A. For $A\in VAR$ this is true by the definition of v'. Boolean combinations of formulas present no difficulties.

Case 1. $A=[\equiv]B$. (⟶). We will proceed by contraposition. Suppose that $\Gamma(\alpha)\nVdash_{v}[\equiv]B$. We have to show that $\alpha\nVdash_{v'}[\equiv]B$.

Case 1.1. $\Gamma(\alpha)\nVdash_{v} B$. Then by the induction hypothesis /i.h./ $\alpha\nVdash_{v'} B$ and since $\alpha\equiv'\alpha$ we obtain $\alpha\nVdash_{v}[\equiv]B$.

Case 1.2. $\Gamma(\alpha)\Vdash_{v} B$ and there is a $\equiv_0$-path $\beta\equiv_0\beta_1\equiv_0\ldots\equiv_0\alpha$ such that $\Gamma(\beta)\nVdash_{v} B$. Then $\alpha\equiv'\beta$ and by i.h. $\beta\nVdash_{v'}B$. This yields $\alpha\nVdash_{v'}[\equiv]B$.

Case 1.3. $\Gamma(\alpha)\Vdash_{v} B$ and there is no $\equiv_0$-path $\beta_0\equiv_0\beta_1\equiv_0\ldots\equiv_0\alpha$ such that $\Gamma(\beta_0)\nVdash_{v} B$. In this case $[\equiv]B\in S(\alpha,[\equiv])$ and by the construction of the tree there exists β such that $\alpha\equiv_0\beta$ and $\Gamma(\beta)\nVdash_{v} B$. Then we have $\alpha\equiv'\beta$ and by i.h. $\beta\nVdash_{v'} B$, which implies $\alpha\nVdash_{v'}[\equiv]B$.

(⟵) Suppose $\Gamma(\alpha)\Vdash_{v}[\equiv]B$ and $\alpha\equiv'\beta$. We have to show that $\beta\Vdash_{v'} B$. From $a\equiv'\beta$ we obtain by lemma 2.5 that $\Gamma(\alpha)\equiv\Gamma(\beta)$, which by $\Gamma(\alpha)\Vdash_{v}[\equiv]B$ gives $\Gamma(\beta)\Vdash_{v} B$. This by i.h. implies $\beta\Vdash_{v'} B$.

Case 2. $A=[\leq]B$. (⟶). Suppose, for the sake of contraposition, that $\Gamma(\alpha)\nVdash_{v}[\leq]B$. We have to show that $\alpha\nVdash_{v'}[\leq]B$.

Case 2.1. $\Gamma(\alpha)\Vdash_{v}[\equiv]B$. Then $[\leq]B\in S(\alpha,[\leq])$ and by the construction of the tree there exists β such that $\alpha\leq_0\beta$ and $\Gamma(\beta)\nVdash_{v} B$. Then we have $\alpha\leq'\beta$ and by i.h. $\beta\nVdash_{v} B$. From here we obtain $\alpha\nVdash_{v'}[\leq]B$.

Case 2.2. $\Gamma(\alpha)\nVdash_{v}[\equiv]B$. Since $[\equiv]B\in\Delta$ then, by case 1 of this proof we have $\alpha\nVdash_{v'}[\equiv]B$. Then there exists β such that $\alpha\equiv'\beta$ and $\beta\nVdash_{v'} B$. From $\alpha\equiv'\beta$ we obtain $\alpha\leq'\beta$ and by $\beta\nVdash_{v'} B$ we get $\alpha\nVdash_{v'}[\leq]B$.

(⟵) Suppose $\Gamma(\alpha)\Vdash_{v}[\leq]B$ and $\alpha\leq'\beta$. Then by lemma 2.5. $\Gamma(\alpha)\leq'\Gamma(\beta)$, which implies $\Gamma(\beta)\Vdash\!\frac{}{v}\!B$. Then by the i.h. we obtain $\beta\Vdash\!\frac{}{v}\!B$.■

Now we complete the proof of theorem 2.1. in the following way. We have $\Gamma(\alpha_0)=x_0$ and $x_0\Vdash\!\frac{}{v}\!\!\not\,A_0$. Then by lemma 2.6 we obtain that $\alpha_0\Vdash\!\frac{}{v'}\!\!\not\,A_0$. This shows that A_0 is not true in the finite qoset $\underline{U}'=(U',\leq',\equiv')$.■

3 Concluding remarks

I'll give some comments about the method of the proof of theorem 1.1. The most important part of this theorem is that, which states the completeness of S4+5 with respect to the desired structures - QO_{inf}. This is done by proving first the completeness of S4+5 with respect to a non-desired class of structures - gQO -, and then, applying some "repairing construction" to gQO models, we obtain the completeness with respect to QO_{inf}. The first kind of such "repairing constructions" is the Segerberg's Buldozer Theorem in [1]. In theorem 1.1 I use an idea of "repairing" the undesired model by constructing certain union of copies of it, and defining the accesibility relations in this union in an apropriate way. This suggests to name such constructions "copying method". I met similar construction, reading the proof in [3] that K is complete in the class of all irreflexive structures. Then I recognized soon that this method can be successfully applied in the axiomatization of polymodal logics under the classes of frames of the form $(U, R_1, R_2, R_1\cap R_2)$, $(U, R, -R)$, classes of frames in which the accesibility relations form Boolean algebra, some examples of monomodal logics, as for e.g. the logic of the classes of the frames (U, R) in which -R is quasi-ordering /the "complement" of S4/, or R is the inequality relation ≠ /the logic of difference/, and so on. For instance the modalities $[R_1]$, $[R_2]$ and $[R_1\cap R_2]$ can be easily axiomatized by $K3+\{[R_i]A\Rightarrow[R_1\cap R_2]\}_{i=1,2}$. A very simple axiomatization has the modality $[\neq]$:

$$K + A\vee[\neq]\neg[\neq]A + A\wedge[\neq]A\Rightarrow[\neq][\neq]A.$$

In this two examples the "undesired" models can be "repaired" by making only two copies.

Many applications of the "copying method" were made by a student of mine - V. Goranko, - in his dissertation [5]. . Another application is given in [6] for the axiomatization of some modal logics, arising from the theory of knowledge representation in Artificial Intelligence. For the axiomatization of [R] and [-R] by the same method see [7]. Another completeness proof of this fact is given by Humberstone [8] by a method, refered to Cresswell [9], which is, probably, the origin of the "subordination method", described in [3] and the similar "unraveling method", given by Sahlqvist in [10]. Both mentioned methods can be applied to obtain other completeness proofs of theorem 1.1. After the reading the manuscript of this paper, Cresswell [11] had sent to me such a proof. He also found [11] a very elegant proof of theorem 2.1. Since this proof is based on a completely different and new idea, I give it, with author's kind permit, in a supplement of the paper.

In [11] Cresswell formulated the following problem. It is a well known fact that the logic K4 is complete in the class of all transitive frames, and in the class of all finite transitive frames as well. So in a monomodal language we cannot distinguish the classes of finite and infinite transitive frames. The question is: is it possible to distinguish finite and infinite transitive frames in a bi-modal language with modalities [R] and [S], such that [R] corresponds to a transitive relation R and [S] corresponds to a relation S, determined by R in the following way:

(*) xSy iff xRy and yRx.

The answer is positive. The description of the corresponding logics is the following. Denote by Tr_{inf} and Tr_{fin} the classes of all infinite, resp. all finite frames of the form (U, R, S) in which R is transitive relation and S is defined by (*). By a K4+5 I will understand the logic having the following axioms and rules:

(Bool), (Mod) $[Q](A\to B)\to([Q]A\to[Q]B)$, $Q\in\{R,S\}$, $[R]A\to[R][R]A$, $A\vee[S]\neg[S]A$, $[S]A\to[S][S]A$, $[R]A\to[S]A$. (MP), (N,R) $A/[R]A$.

Let $K4+5gGrz^*$ be an extension of K4+5 with the following axiom

$gGrz^*$ $[R]([R]([S]A\to[R]A)\to A)\to[R]A$.

Then we have the following results. The logic K4+5 is complete in the class Tr_{inf} but incomplete in the class Tr_{fin}. The logic $K4+5gGrz^*$ is complete in Tr_{fin} but incomplete in the class Tr_{inf}. The proofs of these results can be given in a similar way as the proofs for S4+5 and S4+5gGrz in this paper. As a side result we can obtain the following completeness result in the case of monomodal logics.

The logic K4 is complete in the class of all infinite frames (U, R) with transitive and antisymmetric R, but it is not complete in the class of all finite ones. The logic $K4Grz^*$, which is an extension of K4 with the following axiom Grz^* $\Box(\Box(p\to\Box p)\to p)\to\Box p$, is complete in the class of all finite transitive and antisymmetric frames but not in the class of infinite ones.

4. Supplement. A Cresswell's proof of Theorem 2.1

Suppose A_0 is not a theorem of the logic L_2=S4+5gGrz and let U^* be the set of all maximal L_2-consistent sets of formulas in L_2. Then there exists $x_0 \in U^*$ such that $A_0 \notin x_0$. Now we shall construct a finite qoset in which A_0 will not be true.

Let Φ be the smallest set of formulas with the following properties: (Φi) $A_0 \in \Phi$, (Φii) If $A \in \Phi$ and B is a subformula of A then $B \in \Phi$, (Φiii) If $[\leq]A \in \Phi$ then $[\equiv]A \in \Phi$. Clearly Φ is a finite set. Define

$\Phi^+ = \Phi \cup \{[\leq]A /$ A is a Boolean combination of formulas from $\Phi\}$. The set Φ^+ is not finite, but it is logically finite in a sence that there is a collection of formulas $\{A_1, \ldots, A_n\}$ such that $A \in \Phi$ iff $A \leftrightarrow A_i \in L_2$ for some $1 \leq i \leq n$.

For $x, y \in U^*$ define $x \sim y$ iff $(\forall A \in \Phi^+)(A \in x$ iff $A \in y)$. For $x \in U^*$ let $|x| = \{y / x \sim y\}$ and let $U = \{|x| / x \in U^*\}$. For $A \in VAR$ define $v(A) = \{|x| / A \in x\}$. Define the following relations:

(1) $|x| \leq_1 |y|$ iff $(\forall [\leq]A \in \Phi^+)([\leq]A \in x \rightarrow [\leq]A \in y)$,

(2) $|x| \leq_2 |y|$ iff $(\forall [\equiv]A \in \Phi)([\equiv]A \in x \rightarrow [\equiv]A \in y)$,

(3) $|x| \leq |y|$ iff $|x| \leq_1 |y|$ & $(|y| \leq_1 |x| \rightarrow |x| \leq_2 |y|)$,

(4) $|x| \equiv |y|$ iff $|x| \leq |y|$ & $|y| \leq |x|$.

<u>Lemma 4.1.</u> (i) $(U, \leq)$ is a qoset,

(ii) $|x| \equiv |y|$ iff $|x| \leq_1 |y|$ & $|y| \leq_1 |x|$ & $|x| \leq_2 |y|$ & $|y| \leq_2 |x|$.

<u>Lemma 4.2.</u> The following equivalences are true in S4+5 and cosequently in S4+5gGrz: $[\equiv][\leq]A \leftrightarrow [\leq]A$, $[\equiv]\neg[\leq]A \leftrightarrow \neg[\leq]A$, $[\equiv][\equiv]A \leftrightarrow [\equiv]$, $[\equiv]\neg[\equiv]A \leftrightarrow \neg[\equiv]A$, $[\leq][\leq]A \leftrightarrow [\leq]A$.

<u>Lemma 4.3.</u> For $[\equiv]A \in \Phi$ and $x \in U^*$:

$[\equiv]A \in x$ iff $(\forall y \in U^*)(|x| \equiv |y| \rightarrow A \in y)$.

<u>Proof.</u> $(\rightarrow)$. Suppose $[\equiv]A \in x$ and $|x| \equiv |y|$. Then by lemma 4.1(ii) we have $|x| \leq_2 |y|$ and hence $[\equiv]A \in y$. Then by the axiom $[\equiv]A \Rightarrow A$ we get $A \in y$.

$(\leftarrow)$. Suppose $[\equiv]A \notin x$. We have to find a $y \in U^*$ such that $|x| \equiv |y|$ and $A \notin y$. For that purpose it is sufficient to show that the following set of formulas is L_2-consistent:

$\Lambda = \{[\leq]B / [\leq]B \in x\} \cup \{\neg[\leq]C / \neg[\leq]C \in x\} \cup \{[\equiv]D / [\equiv]D \in x\} \cup \{\neg[\equiv]E / \neg[\equiv]E \in x\} \cup \{\neg A\}$

If Λ is not consistent then for some $[\leq]B_1, \ldots, [\leq]B_i$, $\neg[\leq]C_1, \ldots \neg[\leq]C_j$, $[\equiv]D_1, \ldots, [\equiv]D_k$, $\neg[\equiv]E_1, \ldots, \neg[\equiv]E_l$ from x we have that the following formula is a theorem of L_2:

$$([\leq]B_1 \wedge \ldots \wedge [\leq]B_i) \wedge (\neg[\leq]C_1 \wedge \ldots \wedge \neg[\leq]C_j) \wedge$$
$$\wedge([\equiv]D_1 \wedge \ldots \wedge [\equiv]D_k) \wedge (\neg[\equiv]E_1 \wedge \ldots \wedge \neg[\equiv]E_l) \Rightarrow A.$$

Then, by the principles of K for $[\equiv]$ we have:

$$([\equiv][\leq]B_1 \wedge \ldots \wedge [\equiv][\leq]B_i) \wedge ([\equiv]\neg[\leq]C_1 \wedge \ldots \wedge [\equiv]\neg[\leq]C_j) \wedge$$
$$\wedge([\equiv][\equiv]D_1 \wedge \ldots \wedge [\equiv][\equiv]D_k) \wedge ([\equiv]\neg[\equiv]E_1 \wedge \ldots \wedge [\equiv]\neg[\equiv]E_l) \Rightarrow [\equiv]A \in L_2.$$

Now, apllying lemma 4.2 we obtain:

$$([\leq]B_1 \wedge \ldots \wedge [\leq]B_i) \wedge (\neg[\leq]C_1 \wedge \ldots \wedge \neg[\leq]C_j) \wedge$$
$$\wedge([\equiv]D_1 \wedge \ldots \wedge [\equiv]D_k) \wedge (\neg[\equiv]E_1 \wedge \ldots \wedge \neg[\equiv]E_l) \Rightarrow [\equiv]A \in L_2.$$

This implies that $[\equiv]A\in x$, which contradicts the assumption.■

<u>Lemma 4.4.</u> For $[\leq]A\in\Phi$ and $x\in U^*$:

$$[\leq]A\in x \text{ iff } (\forall y\in U^*)(|x|\leq|y| \longrightarrow A\in y).$$

<u>Proof.</u> The case $(\longrightarrow)$ is as in lemma 4.3.

$(\longleftarrow)$. Suppose $[\leq]A\notin x$. We have to show that there exists $y\in U^*$ such that $|x|\leq|y|$ and $A\notin y$. For that purpose it is sufficient to show that there exists a $y\in U^*$ satisfying the following conditions:

(i) $A\notin y$,

(ii) $\{[\leq]B/[\leq]B\in x\}\subseteq y$,

(iii) either (iiia) $\exists[\leq]C\in\Phi^+$ such that $[\leq]C\in y$ and $[\leq]C\notin x$,

or (iiib) $\{[\equiv]D/[\equiv]D\in x\}\subseteq y$.

<u>Case 1:</u> There is a $y\in U^*$ satisfying (i), (ii) and (iiib). Then the assertion holds.

<u>Case 2:</u> There is no $y\in U^*$ satisfying (i), (ii) and (iiib). In this case we shall prove the following assertions:

<u>Assertion 1:</u> $[\equiv]A\in x$.

<u>Assertion 2:</u> there is a $y\in U^*$ satisfying (i), (ii) and (iiia).

<u>Proof of the assertion 1.</u> If there is no $y\in U^*$ satisfying (i), (ii) and (iiib) then the following set of formulas is L_2-inconsistent:

$\{[\leq]B/[\leq]B\in x\}\cup\{[\equiv]D/[\equiv]D\in x\}\cup\{\neg A\}$.

Then for some formulas $[\leq]B_1,\dots,[\leq]B_i$, $[\equiv]D_1,\dots,[\equiv]D_j$ from x we have that the following formula is a theorem of L_2:

$$([\leq]B_1\wedge\dots\wedge[\leq]B_i)\wedge([\equiv]D_1\wedge\dots\wedge[\equiv]D_j) \Rightarrow A$$

Operating as in lemma 4.3 we obtain that $[\equiv]A\in x$.

<u>Proof of the assertion 2.</u> We take $C=[\equiv]A \Rightarrow [\leq]A$. Since $[\leq]A\in\Phi$ then $[\equiv]A$ is also in Φ and $[\equiv]A \Rightarrow [\leq]A$ is a Boolean combination of formulas from Φ. So $[\leq]C=[\leq]([\equiv]A \Rightarrow [\leq]A)\in\Phi^+$. Since $[\leq]A\notin x$ and $[\equiv]A\in x$ then we have that $[\equiv]A \Rightarrow[\leq]A\notin x$ and hence $[\leq]([\equiv]A \Rightarrow [\leq]A)\notin x$. So $[\leq]C\notin x$.

Now to prove the assertion 2 it is sufficient to show that the set $\Lambda=\{[\leq]B/[\leq]B\in x\}\cup\{[\leq]C\}\cup\{\neg A\}$ is L_2-consistent. If Λ is not L_2-consistent then for some formulas $[\leq]B_1,\dots,[\leq]B_n \in x$ the following formula is a theorem of L_2:

$$[\leq]B_1\wedge\dots\wedge[\leq]B_n\wedge[\leq]([\equiv]A \Rightarrow [\leq]A) \Rightarrow A. \quad \text{Then}$$

$$[\leq]B_1\wedge\dots\wedge[\leq]B_n \Rightarrow ([\leq]([\equiv]A \Rightarrow [\leq]A) \Rightarrow A)\in L_2.$$

Since $[\leq]$ is an S4 modality then we have

$$[\leq]B_1\wedge\dots\wedge[\leq]B_n \Rightarrow [\leq]([\leq]([\equiv]A \Rightarrow [\leq]A) \Rightarrow A)\in L_2.$$

Then by the axiome gGrz we obtain that $[\leq]B_1\wedge\dots\wedge[\leq]B_n \Rightarrow A \in L_2$. Again, since $[\leq]$ is an S4 modality we obtain that $[\leq]B_1\wedge\dots\wedge[\leq]B_n \Rightarrow [\leq]A \in L_2$. Since $[\leq]B_1,\dots,[\leq]B_n\in x$ we get $[\leq]A\in x$, cotrary to the assumption. This ends the proof of the lemma.■

Theorem 4.5. For any $A\in\Phi$ and $x\in U^*$: $|x| \Vdash_V A$ iff $A\in x$.

Proof. By induction on the complexity of the formula A and by using lemma 4.3 and lemma 4.4 for the cases when $A=[\equiv]B$ and $A=[\leq]B$.■

Now the proof of Theorem 2.1 can be completed in the following way. We have that $A_0\notin x_0$ for some $x_0\in U^*$. Then by theorem 4.5 we have that $|x_0| \nVdash_V A_0$, so A_0 is not true in the finite qoset $(U, \leq)$.■

Acknowledgements. The author wishes to thank the referees and T. Tinchev for his valuable remarks.

References

1 Segerberg, K. An Essay in Classical Modal Logic, Upsala, 1971.

2 Vakarelov, D. S4 and S5 together - S4+5, in the proc. of 8 Int. Congress of Logic, Methodology and Phil. of Science vol. 5, part 3 pp. 271, Moscow, 1987.

3 Hughes G & M. J. Cresswell, A Companion to Modal Logic, Methuen, London, 1984.

4 Gargov, G & K. Kirov, The Logic of "Strong Box" in IGL is IS4Grz, in Mathematics and Education of Mathematics, proc. of the 11 Spring Conf. of the Union of Bulgarian Mathematicians, April 82, Sofia, 1982.

5 Goranko, V. Definability and Completeness in Polymodal Logics, Ph.D. thesis, Sofia University, Faculty of Mathematics, Sofia, 1988.

6 Vakarelov D. Modal Logics for Knowledge Representation Systems, report in 6-th Symposium on Computation Theory, November 30 - December 4, 1987, Wendisch-Rietz, GDR.

7 Gargov, G, S. Passy & T. Tinchev, Modal Enviroment for Boolean Speculations, in: Mathematical Logic and its Applications, D. Skordev ed., Plenum Press, New York, 1987.

8 Humberstone, I. L. Inaccessible Worlds, Notre Dame Journal of Formal Logic, vol. 24(1983), pp 346-352.

9 Cresswell, M. J. A Henkin Completeness Theorem for T, Notre Dame Journal of Formal Logic, vol. 8(1967), pp. 187-190.

10 Sahlqvist, H, Completeness and Correspondence in the First and the second Order Semantics for Modal Logic, in: Kanger, S.(ed), Proc. of the Third Scandinavian Logic Symp. Amsterdam, North Holland, 1975, pp 110-143.

11 Cresswell, M. J. Private Letter, May 16 1988.

LEAST FIXED POINTS IN PREASSOCIATIVE COMBINATORY ALGEBRAS[*)]

J. Zashev

Laboratory for Applied Logic
Sofia University
Sofia, Bulgaria

The intention of the investigations to which the present work belongs is to find an axiomatic basis of the recursion theory for which the following qualities are desirable to possess: (1) to be algebraically styled; that means we expect to find a suitable algebraic language, which is simple and well working and allows us to obtain a connection between recursion theory and same algebraic structures which may be interesting by themselves; (2) to have as large as possible area of applications, i.e. we expect for those structures to have many and easy constructable models; (3) to allow us to prove all the basic facts of the recursion theory in the abstract case, especially the least fixed point or first recursion theorem.

There are two kinds of structures which seem to be very suitable for those purposes: partially ordered combinatory algebras (i.e. combinatory algebras [1], which are partially ordered a.s. in the sense of [2]): and combinatory spaces [3] and operative spaces [4]. In order to comprise some of the advantages of both some other structures were considered in [2,5,6,7] which are intermediate between those two kinds of structures. Here we shall prove the essential results anounced in [2], which are not directly derivable from the results of [6,7].

PRELIMINARY NOTATIONS

Let $\mathcal{F}$ be a p.o.a.s. in the terms of [2]. For all p.o.a.s. $\mathcal{F}$ in the present work we shall suppose that the least upper bound $\varphi\vee\psi$ exists in $\mathcal{F}$ for all φ, ψ in $\mathcal{F}$, and is distributive in the sense that

(1.1) $\varkappa(\varphi\vee\psi) = (\varkappa\varphi)\vee(\varkappa\psi)$ and $(\varphi\vee\psi)\varkappa = (\varphi\varkappa)\vee(\psi\varkappa)$

for all $\varphi, \psi, \varkappa$ in $\mathcal{F}$. Let $\mathcal{C}$ be a list of operations in $\mathcal{F}$. $\mathcal{C}$ may contain constants also, considered as 0-ary operations. Define $\mathcal{C}$-terms inductively as usual: the variables (x,y,z,...) are $\mathcal{C}$-terms: if t and s are $\mathcal{C}$-terms, then (ts) is a $\mathcal{C}$-term; if f is an n-ary operation in $\mathcal{F}$, $\bar{f}$ is

*Research partially supported by the Ministry of Culture Science and Education. Contract No 933 1988.

Mathematical Logic
Edited by P. P. Petkov
Plenum Press, New York, 1990

a symbol for it, and $t_1,\dots,t_n$ are $\mathscr{C}$-terms, then $\bar{f}(t_1,\dots,t_n)$ is a $\mathscr{C}$-term. An evaluation in $\mathscr{F}$ is a function $\theta: X \to \mathscr{F}$, where X is a finite set of variables. The evaluation θ is defined for a term t iff all variables in t belong to X. If θ is defined for t then the value $\tilde{\theta}(t)$ is defined inductively as follows: $\tilde{\theta}(t) = \theta(t)$ if t is a variable; $\tilde{\theta}((t_1 t_2)) = \tilde{\theta}(t_1)\tilde{\theta}(t_2)$ and $\tilde{\theta}(f(t_1,\dots,t_n)) = f(\tilde{\theta}(t_1),\dots,\tilde{\theta}(t_n))$ for n-ary $f \in \mathscr{C}$. If a term t is closed (does not contain variables) then by $\tilde{t}$ will be denoted the value $\tilde{\theta}(t)$ for any θ. An element $\varphi \in \mathscr{F}$ will be called $\mathscr{C}$-expressible, iff $\varphi = \tilde{t}$ for a suitable closed term t. Similarly an operation $g: \mathscr{F}^n \to \mathscr{F}$ will be called $\mathscr{C}$-expressible, iff there is a term t such that all variables in t are among $x_1,\dots,x_n$, and for every evaluation $\theta: \{x_1,\dots,x_n\} \to \mathscr{F}$ we have $g(\theta(x_1),\dots,\theta(x_n)) = \tilde{\theta}(t)$.

The notion of preassociative combinatory algebra was introduced in [2] (under the name of 'preassociative applicative system'). Here we shall use the corresponding notations from [2,6], especially the letters A, C and D will denote the constants in a preassociative combinatory algebra, satiafying

(1.2) $A\varphi\psi\varkappa = \varphi(\psi\varkappa)$,

and

(1.3) $C\varphi(D\psi\varkappa) = \varphi\psi\varkappa$

for all elements $\varphi, \psi, \varkappa$ of that algebra. The letter $\mathscr{F}$ will denote a p.o.a.s. (satisfying (1.1)), and 0 will be the least element of $\mathscr{F}$.

NATURAL PARTIALLY ORDERED PREASSOCIATIVE COMBINATORY ALGEBRAS

That notion was defined in [2] as a partially ordered preassociative combinatory algebra, in which five more constants $C', Z, Z', S, S' \in \mathscr{F}$ and a subset $\mathscr{N} \subseteq \mathscr{F}$ are given, such that for all $\varphi \in \mathscr{F}$ and $\nu \in \mathscr{N}$:

(2.1) $S\nu \in \mathscr{N}$, $Z \in \mathscr{N}$, and $Z \neq 0$;

(2.2) $C'\varphi(Z'Z) = \varphi$;

(2.3) $S'(S\nu) = \nu$;

and

(2.4) $Z'(S\nu) = S'Z = 0$.

That notion serves two porposes: to code the natural numbers in $\mathscr{F}$ and to imitate definitions by cases. Its role is analogous to the role of the branching operation in [6,7]. The code $\bar{n}$ of the natural number n may be defined inductively by $\bar{o} = Z$ and $\overline{n+1} = S\bar{n}$. Then $\bar{n} \in \mathscr{N}$ for all natural numbers n; and for all natural numbers n and m we have:

(2.5) $0\bar{n} = 0$;

(2.6) $\bar{n} \neq 0$;

and

(2.7) if $\bar{n} = \bar{m}$, then $n = m$.

Indeed, (2.5) follows from (2.4) since $0Z \leq S'Z = 0$ and $0\overline{n+1} \leq Z'(S\bar{n}) = 0$; (2.6) follows by induction on n: $\bar{o} = Z \neq 0$ by (2.1), and if $\overline{n+1} = 0$, then by (2.3) and (2.4) $\bar{n} = S'\overline{n+1} = S'0 \leq S'Z = 0$; (2.7) follows again by induction on n, since $\bar{o} = \overline{n+1}$ is impossible because otherwide $\bar{n} = S'\overline{n+1} = S'\bar{o} = S'Z = 0$ contrary to (2.6), and if $\overline{n+1} = \overline{m+1}$, then $\bar{n} = S'\overline{n+1} = S'\overline{m+1} = \bar{m}$.

We are not intended to consider examples of natural partially ordered preassociative combinatory algebras here, since examples deserve to be considered in detail and the principal purpose of the paper is to present the ideas of the proofs. But some remarks are to be made about the examples listed in [2] in order to show the connection with the classical recursion theory and some modern generalizations.

1) The usual recursion theory (the elementary theory of recursive enumerable sets) is comprised by the results of the present paper through the example 1 in [2] with the set of the natural numbers as M and any primitive recursive pairing as $\langle x,y \rangle$. This example allows us to comprise the positive existential induction in the sense of [8] as well.

2) The example 4 in [2] comprises in some sense the recursion theory in Scott's domains. Namely, Scott's domains [9] are complete f_o-spaces in the sense of Ershov [10], and the essential part of the model of the typed λ-calculus, defined by Scott in [9], (i.e. the part remaining after omiting cartesian products) can be shown to have an isomorphic copy in a suitable factorization of the algebra $\mathcal{F}$ in that example. The isomorphism preserves recursiveness in certain good sense [11]. That factorization can can be avoided by restricting $\mathcal{F}$ to the set of all ideals, i.e. elements $\varphi \in \mathcal{F}$, such that: (a) if $f,g \in C_r$, $f \in \varphi$ and $g \leq f$, then $g \in \varphi$; and (b) if $\psi \subseteq \varphi \cap C_r$ is right directed by $\leq$ and $f = \sup \varphi$, then $f \in \varphi$; and defining the application $\varphi\psi$ as the ideal produced by $\varphi\psi$ in the former sense.

3) Many examples are obtainable from the fact that semigroups are a special case of preassociative combinatory algebras, especially by using the ideas of Skordev [3].

ITERATIVE EXTENSIONS AND INDUCTIVE ELEMENTS

The notion of an extension $\mathcal{F}_1$ of $\mathcal{F}$ (and of an A-extension, etc.) was introduced in [2]. Here for the extensions of $\mathcal{F}$ we shall suppose also that the condition (1.1) is fulfilled in them. That will allow us to simplify the proof of theorem 1 below. However it is not necessary for the validity of the results. An extension $\mathcal{F}_1$ of $\mathcal{F}$ will be called iterative, iff the following conditions are fulfilled:

(a) $\mathcal{F}_1$ is closed under both left and right divisions and the countable infima (for the definitions see [2]);

(b) for all $\varphi,\psi\in\mathcal{F}$, $\varphi\vee\psi$ is the least upper bound of φ and ψ in $\mathcal{F}_1$; and

(c) for all $\varphi,\psi,\varkappa\in\mathcal{F}$ the system of inequalities

(3.1) $\varphi \le \xi\ ,\ \psi\xi\varkappa \le \xi$

has a least solution $\mathbf{I}(\varphi,\psi,\varkappa)$ with respect to ξ in $\mathcal{F}_1$, which belongs to $\mathcal{F}$.

The role of the notion of iterative extension is analogous to the role of the notion of iterativity of a combinatory space [3] or an operative space [4], and to principles like Scott's μ-induction rule. Some general criteria of existence of iterative extensions may be proved which allow us to obtain such extensions in all the examples mentioned in [2] and others. We shall not consider them here, since their formulations and proofs are quite similar to those given in [6].

Till the end of the paper we shal fix a list $\mathcal{C}$ of (symbols for) several constants in $\mathcal{F}$ and we shall suppose that 0 belongs to that list.

Definition. .A system of $\mathcal{C}$-inequalities is an expression of the form

(3.2) $(s_1\le x_1)(s_2\le x_2)\dots(s_n\le x_n)$,

where $s_1,\dots,s_n$ are $\mathcal{C}$-terms, and the variables $x_1,\dots,x_n$ are not necessarily distinct. An evaluation $\theta: X\to\mathcal{F}$ is a solution of a system $\mathcal{S}$ of the form (3.2), iff it is defined for $\mathcal{S}$, i.e. X contains all the variables in $\mathcal{S}$, and for each $i\in\{1,\dots,n\}$, $\tilde\theta(s_i) \le \tilde\theta(x_i)$. A solution θ of (3.2) is $\mathcal{F}_1$-minimal, iff for every solution $\theta': X'\to\mathcal{F}_1$ of (3.2) we have $x\in X'$ and $\theta(x) \le \theta'(x)$ for all $x\in X$, where $\mathcal{F}_1$ is an extension of $\mathcal{F}$. An element $\varphi\in\mathcal{F}$ will be called $\mathcal{F}_1$-inductive in $\mathcal{C}$, iff there is a system $\mathcal{S}$ of $\mathcal{C}$-inequalities and an $\mathcal{F}_1$-minimal solution θ of $\mathcal{S}$, such that $\varphi = \theta(x)$ for suitable variable x.

Note that all $\mathcal{F}_1$-inductive in $\mathcal{C}$ elements are definable by such systems of the form (3.2) , which are canonical in the following sense: each variable in the system occures in the list $x_1,\dots,x_n$ at least once and at most twice:

(3.3) for every system $\mathcal{S}$ there is a canonical system $\mathcal{S}'$, such that all variables in $\mathcal{S}$ occur in $\mathcal{S}'$, and and if θ' is an $\mathcal{F}_1$-minimal solution of $\mathcal{S}'$, then the restriction θ of θ' on the variables in $\mathcal{S}$ is an $\mathcal{F}_1$-minimal solution of $\mathcal{S}$.

Indeed, we may suppose that each variable in $\mathcal{S}$ occurs as a right side of an inequality since otherwise we may add some inequalities of the form $(x \le x)$ without changing the solutions of the system. Then suppose the system $\mathcal{S}$ has the form $(t_1 \le x)(t_2 \le x)\mathcal{S}_1$, where x may occur in the system $\mathcal{S}_1$ as a right side of an inequality. Taking a variable y non occuring in $\mathcal{S}$, define a system $\mathcal{S}_2$ as $(y \le x)(t_1 \le y)(t_2 \le y)\mathcal{S}_1$. It is clear that if θ_2 is an $\mathcal{F}_1$-minimal solution of $\mathcal{S}_2$, then the restriction of θ_2 upon the variables of $\mathcal{S}$ wiil be an $\mathcal{F}_1$-minimal solution of $\mathcal{S}$. Repeating that process of adding new variables at last we shall reach a canonical system.

THE FIRST RECURSION THEOREM

Now we shall fix a canonical system $\mathcal{S}$ of the form (3.2). If $x = x_i$ where $i \in \{1,\dots,n\}$, but $x \notin \{x_1,\dots,x_{i-1}\}$, then s_i will be denoted below by $\mathcal{S}_1(x)$. Similarly if $x = x_i$ and $x \notin \{x_{i+1},\dots,x_n\}$, then define $\mathcal{S}_2(x) = s_i$. Define also $\mathcal{S}_0(x)$ as the symbol for the constant 0. For each $\mathcal{E}$-term t whose variables occur in $\mathcal{S}$ and each $i<3$ let $r_i(t)$ be the result of the replacement of the right-most occurence of a variable (suppose x) in t with $\mathcal{S}_i(x)$. If t is closed then define $r_i(t)$ as t. By $\tilde{0}(t)$ we shall denote the value $\tilde{\theta}_0(t)$, where $\theta_0(t) = 0$ for all variables x in t.

<u>Definition</u>. Let $\mathcal{T}$ be the set of all $\mathcal{E}$-terms whose variables are in $\mathcal{S}$. Let $\mathbb{A}$ be a function $\mathbb{A}: \mathcal{T} \to \mathcal{F}$, $\rho_0, \rho_1^0, \rho_1^1, \rho_1^2 \in \mathcal{F}$, and for all $t \in \mathcal{T}$:

(4.1) $0\mathbb{A}(t) = 0$;

(4.2) $\rho_0 \mathbb{A}(t) = \tilde{0}(t)$;

and

(4.3) $\rho_1^i \mathbb{A}(t) = \mathbb{A}(r_i(t))$ for each $i<3$.

Then $\mathbb{A}$ willl be called a code function for $\mathcal{S}$ with reading elements ρ_0, ρ_1^0, ρ_1^1, ρ_1^2 .

<u>Theorem</u> 1. Let $\mathcal{F}$ be a p.o.a.s. with a constant A satisfying (1.2) and with an A-extension $\mathcal{F}_1$, which satisfies (a) and (b) above. Suppose there is a code function $\mathbb{A}$ in $\mathcal{F}$ for $\mathcal{S}$ with reading elements ρ_0, ρ_1^0, ρ_1^1, ρ_1^2, and define $\rho_1 = \sup_{i<3} \rho_1^i$. Then if $\omega \in \mathcal{F}$ is a least solution of the system

(4.4) $\rho_0 \le \xi$, $A\xi\rho_1 \le \xi$

with respect to ξ in $\mathcal{F}_1$, then the evaluation θ, defined by $\theta(x) = \mathbb{A}(x)$ for each variable x in $\mathcal{S}$ is a $\mathcal{F}_1$-minimal solution of $\mathcal{S}$.

<u>Proof.</u> Till the end of the proof we shall suppose $t, s \in \mathcal{T}$. Since ω is a solution of (4.4), we have

(4.5) $\tilde{0}(t) = \rho_0 \mathbb{A}(t) \le \omega\mathbb{A}(t)$,

and

(4.6) $\omega k(r_i(t)) = \omega(\rho_1^i k(t)) \le \omega(\rho_1 k(t)) = A\omega\rho_1 k(t) \le \omega k(t)$

for all t and i<3. We shall show that if s is closed, then

(4.7) $\omega k(t)\tilde{s} \le \omega k(ts)$.

Fix s. Then using (a) we may define an element $\theta \in \mathcal{F}_1$, such that

$\theta k(t)\tilde{s} \le \omega k(ts)$ for all t,

and θ is the greatest element of $\mathcal{F}_1$ with the last property. Indeed, if $\eta_t = \max\{\eta\in\mathcal{F}_1 : \eta\tilde{s}\le\omega k(ts)\}$, and $\theta_t = \max\{\eta\in\mathcal{F}_1 : \eta k(t)\le\eta_t\}$, then we may define $\theta = \inf\{\theta_t : t\in\mathcal{T}\}$. To justify that definition we must know also that $0\tilde{s} \le \omega k(ts)$ and $0k(t) \le \eta_t$. The last follows from (4.1), and by (4.5)

$\tilde{0}\tilde{s} \le \tilde{0}(t)\tilde{s} = \tilde{0}(ts) \le \omega k(ts)$.

Now we shall show that θ satisfies (4.4): $\rho_0 k(t)\tilde{s} = \tilde{0}(t)\tilde{s} \le \omega k(ts)$, and by the definition of θ we have $\rho_0 \le \theta$; since $\mathcal{F}_1$ is an A-extension,

$$A\theta\rho_1 k(t)\tilde{s} = \theta(\rho_1 k(t))\tilde{s} = \sup_{i<3}\theta(\rho_1^i k(t))\tilde{s} = \sup_{i<3}\theta k(r_i(t))\tilde{s} \le \omega k(ts),$$

whence $A\theta\rho_1 \le \theta$. Since ω is $\mathcal{F}_1$-minimal solution of (4.4), $\omega \le \theta$ and by the definition of θ we obtain (4.7). Next we shall show that

(4.8) $\omega k(t)(\omega k(s)) \le \omega k(ts)$.

Fix t and define θ' as the greatest element of $\mathcal{F}_1$ with the property: $\omega k(t)(\theta' k(s)) \le \omega k(ts)$ for all $s\in\mathcal{T}$. To justify the definition we must know that $\omega k(t)(0k(s)) \le \omega k(ts)$. But by (4.1), (4.7) and (4.6)

$\omega k(t)(0k(s)) \le \omega k(t)\tilde{0}(s) = \omega k(t)\tilde{s}_0 \le \omega k(ts_0) \le \omega k(ts)$,

where s_0 is the result of substitution of (the symbol for) 0 for each variable in s. That shows also that $\rho_0 \le \theta'$:

$\omega k(t)(\rho_0 k(s)) = \omega k(t)\tilde{0}(s) \le \omega k(ts)$.

And by (4.6)

$$\omega k(t)(A\theta'\rho_1 k(s)) = \sup_{i<3}\omega k(t)(\theta'(\rho_1^i k(s)) = \sup_{i<3}\omega k(t)(\theta' k(r_i(s)))$$

$$\le \sup_{i<3}\omega k(tr_i(s)) = \begin{cases} \sup_{i<3}\omega k(ts) , & \text{if s is closed} \\ \sup_{i<3}\omega k(r_i(ts)) & \text{otherwise} \end{cases} \le \omega k(ts).$$

Therefore θ' is a solution of (4.4) and $\omega \le \theta'$, whence by the definition of θ' we obtain (4.8). Then by induction on t we see that $\Theta(t) \le \omega k(t)$. The basis of that induction is by the definition of Θ and (4.5), and the induction step is immediate from (4.8). That shows that Θ is a solution of $\mathcal{S}$ since if $(s \le x)$ is an inequality in $\mathcal{S}$, then $s = r_i(x)$ for i=1 or i=2, and by (4.6) $\tilde{\Theta}(s) \le \omega k(s) = \omega k(r_i(x)) \le \omega k(x) = \Theta(x)$.

To show that Θ is an $\mathcal{F}_1$-minimal solution of $\mathcal{S}$ consider an arbitrary solution Θ' of $\mathcal{S}$ in $\mathcal{F}_1$ and define $\theta''\in\mathcal{F}_1$ as the greatest element of $\mathcal{F}_1$ satisfying the condition: $\theta'' k(t) \le \Theta'(t)$ for all t. The definition is correct by (4.1). And obviously $\rho_0 k(t) = \tilde{0}(t) \le \tilde{\Theta}'(t)$ whence $\rho_0 \le \theta''$.

Since θ' ia a solution of $\mathcal{S}$, then $\tilde{\theta}'(r_i(t)) \le \tilde{\theta}'(t)$ for each $i<3$. Therefore

$$A\theta''\rho_1\mathbb{A}(t) = \sup_{i<3}\theta''(\rho_1^i\mathbb{A}(t)) = \sup_{i<3}\theta''\mathbb{A}(r_i(t)) \le \sup_{i<3}\tilde{\theta}'(r_i(t)) \le \tilde{\theta}(t),$$

whence $A\theta''\rho_1 \le \theta''$, and θ'' is a solution of (4.4). Then $\omega \le \theta''$, whence $\omega\mathbb{A}(t) \le \tilde{\theta}'(t)$. That shows that θ is $\mathcal{F}_1$-minimal, since for each variable x in $\mathcal{S}$ we have $\theta(x) = \omega\mathbb{A}(x) \le \tilde{\theta}'(x) = \theta'(x)$. The theorem is proved.

Now we shall use some results and definitions from [7], section 1. Define $\mathcal{C}' = \{A, C, D, C', Z, Z', S, S'\}$.

Lemma 1. Let $\mathcal{F}$ be a natural partially ordered preassociative combinatory algebra with an iterative extension $\mathcal{F}_1$ and $\varphi 0 = 0$ for all $\varphi\in\mathcal{F}$. Then there are primitive recursive branching $\mathbb{R}_0$, a primitive recursive iteration $\mathbb{R}_1$, and a special primtive recursion $\mathbb{R}_1^*$, which are $\mathcal{C}'\cup\{\vee, \mathbb{I}\}$-expressible.

Proof. Let $\mathbb{R}_0(\varphi,\psi) = (A(C'\varphi)Z')\vee(A\psi S')$. Then by the definitions $\mathbb{R}_0(\varphi,\psi)\bar{0} = (C'\varphi(Z'\bar{0}))\vee(\psi(S'\bar{0})) = \varphi\vee(\psi 0) = \varphi$, and

$$\mathbb{R}_0(\varphi,\psi)\overline{n+1} = (C'\varphi(Z\overline{n+1}))\vee(\psi(S'\overline{n+1})) = (C'\varphi 0)\vee(\psi\bar{n}) = \psi\bar{n}.$$

Define $\mathbb{R}_1(\varphi,\psi) = \mathbb{I}(A(C'\varphi)Z', AA(A\psi), S')$. Then $\mathbb{R}_1(\varphi,\psi)$ is the least solution in $\mathcal{F}$ of the inequality $\mathbb{R}_0(\varphi, A\psi\xi) \le \xi$ with respect to ξ. Therefore $\mathbb{R}_0(\varphi, A\psi\mathbb{R}_1(\varphi,\psi)) = \mathbb{R}_1(\varphi,\psi)$ whence $\mathbb{R}_1(\varphi,\psi)\bar{o} = \varphi$, and $\mathbb{R}_1(\varphi,\psi)n+1 = A\psi\mathbb{R}_1(\varphi,\psi)n = \psi(\mathbb{R}_1(\varphi,\psi)n)$. The proof of the existence of a special primitive recursion in $\mathcal{F}$ is the same as in the proof of proposition 4 in [7].

Lemma 2. Under the suppositions of Lemma 1, there are $\mathcal{C}'\cup\{\vee,\mathbb{I}\}$-expressible elements ρ_0, ρ_1^0, ρ_1^1, ρ_1^2 of $\mathcal{F}$ and a function $\mathbb{A}(\mathcal{S},t)$ defined for all canonical systems $\mathcal{S}$ and $\mathcal{C}$-terms t without other variables than those in $\mathcal{S}$, such that for every fixed $\mathcal{S}$ the function $\lambda t.\mathbb{A}(\mathcal{S},t)$ is a code function for $\mathcal{S}$ in $\mathcal{F}$ with reading elements ρ_0, ρ_1^0, ρ_1^1, ρ_1^2.

Proof. By Lemma 1 there are primitive recursive branching $\mathbb{R}_0$, iteration $\mathbb{R}_1$ and special recursion $\mathbb{R}_1^*$ in $\mathcal{F}$, which are $\mathcal{C}'\cup\{\vee,\mathbb{I}\}$-expressible. We may define a numeration of the $\mathcal{C}$-terms t and the systems $\mathcal{S}$ of $\mathcal{C}$-inequalities in an usual way [12]. Denote the numbers of t and $\mathcal{S}$ by $\ulcorner t\urcorner$ and $\ulcorner\mathcal{S}\urcorner$ respectively, and define $\mathbb{A}(\mathcal{S},t) = \langle\ulcorner\mathcal{S}\urcorner, \ulcorner t\urcorner\rangle$, where $\langle n,m\rangle$ is a primitive recursive pairing of the natural numbers. Then there is a primitive recursive function f_0, such that $f_0(\langle\ulcorner\mathcal{S}\urcorner, \ulcorner t\urcorner\rangle) = \ulcorner t_0\urcorner$, where t_0 is the result of the substitution of the symbol for 0 for all the variables in t. By theorem 1 in [7] we have a $\mathcal{C}'$-expressible element $\varphi_0\in\mathcal{F}$ such that $\varphi_0\bar{n} = \overline{f_0(n)}$ for all n. Then by proposition 1 in [7] there is a $\mathcal{C}'\cup\{\mathbb{R}_0, \mathbb{R}_1, \mathbb{R}_1^*\}$-expressible element $\tau\in\mathcal{F}$, such that $\tau\overline{\ulcorner t\urcorner} = \tilde{t}$ for every closed term t. Therefore

$$\tilde{O}(t) = \tilde{t}_o = \tau\ulcorner \overline{t_o}\urcorner = \tau f_o(\langle \ulcorner \mathscr{S}\urcorner, \ulcorner t\urcorner\rangle) = \tau(\varphi_o \mathbin{\&}(\mathscr{S},t)) = A\tau\varphi_o \mathbin{\&}(\mathscr{S},t),$$

and we may define $\rho_o = A\tau\varphi_o$. The existence of the elements ρ_1^o, ρ_1^1 and ρ_1^2 follows from the representation of the primitive recursive functions (theorem 1 in [7]), since there are primitive recursive functions f_1^o, f_1^1 and f_1^2 such that $f_1^i(\langle \ulcorner \mathscr{S}\urcorner, \ulcorner t\urcorner\rangle) = \langle \ulcorner \mathscr{S}\urcorner, \ulcorner r_i(t)\urcorner\rangle$ for each $i<3$.

By Theorem 1, Lemma 2, and the remark (3.3) we obtain

<u>Theorem</u> 2. Let $\mathscr{F}$ be a natural partially ordered preassociative combinatory algebra with an iterative A-extension $\mathscr{F}_1$, in which $\varphi 0=0$ for all $\varphi\in\mathscr{F}$. Then every system of $\mathscr{C}$-inequalities has an $\mathscr{F}_1$-minimal solution, whose members are $\mathscr{C}'\cup\{\vee,\mathbb{O}\}$-expressible elements of $\mathscr{F}$.

Therefore every $\mathscr{F}_1$-inductive in $\mathscr{C}$ element of $\mathscr{F}$ is $\mathscr{C}'\cup\{\vee,\mathbb{O}\}$-expressible. The parametrisation theorem (proposition 3 in [2]) is easily obtainable by Theorem 1 and Lemma 2 as well.

Finally we shall consider one more variant of the Theorem 2 in which the condition $\varphi 0=0$ is replaced by another one, which is often fulfilled in the examples.

<u>Theorem</u> 3. Let $\mathscr{F}$ be a *-natural AC-p.o.a.s. [2], and let $\mathscr{F}_1$ be an A-extension of $\mathscr{F}$. Then every system of $\mathscr{C}$-inequalities has a $\mathscr{F}_1$-minimal solution in $\mathscr{F}$, whose members are $\mathscr{C}\cup\{\vee,\mathbb{O}'\}$-expressible, where $\mathbb{O}'(\varphi,\psi)$ is the (operation giving the) least solution of the system $\varphi\le\xi$, $\psi\xi\le\xi$ with respect to ξ.

The proof differs from that of the theorem 2 only in matter of coding natural numbers and defining the operation $\mathbb{R}_o$. Namely, the presupposition of *-naturality means that there are constants T,F,T',F' in $\mathscr{F}$, such that $T'T\varphi = F'F\varphi = \varphi$, $T'F = F'T = 0 = 0\varphi$, and $\mathscr{F}$ is untrivial, i.e. there are at least two different elements in $\mathscr{F}$. Since $\mathscr{F}$ is an AC-p.o.a.s., the combinator C in (1.3) may be supposed to satisfy $C\varphi\psi = \psi\varphi$ (and D may be defined as A(C(AAC))(AA(AAC))). Then the elements defined by $Z = DTT$, $S = DF$, $Z' = C(A(CT')T')$, $S' = CF'$, and $C' = C$ are directly chackable to satisfy (1.3), (2.2) – (2.4); $Z \ne 0$ since otherwise $\varphi = C\varphi(Z'Z) = Z'Z\varphi = Z'0\varphi \le Z'(SZ) = 0\varphi = 0$ and $\mathscr{F}$ would be trivial; and (2.1) is obvious with $\mathscr{N}=\mathscr{F}$. Therefore natural numbers may be coded as before. Then defining $P(\varphi)= C(A(DT'\varphi)C)$ and $Q(\varphi) = C(A(DF'\varphi)C)$ we have $P(\varphi)(DT\varkappa) = A(DT'\varphi)CT\varkappa = DT'\varphi(CT)\varkappa = CTT'\varphi\varkappa = T'T\varphi\varkappa = \varphi\varkappa$, and similarly $Q(\varphi)(DF\varkappa) = \varphi\varkappa$ and $P(\varphi)(DF\varkappa) = Q(\varphi)(DT\varkappa) = 0$, whence $P(A(DT'\varphi)C)(DTT) = A(DT'\varphi)CT = \varphi$, and defining $\mathbb{R}_o(\varphi,\psi) = P(A(DT'\varphi)C) \vee Q(\psi)$ we see immediately that $\mathbb{R}_o$ is a primitive recursive branching operation.

REFERENCES

1. H. P. Barendregt, "The Lambda Calculus. Its Syntax and Semantics", North-Holland, Amsterdam (1981).
2. J. A. Zashev, Basic recursion theory in partially ordered models of some fragments of the combinatory logic, C. R.Acad. bulg. Sci., 37:561 (1984).
3. D. G. Skordev, "Combinatory spaces and recursiveness in them" (Russian), BAN, Sofia (1980).
4. L. L. Ivanov,. "Algebraic Recursion Theory", Ellis Horwood, Chichester (1986).
5. J. A. Zashev, Abstract Plotkin's models, in: "Mathematical theory of programming", A. P. Ershov, D.G.Skordev, ed., Novosibirsk (1985).
6. J. A. Zashev, B-combinatory algebras, Serdica Bulg. math. publ., 12:225 (1986).
7. J. A. Zashev. Recursion theory in B-combinatory algebras (Russian), Serdica Bulg. math. publ., 13:210 (1987).
8. P. Aczel, An introduction to inductive definitions, in: "Handbook of Mathematical Logic", J. Barwise, ed., North Holland, Amsterdam (1977).
9. D. S. Scott, Lectures on a mathematical theory of computation, in: "Theoritical Foundations of Programming Methodology", M. Broy and G. Schmidt, ed., D. Reidel Publishing Co, Dordreht (1982).
10. J. L. Ershov, "Theory of numerations" (Russian), Nauka, Moskow (1977).
11. J. A. Zashev, "Recursion theory in partially ordered combinatory models" (Bulgarian), Thesis, Sofia University (1984).
12. S. C. Kleene, "Introduction to Metamathematics", North-Holland, Amsterdam (1952).

PARTICIPANTS, CONTRIBUTORS AND PROGRAMME COMMITTEE MEMBERS

V. Michele Abrusci
Dipart. di Scienze Filosofiche
Universita di Bari
Palazzo Ateneo
70121 Bari
ITALY

Moni Almalech
ul. Lakatica 1
Sofia 4
BULGARIA

Victor Andonov
Inst. for Appl. Math. & Comp. Sci.
P.O.Box 384
Sofia 1000
BULGARIA

Kalin Angelov Angelov
c\o Gorchevi
j.k. Slatina bl 7, vh. 3, ap. #45
Sofia
BULGARIA

Piotr Borowik
Institute of Mathematics
Pedagogical Univ. of Czestochowa
42-201 Czestochowa
POLAND

Melkana Brakalova
Institute of Mathematics
Bulg. Acad. of Scie.
PO Box 373
1090 Sofia
BULGARIA

Douglas Bridges
The University of Buckingham
Buckingham, MK18 1EG
ENGLAND

Anatoly Buda
Sector of Mathematical Logic
Faculty of Math. & Inf.
boul. Anton Ivanov 5
Sofia 1126
BULGARIA

Alexandr Chagrov
Department of Mathematics
Kalinin State University
Zhelyabov str. 33
Kalinin 170013
USSR

Lilia Chagrova
Department of Mathematics
Kalinin State University
Zhelyabov str. 33
Kalinin 170013
USSR

Ivan Chernev
"ITKR" - BAN
boul. "Akademik Bonchev" bl. 29
Sofia
BULGARIA

Elena Chornaya
ul. Sahalinskaja
d. 6, korp. 1 kv. 34
107065 Moscow
USSR

Barry Cooper
School of Mathematics
University of Leeds
Leeds LS2 9JT
ENGLAND

Catherine Copestake
Department of Pure Mathematics
The University of Leeds
Leeds, LS2 9JT
ENGLAND

Giovanna Corsi
Dipartamento di Filosofia
via Bolognese 52
50139 Firenze
ITALY

M. J. Cresswell
Dept. of Philosophy
Victoria Univ. of Wellington
Private Bag
Wellington
NEW ZEALAND

Johan J. de Iongh
Math. Inst.
Catholic University
Driehuierweg 200
Nijmegen
THE NETHERLANDS

Dick H.J. de Jongh
Mathematisch Instituut
Universiteit van Amsterdam
Roetersstraat 15
1018 WB Amsterdam
THE NETHERLANDS

Osvald Demuth
Na vrcholu 17
130 00 Praha 3
CZECHOSLOVAKIA

Justus Diller
Institut fur Mathematische
Logik und Grundlagenforschung
Westfalische Wilhelms-Universitat
Roxeler Strase 64
4400 Munster (Westf.)
BRD

Nikolai Dimitrov
jk. Ljulin 7
ul. 714 bl 725 vh V ap #54
Sofia
BULGARIA

Angel Ditchev
Laboratory for Applied Logic
University of Sofia
boul. Anton Ivanov 5
Sofia 1126
BULGARIA

Dimiter Dimitrov Dobrev
komp. Iztok bl. 51
Sofia
BULGARIA

Kosta Dosen
Matematicki Institut
Knez-Mihailova 35
11000 Beograd
YUGOSLAVIA

Albert Dragalin
Mathematical Institute
Debrecen University
4010 Debrecen
HUNGARY

Rumjana Draganova
ul. Akad. Metodi Popov, 9
1113 Sofia
BULGARIA

Wojciech Dzik
Inst. Matematyki
Uniwersytet Slaski
ul. Bankowa 14
40-007 Katowice
POLAND

Don Faust
Dept. of Math. & Computer Science
Northern Michigan University
Marquette, Michigan 49855-5340
USA

Pavel Filipec
Inst. of Economics
CSAV, 1 odbor
Politickych veznu 7
111 73 Prague 1
CZECHOSLOVAKIA

Kit Fine
Dept. of Philosophy
U. C. L. A.
405 Hilgard Avenue
Los Angeles, CA 90024
USA

Robin O. Gandy
9 Squitchey Lane
Summertown
Oxford OX2 7LD
ENGLAND

Alexandra Gapik
Szk. Pods. Nr 48
ul. Schillera 5
42-224 Czestowa
POLAND

George Gargov
Linguistic Modelling Laboratory
CICT - BAS
25a Acad. G. Bonchev Str.
1113 Sofia
BULGARIA

Nadezhda Georgieva
Mladost I, 16, vh. E
1199 Sofia
BULGARIA

Adriana Georgieva
Inst. for Appl. Math. & Comp. Sci.
P.O.Box 384
Sofia 1000
BULGARIA

Silvio Ghilardi
via Balestra 5
24100 Bergamo
ITALY

Jan Goossenaerts
Dept. of Computer Science
K. U. Leuven
Celestijnenlaan 200 A
B-3030 Heverlee
BELGIUM

Valentin Goranko
Sector of Math. logic
Faculty of Math.
boul. A. Ivanov 5
Sofia 1126
BULGARIA

Peter Goring
Section Mathematik
Humboldt Universitat
PF 1297
Berlin 1086
DDR

J. Daniel Halpern
S 416
1060 Continentals Way
Belmont, CA 94002
USA

David Harel
Department of Applied Mathematics
The Weizmann Institute of Science
Rehovot 76100
ISRAEL

Susumu Hayashi
Res. Inst. for Math. Sci.
Oiwake-cho
Kitashirakawa
Kyoto 606
JAPAN

Shamil Ishmukhametov
Dept. of Mathematics
Uljanovskij Polytechnical Inst.
ul. Severny Venez, 32
432700 Uljanovsk
USSR

Lyubomir Ivanov
Sector of Math. Logic
Faculty of Math. & Inf.
bul. A. Ivanov 5
1126 Sofia
BULGARIA

Mitko Janchev
bul. D. Petkov, 21
bl. 1, vh. A, ap. 64
1309 Sofia
BULGARIA

Vladimir Jotsov
bul. K. Gotvald 56
1504 Sofia
BULGARIA

M. Kanovich
Kalinin State University
ul. Zheljabova 33
170 013 Kalinin
USSR

Alexander S. Kechris
Department of Mathematics 253-37
California Institute of Technology
Pasadena
California 91125
U.S.A.

Valery Khakhanian
Chair "Applicated math. II"
Moscow Institute of
Rail-Transportation Engineering
Obraztsova str. 15
103055 Moscow
USSR

Krassimir Kirov
Corporation SPS
compl. Bukston 207 A
Sofia
BULGARIA

Pertti Koivisto
Univ. of Tampere
Dept. of Math. Sciences
P.O.Box 607, SF-33101
Tampere
FINLAND

Phokion G. Kolaitis
Department of Computer Science
Building 460
Stanford University
Stanford, CA 94305
U.S.A.

Nikolaj Kotov
ul. Kiril Pchelinski, bl. 1 ap. 32
Sofia 1345
BULGARIA

Vladislav Ya. Kreinovich
PO Box 761
197022 Leningrad 22
USSR

Lill Kristiansen
Institute of Mathematics
P.O.Box 1053-Blindern
0316 Oslo 3
NORWAY

Natalia Kurtonina
Shvernika d. 12, k. 3, kv. 47
117449 Moscow
USSR

Boris Kushner
Computing Centre
AS of the USSR
Vavilova str. 40
Moscow 117333
USSR

Azriel Levy
Inst. of Math. and Comp. Sci.
Hebrew University of Jerusalem
Givat Ram
91904 Jerusalem
ISRAEL

Vladimir Lifschitz
Department of Computer Science
Building 460
Stanford University
Stanford, California 94305-2095
USA

Fu-tseng Liu
Department of Philosophy
National Taiwan University
Taipei
TAIWAN
REPUBLIC OF CHINA

Irina Lomazova
Program Systems Institute
AS of the USSR
Pereslavl-Zalesskii 152140
USSR

Rumen Lozanov
Mladost 4, bl. 428, vh. A, ap. 16
Sofia 1715
BULGARIA

Krassimira Lozanova
Mladost 4, bl. 428, vh. A, ap. 16
Sofia 1715
BULGARIA

Vasilij A. Lyubetsky
Rublevskoe shose 42
korpus 2,kv. 164
Moscow 121609
USSR

Andreana Madguerova
Tolbukhin 24
1000 Sofia
BULGARIA

Larisa Maksimova
Institute of Mathematics
Sib. Div. of AS of the USSR
Universitetsky Avenue 4
Novosibirsk 630090
USSR

Sergej Mardaev
Institute of Mathematics
Sib. Div. of AS of the USSR
Universitetsky Avenue 4
Novosibirsk 630090
USSR

Christoph Meinel
Institut fur Mathematik
Akademie der Vissenschaften
1086 Berlin
PE 1304
DDR

Mirko Mihaljinec
Matematicki Odjel PMF
Marulicev TRG 19/1
41000 Zagreb
YUGOSLAVIA

Grigorii Mints
Institute of Cybernetics
Estonian Academy of Sciences
Akadeemia tee 21
200 108 Tallinn
USSR

Marion Mircheva
Inst. of Philosophy, BAS
blvd. P. Evtimii 6
Sofia
BULGARIA

Albert A. Muchnik
Institute of Applied Math.
AS of the USSR
Miusskaya Sq. 4
Moscow 125047
USSR

A. Muravitski
Inst. of Math. with Comp. Centre
ul. Grosula 5
Kishinev 277028
USSR

Valerij Nepomniaschy
Computer Center
SD of AS of the USSR
pr. Lavrentieva 6
Novosibirsk 630090
USSR

Stela Nikolova
Sector of Math. Logic
Faculty of Math. & Inf.
boul. A. Ivanov 5
1126 Sofia
BULGARIA

Lilly Nikolova
ul. Lazar Stanev 25
bl. 245, vh. A, ap. 3
Sofia 1113
BULGARIA

Hirokazu Nishimura
Institute of Mathematics
University of Tsukuba
Sakura-Mura
Niihari-gun Ibaraki, 305
JAPAN

Dag Normann
Department of Mathematics
University of Oslo
P.O.Box 1053
Blindern
0316 Oslo 3
NORWAY

Hiroakira Ono
Faculty of Integrated
Arts and Sciences
Hiroshima University
Hiroshima 730
JAPAN

Ewa Orlowska
Institute of Computer Science
Polish Academy of Sciences
P.O.Box 22
00-901 Warsaw
POLAND

Solomon Passy
Sector of Math. Logic
Faculty of Math. & Inf.
boul. A. Ivanov 5
1126 Sofia
BULGARIA

Wojciech Penczek
Institute of Computer Science
Polish Academy of Sciences
PKiN, P.O. Box 22
Warsaw
POLAND

Petio Petkov
Sector of Math. Logic
Faculty of Math. & Inf.
boul. A. Ivanov 5
1126 Sofia
BULGARIA

Assen Petkov
Opalchenska 78
1303 Sofia
BULGARIA

Veselin Petrov
Studentski grad, 19 bl., 704
1100 Sofia
BULGARIA

Karl M. Podnieks
Computing Centre
Latvian State University
VC LGU
bul. Raina 19
Riga 226250
USSR

Irina Popova-Atanasova
ul. Borjana 23
Sofia 1618
BULGARIA

Gong Qirong
Philosophy Department
Guizhou University
Guiyang
Guizhou Province
PEOPLE'S REPUBLIC OF CHINA

Slavjan Radev
Sector of Math. Logic
Faculty of Math. & Inf.
boul. A. Ivanov 5
1126 Sofia
BULGARIA

Ivan Raitchev
blvd. Lenin, bl. 6, ap. 9
4700 Smolian
BULGARIA

Vladimir Rybakov
Kafedra of Algebra
Krasnojarsk State University
pr. Svobodnyi 79
Krasnojarsk 660062
USSR

Gerald E. Sacks
Harvard University
Department of Mathematics
Science Center
One Oxford Street
Cambridge, Massachusetts 02138
USA

J. Sakalauskaite
Inst. of Math. & Cybernetics
Acad. of Sci. of Lithuanian SSR
Akademijos str. 4
232 021 Vilnius
Lithuanian SSR
USSR

V. Sazonov
Institute of Programme Systems
AS of the USSR
152140 Pereslavl-Zalessky
USSR

Kajetan Seper
Katedra za matematicu
Faculty of Mechanical Engineering
University of Osijek
Trg m. Tita 18
55000 Slavonski Brod
YUGOSLAVIA

J. C. Shepherdson
School of Mathematics
University of Bristol
University Walk
Bristol, BS8 1TW
ENGLAND

Dimiter Skordev
Sector of Math. Logic
Faculty of Math. & Inf.
boul. A. Ivanov 5
1126 Sofia
BULGARIA

Craig Smorynski
State University of Utrecht
Mathematical Institute
Budapestlaan 6
P.O.Box 80.010
3508 TA Utrecht
THE NETHERLANDS

Robert I. Soare
Department of Mathematics
University of Chicago
5734 University Avenue
Chicago, IL 60637
USA

Kornel Solt
Kinizsi u. 22.IV.5
H-1092 Budapest
HUNGARY

Ivan Soskov
Laboratory for Applied Logic
University of Sofia
boul. Anton Ivanov 5
Sofia 1126
BULGARIA

Alexandra Soskova
Sector of Math. Logic
Faculty of Math. & Inf.
boul. A. Ivanov 5
1126 Sofia
BULGARIA

V. B. Shehtman
2-raya Brestskaya ul., 2/14
Inst. Genplana goroda Moskvi
125047 Moscow
USSR

D. P. Skvortsov
Inst. of Sci. & Technical Infor.
AS of the USSR
Baltijskaya str. 14
Moscow 125219
USSR

Vladimir Sotirov
Sector of Math. Logic
Faculty of Math. & Inf.
boul. A. Ivanov 5
1126 Sofia
BULGARIA

Maria Stambolieva
Linguistic Modelling Laboratory
CICT - BAS
25a Acad. G. Bonchev Str.
1113 Sofia
BULGARIA

Viggo Stoltenberg-Hansen
Dept. of Mathematics
Univ. of Uppsala
Thunbergsvagen 3
S-752 38 Uppsala
SWEDEN

Stoyan Georgiev Stoyanov
Mladost bl 97, vh. B, et. 6 ap. 38
Sofia 1156
BULGARIA

Martin Tabakov
Inst. of Philosophy/BAS
blvd. P. Evtimii 6
Sofia 1000
BULGARIA

Gaisi Takeuti
Departement of Mathematics
Univ. of Illinois at Urbana-Champaign
273 Altgeld Hall
1409 West Green Street
Urbana, IL 61801
USA

Tinko Tinchev
Laboratory for Applied Logic
University of Sofia
boul. Anton Ivanov 5
Sofia 1126
BULGARIA

Jerzy Tiuryn
Institute of Mathematics
Department of Mathematics
University of Warsaw
PKiN IXp.
00-901 Warsaw
POLAND

Antoni Tenev Tonchev
SD "OMIR"
ul. Hadji Dimitar 14
1000 Sofia
BULGARIA

Dimka Topalova
TTK "IST"
PO Box 191
Sofia 1618
BULGARIA

Boris A. Trakhtenbrot
Dept. of Comp. Sci.
Tel Aviv Univ.
Ramat Aviv
ISRAEL

A. S. Troelstra
Mathematical Institute
Univ. of Amsterdam
Roetersstraat 15
1018 WB Amsterdam
THE NETHERLANDS

Dimiter Vakarelov
Sector of Math. Logic
Faculty of Math. & Inf.
boul. A. Ivanov 5
1126 Sofia
BULGARIA

Mars Valiev
Institute for Systems Studies
AS of the USSR
Prospekt 60-letiya Oktyabrya, 9
Moscow 117312
USSR

Dirk van Dalen
State University of Utrecht
Mathematical Institute
Budapestlaan 6
P.O.Box 80.010
3508 TA Utrecht
THE NETHERLANDS

Vasil Vasilev
Inst. for Tech. Cyber. & Robotics
Akad. G. Bonchev, 2
1113 Sofia
BULGARIA

Wim H. M. Veldman
Mathematisch Instituut
Faculteit der
Wiskunde en Natuurwetenschappen
Katholieke Univers. Nijmegen
Toernooiveld
6525 ED Nijmegen
THE NETHERLANDS

Yuri Velinov
Inst. for Appl. Math. & Comp. Sci.
P.O.Box 384
Sofia 1000
BULGARIA

Frank Veltman
Department of Philosophy
University of Amsterdam
Amsterdam
THE NETHERLANDS

Yurij Ventsov
Institute of Mathematics
Sib. Div. of AS of the USSR
Universitetsky Avenue 4
Novosibirsk 630090
USSR

Albert Visser
State University of Utrecht
Central Interfaculty
Department of Philosophy
Heidelberglaan 2
Postbox 80.103
3508 TC Utrecht
THE NETHERLANDS

Sergey Vorobyov
Program Systems Institute
Acad. of Sciences of the USSR
PO Box 11
152140 Pereslavl-Zalessky
USSR

Djordje Vukomanovic
Faculty of Civil Engineering
Belgrade University
Bulevar Revolucije 73/I
11000 Belgrade
YUGOSLAVIA

Stanley Wainer
Centre for Theoretical Comp. Sci.
University of Leeds
Leeds LS2 9JT
ENGLAND

Henrich Wansing
Halbauer Weg 19
1000 Berlin 46
WEST BERLIN

Martin Weese
Section Mathematik
Humboldt Universitat
PF 1297
Berlin 1086
DDR

Piotr Wojtylak
Institute of Mathematics
Silesian University
Bankowa 14
40-007 Katowice
POLAND

David Otway Wray
511 Bay St
Texas City, TX 77590
USA

Ognjan Zahariev
"Interprograma"
boul. Al. Stambolijski 62-64
Sofia
BULGARIA

Jordan Zashev
Laboratory for Applied Logic
University of Sofia
boul. Anton Ivanov 5
Sofia 1126
BULGARIA

Nelli Zlatareva
Inst. for Tech. Cyber. & Robotics
Akad. G. Bonchev 2
1113 Sofia
BULGARIA

INDEX